Hydrogen and Syngas Production and Purification Technologies

Hydrogen and Syngas Production and Purification Technologies

Edited by

Ke Liu
GE Global Research Center

Chunshan Song
Pennsylvania State University

Velu Subramani
BP Products North America, Inc.

A John Wiley & Sons, Inc., Publication

Published by John Wiley & Sons, Inc., Hoboken, New Jersey
Published simultaneously in Canada

For general information on our other products and services or for technical support, please contact our Customer Care Department within the United States at (800) 762-2974, outside the United States at (317) 572-3993 or fax (317) 572-4002.

Wiley also publishes its books in a variety of electronic formats. Some content that appears in print may not be available in electronic formats. For more information about Wiley products, visit our web site at www.wiley.com.

Library of Congress Cataloging-in-Publication Data:
Hydrogen and syngas production and purification technologies / edited by Ke Liu, Chunshan Song, Velu Subramani.
     p. cm.
    Includes index.
    ISBN 978-0-471-71975-5 (cloth)
  1. Hydrogen as fuel. 2. Synthesis gas. 3. Coal gasification. I. Liu, Ke, 1964– II. Song, Chunshan. III. Subramani, Velu, 1965–
    TP359.H8H8434 2010
    665.8'1–dc22

                       2009022465

Printed in the United States of America
10 9 8 7 6 5 4 3 2 1

Contents

6. Water-Gas Shift Technologies   **311**

Alex Platon and Yong Wang

7. Removal of Trace Contaminants from Fuel Processing Reformate: Preferential Oxidation (Prox)   **329**

Marco J. Castaldi

Preface

Hydrogen and synthesis gas (syngas) are indispensable in chemical, oil, and energy industries. They are important building blocks and serve as feedstocks for the production of chemicals such as ammonia and methanol. Hydrogen is used in petroleum refineries to produce clean transportation fuels, and its consumption is expected to increase dramatically in the near future as refiners need to process increasingly heavier and sour crudes. In the energy field, the developments made recently in IGCC (Integrated Gasification Combined Cycle) and fuel cell technologies have generated a need to convert the conventional fuels such as coal or natural gas to either pure hydrogen or syngas for efficient power generation in the future. In addition, the dwindling supply of crude oil and rising demand for clean transportation fuels in recent years led to intensive research and development worldwide for alternative sources of fuels through various conversion technologies, including gas-to-liquid (GTL), coal-to-liquid (CTL) and biomass-to-liquid (BTL), which involve both hydrogen and syngas as key components.

The purpose of this multi-authored book is to provide a comprehensive source of knowledge on the recent advances in science and technology for the production and purification of hydrogen and syngas. The book comprises chapters on advances in catalysis, chemistry and process for steam reforming and catalytic partial oxidation of gaseous and liquid fuels, and gasification of solid fuels for efficient production of hydrogen and syngas and their separation and purification methods, including water-gas-shift, pressure swing adsorption, membrane separations, and desulfurization technologies. Furthermore, the book covers the integration of hydrogen and syngas production with future energy systems, as well as advances in coal-to-liquids and syngas-to-liquids (Fischer-Tropsch) processes. All the chapters have been contributed by active and leading researchers in the field from industry, academia, and national laboratories. We hope that this book will be useful to both newcomers and experienced professionals, and will facilitate further research and advances in the science and technology for hydrogen and syngas production and utilization toward clean and sustainable energy in the future.

We sincerely thank all the authors who spent their precious time in preparing various chapters for this book. We would like to express our sincere gratitude to our family members and colleagues for their constant support and patience while we completed the task of preparing and editing this book. We are also grateful to all the staff members at John Wiley & Sons for their great and sincere efforts in editing and publishing this book.

KE LIU
Energy and Propulsion Technologies
GE Global Research Center

CHUNSHAN SONG
EMS Energy Institute
Pennsylvania State University

VELU SUBRAMANI
Refining and Logistics Technology
BP Products North America, Inc.

Contributors

Anders Bitsch-Larsen, Department of Chemical Engineering & Materials Science, University of Minnesota, Minneapolis, MN

Marco J. Castaldi, Department of Earth and Environmental Engineering, Columbia University, New York, NY. E-mail: mc2352@columbia.edu

Wei Chen, Energy and Propulsion Technologies, GE Global Research Center, Irvine, CA

Zhe Cui, Energy and Propulsion Technologies, GE Global Research Center, Irvine, CA

Gregg Deluga, Energy and Propulsion Technologies, GE Global Research Center, Irvine, CA

David Edlund, Azur Energy, La Verne, CA. E-mail: dedlund@azurenergy.com

Thomas H. Fletcher, Department of Chemical Engineering, Brigham Young University, Provo, UT

Timothy C. Golden, Air Products and Chemicals, Inc., Allentown, PA. E-mail: goldentc@airproducts.com

W.S. Winston Ho, William G. Lowrie Department of Chemical and Biomolecular Engineering, Department of Materials Science and Engineering, Ohio State University, Columbus, OH. E-mail: ho@chbmeng.ohio-state.edu

Jin Huang, William G. Lowrie Department of Chemical and Biomolecular Engineering, Department of Materials Science and Engineering, Ohio State University, Columbus, OH. E-mail: jhuang@osisoft.com

Parag Kulkarni, Energy and Propulsion Technologies, GE Global Research Center, Irvine, CA

Ke Liu, GE Global Research Center, Energy and Propulsion Technologies, Irvine, CA. E-mail: liuk@research.ge.com

Xiaoliang Ma, EMS Energy Institute, and Department of Energy and Mineral Engineering, Pennsylvania State University, University Park, PA. E-mail: mxx2@psu.edu

Alex Platon, Institute for Interfacial Catalysis, Pacific Northwest National Laboratory, Richland, WA

Lanny Schmidt, Department of Chemical Engineering & Materials Science, University of Minnesota, Minneapolis, MN

Pradeepkumar Sharma, Center for Energy Technology, Research Triangle Institute, Research Triangle Park, NC

Shivaji Sircar, Department of Chemical Engineering, Lehigh University, Bethlehem, PA

Chunshan Song, EMS Energy Institute, and Department of Energy and Mineral Engineering, Pennsylvania State University, University Park, PA. E-mail: csong@psu.edu

Velu Subramani, Refining and Logistics Technology, BP Products North America, Inc., Naperville, IL. E-mail: velu.subramani@bp.com

Yong Wang, Institute for Interfacial Catalysis, Pacific Northwest National Laboratory, Richland, WA. E-mail: yongwang@pnl.gov

Wei Wei, Energy and Propulsion Technologies, GE Global Research Center, Irvine, CA

Lingzhi Zhang, Energy and Propulsion Technologies, GE Global Research Center, Irvine, CA

Jian Zou, William G. Lowrie Department of Chemical and Biomolecular Engineering, Department of Materials Science and Engineering, Ohio State University, Columbus, OH

Chapter 1

Introduction to Hydrogen and Syngas Production and Purification Technologies

Clean Fuels and Catalysis Program, EMS Energy Institute, and Department of Energy and Mineral Engineering, Pennsylvania State University

1.1 IMPORTANCE OF HYDROGEN AND SYNGAS PRODUCTION

Clean energy and alternative energy have become major areas of research worldwide for sustainable energy development. Among the important research and development areas are hydrogen and synthesis gas (syngas) production and purification as well as fuel processing for fuel cells. Research and technology development on hydrogen and syngas production and purification and on fuel processing for fuel cells have great potential in addressing three major challenges in energy area: (a) to supply more clean fuels to meet the increasing demands for liquid and gaseous fuels and electricity, (b) to increase the efficiency of energy utilization for fuels and electricity production, and (c) to eliminate the pollutants and decouple the link between energy utilization and greenhouse gas emissions in end-use systems.[1]

The above three challenges can be highlighted by reviewing the current status of energy supply and demand and energy efficiency. Figure 1.1 shows the energy supply and demand (in quadrillion BTU) in the U.S. in 2007.[2] The existing energy system in the U.S. and in the world today is largely based on combustion of fossil fuels—petroleum, natural gas, and coal—in stationary systems and transportation vehicles. It is clear from Figure 1.1 that petroleum, natural gas, and coal are the three largest sources of primary energy consumption in the U.S. Renewable energies

Hydrogen and Syngas Production and Purification Technologies, Edited by Ke Liu, Chunshan Song and Velu Subramani
Copyright © 2010 American Institute of Chemical Engineers

2

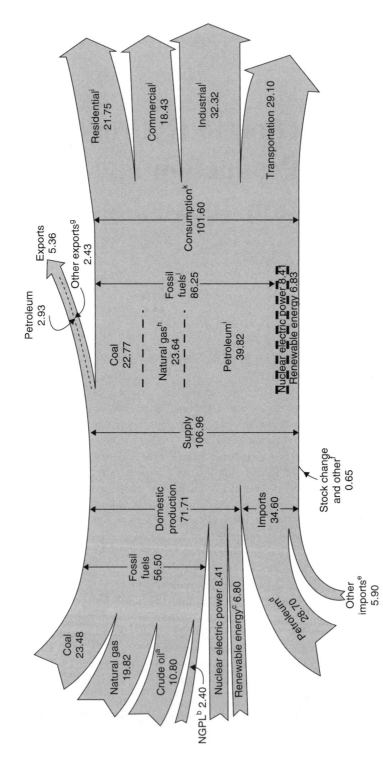

Figure 1.1. Energy supply by sources and demand by sectors in the U.S. in 2007 (in quadrillion BTU).[2]

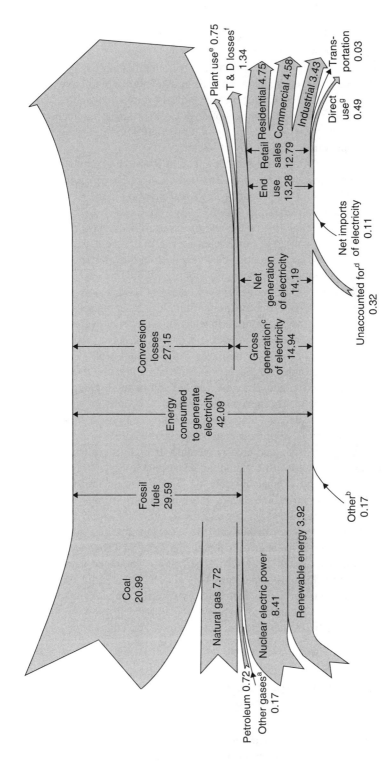

Figure 1.2. Energy consumption for electricity generation in the U.S. in 2007 (in quadrillion BTU).[2]

3

are important but are small parts (6.69%) of the U.S. energy flow, although they have potential to grow.

Figure 1.2 illustrates the energy input and the output of electricity (in quadrillion BTU) from electric power plants in the U.S. in 2007.[2] As is well known, electricity is the most convenient form of energy in industry and in daily life. The electric power plants are the largest consumers of coal. Great progress has been made in the electric power industry with respect to pollution control and generation technology with certain improvements in energy efficiency.

What is also very important but not apparent from the energy supply–demand shown in Figure 1.1 is the following: The energy input into electric power plants represents 41.4% of the total primary energy consumption in the U.S., but the electrical energy generated represents only 35.5% of the energy input, as can be seen from Figure 1.2. The majority of the energy input into the electric power plants, over 64%, is lost and wasted as conversion loss in the process. The same trend of conversion loss is also applicable for the fuels used in transportation, which represents 28.6% of the total primary energy consumption. Over 70% of the energy contained in the fuels used in transportation vehicles is wasted as conversion loss. This energy waste is largely due to the thermodynamic limitations of heat engine operations dictated by the maximum efficiency of the Carnot cycle.

Therefore, the current energy utilization systems are not sustainable in multiple aspects, and one aspect is their wastefulness. Fundamentally, all fossil hydrocarbon resources are nonrenewable and precious gifts from nature, and thus it is important to develop more effective and efficient ways to utilize these energy resources for sustainable development. The new processes and new energy systems should be much more energy efficient, and also environmentally benign. Hydrogen and syngas production technology development represent major efforts toward more efficient, responsible, comprehensive, and environmentally benign use of the valuable fossil hydrocarbon resources, toward sustainable development.

Hydrogen (H_2) and syngas (mixture of H_2 and carbon monoxide, CO) production technologies can utilize energy more efficiently, supply ultraclean fuels, eliminate pollutant emissions at end-use systems, and significantly cut emissions of greenhouse gases, particularly carbon dioxide, CO_2. For example, syngas production can contribute to more efficient electrical power generation through advanced energy systems, such as coal-based Integrated Gasification Combined Cycle (IGCC), as well as syngas-based, high-temperature fuel cells such as solid oxide fuel cells (SOFCs)[3] and molten carbonate fuel cells (MCFCs). Syngas from various solid and gaseous fuels can be used for synthesizing ultraclean transport fuels such as liquid hydrocarbon fuels, methanol, dimethyl ether, and ethanol for transportation vehicles.

1.2 PRINCIPLES OF SYNGAS AND HYDROGEN PRODUCTION

With gaseous and liquid hydrocarbons and alcohols as well as carbohydrate feedstock, there are many process options for syngas and hydrogen production. They are steam reforming, partial oxidation, and autothermal reforming or oxidative steam reforming. With solid feedstock such as coal, petroleum coke, or biomass, there are various gasification processes that involve endothermic steam gasification and exothermic oxidation reaction to provide the heat *in situ* to sustain the reaction process.

The following equations represent the possible reactions in different processing steps involving four representative fuels: natural gas (CH_4) and liquefied propane gas (LPG) for stationary applications, liquid hydrocarbon fuels (C_mH_n) and methanol (MeOH) and other alcohols for mobile applications, and coal gasification for large-scale industrial applications for syngas and hydrogen production. Most reactions (Eqs. 1.1–1.14 and 1.19–1.21) require (or can be promoted by) specific catalysts and process conditions. Some reactions (Eqs. 1.15–1.18 and 1.22) are undesirable but may occur under certain conditions.

- Steam reforming

$$CH_4 + H_2O = CO + 3H_2 \tag{1.1}$$
$$C_mH_n + m\,H_2O = m\,CO + (m + n/2)H_2 \tag{1.2}$$
$$CH_3OH + H_2O = CO_2 + 3H_2 \tag{1.3}$$

- Partial oxidation

$$CH_4 + O_2 = CO + 2H_2 \tag{1.4}$$
$$C_mH_n + m/2\,O_2 = m\,CO + n/2\,H_2 \tag{1.5}$$
$$CH_3OH + 1/2\,O_2 = CO_2 + 2H_2 \tag{1.6}$$
$$CH_3OH = CO + 2H_2 \tag{1.7}$$

- Autothermal reforming or oxidative steam reforming

$$CH_4 + 1/2\,H_2O + 1/2\,O_2 = CO + 5/2\,H_2 \tag{1.8}$$
$$C_mH_n + m/2\,H_2O + m/4\,O_2 = m\,CO + (m/2 + n/2)\,H_2 \tag{1.9}$$
$$CH_3OH + 1/2\,H_2O + 1/4\,O_2 = CO_2 + 2.5H_2 \tag{1.10}$$

- Gasification of carbon (coal, coke)

$$C + H_2O = CO + H_2 \tag{1.11}$$
$$C + O_2 = CO_2 \tag{1.12}$$
$$C + 0.5O_2 = CO \tag{1.13}$$
$$C + CO_2 = 2CO \tag{1.14}$$

- Carbon formation

$$CH_4 = C + 2H_2 \tag{1.15}$$
$$C_mH_n = x\,C + C_{m-x}H_{n-2x} + x\,H_2 \tag{1.16}$$
$$2CO = C + CO_2 \tag{1.17}$$
$$CO + H_2 = C + H_2O \tag{1.18}$$

- Water-gas shift

$$CO + H_2O = CO_2 + H_2 \tag{1.19}$$
$$CO_2 + H_2 = CO + H_2O \text{ (reverse water-gas shift [RWGS])} \tag{1.20}$$

- Selective CO oxidation

$$CO + O_2 = CO_2 \qquad (1.21)$$
$$H_2 + O_2 = H_2O \qquad (1.22)$$

Reforming or gasification produces syngas whose H_2/CO ratio depends on the feedstock and process conditions such as feed steam/carbon ratio and reaction temperature and pressure. Water-gas shift reaction can further increase the H_2/CO ratio of syngas produced from coal to the desired range for conversion to liquid fuels. This reaction is also an important step for hydrogen production in commercial hydrogen plants, ammonia plants, and methanol plants that use natural gas or coal as feedstock.

1.3 OPTIONS FOR HYDROGEN AND SYNGAS PRODUCTION

Both nonrenewable and renewable energy sources are important for hydrogen and syngas production. As an energy carrier, H_2 (and syngas) can be produced from catalytic processing of various hydrocarbon fuels, alcohol fuels, and biofuels such as oxygenates. H_2 can also be produced directly from water, the most abundant source of hydrogen atom, by electrolysis, thermochemical cycles (using nuclear heat), or photocatalytic splitting, although this process is in the early stage of laboratory research.

As shown in Table 1.1, by energy and atomic hydrogen sources, hydrogen (and syngas in most cases) can be produced from coal (gasification, carbonization), natural gas, and light hydrocarbons such as propane gas (steam reforming, partial oxidation, autothermal reforming, plasma reforming), petroleum fractions (dehydrocyclization and aromatization, oxidative steam reforming, pyrolytic decomposition), biomass (gasification, steam reforming, biologic conversion), and water (electrolysis, photocatalytic conversion, chemical and catalytic conversion). The relative competitiveness of different options depends on the economics of the given processes, which in turn depend on many factors such as the efficiency of the catalysis, the scale of production, H_2 purity, and costs of the feed and the processing steps, as well as the supply of energy sources available.

Among the active ongoing energy research and development areas are H_2 and syngas production from hydrocarbon resources including fossil fuels, biomass, and carbohydrates. In many H_2 production processes, syngas production and conversion are intermediate steps for enhancing H_2 yield where CO in the syngas is further reacted with water (H_2O) by water-gas shift reaction to form H_2 and CO_2.

Current commercial processes for syngas and H_2 production largely depends on fossil fuels both as the source of hydrogen and as the source of energy for the production processing.[4] Fossil fuels are nonrenewable energy resources, but they provide a more economical path to hydrogen production in the near term (next 5–20 years) and perhaps they will continue to play an important role in the midterm (20–50 years from now). Alternative processes need to be developed that do not

Table 1.1.   Options of Hydrogen (and Syngas) Production Processing regarding Atomic Hydrogen Source, Energy Source for Molecular Hydrogen Production, and Chemical Reaction Processes

Hydrogen Source	Energy Source	Reaction Processes
1. Fossil hydrocarbons	1. Primary	1. Commercialized process
Natural gas[a]	Fossil energy[c]	Steam reforming[d]
Petroleum[b]	Biomass	Autothermal reforming[d]
Coal[a,b]	Organic waste	Partial oxidation[d]
Tar sands, oil shale	Nuclear energy	Catalytic dehydrogenation[e]
Natural gas hydrate	Solar energy	Gasification[d]
		Carbonization[d]
2. Biomass	Photovoltaic	Electrolysis[f]
3. Water (H_2O)	Hydropower	2. Emerging approaches
4. Organic/animal waste	Wind, wave, geothermal	Membrane reactors
5. Synthetic fuels	2. Secondary	Plasma reforming
MeOH, FTS liquid, etc.		
6. Specialty areas	Electricity	Photocatalytic
Organic compound	H_2, MeOH, etc.	Solar thermal chemical
		Solar thermal catalytic
Metal hydride, chemical	3. Special cases	Biologic
complex hydride		
Ammonia, hydrazine	Metal bonding energy	Thermochemical cycling
Hydrogen sulfide	Chemical bonding energy	Electrocatalytic
7. Others	4. Others	3. Others

[a]Currently used hydrogen sources for hydrogen production.
[b]Currently used in chemical processing that produces H_2 as a by-product or main product.
[c]Currently used as main energy source.
[d]Currently used for syngas production in conjunction with catalytic water-gas shift reaction for H_2 production.
[e]As a part of industrial naphtha reforming over Pt-based catalyst that produces aromatics.
[f]Electrolysis is currently used in a much smaller scale compared with steam reforming.

depend on fossil hydrocarbon resources for either the hydrogen source or the energy source, and such alternative processes need to be economical, environmentally friendly, and competitive. H_2 separation is also a major issue as H_2 coexists with other gaseous products from most industrial processes, such as CO_2 from chemical reforming or gasification processes. Pressure swing adsorption (PSA) is used in current industrial practice. Several types of membranes are being developed that would enable more efficient gas separation. Overall, in order for hydrogen energy to penetrate widely into transportation and stationary applications, the costs of H_2 production and separation need to be reduced significantly from the current technology, for example, by a factor of 2.

1.4 HYDROGEN ENERGY AND FUEL CELLS

The main drivers for hydrogen energy and fuel cells development are listed in Table 1.2. Hydrogen production has multiple application areas in chemical industry, food industry, and fuel cell systems. Due to the major advantages in efficiency and in environmental benefits, hydrogen energy in conjunction with fuel cells has attracted considerable attention in the global research community. H_2 production is a major issue in hydrogen energy development. Unlike the primary energy sources such as petroleum, coal, and natural gas, hydrogen energy is a form that must be produced first from the chemical transformation of other substances. Development of science and technology for hydrogen production is also important in the future for more efficient chemical processing and for producing ultraclean fuels.

The development of H_2-based and syngas-based energy systems require multi-faceted studies on hydrogen sources, hydrogen production, hydrogen separation, hydrogen storage, H_2 utilization and fuel cells, H_2 sensor, and safety aspects, as well

Table 1.2.   Drivers for Hydrogen Energy and Fuel Cell System Development

Category	Drivers	Remarks
Basic reaction	$H_2 + 1/2\ O_2 = H_2O$ $\Delta H = -241.8\,\text{kJ/mol (Gw, LHV)}$ $\Delta H = -285.8\,\text{kJ/mol (Lw, HHV)}$	LHV refers to the reaction with H_2O as vapor
Technical	Efficiency—major improvement potential with fuel cells Environmental advantage—no emissions of pollutants and CO_2	Overcome the thermodynamic limitations of combustion systems
Sustainability	Bridge between nonrenewable (fossil) and renewable (biomass) energy utilization Sustainable in terms of hydrogen atom sources	Hydrogen atom from H_2O
Political and regional	Energy security and diversity Dependence on import of oils	Wide range of resources can be used
Economical	New business opportunities Niche application/market development Potential role and domain for new players	Gas producers and other industrial and small business organizations
Specific applications	Portable power sources Quiet power sources Remote power sources Space explorations Military applications	On-site or on-board fuel cells for stationary, mobile, and portable systems

as infrastructure and technical standardization. The production and utilization of hydrogen energy is also associated with various energy resources, fuel cells, CO_2 emissions, and safety and infrastructure issues. Hydrogen energy and fuel cell development are closely related to the mitigation of CO_2 emissions. Fuel cells using hydrogen allow much more efficient electricity generation; thus, they can decrease CO_2 emission per unit amount of primary energy consumed or per kilowatt-hour of electrical energy generated.

1.5 FUEL PROCESSING FOR FUEL CELLS

Hydrogen and syngas production process concepts can be applied to fuel processing for fuel cells, as outlined in Figure 1.3.[5] In general, all the fuel cells operate without combusting fuel and with few moving parts, and thus they are very attractive from both energy and environmental standpoints. A fuel cell is two to three times more efficient than an internal combustion (IC) engine in converting fuel to electricity.[6] On the basis of the electrolyte employed, there are five types of fuel cells. They differ in the composition of the electrolytes and in operating temperature ranges and are in different stages of development. They are alkaline fuel cells (AFCs), phosphoric acid fuel cells (PAFCs), proton exchange membrane fuel cells (PEMFCs), MCFCs, and SOFCs. In all types, there are separate reactions at the anode and the cathode, and charged ions move through the electrolyte, while electrons move round an external circuit. Another common feature is that the electrodes must be porous,

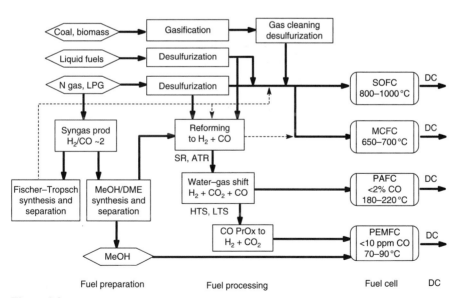

Figure 1.3. Fuel processing of gaseous, liquid, and solid fuels for syngas and hydrogen production for different fuel cells (modified after Song[5]).

because the gases must be in contact with the electrode and the electrolyte at the same time.

A simplified way to illustrate the efficiency of energy conversion devices is to examine the theoretical maximum efficiency.[7] The efficiency limit for heat engines such as steam and gas turbines is defined by the Carnot cycle as maximum efficiency $= (T_1 - T_2)/T_1$, where T_1 is the maximum temperature of fluid in a heat engine and T_2 is the temperature at which heated fluid is released. All the temperatures are in kelvin ($K = 273 +$ degrees Celsius), and therefore, the lower temperature T_2 value is never small (usually >290K). For a steam turbine operating at 400 °C, with the water exhausted through a condenser at 50 °C, the Carnot efficiency limit is $(673 - 323)/673 = 0.52 = 52\%$. (The steam is usually generated by boiler based on fossil fuel combustion, and so the heat transfer efficiency is also an issue in overall conversion.) For fuel cells, the situation is very different. Fuel cell operation is a chemical process, such as hydrogen oxidation to produce water ($H_2 + 1/2O_2 = H_2O$), and thus involves the changes in enthalpy or heat (ΔH) and changes in Gibbs free energy (ΔG). It is the change in Gibbs free energy of formation that is converted to electrical energy.[7] The maximum efficiency for fuel cell can be directly calculated as maximum fuel cell efficiency $= \Delta G/(-\Delta H)$. The ΔH value for the reaction is different depending on whether the product water is in vapor or in liquid state. If the water is in liquid state, then $(-\Delta H)$ is higher due to release of heat of condensation. The higher value is called higher heating value (HHV), and the lower value is called lower heating value (LHV). If this information is not given, then it is likely that the LHV has been used because this will give a higher efficiency value.[7]

Hydrogen, syngas or reformate (hydrogen-rich syngas from fuel reforming), and methanol are the primary fuels available for current fuel cells. Reformate can be used as a fuel for high-temperature fuel cells such as SOFC and MCFC, for which the solid or liquid or gaseous fuels need to be reformed.[5,8,9] Hydrogen is the real fuel for low-temperature fuel cells such as PEMFC and PAFC, which can be obtained by fuel reformulation on-site for stationary applications or on-board for automotive applications. When natural gas or other hydrocarbon fuel is used in a PAFC system, the reformate must be processed by water-gas shift reaction. A PAFC can tolerate about 1%–2% CO.[10] When used in a PEMFC, the product gas from the water-gas shift must be further processed to reduce CO to <10 ppm.

1.6   SULFUR REMOVAL

Sulfur is contained in most hydrocarbon resources including petroleum, natural gas, and coal. Desulfurization of fuels, either before or after reforming or gasification, is important for syngas and hydrogen production and for most fuel cell applications that use conventional gaseous, liquid, or solid fuels.[5,11] Sulfur in the fuel can poison the fuel processing catalysts such as reforming and water-gas shift catalysts. Furthermore, even trace amounts of sulfur in the feed can poison the anode catalysts in fuel cells. Therefore, sulfur must be reduced to below 1 ppm for most fuel cells, preferably below 60 ppb.

1.7 CO₂ CAPTURE AND SEPARATION

CO_2 capture and separation have also become an important global issue in the past decade, not only for H_2 and syngas purification, but also for the greenhouse gas control. When syngas is used for making liquid fuels, CO_2 may be recovered and added to the feed gas for reforming to adjust the H_2/CO ratio. A new process concept called tri-reforming has been proposed[12] and established for using CO_2 in reforming for producing industrially useful syngas with desired H_2/CO ratios for the Fischer–Tropsch synthesis and methanol synthesis. CO_2 utilization and recycling as fuels and chemicals are also important long-term research subjects. Many recent publications have discussed the CO_2 issues including new ways to capture CO_2 by solid sorbents.[1,13,14]

1.8 SCOPE OF THE BOOK

To facilitate the advances in science and technology development for hydrogen and syngas production and purification as well as fuel processing for fuel cells, this book was developed based on the contributions from many active and leading researchers in industry, academia, and national laboratory. Following Chapter 1 as an introduction and overview, Chapters 2–5 deal with the production of syngas and subsequent syngas conversion to hydrogen. In Chapter 2, catalytic steam reforming technologies are reviewed by Velu Subramani of BP, Pradeepkumar Sharma of RTI, and Lingzhi Zhang and Ke Liu of GE Global Research. This is followed by the discussion on catalytic partial oxidation and autothermal reforming in Chapter 3 by Ke Liu and Gregg Deluga of GE Global Research, and Lanny Schmidt of the University of Minnesota. These two chapters collectively cover the production technologies using gaseous and liquid feedstocks. In Chapter 4, coal gasification is reviewed as a solid-feed-based hydrogen and syngas production approach by Ke Liu and Zhe Cui of GE Global Research and Thomas H. Fletcher of Brigham Young University. Coal gasification technology development is also an area of research and development programs of the U.S. Department of Energy.[15,16] It should be mentioned that the basic processing approach of coal gasification is also applicable in general to the gasification of petroleum coke and biomass. Since the hydrocarbon resources including gaseous, liquid, and solid fuels all contain sulfur, which is environmentally harmful and poisonous to process catalysts, Chapter 5 is devoted to a review of desulfurization technologies for various sulfur removal options from liquid and gaseous fuels by Chunshan Song and Xiaoliang Ma of Pennsylvania State University. The step in the hydrogen production process following reforming or gasification and desulfurization is the water-gas shift, which is covered in Chapter 6 by Alex Platon and Yong Wang of Pacific Northwest National Laboratory.

Chapters 7–10 cover the syngas purification and separation. When reforming and water-gas shift are applied to PEMFC systems, trace amounts of CO in the gas that poisons anode catalyst must be removed. This is achieved by preferential CO oxidation, which is covered in Chapter 7 by Marco J. Castaldi of Columbia

University. Membrane development is a promising approach for efficient gas separation in various applications. Chapter 8 provides an overview on hydrogen membrane separation and application in fuel processing by David Edlund of IdaTech. In Chapter 9, CO_2-selective membrane development is reviewed by Jin Huang, Jian Zou, and W.S. Winston Ho of Ohio State University. The CO_2 membrane application for fuel processing is also discussed. For the commercial hydrogen production technologies, PSA is an important technology, for which the state of the art is reviewed by Shivaji Sircar of Lehigh University and Timothy C. Golden of Air Products and Chemicals.

For practical applications, integrated production technologies are highly desired and often provide more efficient and also flexible processing options in response to demands. Chapter 11 focuses on the integration of H_2/syngas production technologies with future energy systems, which is discussed by Wei Wei, Parag Kulkarni, and Ke Liu of GE Global Research.

One of the most important applications of syngas is the synthesis of liquid fuels and chemicals. It is well known that syngas with different H_2/CO ratios can be used for the Fischer–Tropsch synthesis of liquid hydrocarbon fuels for the synthesis of methanol and dimethyl ether, as well as ethanol and higher alcohols. Chapter 12 provides an overview of coal and syngas to liquid technologies, which is authored by Ke Liu, Zhe Cui, Wei Chen, and Lingzhi Zhang of GE Global Research. The indirect coal-to-liquids (CTL) technology via syngas conversion has its root in Germany as reflected by the well-known Fischer–Tropsch synthesis, which can also be applied to natural gas-to-liquids (GTL) and biomass-to-liquids (BTL) development.

We hope this book will provide the balanced overview of science and technology development that will facilitate the advances of hydrogen and syngas production for clean energy and sustainable energy development.

ACKNOWLEDGMENTS

We wish to thank all the authors for their contributions and for their patience in the long process of manuscript preparation, editing, and book production. We also gratefully acknowledge the acquisition editors and editorial office of Wiley publisher for their support of the book project and for their editorial assistance. Finally, we wish to thank the Pennsylvania State University, GE Global Research, and BP Refining Technology for their support of the efforts by the editors for contributing to and editing this book.

REFERENCES

1. SONG, C.S. Global challenges and strategies for control, conversion and utilization of CO_2 for sustainable development involving energy, catalysis, adsorption and chemical processing. *Catalysis Today*, **2006**, 115, 2.
2. EIA/AER. Annual Energy Review 2007. Energy Information Administration, US Department of Energy, Washington, DC. *DOE/EIA-0384(2007)*, June **2008**.

3. WILLIAMS, M.C., STRAKEY, J.P., SURDOVAL, W.A., and WILSON, L.C. Solid oxide fuel cell technology development in the U.S. *Solid State Ionics*, **2006**, 177, 2039.
4. GUNARDSON, H. In *Industrial Gases in Petrochemical Processing*. New York: Marcel Dekker, p. 283, **1998**.
5. SONG, C.S. Fuel processing for low-temperature and high-temperature fuel cells. Challenges and opportunities for sustainable development in the 21st century. *Catalysis Today*, **2002**, 77, 17.
6. THOMAS, S. and ZALBOWITZ, X. Fuel cells. Green power. Los Alamos, NM: Los Alamos National Laboratory. *Publication No. LA-UR-99-3231*, **2000**.
7. LARMINIE, J. and DICKS, A. In *Fuel Cell Systems Explained*. New York: John Wiley, p. 308, **2000**.
8. GHENCIU, A.F. Review of fuel processing catalysts for hydrogen production in PEM fuel cell systems. *Current Opinion in Solid State and Materials Science*, **2002**, 6 (5), 389.
9. FARRAUTO, R.J. From the internal combustion engine to the fuel cell: Moving towards the hydrogen economy. *Studies in Surface Science and Catalysis*, **2003**, 145, 21.
10. HIRSCHENHOFER, J.H., STAUFFER, D.B., ENGLEMAN, R.R., and KLETT, M.G. *Fuel Cell Handbook, DOE/FETC-99/1076*, 4th edn. Morgantown, WV: U.S. Department of Energy, Federal Energy Technology Center, November **1998**.
11. SONG, C.S. An overview of new approaches to deep desulfurization for ultra-clean gasoline, diesel fuel and jet fuel. *Catalysis Today*, **2003**, 86 (1–4), 211.
12. SONG, C.S. and PAN, W. Tri-reforming of methane: A novel concept for catalytic production of industrially useful synthesis gas with desired H_2/CO ratios. *Catalysis Today*, **2004**, 98 (4), 463.
13. XU, X.X., SONG, C.S., ANDRESEN, J.M., MILLER, B.G., and SCARONI, A.W. Preparation and characterization of novel CO_2 "molecular basket" adsorbents based on polymer-modified mesoporous molecular sieve MCM-41. *Microporous and Mesoporous Materials*, **2003**, 62, 29.
14. MA, X.L., WANG, X.X., and SONG C.S. Molecular basket sorbents for separation of CO_2 and H_2S from various gas streams. *Journal of the American Chemical Society*, **2009**, 131 (16), 5777.
15. STIEGEL, G.J. and RAMEZAN, M. Hydrogen from coal gasification: An economical pathway to a sustainable energy future. *International Journal of Coal Geology*, **2006**, 65, 173–190.
16. DOE. Clean coal & natural gas power systems-gasification technology R&D. US Department of Energy. http://www.fossil.energy.gov/programs/powersystems/gasification/index.html (accessed March 1, 2009).

Chapter 2

Catalytic Steam Reforming Technology for the Production of Hydrogen and Syngas

VELU SUBRAMANI,[1] PRADEEPKUMAR SHARMA,[2]
LINGZHI ZHANG,[3] AND KE LIU[3]

[1]*Refining and Logistics Technology, BP Products North America, Inc.*
[2]*Center for Energy Technology, Research Triangle Institute*
[3]*Energy & Propulsion Technologies, GE Global Research Center*

2.1 INTRODUCTION

Hydrogen (H_2) has a long tradition as an energy carrier as well as an important feedstock in chemical industries and in refineries. It has a very high energy density. As shown in Table 2.1, 1 kg of H_2 contains the same amount of energy as 2.6 kg of natural gas/methane (CH_4) or 3.1 kg of gasoline. This makes H_2 an ideal fuel in applications where weight rather than volume is an important factor, such as providing lift for balloons or zeppelins and recently as a fuel for spacecraft.

The use of H_2-rich gas, known as "town gas," produced from coal and containing about 50% H_2 with the rest mostly CH_4 and carbon dioxide (CO_2), for lighting and heating began in early 1800s and continued until mid-1900s.[1] Town gas was celebrated as a wonder, bringing light and heat to the civilized world. Later, the discovery of oil and natural gas reserves slowly displaced the supply of town gas. The use of H_2 as a feedstock for the production of ammonia and fertilizer began in 1911. Today, ammonia synthesis has become one of the major uses of H_2.

The worldwide H_2 production at present has been estimated to be about 12 trillion standard cubic feet (SCF)/year, including about 1.7 trillion SCF/year of merchant H_2.[2] Most of this H_2 is consumed for the synthesis of ammonia and methanol. A significant portion is also used in refineries for upgrading crude oils by

Hydrogen and Syngas Production and Purification Technologies, Edited by Ke Liu, Chunshan Song and Velu Subramani

Table 2.1. Energy Density and Hydrogen to Carbon Ratio of Various Hydrocarbon and Alcohol Fuels

Fuel	Major Chemical Compound	Energy Density (MJ/kg)	H/C Ratio
Hydrogen	H_2	142.0	–
Natural gas	CH_4	55.5	4
Biogas-I[a]	CH_4, CO_2	28–45	2–3.2
Biogas-II[b]	CH_4, H_2, CO_2, CO	4–14	0.7–2.0
LPG	C_3–C_4	50.0	2.5–2.7
Methanol	CH_3OH	22.5	4
Ethanol	C_2H_5OH	29.7	3
Gasoline	C_4–C_{12}	45.8	1.6–2.1
Jet fuel	Up to C_{25}	46.3	1.6–2.0
Diesel	C_9–C_{24}	45.3	1.8–2.3

[a]Biogas from anaerobic digester.

[b]Biogas from gasifier.

processes such as hydrocracking and hydrotreating to produce gasoline and diesel. Pure H_2 streams have also been used in a number of hydrogenation reactions, including hydrogenation of edible oils, aromatics, hydrocarbons, aldehydes, and ketones for the production of vitamins, cosmetics, semiconductor circuits, soaps, lubricants, margarine, and peanut butter.

There is a growing worldwide demand for H_2 in refineries because of the need to process heavier and dirtier feedstocks, combined with the desire to produce much cleaner transportation fuels that are almost free from sulfur to meet the stringent environmental regulations imposed in several countries.[3,4] Processing of heavier and higher-sulfur crude oils will require a greater H_2 stream. In addition, the evolving interest in using H_2 as a future energy carrier, especially in the automotive sector, will result in a large demand for H_2 in the future.

H_2 can be produced from a variety of feedstocks, including fossil fuels such as natural gas, oil, and coal and renewable sources such as biomass and water with energy input from sunlight, wind, hydropower, and nuclear energy. H_2 production from fossil fuels and biomass involves conversion technologies such as reforming (hydrocarbons, oils and alcohols), gasification, and pyrolysis (biomass/coal), while other conversion technologies such as electrolysis and photolysis are used when the source of H_2 is water (Fig. 2.1). The former processes produce syngas, which is a mixture of H_2 and CO with a H_2/CO ratio dictated by the type of fuel source and the conversion technology used. The syngas obtained is subjected to several downstream processes, which produce pure H_2. The discussion in this chapter will focus on reforming of fossil fuels and biofuels for the production of syngas, which does not involve downstream gas cleanup and conditioning.

Reforming occurs when a hydrocarbon or alcohol fuel and steam and/or oxygen is passed through a catalyst bed under optimum operating conditions. Depending

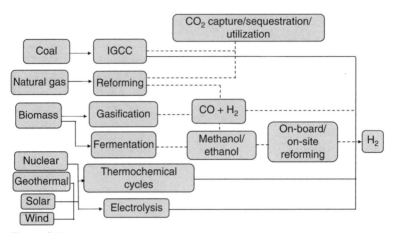

Figure 2.1.   Technological options for the production of hydrogen from various carbon-containing feedstocks. IGCC, Integrated Gasification Combined Cycle.

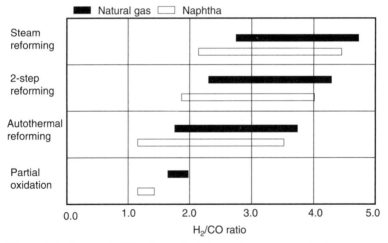

Figure 2.2.   Possible H_2/CO ratios obtained from various syngas production processes. Adapted from Rostrup-Nielsen et al.[10]

upon whether steam or oxygen or a mixture of steam and oxygen is used, the reforming technology is termed "steam reforming," "partial oxidation," and "autothermal reforming (ATR)," respectively. Reforming of natural gas with CO_2, also known as "dry reforming," has also been reported in recent years.[5–10] Among the reforming technologies, steam reforming is the preferred process for hydrogen and syngas today because it offers relatively a higher H_2/CO ratio (close to 3) since a part of hydrogen comes from water. The H_2/CO ratio can be varied over a wide range as shown in Figure 2.2, as the reforming reactions are coupled with the shift reaction at the downstream.[10]

Catalytic steam reforming (CSR) involves the extraction of H_2 molecules from a hydrocarbon or alcohol fuel and water over a base metal or noble metal-supported catalysts. CSR is widely employed to produce H_2-rich gas from various gaseous and liquid hydrocarbon fuels. Steam reforming of hydrocarbon fuels, especially steam reforming of natural gas containing methane, is a well-developed technology and practiced commercially for large-scale H_2 production.[3–6,8–10] Research in this area is still being pursued actively to further improve process efficiency. Knowledge gained from natural gas reforming is applied to the reforming of higher hydrocarbons, alcohols, and biofuels for the manufacture of H_2 or syngas depending on the end use. The chemistry, thermodynamics, catalysts, kinetics, reactions mechanisms, and technology developments in the CSR of various hydrocarbon and alcohol fuels for H_2 or syngas production are discussed in detail in the following sections.

2.2 STEAM REFORMING OF LIGHT HYDROCARBONS

2.2.1 Steam Reforming of Natural Gas

2.2.1.1 Chemistry

Natural gas is an odorless and colorless naturally occurring mixture of hydrocarbon and nonhydrocarbon gases found in porous geologic formations beneath the earth's surface, often in association with petroleum or coal. The principal constituent is methane (CH_4) and its composition is regionally dependent. Table 2.2 summarizes the composition of natural gas by region.[8]

Methane reacts with steam in the presence of a supported nickel catalyst to produce a mixture of CO and H_2, also known as synthesis gas or syngas as represented by Equation 2.1. This reaction is also referred to as steam methane reforming (SMR) and is a widely practiced technology for industrial production of H_2. However, the SMR is not really just one reaction as indicated in Equation 2.1 but involves contributions from several different catalyzed reactions such as water-gas shift

Table 2.2.   Composition of Natural Gas by Region[8]

Region	Methane	Ethane	Propane	H_2S	CO_2
U.S./California	88.7	7.0	1.9	–	0.6
Canada/Alberta	91.0	2.0	0.9	–	–
Venezuela	82.0	10.0	3.7	–	0.2
New Zealand	44.2	11.6 (C_2–C_5)	–	–	44.2
Iraq	55.7	21.9	6.5	7.3	3.0
Libya	62.0	14.4	11.0	–	1.1
U.K./Hewett	92.6	3.6	0.9	–	–
U.R.S.S./Urengoy	85.3	5.8	5.3	–	0.4

(WGS), reverse water-gas shift (RWGS), CO disproportionation (Boudouard reaction), and methane decomposition reactions as described in Equations 2.2–2.5:

$$CH_4(g) + H_2O(g) \rightarrow 3H_2(g) + CO(g) \qquad \Delta H^\circ_{298} = +205.9 \text{ kJ/mol}, \qquad (2.1)$$

$$CO(g) + H_2O(g) \rightarrow CO_2(g) + H_2(g) \qquad \Delta H^\circ_{298} = -41 \text{ kJ/mol}, \qquad (2.2)$$

$$2CO(g) \rightarrow CO_2(g) + C \quad \Delta H^\circ_{298} = -172.4 \text{ kJ/mol}, \qquad (2.3)$$

$$CH_4(g) \rightarrow C + 2H_2(g) \quad \Delta H^\circ_{298} = +74.6 \text{ kJ/mol}, \qquad (2.4)$$

$$C + H_2O(g) \rightarrow CO + H_2(g) \quad \Delta H^\circ_{298} = +131.3 \text{ kJ/mol}, \qquad (2.5)$$

$$CO(g) + 0.5O_2(g) \rightarrow CO_2(g) \qquad \Delta H^\circ_{298} = -283 \text{ kJ/mol}, \qquad (2.6)$$

$$CO(g) + 3H_2(g) \rightarrow CH_4(g) + H_2O(g) \qquad \Delta H^\circ_{298} = -205.9 \text{ kJ/mol}. \qquad (2.7)$$

The steps involved in the SMR process for the production of pure H_2 can be divided into (a) feed pretreatment, (b) steam reforming, (c) CO shift conversion, and (d) hydrogen purification. For natural gas, the only pretreatment required is desulfurization, which usually consists of a hydrogenator for the conversion of sulfur-containing species into H_2S followed by a zinc oxide bed for H_2S scrubbing. After desulfurization, the natural gas is fed into a reformer reactor, where it reacts with steam to produce H_2, CO, and CO_2 through reactions represented by Equations 2.1–2.5. The reformer is comprised of several reactor tubes filled with reforming catalysts and kept in a furnace that provides heat necessary for the endothermic reaction and operated in the temperature range between 500 and 900 °C and pressure above 20 atm.[3-6,8-10] Since the reaction produces an increase in the net number of product molecules, additional compression of the product would be necessary if the reaction were run at <20 atm. Although the stoichiometry for Equation 2.1 suggests that only 1 mol of H_2O is required for 1 mol of CH_4, the reaction in practice is being performed using high steam-to-carbon (S/C) ratio, typically in the range 2.5–3 in order to reduce the risk of carbon deposition on the catalyst surface. The gas exiting the reformer is cooled to about 350 °C and then subjected to the WGS reaction in a high-temperature shift (HTS) converter. The current process for the industrial production of pure H_2 (over 99.99%) employs pressure swing adsorption (PSA) for the purification of H_2 after the shift reaction. The PSA off-gas, which contains CO, CO_2 unreacted CH_4, and unrecovered H_2 is used to fuel the reformer.

Alternative technologies to the PSA process for H_2 purification include, after the HTS reaction, a low-temperature shift (LTS) reaction followed by CO_2 scrubbing (e.g., monoethanolamine or hot potash).[11] The LTS reaction can increase the H_2 yield slightly. However, the product stream, after the HTS, needs to be cooled to about 220 °C. Preferential oxidation (Prox) and/or methanation reaction as shown in Equations 2.6 and 2.7, respectively, removes the traces of CO and CO_2. The product H_2 has a purity of over 97%.

2.2.1.2 Thermodynamics

As shown in Equation 2.1, the SMR reaction results in gas volume expansion and is strongly endothermic ($\Delta H^\circ_{298} = +205.9 \text{ kJ/mol}$). Therefore, the reaction is thermodynamically favorable under low pressure and high temperatures. The changes in enthalpy (ΔH) and Gibbs free energy (ΔG) during the reaction can be calculated, along with the corresponding equilibrium constants (shown in Table 2.3). The thermodynamic data presented in the table provides knowledge in identifying operation conditions and feasibility. The reaction requires certain temperatures to achieve sufficient activity. Figure 2.3 shows the variation of ΔG as a function of temperature in the form of an Ellingham-type diagram for three representative reactions during the SMR process: SMR, methane decomposition, and carbon gasification. ΔG declines as temperature increases for all three reactions, again reflecting the endothermic nature of those reactions. It can be seen that methane decomposition (line a), which leads to coke deposition, occurs at relatively low temperature, around 500 °C. However, the SMR and carbon gasification reactions require fairly high temperatures (>700 °C) to move forward. This makes heat transfer a critical reactor design component. It also puts stringent thermal requirement for materials used for reactor and pipeline manufacture.

The equilibrium methane conversions with increasing temperature calculated at different S/C ratios and pressure between 1 and 20 bar are shown in Figure 2.4. The methane conversion increases with higher S/C ratios (S/C varies from 1 to 5) and decreases with increasing pressures (1–20 bar pressures were studied). A complete

Table 2.3. Thermodynamic Data for the Steam Methane Reforming (SMR) Reaction

Temperature (°C)	$\Delta H°$ (kJ/mol)	$\Delta G°$ (kJ/mol)	Log K
25	205.885	141.932	−24.868
75	208.156	131.025	−19.66
125	210.269	119.801	−15.718
175	212.225	108.32	−12.626
225	214.026	96.629	−10.133
275	215.675	84.764	−8.078
325	217.179	72.755	−6.354
375	218.541	60.626	−4.886
425	219.769	48.397	−3.621
475	220.867	36.085	−2.52
525	221.841	23.703	−1.551
575	222.698	11.264	−0.694
625	223.442	−1.222	0.071
675	224.077	−13.747	0.757
725	224.608	−26.303	1.377
775	225.037	−38.883	1.938
825	225.369	−51.481	2.449
875	225.608	−64.092	2.916
925	225.76	−76.711	3.345
975	225.831	−89.335	3.739
1000	225.838	−95.648	3.925

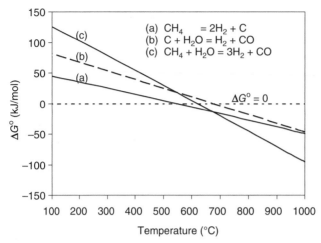

Figure 2.3. Variation of ΔG as a function of temperature in the form of an Ellingham-type diagram for the SMR process.

conversion of methane could be achieved around 700 °C at 1 bar pressure and the S/C ratio of above 2.5, while a temperature of above 900 °C would be required to achieve the complete methane conversion at 20 bar pressure. It has been reported that all currently available steam reforming catalysts promote carbon formation to different extents. Presence of excess stream can suppress carbon deposition and avoid plant shutdown caused by catalyst deactivation. Therefore, although stoichiometrically only S/C = 1 is needed for the SMR reaction, a 3.0 to 3.5 ratio is commonly used in practical applications.[12] In modern H_2 plants, driven by economic and efficiency considerations, reactor and process designs are improved to reduce the steam consumption, with a typical ratio of S/C = 2.5.[10]

Equilibrium H_2 and CO compositions can also be derived thermodynamically. Depending on the ultimate application for the gas product, H_2/CO ratio can be further tailored by integrating with secondary reactor stage (e.g., WGS) or by optimizing catalysts or operating conditions.

2.2.1.3 Catalyst

Natural gas steam reforming has been widely practiced in the industry, and a large body of catalyst development research can be found in literature. This section is not meant to be a comprehensive literature review on steam reforming catalysis, but outlining major research aspects in steam reforming catalyst development. Contributions from the following authors on steam reforming literature reviews are highly acknowledged: Trimm,[13] Bartholomew,[14] Rostrup-Nielsen,[15] Twigg,[12] Trimm,[16] and Sehested.[17]

In industrial practice, steam reforming of natural gas has been performed at high temperatures over Ni-based catalysts. Ni has been the favored active metal because

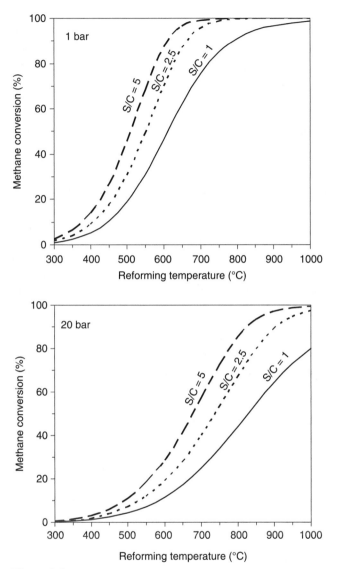

Figure 2.4.   Equilibrium methane conversions at different temperatures, steam/carbon ratios, and pressures obtained by thermodynamic calculations.

of its sufficient activity and low cost. Ni is typically supported on alumina, a refractory and highly stable material. These catalysts are shaped into an optimal form, often in the shape of multichannel wheels in order to have a better heat and mass transfer and to minimize the pressure drop under the industrial operating conditions. The catalyst performs in excess of 5 years (>50,000 h) of continuous operation. Potential suppliers of steam reforming catalyst include Haldor Topsoe, Johnson Matthey, Süd-Chemie, and BASF.[6] The Ni-based catalysts suffer from catalyst deactivation by coke formation and sintering of metallic Ni active phase. Research has been undergoing to address these issues employing different approaches, including catalyst preparation, promoter incorporation, and support materials.

Conventional Ni–Al$_2$O$_3$ catalysts are prepared by wet-impregnating Ni onto the Al$_2$O$_3$ support. This method has poor control of metal distribution on the support and yields weak binding between metal and the support. As indicated from literature, weakly attached Ni particles tend to aggregate and form large particles, which catalyze coke formation reactions.[18–20] Catalyst preparation was examined to strengthen the interaction between Ni particles and the support or enhance metal dispersion on the support, aiming to achieve higher stability during steam reforming. As revealed from Fonseca and Assaf's work,[21] in comparison with traditional impregnation technique, catalysts synthesized using hydrotalcite precursors displayed high methane reforming activity and long-term stability. Use of hydrotalcite precursors produces homogeneous dispersion of anions during catalyst synthesis. Ni can be uniformly dispersed in the final calcined catalyst structure. Zhang et al. examined one pot sol-gel technique for Ni–Al$_2$O$_3$ preparation.[22] Compared with conventional impregnation, sol-gel technique yields catalysts with highly dispersed Ni particles on the surface and a strong metal–support interaction. This suppresses the carbon filament formation and filament growth, thereby increasing catalyst stability. Zhang et al. studied synthesized nanocomposite Ni-based catalysts using a novel sol-gel method and obtained highly active and extremely stable reforming catalysts.[23] By dipping presynthesized Mg–Al mixed oxides into Ni nitrate solution, Takehira et al. obtained eggshell-type loaded Ni catalysts.[24] These catalysts showed high and stable reforming activity owing to highly dispersed and stable Ni metal particles concentrated in the catalyst surface layer. Catalysts based on hexa-aluminate-type oxides were prepared to uniformly disperse active species (Ni or other active metals) in the lattice.[25] A new concept in catalyst preparation is to combine catalyst and CO$_2$ sorbent into one material for steam reforming. As described by Satrio et al.,[26] small spherical pellets were prepared in the form of a layered structure, with a CO$_2$ sorbent core enclosed by a porous protective shell made of alumina-supported Ni catalysts. This material offers *in situ* CO$_2$ removal and hydrocarbon reforming, thereby achieving 95% H$_2$ yield.

Trace amount of promoters was reported to markedly suppress coke formation during steam reforming. Presence of promoters can modify Ni ensemble size on the surface and inhibit coke deposition.[16] Alkali metals such as K and alkaline earth metals such as Mg and Ca are frequently used to improve catalyst stability. This was attributed to higher reactivity of carbon formed on the surface and neutralization of acidic sites of the support materials (acidic support catalyzes hydrocarbon cracking and polymerization reactions).[15,27] A small amount of molybdenum or tungsten (0.5 wt% MoO$_3$ or WO$_3$) into Ni catalysts was demonstrated by Borowiecki et al. to increase the coking resistance without loss in catalytic activity.[28–30] Lanthanides (La, Ce, Gd, Sm) emerge as promising promoters for Ni-supported catalysts.[31–34] Noble metals including Rh, Pt, and Pd were examined by Nurunnabi et al. and promoted reforming activity and stability.[35–38] Studies on bimetallic Ni-based catalysts showed high stability for hydrocarbon reforming. Formulations examined include Ni–Co, Ni–Mo, Ni–Re, and Ni–Cu.[34,39,40] Exposure of Ni catalysts to very low concentration of sulfur was found beneficial for coke resistance. It is a debating topic whether this is due to ensemble size control or interference with carbon dissolution during whisker formation process as indicated by Trimm.[16]

The typical Al_2O_3 support for steam reforming catalysts is acidic and favors hydrocarbon cracking and polymerization. In addition to using promoters to modify Al_2O_3 support properties, researchers are seeking for alternative supports. Matsumura and Nakamori compared ZrO_2 with Al_2O_3 as support for Ni catalysts. Ni–ZrO_2 exhibited better performance than Ni–Al_2O_3 catalysts, particularly at low reaction temperatures, which was attributed to more hydroxyls groups formed on the ZrO_2 surface.[41] Compared with Al_2O_3, MgO support promotes surface carbon gasification, which markedly suppresses coke deposition. Ni–MgO solid solution catalysts demonstrated stable performance over long-term operation.[42–44] The solid solution creates a strong ionic environment at the metal particle–support interface and effectively minimizes Ni particle clustering and carbon formation. CeO_2 is a popular support for reforming catalysts because its high oxygen storage capacity prevents coke formation reaction.[34,45] Similarly, the lattice oxygen in perovskites materials facilitates the oxidation of CH_x fragments adsorbed on metallic nickel and reduces coking. Urasaki et al. compared the decoking abilities of a series of perovskite-supported Ni catalyst including $LaAlO_3$, $LaFeO_3$, $SrTiO_3$, $BaTiO_3$, and $La_{0.4}Ba_{0.6}Co_{0.2}Fe_{0.8}O_{3-\delta}$, and high activity and stability were measured in $LaAlO_3$ and $SrTiO_3$ catalytic systems.[46] New formulations have been developed. Ross reported some promising results over Mo and W carbides for natural gas dry reforming with CO_2.[47]

Another major type of steam reforming catalysts is based on noble metals. The serious coking problem with Ni is caused by the formation, diffusion, and dissolution of carbon in the metal.[16] However, carbon does not dissolve in noble metals, yielding much less coking in those systems. Ru, Rh, Pd, Ir, and Pt were examined for their reforming performance. Ru and Rh displayed high reforming activities and low carbon formation rates.[48] However, the cost and availability of noble metals limit their application.

2.2.1.4 Kinetics

There is a vast amount of literature studies on the kinetics of SMR. A variety of kinetics models or rate expressions have been reported. There are no general agreements on the rate equations. Discrepancies exist and some may even contradict. Kinetic parameters are largely influenced by catalysts and operation conditions. Neglecting diffusion and heat transfer limitations may yield misleading results. An overview of steam reforming kinetics has been provided in Twigg's catalysis handbook published in 1989.[12] The table of selected kinetic equations for hydrocarbon reforming from the book has been included here, which is still of high value now.

A few of representative kinetic studies in recent years are discussed in this section. Gokon et al.[49] discussed different kinetics models employed for methane reforming studies, including the Langmuir–Hinshelwood (LH), basic (BA),

Eley–Rideal (ER), and stepwise (SW) mechanisms. The LH model was tested by Mark et al. in the CO_2 reforming of methane.[50] It assumed that both reactant species of CH_4 and CO_2 are adsorbed onto the catalyst active sites separately. Adsorbed reactants then associatively react on the active sites and lead to H_2 and CO product formation. The basic model is established on the basis that the reactant species of CH_4 and CO_2 follow the first-order behavior. In the ER mechanism, one of the two reactants (either CH_4 or CO_2) is adsorbed onto the catalyst surface in adsorption equilibrium. The adsorbed species then react with the other reactant from the gas phase, and H_2 and CO are formed subsequently.[51] The SW mechanism assumes that CH_4 dissociatively adsorbed (active carbon and hydrogen species) on the catalytic surface. The active carbon reacts with CO_2 in the gas phase and produces two equivalents of CO.

Rostrup-Nielsen et al.[10] and Trimm and Önsan[52] described the kinetics of the SMR reaction based on LH rate expressions reported by Xu and Froment.[53] These studies have considered that H_2, CO, and CO_2 are produced in the SMR reaction through methane steam reforming and WGS reactions. Also, CO_2 is formed not only through the WGS reaction, but also by the steam reforming reaction with a higher S/C ratio. The rate expressions for stoichiometric methane steam reforming to syngas, WGS, and methane steam reforming with an excess of steam to produce H_2 and CO_2 are given in Equations 2.8–2.10, respectively,

$$CH_4 + H_2O = 3H_2 + CO; \qquad r_1 = \frac{k_1 \cdot P_{CH_4} \cdot P_{H_2O}}{P_{H_2}^2 \cdot Z^2}(1 - \beta), \qquad (2.8)$$

$$CO + H_2O = H_2 + CO_2; \qquad r_2 = \frac{k_2 \cdot P_{CO} \cdot P_{H_2O}}{P_{H_2} \cdot Z^2}(1 - \beta), \qquad (2.9)$$

$$CH_4 + 2H_2O = 4H_2 + CO_2; \qquad r_3 = \frac{k_3 \cdot P_{CH_4} \cdot P_{H_2O}^2}{P_{H_2}^{3.5} \cdot Z^2}(1 - \beta), \qquad (2.10)$$

where

$Z = 1 + K_{a,CO}P_{CO} + K_{a,H2}P_{H2} + K_{a,CH4}P_{CH4} + K_{a,H2O}(P_{H2O}/P_{H2})$,

β = reaction quotient $(Q_R)/Kp$,

K_a = adsorption constant, and

Kp = equilibrium constant.

The rate constants (k_1, k_2, and k_3) for these three reactions and the adsorption constants (K_a) for CH_4, H_2O, CO, and H_2 determined experimentally by two different research groups are compared in Table 2.4.[10]

Wei and Iglesia[54-57] recently reported isotopic and kinetic studies for the SMR and CO_2 reforming of methane over Ni- and noble metal-based catalysts. They considered the sequence of elementary steps involved in the steam reforming and CO_2 reforming of methane as well as methane decomposition and WGS reactions as shown in Figure 2.5. Accordingly, CH_4 decomposes to chemisorbed carbon (C*) via sequential elementary H-abstraction steps, which becomes faster as H atoms are sequentially abstracted from the CH_4 reactant. This cascade process leads to a low

Table 2.4. Kinetic Constants for the Steam Methane Reforming and Water-Gas Shift Reactions[10,53]

Parameter	Data by Xu and Froment	Data by Avetnisov et al.
k_1	$4.23 \times 10^{15} \exp(-240.1/RT)$	$1.97 \times 10^{16} \exp(-248.9/RT)$
k_2	$2.00 \times 10^{6} \exp(-67.1/RT)$	$2.43 \times 10^{5} \exp(-54.7/RT)$
k_3	$1.02 \times 10^{15} \exp(-243.9/RT)$	$3.99 \times 10^{18} \exp(-278.5/RT)$
$K_{a,CO}$	$8.23 \times 10^{-5} \exp(70.65/RT)$	$3.35 \times 10^{-4} \exp(65.5/RT)$
$K_{a,H2}$	$6.12 \times 10^{-9} \exp(82.90/RT)$	$2.06 \times 10^{-9} \exp(58.5/RT)$
$K_{a,CH4}$	$6.65 \times 10^{-4} \exp(38.28/RT)$	$6.74 \times 10^{-3} \exp(34.1/RT)$
$K_{a,H2O}$	$1.77 \times 10^{5} \exp(-88.68/RT)$	$9.48 \times 10^{4} \exp(-74.9/RT)$

k_1, k_2, and k_3 are rate constants of Equations 2.8–2.10, respectively.

$K_{a,CO}$; $K_{a,H2}$; $K_{a,CH4}$; and $K_{a,H2O}$ are adsorption constants for CO, H_2, CH_4, and H_2O, respectively. Catalyst used was $Ni/MgAl_2O_4$ with an Ni metal surface area of $3\,m^2/g$. Activation energies and heats of adsorption (values in the parentheses) are in kilojoule per mole.

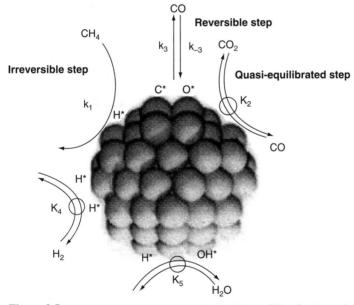

Figure 2.5. Sequence of elementary steps involved in the CH_4 reforming and water-gas shift reactions over Ni-based catalysts. Adapted from Wei and Iglesia.[54]

CHx* coverage and to C* as the most abundant carbon-containing reactive intermediate. Chemisorbed carbon is then removed by steam or CO_2 as a coreactant. These elementary steps are consistent also with kinetic and isotopic measurements on other noble metal-based catalysts such as Pt and Ir studied by them. When exposed metal atoms are the abundant surface species, only the rate constant for the activation of

Table 2.5. Kinetic Parameters for the Steam Methane Reforming over Ni-, Pt-, and Ir-Based Supported Catalysts[54–57]

Catalyst	Turnover Rate (s^{-1})	Rate Constant (s/kPa)	Activation Energy (kJ/mol)	Preexponential Factor (s/kPa)
Ni/MgO	4.0	0.2	105	3.8×10^3
Pt/ZrO$_2$	13.1	0.66	75	2.0×10^2
Ir/ZrO$_2$	12.4	0.62	87	9.9×10^4

Reaction conditions: temperature: 600 °C; CH$_4$ partial pressure: 20 kPa; H$_2$O partial pressure: 25 kPa.

the first C–H bond in CH$_4$ appears in the rate expression, and the reaction rates become first order in CH$_4$ and independent on the concentration of the coreactant, steam, or CO$_2$. The rate equation is shown in Equation 2.11,

$$r_f = kP_{CH4}, \tag{2.11}$$

where r_f is the rate of the forward reaction for the SMR, k is the rate constant, and P_{CH4} is the partial pressure of CH$_4$. The kinetic parameters such as turnover rate, rate constant, activation energy, and the preexponential factor for the SMR over supported Ni, Pt, and Ir catalysts determined under the same experimental conditions are gathered in Table 2.5.[54–57] Additional kinetic equations for the steam reforming of methane and other higher hydrocarbons reported in the literature are gathered in Table 2.6.

2.2.1.5 Mechanism

Methane is a stable and highly symmetrical molecule. The bond energy, which is essentially the average enthalpy change in a gas-phase reaction to break all similar bonds, of C–H bond in methane is 416 kJ/mol. The activation of the rigid C–H bond by dissociative adsorption of methane is the most critical and the rate-determining step (RDS) in the SMR reaction, and this occurs with different rates over Ni- and noble metal-based supported catalysts.[10,54–58] The activated methane molecule then undergoes surface reaction with adsorbed oxygen atom obtained from the dissociation of H$_2$O as described in Equations 2.12–2.20,[10,58]

$$CH_4 + 2* \rightarrow CH_3* + H* \quad (RDS), \tag{2.12}$$
$$CH_3* + * \leftrightarrow CH_2* + H*, \tag{2.13}$$
$$CH_2* + * \leftrightarrow CH* + H*, \tag{2.14}$$
$$CH* + * \leftrightarrow C* + H*, \tag{2.15}$$
$$H_2O + 2* \leftrightarrow OH* + H*, \tag{2.16}$$
$$OH* + * \leftrightarrow O* + H*, \tag{2.17}$$
$$C* + O* \leftrightarrow CO* + *, \tag{2.18}$$
$$CO* \leftrightarrow CO + *, \tag{2.19}$$
$$2H* + 2* \leftrightarrow H_2 + 2*, \tag{2.20}$$

Table 2.6. Selected Kinetic Equations for Hydrocarbon Steam Reforming

Hydrocarbon (hc)	Catalyst	Temperature (°C)	Pressure (bar)	Rate Equation	Remarks
CH_4	Ni	500–900	1–15	$[hc](1 - K'_5/K_5)$	Plant design model
CH_4	Ni	500–900	1–15	$[hc](1 - K'_6/K_6)$	Plant design model
CH_4	Ni	500–900	1–15	$[hc][H_2O]^2(1 - K'_6/K_6)$	CO_2 from CH_4, then reverse shift to give CO
CH_4	Industrial Ni catalyst	500–900	21–41	$\dfrac{[hc]}{(1 - K'_5/K_5)[H_2]}$	Rate constant is pressure dependent because of diffusion
CH_4	Ni foil	470–800	1–41	$\dfrac{[hc][H_2O]}{[H_2O]+a[H_2]^2+b[H_2]^3(1-K'_5/K_5)}$	Rate constant is pressure dependent. Adsorption of hc to form $=CH_2$, then reaction with gas-phase H_2O
CH_4	Ni/α-Al$_2$O$_3$	350–450	1–2	$\dfrac{[hc][H_2O]^2}{1-a[hc](1-K'_6/K_6)}$	Rate-determining step reaction of adsorbed hc and gas-phase H_2O; H_2 inhibits reaction
CH_4	Commercial Ni catalyst	638	1–18	$\dfrac{-K_{CO}([CO]-K[CH_4][H_2O]/[H_2O]^3)}{1+a[H_2O]+b[CH_4][H_2O]/[H_2]^3+c[CH_4][H_2O]^2/[CH_2]^4}$	Equation is for CO formation, similar one for CO_2 formation. Rate-determining step is desorption of CO and CO_2 after the reaction of CH_4 gas with adsorbed H_2O

(*Continued*)

Table 2.6. *Continued*

Hydrocarbon (hc)	Catalyst	Temperature (°C)	Pressure (bar)	Rate Equation	Remarks
CH_4	Ni/Al_2O_3 or SiO_2	670–770	16–26	$[hc][H_2O]^2(1 - K_6'/K_6)$	Rate constant is pressure dependent because of diffusion: $Ea = 38.5$–62 kJ/mol
CH_4	Rh, etc./SiO_2	350–600	1	$[hc]^\circ[H_2O]^{0.5}$	Gaseous products at equilibrium for shift reactions and $CH_4 + H_2O \rightarrow CO + 3H_2$
C_3H_6	Ni/SiO_2 or C	500–750	1	$[hc]^{0.75}[H_2O]^{0.5}$	Two-site mechanism; nondissociative hc adsorption: $Ea = 64$ kJ/mol
C_2H_6	Ni/Cr_2O_3	300–360	1	$\dfrac{[hc]^\circ}{1 + a[hc]/[H_2]}$	Rate-determining step hc adsorption; reaction via $=CH_2$
$n\text{-}C_6H_{14}$, etc.	Ni	500–800	1–30	$[hc]^\circ[H_2O]^\circ$	$Ea = 46$ kJ/mol
$n\text{-}C_7H_{16}$	Rh/γ-Al_2O_3	550–800	1	–	CH_4 and CO greater than thermodynamics involving $CH_4 + H_2O = CO + 3H_2$ and shift reaction. CH_4 direct from hc and from methanation
$n\text{-}C_7H_{16}$	Rh/MgW_4	500	1	$\dfrac{[hc][H_2O]}{1 + a[hc]}$	As above. Zero order for heat high [hc]. CO_2 primary product: $Ea = 78$ kJ/mol

Hydrocarbon	Catalyst	Temperature	Pressure	Rate expression	Comments
n-C_6H_{14}	Ni/K polyaluminate	500–800	1–20	–	CH_4 less than thermodynamics involving $CH_4 + H_2O = CO + 3H_2$ and CO_2 less than shift equilibrium at high H_2O/hc ratio. CO primary product
C_3H_8, etc.	Ni/Al_2O_3 or Mg silicate	450–500	1	–	As above, but CO_2 less than shift equilibrium, CO from reverse shift, CH_4 from CO/$CO_2 + H_2$
Toluene	Rh/γ-Al_2O_3	520	1	$[hc]^{\circ}[H_2O]^{\circ}$	Two-site adsorption. Some gasification via dealkylation, π-bonded ring and $=CH_4$
Toluene	Rh/γ-Al_2O_3	500–600	1	–	Gasification via adsorbed six-membered aromatic and CH_x species
Toluene	Rh/γ-Al_2O_3	400–500	1	$[hc]^{n}[H_2O]^{(1-n/2)}$	$N = 0.1$. Two-site adsorption followed by surface reaction: $Ea = 138\,kJ/mol$
Toluene	Rh/α-Cr_2O_3, etc.	625	1–20	$\dfrac{[hc][H_2O]}{(1+a[hc])+b[H_2O]^2}$	Single-site adsorption. Rate-determining step is surface hc and H_2O reaction: $Ea = 115\,kJ/mol$

Letters a, b, and c are constants.

[x], partial pressure of species x; Ea, activation energy; K, equilibrium constants; K', K calculated from nonequilibrium concentrations.

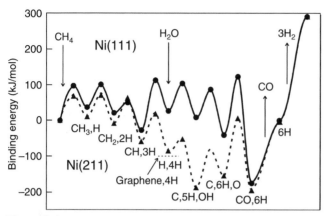

Figure 2.6. Potential energy diagram for the steam methane reforming over Ni(111) and Ni(211) surfaces based on the density functional theory (DFT) study. Adapted from Rostrup-Nielsen et al.[10]

where the asterisks (*) denotes the adsorption sites on the catalyst surface. Wei and Iglesia[54–57] also considered the involvement of similar elementary steps in SMR and CO_2 reforming of methane over Ni- and noble metal-based catalysts and showed the elementary reactions schematically as shown in Figure 2.5 and described above in the kinetics section.

Rostrup-Nielsen and coworkers, based on the density functional theory (DFT) calculations for the SMR over Ni(111) and Ni(211) surfaces, showed the energies of intermediate species formed on the surface and activation barrier separating the intermediates along the reaction path (Fig. 2.6).[10,58] The data indicate that the C or CH species are the most stable intermediates on these surfaces. For the reverse reaction, methanation, the dissociation of CO has a large activation barrier on the perfect Ni(111) surface but is favored on the stepped Ni(211) surface. Hence, the steps on the Ni(111) surface are predicted to be the sites where CO dissociates. The reaction energy, (the final point in the figure) 292 kJ/mol calculated from this DFT study, corresponds to a reaction enthalpy of 230 kJ/mol, and this is in good agreement with the experimental value of 206 kJ/mol.

2.2.1.6 Modeling and Simulation

Process modeling and simulation significantly improves the efficiency of system design and development and economic analysis, which is important for the success of SMR industry operation. The behavior of the process can be simulated with diverse mathematical models. Many parameters have been factored into the modeling process, including reactor type, mass and heat transfer, operation conditions (temperature, pressure, feed composition), kinetic modeling, and flow distribution pattern. Knowledge derived from these studies provides understanding of the complex process and facilitates developmental work. This section introduces examples of process modeling and simulation work available in the literature, most of which were performed in SMR reactors combined with H_2 removal or CO_2 sorption.

Lee et al.[59] simulated a hybrid reaction system of SMR and *in situ* noncatalytic removal of CO_2 by the carbonation of CaO to $CaCO_3$ in a moving bed reactor where reforming catalyst and CaO-based CO_2 acceptor in pellets move concurrently with gaseous reactants. Effects of parameters like feed rates of CaO and CH_4, and the reactor bed temperature on steady-state behavior of the hybrid reaction have been determined. In another paper, these researchers simulated the transient behavior of SMR coupled with simultaneous CO_2 removal by carbonation of CaO pellets in a packed bed reactor for hydrogen production.[60] A mathematical model was developed to describe both the SMR reaction and the CaO carbonation-enhanced SMR reaction at nonisothermal, nonadiabatic, and nonisobaric operating conditions. It has been successfully validated with reaction experiments. Apparent carbonation kinetics of the CaO pellet prepared has been determined using thermogravimetric analysis (TGA) carbonation experiments at various temperatures and has been incorporated into the model. Effects of major operating parameters (bed temperature, pressure, steam to methane feed ratio, and flow rate) on the transient behavior of the CaO carbonation-enhanced SMR have been investigated using the model. Therefore, the optimum operation conditions can be identified to achieve desired CO_2 uptake capacity, to lower CO concentration in the product, and to maximize H_2 yields.

Ochoa-Fernández et al.[61] studied the kinetics of CO_2 sorption on a solid adsorbent, namely lithium zirconate, in an oscillating microbalance. The solid sorbent has been prepared by a novel route resulting in a high capacity, good stability, and much improved sorption rates, making it suitable for its application in sorption enhanced hydrogen production by SMR. A kinetic equation for the sorption as a function of CO_2 partial pressure and temperature has been developed. The hydrogen production by sorption-enhanced reaction process has been simulated by a dynamic one-dimensional pseudo-homogeneous model of a fixed-bed reactor, where a hydrotalcite-derived Ni catalyst has been used as steam reforming catalysts. Simulation results show that hydrogen purer than 95% with a concentration of carbon monoxide lower than 0.2 mol% can be produced in a single step.

Cao et al.[62] mathematically described and experimentally demonstrated the microstructured catalysts used for SMR reaction in microchannel reactors. Porous metal substrates (FeCrAlY) were used to form engineered catalysts with Rh. Two types of structures were evaluated in the microchannel reactors and simulated with the developed heterogeneous reactor model. The modeling technique described in the paper provides a convenient way to evaluate variables in designing more efficient catalysts for SMR. Yu et al.[63] performed a simulation study to investigate the performance of a porous ceramic membrane reactor for hydrogen production through SMR. The results show that the methane conversions much higher than the corresponding equilibrium values can be achieved in the membrane reactor due to the selective removal of products from the reaction zone. The comparison of isothermal and nonisothermal model predictions was made. It was found that the isothermal assumption overestimates the reactor performance and the deviation of calculation results between the two models is subject to the operating conditions. The effects of various process parameters such as the reaction temperature, the reaction side pressure, the feed flow rate, and the steam to methane molar feed ratio, as well as the sweep gas flow rate and the operation modes, on the behavior of membrane reactor were analyzed and discussed.

Rakib and Alhumaizi[64] mathematically studied a bubbling fluidized bed membrane reactor for SMR reaction, with the permselective Pd membranes removing hydrogen from the reaction system to enhance the methane conversion. Oxygen fed into the reaction system can decrease the endothermicity of the overall reaction by the combustion of methane, thereby reducing the need of external firing. Operation at low feed steam:carbon ratios is also possible with the steam required for the reforming reaction being provided as a product from the combustion reactions, although problems related to coking also need to be addressed at very low ones. Because of the high endothermicity and fast kinetics of the steam reforming reactions, to provide heat effectively to the reaction, an *in situ* heat generation by combustion is employed. As an alternative, a higher feed temperature can be used as well. However, since higher oxygen:methane ratios also tend to consume more of the methane itself, this cannot be increased much, and an optimum value exists with respect to the favorable production of pure hydrogen from the reactor permeate side.

Wang and Rodrigues[65] reported the fundamental analysis of the sorption-enhanced steam methane reforming (SE-SMR) process in which the simultaneous removal of carbon dioxide by hydrotalcite-based chemisorbent is coupled. A two-section reactor model has been developed to describe the SE-SMR reactor, decoupling the complexity in process analysis. The model defines two subsequent sections in the reactor: an equilibrium conversion section (upstream) and an adsorption reforming section (downstream). The material balance relationship in the equilibrium conversion section is directly determined by thermodynamic equilibrium calculation, providing an equilibrated atmosphere to the next section. The adsorption reforming section is described using an isothermal multicomponent dynamic model into which the SMR reactions and the high-temperature CO_2 adsorption are embedded. The multiple requirements (including H_2 purity, H_2 productivity, CH_4 conversion enhancement, and carbon oxide concentrations) are taken into account simultaneously so as to analyze and define feasible operation window for producing high-purity hydrogen with ppm-level CO impurity. The performances of the reactors with different dimensions (laboratory scale and pilot scale) are explored, highlighting the importance of operation parameter control to the process feasibility.

Posada and Manousiouthakis[66] discussed the heat and power integration studies for a conventional methane reforming based hydrogen production plant with the purpose of finding minimum hot/cold/electric utility cost. Hot methane and steam were fed into the SMR reformer, where the reversible reactions (r_1), (r_2), and (r_3) were the main global reactions taking place. Heat and power integration results in utility profit due to electricity production in excess of process needs. Heat integration alone resulted in a 36% reduction in utility costs. Operation at the minimum hot/cold or hot/cold/electric utility cost did not require hot utility with a consequent reduction of carbon dioxide emissions of 6.5%.

Takeuchi et al.[67] reported a membrane reactor as a reaction system that provides higher productivity and lower separation cost in chemical reaction processes. In this paper, packed bed catalytic membrane reactor with palladium membrane for SMR reaction has been discussed. The numerical model consists of a full set of partial differential equations derived from conservation of mass, momentum, heat, and chemical species, respectively, with chemical kinetics and appropriate boundary conditions for the problem. The solution of this system was obtained by computational fluid dynamics (CFD). To perform CFD calculations, a commercial solver FLUENT™ has been used, and the selective permeation through the membrane has been modeled by user-defined functions. The CFD simulation results exhibited the flow distribution in the reactor by inserting a membrane protection tube, in addition to the temperature and concentration distribution in the axial and radial directions in the reactor, as reported in the membrane reactor numerical simulation. On the basis of the simulation results, effects of the flow distribution, concentration polarization, and mass transfer in the packed bed have been evaluated to design a membrane reactor system.

Zanfir and Gavriilidis[68] presented a theoretical study of the influence of flow arrangement on the thermal behavior of a catalytic plate reactor (CPR) for SMR reaction using methane catalytic combustion as heat source. A two-dimensional model is presented. CPR performance and the thermal behavior is strongly affected by overall and local balance between heat generated on the exothermic side and heat consumed on the endothermic one, which in turn is influenced by flow arrangement. Simulations for co-current and countercurrent flow were carried out for similar inlet conditions and catalyst loadings. It was found that the reactor is better balanced thermally for concurrent operation. For countercurrent arrangement, higher conversions and better utilization of the overall heat generated in the exothermic process are achieved at the expense of pronounced temperature extremes. Thus, reforming conversion for countercurrent operation is 62.8% compared with 52% for concurrent operation, while maximum transverse temperature difference for concurrent operation is only 16.5K compared with 310K for countercurrent operation. This increases the chances of the reactor runaway and of homogeneous combustion being initiated. Utilization of a nonuniform catalyst distribution can overcome the heat imbalance by inducing favorable reactant depletion along the reactor during countercurrent flow.

Gallucci et al.[69] investigated the SMR reaction from a modeling viewpoint, considering the effect of different parameters on methane conversion. For example, considering the influence of the lumen pressure on methane conversion at constant temperature, it has been found that increasing this parameter increases the equilibrium methane conversion for the membrane reactor, while it decreases for the traditional one. Moreover, in a realistic membrane reactor (i.e., considering a simulation performed using kinetic expressions), the behavior of methane conversion versus lumen pressure at various temperatures shows a minimum value, depending on the membrane thickness, on the reactor length, and on the temperature.

2.2.1.7 Reactor Design and Development

The SMR reaction results in the production of H_2 and CO_2 and is a strongly endo-thermic reaction and favored at high temperatures. Heat transfer is critical to ensure fast kinetics as the reaction moves forward. To enhance SMR reaction, the following approaches can be employed:

- Integration of SMR reaction and products (H_2 or CO_2) separation steps to shift the equilibrium and enhance the reaction rate in the forward direction. This also lowers the reforming temperature to achieve the same conversion level.

- Novel reactor design with improved mass and heat transfer. Efficient heat transfer facilitates reaction kinetics and decreases energy consumption. As such, reactor size and cost can be significantly reduced.

- Alternative heating sources to increase system efficiency. This may lead to additional benefit of reducing CO_2 emissions.

A look into the recent advances taking place in the equipment and reactor designs reveals that much emphasis has been laid on preparing and experimenting with the membrane reactors and CO_2 sorption-enhanced reactors. There are extensive studies on Pd-based membrane for selective separation of H_2. Cost of Pd has been proven to be insignificant among the total cost considering typical membrane thickness of $5-10\,\mu$.[70] With H_2 permeable membrane reactor configuration, ultrapure H_2 can be obtained and high-pressure CO_2 can be readily sequestrated or for other application. As reported by Patil et al.,[71] permselective Pd metallic membranes for H_2 removal (500–600 °C operating temperature) was integrated inside a fluidized bed reactor along with selective O_2 addition through dense perovskite membranes (900–1000 °C operating temperature). CH_4 conversion over 95% could be obtained. Ultrapure H_2 (no more than 10 ppm CO) can be generated from light hydrocarbons such as CH_4, which meets the requirement for feeding proton exchange membrane fuel cells (PEMFCs). Instead of fitting a ceramic substrate to a thin Pd-based layer that lacks mechanical stability, Tong and coworkers performed a series of studies on controlled deposition of Pd or Pd alloy (Pd–Ag, Pd–Ce) composite membranes onto porous stainless steel (PSS) tubes for SMR.[72–75] The deposited membrane exhibited stable homogeneous structure and provided high hydrogen permeable flux and complete selectivity of H_2 versus argon or helium. Under typical operation temperature (~500 °C) and pressure (100–300 kPa), a methane conversion of over 97% was measured. Pd-based membrane supported on PSS was also reported by Ayturk et al.[70] Long-term stability at high temperature with high methane conversion (>95%) has been achieved.

Research on other types of materials for H_2 separation has been motivated by relatively high cost of Pd and possible membrane degradation by acidic gases and carbon as summarized in Tsuru et al.[76] These authors examined microporous silica membranes together with an Ni catalyst layer for SMR reaction. However, this type of membrane allows the permeation of hydrogen as well as other gases in reactants and products, which markedly reduces hydrogen selectivity and limits methane conversion (up to 0.8 under their testing conditions). Improvement in methane conversion and H_2 selectivity was reported in Nomura et al.,[77] who synthesized a catalyst composite silica membrane using a counter diffusion chemical vapor deposition method and observed higher H_2 selectivity (H_2/N_2 permeance ratio over 1000).

There are many studies available in literature on CO_2 sorption-enhanced SMR. Pure CO_2 from the reaction is suitable for sequestration or other use.[78] Primary efforts were in the development of CO_2 sorbent materials and improvement of multicycle stability. Johnsen et al.[79] studied dolomite as a CO_2 sorbent in a bubbling fluidized bed reactor. The operating temperature and pressure are 600 °C and 1 atm, respectively. By cycling between CO_2 absorption and regeneration to reduce prebreakthrough and subsequent conversion decline, an H_2 concentration of >98% could be obtained. Dolomite was found to be a better sorbent than limestone. $Ca(OH)_2$ or $CaOSiO_2$ is another major type of adsorbent that is effective for CO_2 separation. They displayed superior performance during hydrocarbon gasification compared with other metal oxides including MgO, SnO, and Fe_2O_3.[80] Wu et al.[81] characterized the production of hydrogen with a sorption-enhanced steam methane reaction process using $Ca(OH)_2$ as the CO_2 adsorbent. Ninety-four percent H_2 concentration can be reached, which is nearly 96% of the theoretical equilibrium limit, much higher than the equilibrium concentration of 67.5% without CO_2 sorption under the same conditions of 500 °C, 0.2 MPa pressure, and a steam-to-methane ratio of 6. In addition, the residual mole fraction of CO_2 was less than 0.001. Li_2ZrO_3 synthetic sorbents are also reported in literature with better multicycle stability, but the costs are prohibitive.[79,82]

Kusakabe et al.[83] proposed selective CO oxidation membrane concept to facilitate SMR reaction. Yttria-stabilized zirconia (YSZ) membrane was deposited on the surface of a porous alumina support tube by sol-gel procedure. This again was impregnated with Pt and Rh aqueous solution to produce a Pt- or Rh-loaded YSZ membrane. With addition of O_2 in the feed, oxidation of CO can bring CO concentration to the level appropriate for PEMFC (<30 ppm).

A substantial amount of work on new reactor design has been undergoing to enhance mass and heat transfer efficiency and SMR kinetics as a result. Xiu et al.[84] applied the subsection-controlling strategy to design adsorptive SMR reactor. The whole process was divided into four steps within one bed, with adsorbent/catalyst packing ratio and wall temperature separately regulated in each subsection. By maintaining a pressure swing sorption-enhanced SMR cyclic process, a product stream with H_2 purity over 85% purity, CO concentration below 30 ppm, and CO_2 concentration below 300 ppm can be reached.

Roychoudhury et al.[85] reported the use of microlith catalytic reactors (patented by Precision Combustion, Inc. (PCI)) for ATR, WGS, or Prox. The microlith substrates are composed of a series of ultrashort channel length, low thermal mass, catalytically coated metal meshes with very small channel diameters. This design allows fast heat and mass transfer, reduces the reactor size, and improves the overall performance. ATR test showed high resistance to coking, especially at low steam/C ratios.

Chikazawa et al.[86,87] evaluated the feasibility of small sodium-cooled reactor as a diversified power source in terms of economical and safety potential and reviewed the application of this type of reactor for hydrogen generation from natural gas reforming. This type of reactor demonstrates high thermal efficiency and availability, yields a compact reactor design, and simplifies the cooling system. Glockler et al.[88] and Kolios et al.[89] reported the multifunctional reactor concept. This design is capable of integrating exothermic and endothermic reactions in microstructured devices and employing recuperative heat exchange between the process streams. This offers efficient heat recovery. Hoglen and Valentine[90] applied Coriolis meters in SMR technology. Accurate control of S/C ratios provides high operation efficiency.

Velocys, which is a spinout company from Battelle Memorial Institute, has developed microchannel process technology for large-scale chemical processing. Hydrogen can be produced from the compact microchannel unit at high heat and mass transfer rates. As discussed by Tonkovich et al.,[91] a microchannel methane steam reforming reactor integrates catalytic partial oxidation of methane prior to catalytic combustion with low excess air (25%) to generate the required energy for endothermic methane steam reforming in adjacent channels. This design improves process intensification and results in significant capital and operating cost savings for commercial applications. Galvita and Sundmacher[92] proposed a periodically operated two-layer (reduction and reoxidation) reactor design concept for steam reforming of methane. In the reduction phase, methane is converted at "layer 1" into CO and H_2 using partial oxidation. The products are fed to "layer 2" where CO_2 and H_2O are produced. In the reoxidation phase, the reactor is fed with pure steam, which reoxidizes the catalytic materials and is itself converted into CO-free H_2 gas. This process generates high-quality H_2 and eliminates additional H_2 purification units at a reduced cost.

SMR reactor design can also be approached by employing other energy sources. Nozaki et al.[93,94] reported improvement of SMR performance (much higher methane conversion) by combination of catalysts and barrier discharges. Use of barrier discharge greatly reduces energy cost and increases energy efficiency. Solar steam reforming has been examined by Moller et al.[95] Using the solar reformer technology, up to 40% fuel savings can be expected compared with a conventional plant. CO_2 emission reduction is another benefit. Another clean energy source is nuclear power. Fukushima and Ogawa[96] presented a conceptual design of low-temperature hydrogen production and high-efficiency nuclear reactor technology.

2.2.2 Steam Reforming of C_2–C_4 Hydrocarbons

While natural gas reforming is the primary process for the industrial production of H_2, the reforming of other gaseous hydrocarbons such as ethane, propane, and n-butane have been explored for the production of H_2 for fuel cells.[52,97] The reforming of propane and n-butane received particular attention in recent years, because they are the primary constituents of liquefied petroleum gas (LPG), which is available commercially and can be easily transported and stored on-site. LPG could be an attractive fuel for solid oxide fuel cells (SOFCs) and PEMFCs for mobile applications.[98–101] The chemistry, thermodynamics, catalysts, kinetics, and reaction mechanism involved in the reforming of C_2–C_4 hydrocarbons are briefly discussed in this section.

2.2.2.1 Chemistry and Thermodynamics

Steam reforming reaction of ethane, propane, and *n*-butane are represented by Equations 2.21–2.23, while the thermodynamic data for these reactions are summarized in Tables 2.7–2.9, respectively:

$$C_2H_6(g) + 2H_2O(g) \rightarrow 5H_2(g) + 2CO(g); \Delta H^{\circ}_{298} = +374.3 \text{ kJ/mol}, \quad (2.21)$$

$$C_3H_8(g) + 3H_2O(g) \rightarrow 7H_2(g) + 3CO(g); \Delta H^{\circ}_{298} = +497.7 \text{ kJ/mol}, \quad (2.22)$$

$$n\text{-}C_4H_{10}(g) + 4H_2O(g) \rightarrow 9H_2(g) + 4CO(g); \Delta H^{\circ}_{298} = +651.3 \text{ kJ/mol}. \quad (2.23)$$

These are endothermic reactions and the endothermicity increases with increasing carbon number of the hydrocarbon. These reactions are generally carried out in a wide temperature range between 300 and 900 °C over a nickel-based or noble metal-based supported catalyst. Under the reaction operating conditions, other reactions such as cracking into carbon and hydrogen (Eqs. 2.24–2.26) followed by carbon gasification, cracking into methane (Eq. 2.27) followed by steam reforming of methane can also occur. Methane formation can also occur by methanation of carbon oxide and H_2 (Eq. 2.28) formed in the cracking reactions. Similar to the SMR process, the steam reforming of higher hydrocarbons in practice is also performed

Table 2.7. Thermodynamic Data for the Steam Reforming of Ethane: $C_2H_6(g) + 2H_2O(g) \rightarrow 5H_2(g) + 2CO(g)$

Temperature (°C)	ΔH° (kJ/mol)	ΔG° (kJ/mol)	Log K
25	374.3	215.6	−37.782
75	351.3	193.2	−28.995
125	354.8	170.3	−22.344
175	358.1	146.9	−17.126
225	361.0	123.2	−12.92
275	363.7	99.2	−9.454
325	366.0	75.0	−6.548
375	368.1	50.6	−4.075
425	370.0	26.0	−1.944
475	371.6	1.3	−0.09
525	373.1	−23.5	1.539
575	374.3	−48.4	2.98
625	375.4	−73.3	4.266
675	376.3	−98.4	5.419
725	377.1	−123.4	6.458
775	377.7	−148.5	7.401
825	378.2	−173.6	8.258
875	378.5	−198.7	9.042
925	378.7	−223.9	9.761
975	378.8	−249.0	10.422
1000	378.8	−261.6	10.734

Table 2.8. Thermodynamic Data for the Steam Reforming of Propane:
$C_3H_8(g) + 3H_2O(g) \rightarrow 7H_2(g) + 3CO(g)$

Temperature (°C)	$\Delta H°$ (kJ/mol)	$\Delta G°$ (kJ/mol)	Log K
25	497.701	297.577	−52.139
75	503.172	263.571	−39.548
125	508.015	228.822	−30.022
175	512.331	193.495	−22.555
225	516.185	157.711	−16.539
275	519.624	121.56	−11.585
325	522.685	85.112	−7.433
375	525.398	48.422	−3.903
425	527.787	11.535	−0.863
475	529.875	−25.513	1.781
525	531.681	−62.69	4.103
575	533.224	−99.973	6.158
625	534.518	−137.34	7.988
675	535.572	−174.772	9.629
725	536.398	−212.254	11.109
775	537.008	−249.772	12.448
825	537.413	−287.314	13.668
875	537.623	−324.871	14.781
925	537.652	−362.432	15.802
975	537.517	−399.991	16.741
1000	537.392	−418.768	17.183

at higher S/C ratios, typically between 1 and 3, in order to reduce the risk of carbon deposition. While higher operating pressures are preferred for industrial applications, the reactions are generally operated at atmospheric pressure for fuel cell applications. The syngas produced can be fed directly into SOFC. HTS and LTS and/or Prox reactions are also needed in the downstream of the steam reformer to remove CO from the reformed gas for PEMFC applications:

$$C_2H_6(g) \rightarrow 3H_2(g) + 2C; \Delta H_{298}° = +84.7 \text{ kJ/mol}, \quad (2.24)$$
$$C_3H_8(g) \rightarrow 4H_2(g) + 3C; \Delta H_{298}° = +103.8 \text{ kJ/mol}, \quad (2.25)$$
$$n\text{-}C_4H_{10}(g) \rightarrow 9H_2(g) + 4C; \Delta H_{298}° = +126.2 \text{ kJ/mol}, \quad (2.26)$$
$$C_nH_m + xH_2O \rightarrow yH_2 + (n-1)CO + CH_4, \quad (2.27)$$
$$CO + 3H_2 \rightarrow CH_4 + H_2O; \Delta H_{298}° = -206 \text{ kJ/mol}. \quad (2.28)$$

2.2.2.2 Catalysts and Kinetics

The Ni-based catalysts known for the steam reforming of natural gas are also active for the steam reforming of C_2–C_4 hydrocarbons. However, carbon deposition and

Table 2.9. Thermodynamic Data for the Steam Reforming of *n*-Butane:
$C_4H_{10}(g) + 4H_2O(g) \rightarrow 9H_2(g) + 4CO(g)$

Temperature (°C)	$\Delta H°$ (kJ/mol)	$\Delta G°$ (kJ/mol)	Log K
25	651.286	382.595	−67.035
75	658.042	336.987	−50.564
125	664.086	290.456	−38.109
175	669.497	243.202	−28.349
225	674.33	195.375	−20.488
275	678.629	147.089	−14.018
325	682.432	98.432	−8.596
375	685.771	49.476	−3.988
425	688.675	0.278	−0.021
475	691.173	−49.113	3.429
525	693.292	−98.659	6.457
575	695.058	−148.327	9.136
625	696.488	−198.089	11.521
675	697.6	−247.922	13.659
725	698.411	−297.805	15.586
775	698.939	−347.722	17.33
825	699.203	−397.659	18.917
875	699.219	−447.601	20.365
925	699.009	−497.54	21.693
975	698.598	−547.465	22.913
1000	698.326	−572.421	23.487

subsequent catalyst deactivation are the major issues. The industrial methane steam reforming catalysts are suitably modified to reduce the carbon deposition for the reforming of higher hydrocarbons. Approaches used to minimize the carbon formation in Ni-based catalysts include addition of alkali promoters such as K_2O, MgO, CaO, and SrO, to the catalyst formulation. Ni supported on alkali metal oxides such as MgO, CaO, $MgO-Al_2O_3$, and $CaO-Al_2O_3$ mixed oxides have also been reported.[10,42,52,102] The compositions of some of the commercial catalysts known for steam reforming of light hydrocarbons and LPG are included in Table 2.10.[103] Noble metal-supported catalysts are less susceptible for carbon formation compared with Ni-based catalysts. Consequently, the steam reforming of methane, ethane, propane, and butane over Pt, Pd, and Rh, supported on Al_2O_3, SiO_2, $MgO-Al_2O_3$, CeO_2, ZrO_2, and so on, has been reported.[52,97,104] Alloy formation by the addition of a small amount of noble metals such as Au to the Ni-based catalysts has also been reported to suppress the carbon formation.[58,105] Consequently, various bimetallic and multimetallic catalysts containing noble metals and Ni supported on above-mentioned oxides are known.[22,97,105–116] A few catalyst formulations reported in recent years are summarized in Table 2.11.

Table 2.10. Compositions of a Few Commercial Steam Reforming Catalysts[103]

Manufacturer	Trade Name	Hydrocarbon	NiO (wt%)	Promoter (wt%)	Carrier (wt%)	SiO$_2$
Haldor Topsoe	R-67-7H	NG to naphtha	>12	–	MgAl$_2$O$_4$ (balance)	<0.2
	AR-301	NG to light HC (PR)	>30	2–5	MgAl$_2$O$_4$ (balance)	–
	RK 202	Naphtha	>15	K$_2$O, >1	MgAl$_2$O$_4$ (balance)	<0.2
Johnson Matthey	23-4	Naphtha, NG, and light HC	18	–	Al$_2$O$_3$ (balance)	<0.1
	57-4	NG and light HC	18	–	CaAl$_2$O$_4$ (balance)	<0.1
	46-6Q	Naphtha	16	–	CaAl$_2$O$_4$ (balance)	0.5
Süd-Chemie	G-90	NG	14	–	CaAl$_{12}$O$_{19}$ (balance)	–
	G-91	NG/LPG	18	K$_2$O (1.6)	CaK$_2$Al$_{22}$O$_{34}$ (balance)	–
	C11-NK	Naphtha	25	K$_2$O (8.5)	CaAl$_2$O$_4$ (balance)	–
United Catalysts	NGR-615-K	Naphtha	15–20	K$_2$O (1.5–8)	Al$_2$O$_3$ (balance)	0.1–0.2
	NGPR-1	PR	25	Unspecified	MgAl$_2$O$_4$	–
BASF	SG-9301	NG	16.5	MgO (6.0) ReO (3.0)	Al$_2$O$_3$ (balance)	<0.1
	G1-50	Naphtha	20	MgO (11) CaO (16) K$_2$O (7.0)	Al$_2$O$_3$ (32)	14.0

NG, natural gas; HC, hydrocarbons; PR, prereforming; LPG, liquefied petroleum gas.

40

Table 2.11. A Few Selected Catalysts Reported Recently for the Steam Reforming of C_2–C_4 Hydrocarbons

Hydrocarbon	Catalyst	Reaction Conditions Temperature (°C)	S/C Ratio	Results	Reference
Propane	20% Ni/Al$_2$O$_3$	500	1.3	Catalysts prepared by sol-gel technique exhibited a stable activity than that prepared by impregnation.	Zhang et al.[22]
Propane	10% Ni coated on SiO$_2$	600	1.0	SiO$_2$-coated Ni catalyst is more active and stable compared with Ni/SiO$_2$, Ni/MgO, and Ni/Al$_2$O$_3$ catalysts.	Takenaka et al.[112]
Ethane and propane	Ni/Al$_2$O$_3$–CeO$_2$	700–900	1.5	Ni/Al$_2$O$_3$ catalyst doped with 14% CeO$_2$ showed the best performance.	Laosiripojana et al.[111]
n-Butane	1% Pd/CeO$_2$, 1% Pd/Al$_2$O$_3$, 1% Pd/SiO$_2$	300–600	1–2	1% Pd/CeO$_2$ catalyst exhibited much higher activity than other Pd-supported catalysts.	Wang and Gorte[110]
LPG	High surface area CeO$_2$	700–900	1–3	High surface area CeO$_2$ showed better performance compared with an Ni/Al$_2$O$_3$ catalyst.	Laosiripojana and Assabumrungrat[109]
C$_1$ and C$_2$ HC	1% Pt, Rh, or Pd supported on YSZ	450–800	1.2	The order of reactivity on Rh/YSZ: C$_2$H$_6$ > C$_2$H$_4$ > CH$_4$; on Pt/YSZ: CH$_4$ > C$_2$H$_6$ > C$_2$H$_4$. Pt/YSZ is a better coke resistant catalyst.	Graf et al.[108]
n-Butane	Ni/δ-Al$_2$O$_3$, Ni–Pt/δ-Al$_2$O$_3$	300–400	1–4	Ni-Pt/δ-Al$_2$O$_3$ catalyst showed superior performance than the Pt-free catalyst.	Avci et al.[107]
n-Butane	8.8% Ni–0.2% Au/MgAl$_2$O$_4$	450–550	0.7	Ni–Au alloy formation helps in minimizing the carbon formation.	Chin et al.[106]

HC, hydrocarbons; YSZ, yttrium stabilized zirconia; S/C, steam-to-carbon ratio.

41

Takenaka and coworkers[112] reported that silica-coated Ni catalysts prepared by a water-in-oil microemulsion method are more active in the steaming reforming of propane compared with a Ni catalyst just supported on silica without coating. Transmission electron microscope (TEM) analysis of the former catalyst prereduced in H_2 at 700 °C showed small Ni particles with diameters of about 5 nm located at the center of about 50-nm-diameter SiO_2 particles. On the other hand, in Ni/SiO_2 catalyst, the Ni particles of 5–20 nm in diameter are attached to the outer surface of the silica support (Fig. 2.7). Also, the silica-coated Ni catalysts are more active than that supported on MgO or Al_2O_3. They also observed a lower amount of carbon deposition on silica-coated Ni catalyst compared with other catalysts. X-ray absorption studies revealed the existence of a strong metal–support interaction in the silica-coated Ni catalyst, and this interaction prevents the sintering of Ni metal particles during propane steam reforming. Zhang et al.[22] reported that the Ni/Al_2O_3 catalyst prepared by sol-gel method is more stable during steam reforming of propane compared with that prepared by conventional impregnation method.

Wang and Gorte[110] measured the rates of steam reforming of n-butane on 1 wt% Pd supported on CeO_2, SiO_2, or Al_2O_3. Among them, the Pd/CeO_2 catalyst exhibited a stable reaction rate. Also, the catalyst produced a higher CO_2/CO ratio in the reformed product stream. The higher activity of CeO_2-supported catalyst has been attributed to the existence of a dual-function mechanism, in which CeO_2 can be oxidized by H_2O and then supply oxygen to the Pd metal. The participation of CeO_2 in WGS reaction reduces the amount of CO in the product stream thereby increasing

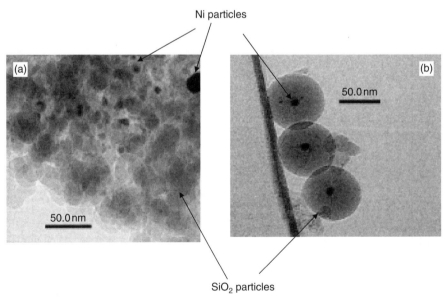

Figure 2.7.   TEM images of (a) Ni/SiO_2 catalyst prepared by impregnation and (b) SiO_2-coated Ni catalyst prepared by water-in-oil microemulsion methods. Adapted from Takenaka et al.[112]

the CO_2/CO ratio. Laosiripojana and Assabumrungrat[109] reported the participation of CeO_2 lattice oxygen in the steam reforming of hydrocarbons. A high surface area CeO_2 with a Brunauer Emmett and Teller (BET) surface area of about $100 m^2/g$ showed a better catalytic performance than that of a low BET surface area CeO_2 with $55 m^2/g$ and conventional Ni/Al_2O_3 catalyst.

Graf et al.[108] performed a comparative study of steam reforming of methane, ethane and ethylene on Pt, Rh, and Pd supported on YSZ. They observed that the reactivity and product distribution depends on the type of noble metal loaded. Over Rh/YSZ catalyst, the reactivity decreased in the order $C_2H_6 > C_2H_4 > CH_4$. On the other hand, over Pt/YSZ, methane reacted much faster than the C_2 hydrocarbons and the order of reactivity is $CH_4 > C_2H_4 > C_2H_6$ (Fig. 2.8). The higher reactivity of Rh

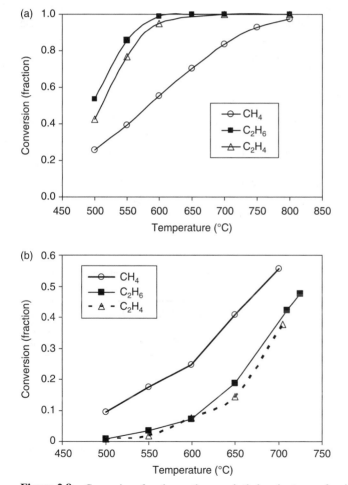

Figure 2.8. Conversion of methane, ethane, and ethylene in steam reforming reaction over (a) Rh/YSZ and (b) Pt/YSZ. Adapted from Graf et al.[108]

has been attributed to a higher binding strength of Rh toward carbon atoms compared with that of Pt. In other words, Rh has a higher tendency to break the C–C bond of adsorbed hydrocarbon than Pt metal, and hence, the Rh/YSZ catalyst exhibits a higher reactivity in the steam reforming of C_2H_6 than Pt/YSZ catalyst.

Avci and coworkers[107] observed an improved n-butane conversion and higher H_2/CH_4 ratio in the steam reforming of n-butane over a 15 wt% Ni/δ-Al$_2$O$_3$ catalyst promoted with 0.2 wt% of Pt compared with the Pt-free catalyst. Around 400 °C, the Pt-promoted Ni/δ-Al$_2$O$_3$ catalyst exhibited close to 100% n-butane conversion, while the Pt-free catalyst showed only about 67% conversion. The higher catalytic activity of the bimetallic catalyst has been attributed to the existence of a synergistic interaction between Pt and Ni metals. An empirical power-function expression describing the rate of n-butane conversion has been used to determine the kinetic parameters. The reaction orders estimated are 1.20 and −0.18 with respect to n-butane and steam, respectively. The activation energy for the reaction was found to be around 81 kJ/mol. These kinetic parameters are in close agreement with those reported over Ni/MgO, Ni/Al$_2$O$_3$, and Ni/SiO$_2$ catalysts in the steam reforming of C_2–C_4 hydrocarbons as summarized in Table 2.12.[52,107]

2.2.2.3 Mechanism

The mechanism of steam reforming of higher hydrocarbons over Ni-based catalysts has been reported to involve adsorption of the hydrocarbon on the catalyst surface followed by the formation of C_1 species by the successive α-cleavage of the C–C bonds. Similar to that proposed for the SMR reaction (Eqs. 2.12–2.20), the adsorbed C_1 species are then dehydrogenated stepwise into adsorbed carbon atoms, which may dissolve in the Ni crystal. When the concentration of carbon is above saturation, a carbon whisker will nucleate. These reactions compete with the reaction of C_1

Table 2.12. Kinetic Parameters for the Steam Reforming of Gaseous Light Hydrocarbons over Various Supported Catalysts[52,107]

Hydrocarbon	Catalyst	Temperature Range (°C)	Pressure (MPa)	Order with Respect to		Activation Energy (kJ/mol)
				Hydrocarbon	Steam	
CH$_4$	Ni/MgO	350–400	0.1	0.96	−0.17	60
C$_2$H$_6$	Ni/MgO	450–550	0.1	0.60	−0.40	76
C$_2$H$_6$	Ni/MgO	300–350	0.1	0.95	−0.46	81
C$_3$H$_8$	Ni/α-Al$_2$O$_3$	500–640	0.1	0.75	+0.60	67
C$_3$H$_8$	Ni/MgO	300–350	0.1	0.93	−0.53	45
C$_4$H$_{10}$	Ni/γ-Al$_2$O$_3$	425–475	3.0	0.00	1.00	54
C$_4$H$_{10}$	Ni/SiO$_2$	400–450	0.1	0.00	1.00	–
C$_4$H$_{10}$	Pt–Ni/ δ-Al$_2$O$_3$	330–395	0.1	1.20	−0.18	81

species with adsorbed oxygen atoms to gaseous products. The concentration of adsorbed oxygen depends on the steam adsorption on the catalyst. The steps involved in the simplified mechanism for the steam reforming of higher hydrocarbons are given below in Equations 2.29–2.37,[117]

$$C_nH_m + *2 \rightarrow *2C_nH_y, \tag{2.29}$$

$$*2C_nH_y + * \rightarrow *CH_x + *2C_n - 1H_z, \tag{2.30}$$

$$*2C_nH_y \rightarrow n*C, \tag{2.31}$$

$$*CH_x \rightarrow *C, \tag{2.32}$$

$$*C \leftrightarrow [C, Ni]bulk \rightarrow whisker\ carbon, \tag{2.33}$$

$$H_2O + * \leftrightarrow *O + H_2, \tag{2.34}$$

$$*CH_x + *O \rightarrow gas, \tag{2.35}$$

$$*C + *O \rightarrow gas, \tag{2.36}$$

$$H_2 + 2* \leftrightarrow 2*H, \tag{2.37}$$

where the asterisks (*) denotes the adsorption sites on the catalyst surface.

The mechanism for the steam reforming and dry reforming of propane, over Ni and noble metal-based supported catalysts, has been recently studied and a reaction sequence similar to the one described in Equations 2.38–2.46 has been proposed.[40,115,118,119] Unlike SMR reaction, wherein the activation of the C–H bond of CH_4 is the critical and RDS (Eq. 2.12), both C–C bond rupture in propane (Eq. 2.40) and oxidation of adsorbed CH_n species (Eq. 2.43) have been proposed to be the RDSs in the reforming of propane. Based on the kinetic isotopic effect and isotopic tracer studies, Jensen et al.[118] suggested that the C–C bond rupture (Eq. 2.40) is the RDS, because they observed the presence of C_3 species as the main hydrocarbon species on the catalyst surface. The adsorbed CH_n species reacts with adsorbed oxygen species generated either from H_2O in the steam reforming reaction or from CO_2 in the dry reforming reaction. The formation of adsorbed oxygen species from H_2O could be similar to that shown in Equations 2.16 and 2.17. The adsorbed CH_n species may also undergo hydrogenolysis reaction to produce CH_4 as shown in Equation 2.44. Graf et al.[108] also reported the formation of CH_4 primarily via hydrogenolysis rather than methanation reaction in the steam reforming of ethane over Rh supported on YSZ catalyst. Similar elementary reaction steps have also been proposed for the steam reforming of ethane:[40]

$$C_3H_8(g) + 2* \leftrightarrow C_3H_7* + H*, \tag{2.38}$$

$$C_3H_7* + * \leftrightarrow C_3H_6* + H*, \tag{2.39}$$

$$C_3H_7* + (9-n)* \rightarrow 3CH_n* + (7-n)H*\ \ (RDS), \tag{2.40}$$

$$H_2O + 2* \leftrightarrow O* + 2H*\ \ (in\ steam\ reforming\ reaction), \tag{2.41}$$

$$CO_2(g) + 2* \leftrightarrow O* + CO*\ \ (in\ dry\ reforming\ reaction), \tag{2.42}$$

$$CH_n* + O* + n* \leftrightarrow CO* + nH* + *, \tag{2.43}$$

$$CH_n* + (4-n)H* \leftrightarrow CH_4(g) + (5-n)*\ \ (hydrogenolysis), \tag{2.44}$$

$$CO* \leftrightarrow CO(g) + *, \tag{2.45}$$

$$2H* \leftrightarrow H_2(g) + 2*. \tag{2.46}$$

2.3 STEAM REFORMING OF LIQUID HYDROCARBONS

Steam reforming of liquid hydrocarbons such as naphtha was introduced to the chemical industry during the 1960s, and the process served as a major source of H_2 for ammonia synthesis as well as for petroleum refining.[120] Indeed, higher hydrocarbons became the preferred feedstock for the reforming process when natural gas was not available. Many refineries benefit from feedstock flexibility, taking advantage of the surplus of various hydrocarbon streams they produce. Recently, steam reforming of liquid hydrocarbons, especially gasoline, diesel, and jet fuel, has been considered as a potential process for near-term H_2 production for fuel cells because of their readily existing infrastructure and availability.[4–9,103,121,122] Moreover, diesel and jet fuel are the preferred logistic fuels to produce H_2-rich gas for portable fuel cells for military applications. Consequently, a significant progress in catalyst development and reactor design has been made in recent years. The chemistry involved, thermodynamics, type of catalysts, and process development in the steam reforming of these liquid hydrocarbon fuels are discussed in this section.

2.3.1 Chemistry

Typical compositions of gasoline, diesel, and jet fuel are summarized in Table 2.13[123,124] while their energy densities and the H/C ratios are included in Table 2.1. Steam reforming of these liquid hydrocarbons can be expressed by a general equation as shown in Equation 2.47:

$$C_nH_m + nH_2O \rightarrow ((m+2n)/2)H_2 + nCO. \tag{2.47}$$

However, since these transportation fuels are complex mixtures containing a wide range of hydrocarbons, aromatics, and olefins, simple hydrocarbons such as isooctane (C_8 hydrocarbon), dodecane (C_{12} hydrocarbon), and hexadecane (C_{16} hydrocarbon) have been used as representative surrogate fuels for gasoline, jet fuel, and diesel, respectively, in order to better understand the chemistry involved and

Table 2.13. Typical Compositions of Gasoline, Diesel, and Jet Fuel Available in the U.S.[123,124]

Component	Composition (vol%)		
	Gasoline (Regular Unleaded)	Diesel(#2)	JP-8 Jet Fuel
Paraffins	10.2	55.0	71.0
Isoparaffins	37.6	–	–
Aromatics	38.7	24.0	19.0
Naphthalenes	6.0	12.0	6.2
Olefins	7.1	5.0	3.5

thermodynamics of the reforming process better. Steam reforming reactions of these surrogate fuels are represented by Equations 2.48–2.50, respectively. The use of other liquid hydrocarbons, including h-heptane, n-decane, and n-tetradecane, as well as surrogates containing a mixture of a range of hydrocarbons and aromatics, is also known:[125–128]

$$\text{Iso-}C_8H_{18}(g) + 8H_2O(g) \rightarrow 17H_2(g) + 8CO(g); \Delta H^\circ_{298} = +1273 \text{ kJ/mol}, \quad (2.48)$$

$$C_{12}H_{26}(g) + 12H_2O(g) \rightarrow 25H_2(g) + 12CO(g); \Delta H^\circ_{298} = +1866.2 \text{ kJ/mol}, \quad (2.49)$$

$$C_{16}H_{34}(g) + 16H_2O(g) \rightarrow 33H_2(g) + 16CO(g); \Delta H^\circ_{298} = +2473.9 \text{ kJ/mol}. \quad (2.50)$$

These reactions are highly endothermic and occur at relatively higher temperatures, between 700 and 1000 °C over a nickel-based or noble metal-based supported catalyst. Under the reaction operating conditions, other reactions such as cracking into carbon and hydrogen followed by carbon gasification as mentioned in the SMR process, cracking into lower hydrocarbons such as methane, ethane, and ethylene, followed by steam reforming of these cracked hydrocarbons, and WGS reaction (Eq. 2.2) to convert CO into CO_2 and additional H_2 could also occur. The overall reactions for the steam reforming of these liquid fuels after combining the WGS reaction (equimolar concentration of CO and H_2O) are given in Equations 2.51–2.53, respectively. Similar to the SMR process, the steam reforming of higher hydrocarbons in practice is also performed at higher S/C ratios, typically between 1 and 5, in order to reduce the risk of carbon deposition.[129–132] While higher operating pressures are preferred for industrial applications, the reactions are generally operated at atmospheric pressure for fuel cell applications. HTS and LTS and/or Prox reactions are used in the downstream of the steam reformer to remove CO from the reformed gas and to produce H_2-rich gas suitable for fuel cells:

$$\text{Iso-}C_8H_{18}(g) + 16H_2O(g) \rightarrow 25H_2(g) + 8CO_2(g); \Delta H^\circ_{298} = +944.1 \text{ kJ/mol}, \quad (2.51)$$

$$C_{12}H_{26}(g) + 24H_2O(g) \rightarrow 37H_2(g) + 12CO_2(g); \Delta H^\circ_{298} = +1372.6 \text{ kJ/mol}, \quad (2.52)$$

$$C_{16}H_{34}(g) + 32H_2O(g) \rightarrow 49H_2(g) + 16CO_2(g); \Delta H^\circ_{298} = +1816 \text{ kJ/mol}. \quad (2.53)$$

Because gasoline, jet fuel, and diesel are complex hydrocarbon mixtures, reforming of these liquid hydrocarbon fuels is relatively more complicated than natural gas reforming. The catalyst used should be capable of breaking several C–C bonds, in addition to activating the C–H bonds. A prereformer may thus be used for cracking the long-chain hydrocarbons into mainly C_1 products, which then can be reformed in the main reformer. The presence of fuel additives also affects the reforming reaction.[132] Furthermore, since these fuels have a significant amount of sulfur compounds, a desulfurization step is also required to reduce the sulfur content to below tolerable level, usually below 1 ppmw, before reforming. Desulfurization of these transportation fuels is discussed separately in this book.

2.3.2 Thermodynamics

The thermodynamic parameters ΔH, ΔG, and equilibrium constant (Keqm) for the steam reforming of isooctane, dodecane, and hexadecane are given in

Table 2.14. Thermodynamic Data for the Steam Reforming of Isooctane:
Iso-C_8H_{18}(g) + 8H_2O(g) → 17H_2(g) + 8CO(g)

Temperature (°C)	$\Delta H°$ (kJ/mol)	$\Delta G°$ (kJ/mol)	Log K
25	1273.176	714.351	−125.162
75	1285.783	619.609	−92.971
125	1296.784	523.163	−68.641
175	1306.358	425.42	−49.589
225	1314.648	326.676	−34.257
275	1321.769	227.156	−21.648
325	1327.803	127.035	−11.095
375	1332.879	26.449	−2.132
425	1337.096	−74.496	5.574
475	1340.533	−175.716	12.269
525	1343.253	−277.142	18.139
575	1345.311	−378.718	23.326
625	1346.74	−480.397	27.941
675	1347.566	−582.14	32.073
725	1347.812	−683.91	35.793
775	1347.501	−785.679	39.158
825	1346.655	−887.421	42.215
875	1345.29	−989.112	45.003
925	1343.431	−1090.733	47.556
975	1341.109	−1192.267	49.9
1000	1339.785	−1242.996	51.002

Tables 2.14–2.16, respectively. The $\Delta G°$ values for these reactions become negative above 425 °C, implying that the reaction is thermodynamically favorable above this temperature. However, cracking of these liquid hydrocarbons into CH_4 and carbon formation shown in Equations 2.54–2.58 are highly facile reactions with negative $\Delta G°$ values. Also, the $\Delta G°$ value becomes more negative with increasing carbon number of the fuel, indicating that carbon deposition becomes serious at low reaction temperature and with increasing carbon number of the liquid hydrocarbon fuel:[133,134]

- Isooctane cracking

$$\text{Iso-}C_8H_{18}(g) \rightarrow 4CH_4(g) + H_2(g) + 4C(s)$$
$$\Delta H_{298}° = -75.5 \text{ kJ/mol} \qquad (2.54)$$
$$\Delta G_{298}° = -219 \text{ kJ/mol}$$
$$\text{Iso-}C_8H_{18}(g) \rightarrow 9H_2(g) + 8C(s)$$
$$\Delta H_{298}° = +222.9 \text{ kJ/mol} \qquad (2.55)$$
$$\Delta G_{298}° = -16.9 \text{ kJ/mol}$$

Table 2.15. Thermodynamic Data for the Steam Reforming of Dodecane:
$C_{12}H_{26}(g) + 12H_2O(g) \rightarrow 25H_2(g) + 12CO(g)$

Temperature (°C)	$\Delta H°$ (kJ/mol)	$\Delta G°$ (kJ/mol)	Log K
25	1866.238	1046.116	−183.29
75	1885.608	907.034	−136.098
125	1904.646	765.187	−100.396
175	1923.502	620.954	−72.382
225	1942.268	474.618	−49.771
275	1961.001	326.398	−31.106
325	1979.739	176.466	−15.412
375	1998.507	24.966	−2.012
425	2017.325	−127.986	9.577
475	2036.207	−282.289	19.711
525	2055.161	−437.856	28.658
575	2074.196	−594.615	36.623
625	2093.296	−752.498	43.767
675	2112.441	−911.447	50.217
725	2131.615	−1071.406	56.073
775	2150.801	−1232.327	61.418
825	2169.986	−1394.163	66.32
875	2189.15	−1556.872	70.835
925	2208.287	−1720.416	75.01
975	2227.404	−1884.758	78.883
1000	2236.958	−1967.219	80.718

- Dodecane cracking

$$C_{12}H_{26}(g) \rightarrow 6CH_4(g) + H_2(g) + 6C(s)$$
$$\Delta H°_{298} = -156.8 \text{ kJ/mol} \quad (2.56)$$
$$\Delta G°_{298} = -353.9 \text{ kJ/mol}$$
$$C_{12}H_{26}(g) \rightarrow 13H_2(g) + 12C(s)$$
$$\Delta H°_{298} = +290.8 \text{ kJ/mol} \quad (2.57)$$
$$\Delta G°_{298} = -50.8 \text{ kJ/mol}$$

- Hexadecane cracking

  Equation 2.26:

$$C_{16}H_{34}(g) \rightarrow 8CH_4(g) + H_2(g) + 8C(s)$$
$$\Delta H°_{298} = -223.5 \text{ kJ/mol}$$
$$\Delta G°_{298} = -488.4 \text{ kJ/mol}$$
$$C_{16}H_{34}(g) \rightarrow 17H_2(g) + 16C(s) \quad (2.58)$$
$$\Delta H°_{298} = +373.3 \text{ kJ/mol}$$
$$\Delta G°_{298} = -81.2 \text{ kJ/mol}$$

Table 2.16. Thermodynamic Data for the Steam Reforming of Hexadecane: $C_{16}H_{34}(g) + 16H_2O(g) \rightarrow 33H_2(g) + 16CO(g)$

Temperature (°C)	$\Delta H°$ (kJ/mol)	$\Delta G°$ (kJ/mol)	Log K
25	2473.894	1378.25	−241.484
75	2497.996	1192.548	−178.939
125	2519.125	1003.582	−131.674
175	2537.691	812.114	−94.665
225	2553.981	618.695	−64.88
275	2568.211	423.74	−40.383
325	2580.561	227.57	−19.875
375	2591.186	30.437	−2.453
425	2600.224	−167.456	12.53
475	2607.805	−365.945	25.552
525	2614.049	−564.896	36.973
575	2619.074	−764.201	47.068
625	2622.964	−963.769	56.056
675	2625.799	−1163.525	64.105
725	2627.659	−1363.404	71.355
775	2628.625	−1563.355	77.916
825	2628.783	−1763.332	83.882
875	2628.209	−1963.3	89.327
925	2626.996	−2163.228	94.316
975	2625.258	−2363.095	98.903

The effect of temperature on equilibrium compositions in the steam reforming of hexadecane is shown in Figure 2.9 as an example. At a given S/C ratio of 2.5, the H_2 yield generally increases with increasing temperature, but the RWGS reaction dominates at elevated temperatures, above 775 °C, which decreases the H_2 yield with further increasing temperature. Under this high S/C ratio, while the carbon formation disappears above around 250 °C, the formation of methane continues until around 700 °C. These data suggest that the ideal temperature range for steam reforming of hexadecane should be between 700 and 800 °C at the S/C ratio of 2.5.

The effect of S/C ratio on equilibrium compositions in the steam reforming of hexadecane at 800 °C is shown in Figure 2.10. The reaction is thermodynamically favorable without any carbon formation at an S/C ratio above 1.1. Although the H_2 yield increases with increasing S/C beyond 1.1, the reaction at higher S/C ratios will negatively impact the overall efficiency of the process because of the higher heat input required for the system. Therefore, an S/C ratio between 2 and 3 is typically used for the steam reforming of liquid hydrocarbons.

Since the steam reforming reaction proceeds with increasing number of moles, the reaction is favored at low pressure. Increasing reaction pressure decreases the

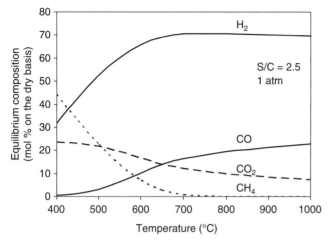

Figure 2.9.   Effect of temperature on equilibrium compositions in steam reforming of *n*-hexadecane; S/C = 2.5, 1 atm pressure.

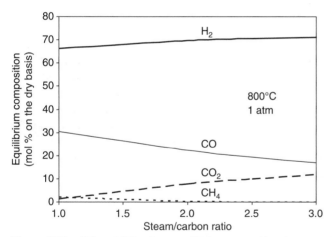

Figure 2.10.   Effect of S/C ratio on equilibrium compositions in steam reforming of *n*-hexadecane at 800 °C and 1 atm pressure.

H_2 yield as shown in Figure 2.11. Also, the yields of CO and CO_2 decline while that of methane increases due to the methanation reaction at higher pressures. Although thermodynamics dictates the use of low pressure, industrial steam reforming reactions are generally carried out at elevated pressures (15–35 atm) because much of the H_2 produced is supplied to methanol and ammonia synthesis plants where higher pressures facilitate a better heat recovery and result in compression energy savings.[10,103,120]

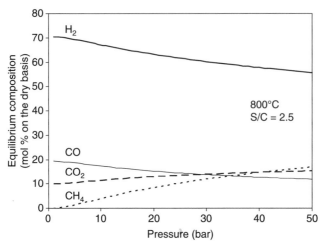

Figure 2.11. Effect of pressure on equilibrium compositions in steam reforming of n-hexadecane at 800 °C; S/C ratio = 2.5.

2.3.3 Catalyst

The type of catalysts used for the steam reforming of liquid hydrocarbons can be broadly classified into at least three major categories: (a) Ni-based base metal catalysts, (b) noble metal-based catalysts, and (c) metal carbides. The Ni-based and noble metal catalysts are usually dispersed on various supports including γ-alumina, α-alumina, magnesium aluminate, calcium aluminate spinels, and pervoskites, as well as oxide ion-conducting substrate such as CeO_2 and ZrO_2. The catalysts known for steam reforming of liquid hydrocarbons were reviewed recently by Cheekatamarla and Finnerty[97] and Shekhawat et al.[103] The performance of some of the most recently reported catalysts are summarized in Table 2.17[125,135–141] and briefly discussed in this section.

The Ni-based supported catalysts known for the steam reforming of natural gas and other gaseous hydrocarbons should be active for the steam reforming of gasoline, jet fuel, and diesel as well. However, as discussed above in the thermodynamics section, carbon formation is a serious issue in the reforming of liquid hydrocarbons. Thus, efforts on designing new catalysts for the reforming of these liquid fuels are focused on avoiding carbon formation. One approach to minimize the carbon formation in Ni-based catalysts is to use alkali promoters such as K_2O, MgO, CaO, and SrO. In some cases, a small amount of SiO_2 was also used to prevent the evaporation of alkali oxides such as K_2O. The compositions of commercial Ni-based catalysts known for the reforming of naphtha are included in Table 2.10.

InnovaTech, Inc. of the U.S. has recently reported proprietary Inniva Tech Catalyst (ITC) catalyst series for the steam reforming of various hydrocarbons, including natural gas, gasoline, and diesel at atmospheric pressure.[130] The detailed information on catalyst compositions is not available. It appears that the catalyst

Table 2.17. Steam Reforming of Liquid Hydrocarbons over Some of the Most Recently Reported Catalyst Compositions

Hydrocarbon	Catalyst	Reaction Conditions			Results	Reference
		Temperature (°C)	S/C Ratio	GHSV (h⁻¹)		
Isooctane	Bulk Mo_2C	1000	1–3	1000–7000	S/C of 1.3 produces a maximum H_2, avoids preformer, negligible coke formation	Cheekatamarla and Thomson[139]
Hexadecane	Bulk Mo_2C	965	1–2	4000–6000	Higher H_2 yield at a lower S/C ratio of 1.18	Cheekatamarla and Thomson[138]
Isooctane	Mo_2C (commercial)	650–1000	0.5–2.0	0.5–2.0	Reaction onset temperature is 850 °C. Higher H_2 yields were obtained at an S/C of 1 and low WHSV of below 1.8 h⁻¹	Flores and Ha[140]
Jet fuel (surrogate and real)	2% Rh–10% Ni/Al_2O_3–CeO_2	515	3	2000–3000	Capable of reforming the real JP-8 jet fuel containing 22 ppm sulfur. Ni–Rh synergism plays a role for improved catalytic activity	Strohm et al.[137]
Hexadecane	Pd/ZrO_2 coated on metal foil from Catacel Corp.	600–900	3–6	22,000	Improved catalytic performance has been observed at higher S/C ratio, higher temperature, and feed containing lower sulfur content	Goud et al.[141]
Hexadecane	0.3%–1% Rh/ NiMgAl hydrotalcite	700–950	3	10,000–100,000	Loading of 0.3 wt% Rh promotes the NiMgAl catalyst for sustained activity	Kim et al.[136]
Isooctane	1.5% Pt/ $Ce_{0.8}Gd_{0.2}O_{1.9}$	750	3	0.5–3.5	Catalyst calcination temperature plays a crucial role for catalytic activity and sulfur tolerance. Catalyst calcined at 800 °C exhibit a better performance and sulfur tolerance due to improved catalyst stability and synergism	Lu et al.[135]
n-Heptane + n-hexane	NiMgAl hydrotalcite having 15% Ni	550	3	225,160	Catalyst preparation methods, physicochemical properties, and catalytic properties were studied. Stronger metal–support interaction with increasing Mg, greater specific surface area with increasing Al	Melo and Morlanés[125]

WHSV, weight hourly space velocity.

53

consisted of a noble metal-based bimetallic compound supported on high surface area Al_2O_3 treated with an oxide having oxygen ion-conducting properties, probably CeO_2 or ZrO_2. A 300-h continuous test has shown that the catalyst has a very stable performance for steam reforming of isooctane at 800 °C with an S/C ratio of 3.6. The same catalyst was also tested in the steam reforming of hexadecane for 73 h as well as natural gas for over 150 h continuously, without deactivation or carbon deposition. The catalyst has been reported to be tolerant to sulfur up to about 100 ppmw present in gasoline.

Melo and Morlanés recently reported mixed oxide catalysts containing Ni, Mg, and Al obtained by thermal decomposition of hydrotalcite or layered double hydroxide material for the steam reforming of a mixture of *n*-heptane and *n*-hexane as a surrogate for naphtha.[125] They studied the effect of catalyst synthesis parameters such as method of introduction of active species, calcinations, and reduction temperatures on the catalytic performance. Murata et al.[142] reported the steam reforming of isooctane and methylcyclohexane (MCH) over $FeMg/Al_2O_3$ and Ni/ZrO_2 catalysts. They observed that the rate of H_2 production over Ni/ZrO_2 catalyst was seven times higher than that over $FeMg/Al_2O_3$ catalyst. Furthermore, the stability of Ni/ZrO_2 catalyst was improved by the addition of Sr. During the 100-h time-on-stream experiments, the rate of H_2 production dropped only by 10 wt% over $NiSr/ZrO_2$ catalyst compared with a 50% drop in H_2 production rate observed over Sr-free Ni/ZrO_2 catalyst. The improved stability of the catalyst has been attributed to the formation of mixed oxide phase containing Ni, Sr, and Zr.

Flytzani-Stephanopoulos and Voeck[143] reported honeycomb monolithic supports as promising for steam reforming catalysts, improving radial heat transfer in the reactor and hence to achieve a uniform reactor temperature zone, which could minimize the carbon formation on the catalyst bed associated with hot spots. Additionally, monolith can provide faster response during transients in fuel cell applications due to their better heat transfer properties. They performed the steam reforming of *n*-hexane over conventional Ni-based supported catalysts and two monolithic supported catalysts, a ceramic monolith of Ni impregnated on γ-alumina washcoated cordierite, and a metal monolith of Ni supported on γ-alumina washcoated Kanthal (Fe–Cr–Al alloy). A hybrid monolith comprised of metal monolith at the top and ceramic monolith at the bottom provided higher H_2 yields and *n*-hexane conversions compared with the conventional pellets at an S/C ratio of 2.5, temperature of 927 °C, and space velocity of 4000 h^{-1}.

Noble metals and hybrid of noble metal- and base metal-supported catalysts have also been developed in order to minimize the carbon formation and sulfur tolerance in the steam reforming of liquid hydrocarbons. Idemitsu Co., Ltd. in Japan has reported Ru-based naphtha reforming catalyst (ISR-7G) for the steam reforming of kerosene,[134] whose chemical composition is more or less similar to that of JP-8 jet fuel as described in Table 2.13. The catalyst exhibited stable activity over 12,000 h with 100% kerosene conversion and stable outlet gas composition. The catalyst was expected to meet the target lifetime of about 40,000 h, which corresponds to 10 years on the presupposition that the fuel cell would be operated in daytime only.

Wang and Gorte[144] measured the differential rates of steam reforming of methane, ethane, *n*-butane, *n*-hexane, 2-4-dimethylhexane, *n*-octane, cyclohexane, benzene, and toluene over 1 wt % Pt or Pd supported on CeO_2 and Al_2O_3. They have reported that the Pd/CeO_2 or Pt/CeO_2 is much more active in the steam reforming of hydrocarbon fuels than that supported on Al_2O_3. Their finding indicated that the primary role of CeO_2 is to dissociate water and transfer the oxygen produced by that reaction to the supported metal. This transfer of oxygen from CeO_2 to the metal also helps to maintain a carbon-free catalyst surface thereby retaining the catalytic activity for a longer period of time. Fourier transform infrared (FTIR) spectroscopic results showed that CeO_2 was covered by a carbonate layer under steam reforming conditions, and this could limit the rate at which CeO_2 can be reoxidized. Since CeO_2–ZrO_2 mixed oxides are more reducible than pure CeO_2, the carbonate formation could be suppressed over these mixed oxides. Consequently, higher reforming rates could be achieved over Pt or Pd supported on CeO_2–ZrO_2 mixed oxides. The authors also reported that in a series of alkanes from methane to *n*-hexane, the rate of formation of CO and CO_2 increased with increasing carbon numbers (Fig. 2.12). For molecules larger than C_6, there is a tendency to form cyclohexane, which further reacts with benzene. Also, aromatics are less susceptible to reforming than *n*-hexane.

Suzuki et al.[145] carried out a long-term stability of a highly dispersed Ru/CeO_2–Al_2O_3 catalyst containing 2 wt % Ru, 20 wt % CeO_2, and balance alumina in the steam reforming of desulfurized kerosene ($C_{10}H_{22}$) containing below 0.1 ppm sulfur at 800 °C and an S/C ratio of 3.5. They observed a stable activity with 97%–99% kerosene conversion with about 70% H_2 in the product stream over 8000 h of continuous operation. They also observed that the presence of CeO_2 in the catalyst formulation acts as a promoter to improve the sulfur tolerance. The kerosene conversion dropped to about 85% in 25 h onstream when the kerosene containing about 50 ppm sulfur was used. On the other hand, the conversion dropped to about 60% during the same period of time (25 h) when Ru-based catalyst without CeO_2 was used. The same

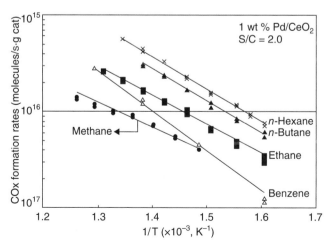

Figure 2.12. Arrhenius plots for the steam reforming of methane, ethane, *n*-butane, *n*-hexane, and benzene over Pd/CeO_2 catalyst. Water partial pressure is 100 torr. Adapted from Wang and Gorte.[144]

group also evaluated the La_2O_3 supported Ru, Rh, and Ir catalysts in the steam reforming of kerosene in the temperature range between 500 and 800 °C. At 600 °C and an S/C ratio of 3, Rh- and Ru-supported catalysts produced higher hydrocarbon conversions of about 95% and H_2 yield of close to 100% compared with Ir-supported catalyst, which showed only about 60% conversion and 80% H_2 yield under the same operating conditions.

Murata et al.[142] compared the performance of various bimetallic catalysts containing a noble metal and a base metal for the steam reforming of n-octane and MCH with noble metal-based monometallic catalysts and $FeMg/Al_2O_3$ catalyst. They noted that the $FeMg/Al_2O_3$ catalyst produced a higher rate of H_2 formation compared with noble metal-based supported catalysts. The highest H_2 production rate has been observed over $RhFe/Al_2O_3$ bimetallic catalyst (Fig. 2.13). Kim et al.[136] also observed that loading of about 0.3 wt % Rh on NiMgAl mixed oxide catalysts derived from hydrotalcite precursors improved the catalytic activity in the steam reforming of n-hexadecane. Added Rh acts as a promoter, reduces the carbon formation on the catalyst surface, and prevents Ni metal from sintering.

Strohm et al.[137] reported that a bimetallic catalyst containing 2 wt % Rh and 10 wt % Ni supported on CeO_2–Al_2O_3 is an excellent catalyst for the low temperature steam reforming of a real jet fuel containing about 22 ppmw sulfur. The catalyst exhibited a stable activity with a jet fuel conversion close to 100% during the first 45 h onstream, after which it decreased slightly but exhibited around 95% conversion even after 72 h of onstream (Fig. 2.14). It has been proposed that the added Ni closely interacts with Rh metal and protects Rh from sulfur poisoning through two possible mechanisms: (a) when Ni reacts preferentially with sulfur, the Rh can remain active for steam reforming and (b) when some Rh atoms remain unreactive toward sulfur, such sulfur atoms can be transferred from Rh to those Ni atoms that are present in close vicinity to Rh. In other words, S in RhSx may be transferred to the Ni metal

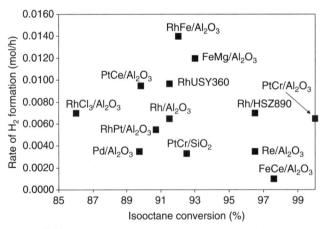

Figure 2.13.   Correlation between conversion and rate of H_2 formation in steam reforming of isooctane over various monometallic- and bimetallic-supported catalysts. Adapted from Murata et al.[142]

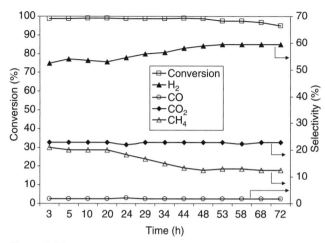

Figure 2.14. Performance of 2 wt% Rh–10 wt% Ni-supported CeO_2–Al_2O_3 catalyst in the steam reforming of a pre-desulfurized jet fuel containing about 22 ppmw sulfur. Adapted from Strohm et al.[137]

through sulfur spillover with or without the aid of gas-phase H_2. In either case, a close interaction between the added Ni and Rh is necessary.

The authors also studied the effect of Ni loading and sulfur concentration in the fuel on the performance of a Rh/CeO_2–Al_2O_3 catalyst using a surrogate of jet fuel called NORPAR-13, an industrial solvent obtained from ExxonMobile comprising only normal paraffins with an average carbon number of 13 and containing about 4 ppmw sulfur and spiked with 3-methylbenzothiophene to achieve different levels of sulfur. The results revealed that an effective interaction between Ni and Rh could be achieved by coimpregnating Ni and Rh on CeO_2–Al_2O_3 support rather than impregnating Rh on pre-Ni-impregnated CeO_2–Al_2O3 catalyst (Fig. 2.15).[137] Also, the data show that an optimum Ni loading is around 10 wt% to achieve a relatively stable activity. The monometallic Rh/CeO_2–Al_2O_3 catalyst deactivates rapidly by S poisoning. It has also been reported that the catalytic activity rapidly decreased when the amount of sulfur in the fuel flown over the catalyst reaches the level corresponding to an S_{fuel} : Rh_{surf} atomic ratio of 0.28:0.30. Ni surface saturation of sulfur was found to occur at an S_{fuel} : Ni_{surf} atomic ratio of 0.59:0.60, and this corresponds to an S_{fuel} : Rh_{surf} atomic ratio of 1:1 for the catalyst 2% Rh–10% Ni/CeO_2–Al_2O_3 with over 95% fuel conversion.

Metal carbides have attracted much interest in recent years as promising catalysts in the steam reforming of hydrocarbons and WGS reaction as these materials are more tolerant toward carbon deposition and sulfur poisoning.[146–150] Bulk molybdenum carbide (Mo_2C) has been employed as catalysts in the steam reforming and oxidative steam reforming of gasoline and diesel for hydrogen production.[97] Since these catalysts are carbon tolerant, fuel prereforming is not required. Also, the steam reforming reactions could be conducted with a much lower S/C ratio of around 1

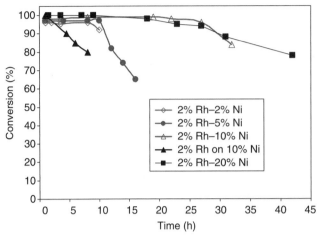

Figure 2.15. Effect of Ni loading on catalytic performance of 2% Rh/CeO_2–Al_2O_3 catalysts in the steam reforming of NORPAR-13 surrogate jet fuel. Adapted from Strohm et al.[137] See color insert.

compared with that employed on the Ni-based supported catalysts, which require an S/C ratio of at least 3. A few recent reports on the steam reforming of liquid hydrocarbons on Mo_2C catalysts are included in Table 2.17.[138–140]

Flores and Ha[140] recently reported that the catalytic activity of Mo_2C is initiated by thermal decomposition of the carbide phase and propagated by the carburization of the dioxide phase with isooctane as shown below in Equations 2.59 and 2.60:

$$Mo_2C + 5H_2O \leftrightarrow 2MoO_2 + CO + 5H_2, \tag{2.59}$$

$$2MoO_2 + 0.51C_8H_{18} \leftrightarrow 2Mo_2C + 4.59H_2 + 2.16CO + 0.92CO_2. \tag{2.60}$$

Thus, the reforming process becomes stable as long as these reactions are continuously taking place. The onset temperature of the reforming process around 750 °C has been directly related to the temperature at which Mo_2C is able to produce Mo metal.

2.3.4 Kinetics

Since the liquid hydrocarbons are complex mixtures of hundreds of components, and each one could undergo several different reactions under the reforming conditions, the determination of a detailed reaction model for reforming them is very complicated. For example, in addition to the reactions associated with reforming, other reactions such as WGS, cracking followed by carbon formation, methanation, hydrocracking, dehydrocyclization, dehydrogenation, ring opening, hydrogenation, and so on may take place. For simplification, the experimental rate data for such complex reactions are fit into power law or even first-order rate expressions. Unfortunately, these rate expressions tend to be limited to a specific catalyst, fuel composition, and operating conditions. It would be desirable to develop predictive models to account for variations in these parameters. However, only limited information is available on the kinetics of liquid hydrocarbon reforming for H_2 production. Literature is limited mostly to kinetic studies of steam reforming of single paraffinic components.

For the steam reforming of higher hydrocarbons, Rostrup-Nielsen and Tottrup postulated a Langmuir–Hinshelwood–Houghen–Watson (LHHW) kinetic model.[15,151,152] They assumed that the hydrocarbon chemisorbs on a duel catalytic site, followed by successive α-scission of the C–C bond. The resulting C_1 species react with adsorbed steam to form H_2 and CO. The rate expressions were generated by fitting the data for steam reforming of n-heptane (C_7H_{16}) on an Ni/MgO catalyst at 500 °C. The overall rate expression is given in Equation 2.61:

$$r = \frac{240 \exp(-8150/T) pC_7H_{16}}{[1 + 25.2 pC_7H_{16}(pH_2/pH_2O) + 0.08(pH_2O/pH_2)]^2}. \tag{2.61}$$

This rate expression was reduced to a power law-type model shown in Equation 2.62,

$$r = k_0 e^{(-8150/T)} p_{C_7H_{16}}^{0.2} p_{H_2O}^{-0.2} p_{H_2}^{0.4}, \tag{2.62}$$

wherein r is the rate of the reaction in kmol/m²(Ni)/h and the partial pressures p are in megapascal units. A very low reaction order with respect to C_7H_{16} suggests a strong adsorption of the hydrocarbon on the catalyst surface. A retarding effect of aromatics and higher hydrocarbons on the reaction rate has been observed. Due to the presence of π-electrons, aromatics can strongly adsorb onto the catalyst surface and can cause the reaction order to approach zero. The reaction order with respect to steam was very sensitive to temperature and was found to decrease from −0.6 at 450 °C to −0.2 at 550 °C. The steam-surface interactions weaken with increasing temperature, and hence, the reaction order with respect to steam approaches zero due to weak adsorption at high temperatures.

Praharso et al.[153] also developed an LH type of kinetic model for the steam reforming of isooctane over a Ni-based catalyst. In their model, they assumed that both the hydrocarbon and steam are dissociatively chemisorbed on two different dual sites on the catalyst surface. The bimolecular surface reaction between dissociated adsorbed species was proposed as the RDS. The generalized rate expression they proposed is given in Equation 2.63,

$$r = \frac{k_{rxn}\sqrt{p_{isooctane} p_{H_2O}}}{\left(1 + \sqrt{K_{isooctane} P_{isooctane}}\right)\left(1 + \sqrt{K_{H_2O} p_{H_2O}}\right)}, \tag{2.63}$$

where $K_{isooctane}$ and K_{H2O} are the adsorption constants for isooctane and steam, respectively. The power law rate expression based on the regression of the rate data over a temperature range of 310–350 °C is given in Equation 2.64:

$$r = k_0 e^{(-5300/T)} p_{isooctane}^{0.2} p_{H_2O}^{0.5}. \tag{2.64}$$

As noted above, the very low reaction order with respect to isooctane suggests a strong adsorption of the hydrocarbon on the catalyst surface.

Shi et al.[154] recently studied the steam reforming of isooctane in a monolithic type reactor simulated by a three-dimensional CFD model. They considered global reactions to represent steam reforming of isooctane, which include steam reforming of isooctane to syngas as expressed in Equation 2.48, WGS reaction as shown in Equation 2.2, and the net reaction by combining these two reactions to produce H_2 and CO_2 as shown in Equation 2.51:

Isooctane to syngas (Eq. 2.48): Iso-C_8H_{18}(g) + 8H_2O(g) → 17H_2(g) + 8CO(g),

WGS (Eq. 2.2): CO(g) + H_2O(g) → CO_2(g) + H_2(g),

   Isooctane to H_2 and CO_2 (Eq. 2.51):
Iso-C_8H_{18}(g) + 16H_2O(g) → 25H_2(g) + 8CO_2(g).

The rate expressions for these three reactions are given in Equations 2.65–2.67, respectively:

$$r_1 (\text{isooctane to syngas}) = \frac{k_1 \left(P_{C_8H_{18}} - P_{H_2}^3 P_{CO} / P_{H_2O} K_{e1} \right)}{P_{H_2O}^{0.6}}$$
$$\left(\text{kmol } C_8H_{18} \text{ reacted} / m^2 / s \right), \qquad (2.65)$$

$$r_2 (\text{WGS}) = k_2 \left(P_{CO} - P_{H2} P_{CO2} / P_{H2O} k_{e2} \right)$$
$$\left(\text{kmol CO reacted} / m^2 / s \right), \qquad (2.66)$$

$$r_2 (\text{isooctane to } H_2 \text{ and } CO_2) = \frac{k_3 \left(P_{C_8H_{18}} - P_{H_2}^4 P_{CO_2} / P_{H_2O}^2 K_{e3} \right)}{P_{H_2O}^{0.3}}$$
$$\left(\text{kmol } C_8H_{18} \text{ reacted} / m^2 / s \right). \qquad (2.67)$$

They have calculated the equilibrium constants for these reactions by fitting the preexponential factors and activation energy for each reaction shown in Table 2.18 in Arrhenius equation (Eq. 2.68). The variation of equilibrium constant with temperature for these reactions is given in Equations 2.69–2.71, respectively:

$$k_i = A \exp(-Ei/RT), \qquad (2.68)$$

$$K \text{ (isooctane to syngas)} = 8.10 \times 10^6 (T)^{3.03} \exp(-202.3 \text{ kJ/mol}/RT)(\text{bar}^2), \quad (2.69)$$

$$K \text{ (WGS)} = 9.01 \times 10^{-6} (T)^{0.968} \exp(-43.6 \text{ kJ/mol}/RT), \qquad (2.70)$$

$$K \text{ (isooctane to } H_2 \text{ and } CO_2) = 4.93 \times 10^5 (T)^{2.85} \exp(-166.38 \text{ kJ/mol}/RT)(\text{bar}^2).$$
$$(2.71)$$

Table 2.18.   Activation Energy and Preexponential Factor for the Steam Reforming of Isooctane Used in the Simulation

Reaction	Preexponential Factor (A)	Activation Energy (kJ/mol)
Iso-C_8H_{18} + 8H_2 → 17H_2 + 8CO	1.4×10^{11} (kmol/m²/s/bar$^{0.4}$)	240.1
CO + H_2O → CO_2 + H_2	25 (kmol/m²/s/bar)	67.1
Iso-C_8H_{18} + 16H_2O → 25H_2 + 8CO_2	1.0×10^{11} (kmol/m²/s/bar$^{0.7}$)	243.9

Goud et al.[141] studied the kinetics of deactivation of Pd/ZrO_2 catalyst in the steam reforming of n-hexadecane. A first-order kinetic model, with first-order deactivation rate, was used to obtain the best fit values for the reaction rate constant and the deactivation rate constant as a function of S/C ratio, temperature, and sulfur loading. They noticed that the reaction rate was enhanced by an increase in temperatures and S/C ratios, but decreased by the presence of sulfur. The catalyst deactivation was more rapid in the presence of sulfur, at low S/C ratios and at lower temperatures.

The kinetics of partial oxidation, ATR, and dry reforming of liquid hydrocarbons have also been reported recently.[103,155] Pacheco et al.[155] developed and validated a pseudo-homogeneous mathematical model for the ATR of isooctane and the subsequent WGS reaction, based on the reaction kinetics and intraparticle mass transfer resistance. They regressed the kinetic expressions from the literature for partial oxidation and steam reforming reactions to determine the kinetics parameters for the ATR of isooctane on Pt/ceria catalyst. The rate expressions used in the reformer modeling and the parameters of these rate expressions are given in Tables 2.19 and 2.20, respectively.

Although the kinetics of reforming of pure paraffins may be satisfactorily represented by LHHW-type models, such models are difficult to apply for real fuels, such as diesel or gasoline. For such complex systems, it may be more practical to use pseudo-homogeneous kinetics. For example, hydrocarbon fuel components can be lumped in groups with similar properties and kinetic behaviors; for example, paraffins are also grouped into lumped reactions. However, the levels of simplification must be carefully evaluated to make them consistent with the final aim of the kinetic model.

2.3.5 Mechanism

The mechanism of steam reforming of higher hydrocarbons discussed in Section 2.2.2.3 is also valid for the steam reforming of liquid hydrocarbons as well. As noted, the mechanism involves adsorption of the hydrocarbon on the catalyst surface followed by the formation of C_1 species by the successive α-cleavage of the C–C bonds. The steps involved in the steam reforming of higher hydrocarbons are also given in Equations 2.29–2.37.[120]

2.3.6 Prereforming

Since liquid hydrocarbons consist of a complex mixture of heavier hydrocarbons, aromatics, and olefins, they undergo severe carbon deposition and catalyst deactivation during reforming. The aromatics present in the fuel could hinder the overall reforming rate by occupying the catalytic sites for a longer time due to π-complexation between the d-electrons of the metal and π-electrons of the aromatics. One way to minimize the carbon deposition is to prereform the liquid hydrocarbons to produce lighter gaseous mixture containing primarily CH_4, H_2, CO, CO_2, and some C_2–C_4

Table 2.19. Rate Expressions Used by Pacheco et al. for Isooctane Reformer Modeling[155]

Reaction	Rate Expression	Catalyst	Temperature (°C)
$C_8H_{18} + 16O_2 \rightarrow 9H_2O + 8CO_2$	$r_1 = k_1 P_{C8H18} P_{O2}$	Ni/Al_2O_3	800–900
$C_8H_{18} + 8H_2O \rightarrow 17H_2 + 8CO$	$r_2 = \dfrac{k_2}{P^{2.5}_{H_2}} \dfrac{P_{C8H18} P_{H2O} - P^3_{H_2} P_{CO}/K_1}{1+(K_{CO}P_{CO} + K_{H_2}P_{H_2} + K_{C8H18}P_{C8H18} + K_{H2O}P_{H2O}/pH_2)^2}$	$Ni/Mg–Al_2O_3$	500–750
$C_8H_{18} + 8CO_2 \rightarrow 9H_2 + 16CO$	$r_3 = k_3 P_{C8H18} P_{CO2}\left(1 - \dfrac{P^2_{CO} P^2 H_2}{K_3 P_{C8H18} P_{CO2}}\right)$	Ni/Al_2O_3	800–900
$C_8H_{18} + 16H_2O \rightarrow 25H_2 + 8CO_2$	$r_4 = \dfrac{k_4}{P^{3.5}_{H_2}}\left(\dfrac{P_{C8H18} P^2_{H2O} - P^4_{H_2} P_{CO_2}/K_4}{1+(K_{CO}P_{CO} + K_{H_2}P_{H_2} + K_{C8H18}P_{C8H18} + K_{H2O}\,p_{H2O}/p_{H_2})^2}\right)$	$Ni/Mg–Al_2O_3$	500–750

Table 2.20. Regression of Kinetic Parameters for the Rate Equations Given in Table 2.19[155]

Parameter	Preexponential Factor (A)	Activation Energy (kJ/mol)
k_1 (mol/gcat s bar^2)	2.58×10^8	166.0
k_2 (mol bar$^{0.5}$/gcat s)	2.61×10^9	240.1
k_3 (mol/gcat s bar^2)	2.78×10^{-5}	23.7
k_4 (mol bar$^{0.5}$/gcat s)	1.52×10^7	243.9
K_{H2O} (dimensionless)	1.57×10^4	88.7 (heat of adsorption of water)

hydrocarbons and then reform the lighter gas mixture in the main reformer. This preforming process is typically carried out in lower temperatures between 400 and 550 °C in an adiabatic fixed-bed reactor; that is, there is no external heat supply during the reaction, upstream of the main reformer. The major reactions occurring in the prereformer include an endothermic irreversible steam reforming reaction that breaks the heavier hydrocarbons and aromatics into CH_4, H_2, and CO (Eq. 2.72), followed by exothermic and equilibrium-driven exothermic methanation (Eq. 2.28) and WGS reaction (Eq. 2.2), with the overall reaction being close to thermally neutral:

$$CnHm + H_2O \rightarrow H_2 + CO + CH_4 + CyHn. \qquad (2.72)$$

The major advantages of incorporating an adiabatic prereforming step in reforming of high hydrocarbons include the following:

- The probability of carbon formation on the catalyst surface is significantly decreased since no higher hydrocarbons are present in the feed.
- The main reformer can be operated at lower S/C ratios without any catalyst deactivation due to coking. This can significantly reduce the energy necessary for steam generation.
- The prereformer allows for greater feed temperatures to the main reformer, thereby reducing the size of the latter.
- It provides feedstock flexibility because the prereformer can crack the heavier hydrocarbons into lighter gaseous products to feed into the main reformer.
- The Ni-based catalyst typically used in the prereforming process can act as a sulfur trap to remove traces of sulfur in the feed. This can improve the lifetime of the catalyst in the main reformer and also the WGS catalyst used in the downstream.
- Help using a cheaper catalyst in the main reformer.
- Because of all the above benefits, it could lower the capital investments.

A wide variety of fuels such as diesel, jet fuel, and NATO F-76 have been reformed using adiabatic prereformers.[103] Since the prereforming process is generally carried out at low temperatures, the catalyst must be highly active. Industries

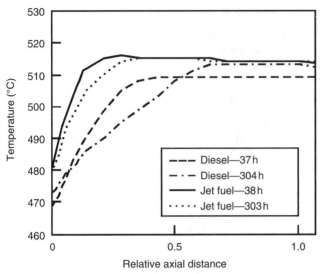

Figure 2.16. Temperature profile for the adiabatic prereforming of jet fuel and diesel over Topsoe RKNGR catalyst. Adapted from Rostrup-Nielsen et al.[10]

use the Ni-based catalysts typically for steam reforming of hydrocarbons for prereforming as well. Haldor Topsoe reported the use of the natural gas steam reforming catalyst R-67-GR and the naphtha steam reforming catalyst RKNGR for the prereforming of diesel and jet fuel.[10] Over RKNGR catalyst, the deactivation rate when using diesel was almost twice as that when using jet fuel because of a combined gum formation and sulfur poisoning, while the deactivation when using jet fuel has been related only to sulfur poisoning (Fig. 2.16). Christensen's studies also reported a similar observation on the prereforming of jet fuel and diesel over high surface area Ni catalyst supported on MgO.[156]

Noble metal-based supported catalysts have also been explored for the prereforming reactions. Suzuki and coworkers have reported the prereforming of kerosene and LPG using Ru supported on Al_2O_3, La_2O_3, and CeO_2–Al_2O_3.[145,157] Among them, the CeO_2-containing supports have been demonstrated to be superior than unmodified Al_2O_3 support since they can further enhance catalyst activity and reduce carbon formation.

Zheng et al.[158] recently compared the catalytic performance of a commercial Ni-based catalyst and a homemade Rh/CeO_2–Al_2O_3 containing about 2 wt% Rh in the prereforming of a surrogate jet fuel containing 80 wt% dodecane and 20 wt% aromatics. During 11 h of onstream experiments on Ni-based catalyst, they observed a drop in dodecane conversion from about 60% to about 15% within 3 h. They also observed a gradual increase in system pressure due to carbon deposition on the catalyst surface. On the other hand, the Rh/CeO_2–Al_2O_3 catalyst exhibited a sustained activity of close to 100% conversion over a period of 20 h and then started dropping slowly to about 90% after about 70 h of onstream. They have also noted distinctly different product distributions between these two different catalyst systems with more H_2 over the Ni-based catalyst than over Rh-based catalyst. Based on these observations, it has been concluded that the Ni-based catalyst catalyzes not only the steam reforming reaction in the prereformer, but also cracking of dodecane or aromatics, which leads to carbon deposition and more hydrogen in the product.

2.4 STEAM REFORMING OF ALCOHOLS

Methanol and ethanol have been considered as promising fuels for generating H_2, especially for on-board fuel cell applications due to their easy availability, ability to transport, and reaction simplicity.[52,121,159–169] For example, both alcohols have high H_2-to-carbon ratio (H/C) of 4 and 3, respectively (Table 2.1). They could be synthesized from renewable sources such as biomass and thus the ability to close the carbon cycle.[161,166] Unlike hydrocarbon fuels, methanol and ethanol are free from sulfur, and this avoids additional sulfur removal step in the fuel processing. In addition, methanol can be reformed at a lower temperature, around 300 °C, and this makes the fuel processing relatively simple and less complicated. Furthermore, unlike natural gas, which produces primarily syngas, reforming of methanol and ethanol can in principle produce a mixture of H_2 and CO_2, and this would also simplify the downstream CO cleanup for fuel cells such as PEMFCs where CO is a poison.

As discussed above in the reforming of hydrocarbon fuels, H_2 can be produced from alcohol fuels by at least three major catalytic processes, namely steam reforming, partial oxidation and ATR or oxidative steam reforming. The chemistry, thermodynamics, and recent developments in catalysis of methanol and ethanol reforming with steam for H_2 production will be discussed in this section.

2.4.1 Steam Reforming of Methanol (SRM)

2.4.1.1 Chemistry

Hydrogen can be produced from methanol by decomposition, steam reforming, partial oxidation, and oxidative steam reforming as shown in Equations 2.73–2.76, respectively. Among them, the SRM is a well-developed process and known for the past 20–30 years. The reaction is generally performed using over Cu-based catalysts such as Cu–ZnO–Al$_2$O$_3$ methanol synthesis catalyst or noble metal-based supported catalysts in the temperature range between 200 and 300 °C at atmospheric pressure. An S/C ratio between 1 and 2 is typically used. Higher S/C ratio could favor the WGS, which can reduce CO concentration in the product stream and can also help in suppressing the carbon formation on the catalyst surface. The SRM process is highly efficient as H_2 is coming from both methanol and water (Eq. 2.76). Yields are typically high, with H_2 concentrations of 70%–75% in the product stream. However, the reaction is highly endothermic and requires external heat supply. This can be avoided by adding a stoichiometric amount of oxygen to the steam reforming process. In the presence of oxygen, the process combines the endothermic steam reforming with exothermic partial oxidation reactions, thereby making the overall process thermally neutral. Due to the flexibility to operate at a wide range of operating conditions, spanning from endothermic to exothermic by simply varying the ratio of oxygen/steam (O/S) as shown in Equation 2.77, the reaction has been referred to as oxidative steam reforming of methanol (OSRM).[160,163,169–176] The process can therefore be easily adapted to the requirements of the system at any given point of operation. The low operating temperature and pressure and high H_2 production rate are the unique features of the OSRM process, and consequently, this hydrogen generation technique has recently gained an increased interest for use in mobile applications.

1. Decomposition of methanol

$$CH_3OH(g) \rightarrow 2H_2(g) + CO(g);$$
$$\Delta H^\circ_{298} = +90.5 \text{ kJ/mol} \tag{2.73}$$

2. SRM

$$CH_3OH(g) + H_2O(g) \rightarrow 3H_2(g) + CO_2(g);$$
$$\Delta H^\circ_{298} = +49.3 \text{ kJ/mol} \tag{2.74}$$

3. Partial oxidation of methanol

$$CH_3OH(g) + 0.5O_2(g) \rightarrow 2H_2(g) + CO_2(g);$$
$$\Delta H^\circ_{298} = -192.5 \text{ kJ/mol} \tag{2.75}$$

4. OSRM
   - Using lower heating values (LHVs)

$$CH_3OH(g) + 0.8H_2O(g) + 0.1O_2(g) \rightarrow 2.8H_2(g) + CO_2(g);$$
$$\Delta H^\circ_{298} = +1.0 \text{ kJ/mol} \tag{2.76}$$

   - Using higher heating values (HHVs)

$$CH_3OH(\ell) + (1-2\delta)H_2O(\ell) + \delta O_2(g) \rightarrow (3-2\delta)H_2(g) + CO_2(g) \tag{2.77}$$

If $\delta = 0.23$, $\Delta H^\circ_{298} = -0.5 \text{ kJ/mol}$.

2.4.1.2 *Thermodynamics*

The thermodynamic parameters ΔH, ΔG, and the equilibrium constant (K) for SRM and OSRM reactions are given in Tables 2.21 and 2.22, respectively. The ΔG° values for these reactions are negative at all temperatures, implying that both steam reforming and oxidative steam reforming reactions are thermodynamically favorable even at room temperature. While the SRM reaction is endothermic, requiring external heat supply, the OSRM reaction can be made thermally neutral or slightly endothermic or slightly exothermic by adjusting the O/S ratio. The data shown in Table 2.23 obtained by using HHVs of methanol and water indicate that under

Table 2.21. Thermodynamic Data for the Steam Reforming of Methanol: $CH_3OH(g) + H_2O(g) \rightarrow 3H_2(g) + CO_2(g)$

Temperature (°C)	$\Delta H°$ (kJ/mol)	$\Delta G°$ (kJ/mol)	Log K
25	49.321	−3.484	0.61
50	50.471	−7.959	1.287
75	51.593	−12.522	1.879
100	52.683	−17.164	2.403
125	53.74	−21.879	2.871
150	54.763	−26.658	3.291
175	55.752	−31.498	3.672
200	56.707	−36.391	4.018
225	57.628	−41.334	4.335
250	58.516	−46.323	4.626
275	59.37	−51.353	4.894
300	60.191	−56.421	5.142
325	60.981	−61.525	5.373
350	61.739	−66.661	5.588
375	62.467	−71.827	5.789
400	63.166	−77.02	5.977
425	63.835	−82.239	6.154
450	64.477	−87.481	6.319
475	65.091	−92.745	6.476
500	65.679	−98.029	6.623

Table 2.22. Thermodynamic Data Using Lower Heating Values for the Oxidative Steam Reforming of Methanol: $CH_3OH(g) + 0.8H_2O(g) + 0.1O_2(g) \rightarrow 2.8H_2(g) + CO_2(g)$

Temperature (°C)	$\Delta H°$ (kJ/mol)	$\Delta G°$ (kJ/mol)	Log K
25	0.956	−49.2	8.62
50	2.056	−53.451	8.641
75	3.128	−57.786	8.671
100	4.17	−62.197	8.707
125	5.179	−66.676	8.748
150	6.155	−71.218	8.792
175	7.097	−75.817	8.838
200	8.006	−80.468	8.884
225	8.881	−85.165	8.931
250	9.723	−89.906	8.978
275	10.532	−94.686	9.024
300	11.309	−99.502	9.069
325	12.055	−104.352	9.114
350	12.77	−109.232	9.157
375	13.455	−114.14	9.199
400	14.111	−119.074	9.241
425	14.739	−124.032	9.281
450	15.339	−129.012	9.32
475	15.913	−134.013	9.357
500	16.461	−139.032	9.394

Table 2.23. Thermodynamic Data Using Higher Heating Values for the Oxidative Steam Reforming of Methanol: $CH_3OH(l) + 0.54H_2O(l) + 0.23O_2(g) \rightarrow 2.54H_2(g) + CO_2(g)$

Temperature (°C)	$\Delta H°$ (kJ/mol)	$\Delta G°$ (kJ/mol)	Log K
25	−0.46	−100.055	17.531
50	−0.942	−108.387	17.521
75	−1.56	−116.677	17.507
100	−2.354	−124.917	17.488
125	−3.366	−133.097	17.463
150	−4.64	−141.205	17.432
175	−6.222	−149.229	17.395
200	−8.158	−157.155	17.351
225	−10.498	−164.969	17.3
250	−13.293	−172.654	17.24
275	−16.603	−180.194	17.173
300	−20.49	−187.57	17.096
325	−25.059	−194.763	17.009
350	−30.344	−201.749	16.913
375	−36.249	−208.51	16.805
400	−42.815	−215.032	16.687
425	−50.084	−221.297	16.559
450	−58.096	−227.288	16.419
475	−66.895	−232.989	16.268
500	−76.522	−238.383	16.107

adiabatic operating conditions, a higher O/S ratio would be required to make the overall process thermally neutral. The effect of temperature on equilibrium compositions in SRM and OSRM are shown in Figure 2.17. Higher temperatures favor the formation of large amount of CO probably due to the RWGS reaction.

2.4.1.3 Catalysts

The catalysts used for the SRM can be broadly classified into (a) Cu-based catalysts and (b) noble metal-based catalysts.[52,97,163,164,169] Since the methanol reforming reaction is the reverse of the methanol synthesis reaction from syngas, the commercial Cu/ZnO–Al$_2$O$_3$ methanol synthesis and WGS catalysts are active for the methanol reforming reaction as well. Because metallic copper is the active site in these reactions, a good catalyst formulation should contain highly dispersed Cu metal particles. The ZnO in the catalyst formulation is to act as a textural support in segregating Cu, which is highly susceptible for sintering. Al$_2$O$_3$ serves as a high surface area support for Cu. It helps in improving Cu metal dispersion and decreases the susceptibility of copper sintering. The performance of the Cu/ZnO–Al$_2$O$_3$ catalysts in methanol reforming could be further improved by improving the CuO reducibility and Cu

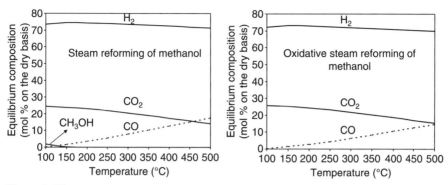

Figure 2.17. Effect of temperature on equilibrium compositions in steam reforming and oxidative steam reforming of methanol for H_2 production.

metal dispersions that could be achieved by changing the way the catalysts are synthesized and also by modifying the chemical compositions. As an example, Velu et al.[160,177,178] showed that $Cu/ZnO–Al_2O_3$ catalysts prepared via hydrotalcite (also known as layered double hydroxide or anionic clay) precursors by coprecipitation is highly active in methanol reforming reactions. Later, Murcia-Mascarós et al.[172] also reported the use of $Cu/ZnO–Al_2O_3$ mixed oxide catalysts derived from hydrotalcite precursors prepared by urea hydrolysis for the OSRM. Shishido et al.[179] prepared similar $Cu/ZnO–Al_2O_3$ catalysts by homogeneous precipitation method by urea hydrolysis. The high activity of the catalyst in SRM has been attributed to improvements in Cu metal dispersions with high accessibility to methanol and steam.

Substitution of Al_2O_3 in the $Cu/ZnO–Al_2O_3$ formulation by an oxide ion-conducting material such as ZrO_2, CeO_2, La_2O_3, and Y_2O_3 has been found to improve the catalytic performance in SRM and OSRM reactions.[160,177,180–188] Table 2.24 summarizes the compositions of CuZnAlZr oxide ($Cu/ZnO–Al_2O_3$) catalysts with various $Cu:Zn:Al:Zr$ ratios synthesized by coprecipitation via hydrotalcite-like precursors. The results indicate that the Cu metal surface area and dispersion increased when Al in the $Cu/ZnO–Al_2O_3$ formulation (CuZnAl-4) is either partially (CuZnAlZr-6) or completely replaced by Zr (CuZnZr-7). Substitution of Al by Zr also improves the CuO reducibility as can be seen in Figure 2.18. The improved CuO reducibility, Cu metal surface area, and dispersion achieved by substituting Al by Zr in the $Cu/ZnO–Al_2O_3$ formulation also improves the catalytic performance (improved methanol conversion and hydrogen production rate and a lower CO selectivity) in OSRM for H_2 production as illustrated in Figure 2.19. X-ray absorption near edge structure (XANES) and extended X-ray absorption fine structure (EXAFS) studies have shown the formation of a Cu–Zr bond (Fig. 2.20).[189] Such $CuO–ZrO_2$ interaction has been found to improve the CuO reducibility, which in turn improved the catalytic performance of ZrO_2-containing catalysts. Time-on-stream experimental results, however, indicated that the catalyst CuZnAlZr-6 containing both Al and Zr is much more stable compared with the CuZnZr-7 without Al, which undergoes slow

Table 2.24.   Physicochemical Properties of CuZnAl(Zr) Oxide Catalysts Tested in the Oxidative Steam Reforming of Methanol[177]

Catalyst	Metal Composition (wt %)[a]				BET Surface Area (m²/g)	Characteristics of Metallic Copper[b]		
	Cu	Zn	Al	Zr		SA (m²/g)	t (nm)	D (%)
CuZnAl-1	35.3	44.1	20.6	0.0	56	24.9	14.5	10.8
CuZnAl-2	36.7	48.0	15.3	0.0	71	32.0	11.8	13.4
CuZnAl-3	39.3	48.9	11.8	0.0	84	34.0	11.9	13.3
CuZnAl-4	37.6	50.7	11.7	0.0	108	34.4	11.2	14.1
CuZnAl-5	37.9	53.0	9.1	0.0	54	37.3	10.4	15.1
CuZnAlZr-6	35.9	44.7	5.1	14.3	65	41.0	9.0	17.5
CuZnZr-7	32.3	40.5	0.0	27.2	59	43.8	7.6	20.8

[a]Determined by X-ray fluorescence spectroscopy.

[b]Determined by CO chemisorption on reduced catalysts.

SA, surface area; t, crystallite size; D, dispersion.

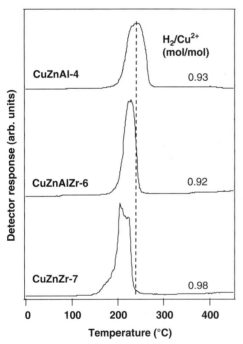

Figure 2.18.   Temperature-programmed reduction profiles of CuZnAlZr oxide catalysts for oxidative steam reforming of methanol. Note that a gradual shift in peak maximum toward lower temperatures, when Al is substituted by Zr, indicates an improvement in CuO reducibility upon Al substitution by Zr in the Cu/ZnO–Al₂O₃ formulation. Adapted from Velu et al.[177]

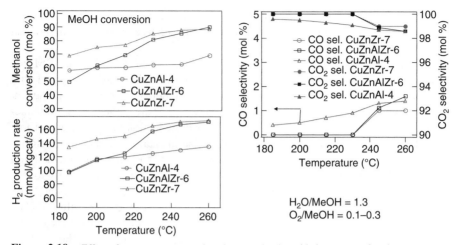

Figure 2.19. Effect of temperature on methanol conversion in oxidative steam reforming of methanol over CuZnAlZr oxide catalysts. Note an increase in methanol conversion and H_2 production rate upon substitution of Al by Zr in the Cu/ZnO–Al₂O₃ formulation. Adapted from Velu et al.[177]

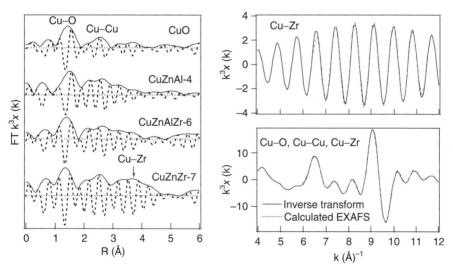

Figure 2.20. Nonphase shift corrected Fourier transform EXAFS spectra (left panel) of Cu/ZnO–Al₂O₃ catalysts containing ZrO₂. Examples for the best fitting of Fourier back-filtered EXAFS oscillations for Cu–Zr scattering (third cell) alone (top right) and curve fitting of the inverse Fourier transform EXAFS over the whole range using scattered parameters of Cu–O, Cu–Cu, and Cu–Zr together (bottom right). Adapted from Velu et al.[189]

deactivation. Breen and Ross[187] and Navarro et al.[169] also observed similar results. The added Al stabilizes the zirconia in its amorphous phase under the experimental conditions.

Agrell et al.[183] observed a higher turnover frequency (TOF) for the SRM and OSRM for H_2 production over Cu/ZnO catalysts promoted by ZrO_2 compared with the catalysts without ZrO_2 promotion (Fig. 2.21). Matter et al. reported that the Cu/ ZnO catalyst promoted by ZrO_2 could be used in the SRM without even prereduction, while the Al_2O_3-promoted catalyst requires prereduction to achieve an optimal activity.[186] The ability to use the unreduced Cu-based catalysts in methanol reforming can avoid the issue of phyrophorisity. Oguchi et al.[182] reported that the addition of ZrO_2 in Cu/CeO_2 catalyst improves the catalytic performance and stability for the SRM for H_2 production. The best weight ratio of $CuO/CeO_2/ZrO_2$ was found to be 8:1:1. According to these authors, the added ZrO_2 improves the reducibility of CuO to form Cu_2O and the existence of such synergistic interaction improves the catalytic performance of these catalysts. Clancy et al.[181] recently reported that the catalytic activity and stability of Cu/ZrO_2 catalyst in the SRM can be further improved by doping $20 \, mol\%$ of Y_2O_3. A threefold increase in Cu metal surface area has been observed by Y_2O_3 doping. Although the Cu-based catalysts are highly active in methanol reforming reactions, they are phyrophoric and susceptible for thermal sintering leading to catalyst deactivation during onstream operation.

Pd/ZnO catalysts have also been reported recently as an alternative to Cu-based catalysts for the reforming of methanol for H_2 production because they have been

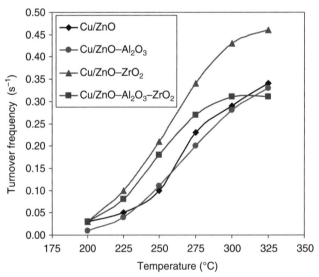

Figure 2.21. Turnover frequencies (TOFs) for the steam reforming of methanol over Cu/ZnO catalysts promoted by ZrO_2 and/or Al_2O_3. Note that the catalyst promoted by ZrO_2 exhibits the highest TOF. Adapted from Agrell et al.[183]

considered to be less susceptible for thermal sintering.[164,170,171,175,190–193] Iwasa et al.[194–196] and Takezawa and Iwasa[197] first reported that Pd supported on ZnO and prereduced above 300 °C has exceptionally high activity and selectivity to H_2 and CO_2. Spectroscopic studies have shown the formation of PdZn alloy phase under reduction conditions over 300 °C in these catalysts and the formation of such an alloy phase is critical for the selectivity of the methanol reforming reaction to produce H_2 and CO_2. Studies on other group VIII metals such as Co, Ni, Ru, Ir, and Pt on various supports such as ZnO, InO_2, and Ga_2O_3 have indicated that only Pd and Pt can form alloy on these supports and that the PdZnO catalyst system is the most active and selective for methanol reforming for H_2 production than the PtZnO system.[195,196]

The catalyst preparation method, Pd loading, and the type of support have major impact on the physicochemical properties and activity of these catalysts.[117,163,170,171,175,198–208] The PdZnO catalysts are generally prepared by impregnation of Pd salts on ZnO. The use of highly acidic Pd nitrate aqueous precursors alters the porosity and crystalline structure. This can be avoided by using organic precursors such as Pd acetate. The Pd loading in these catalyst systems is over 5 wt %. Increased methanol conversion and decreased CO selectivity is generally observed with increasing Pd loading. This is due to the formation of PdZn alloy phase, which is prevalent at higher Pd loading. Liu et al.[170,171,175] compared the catalytic performance of PdZnO catalysts prepared by impregnation and coprecipitation. They prepared PdZnO catalysts with Pd loadings between 1 and 45 wt % by these two methods and found that coprecipitation method offers better catalytic performance when the Pd loading exceeds 5 wt %. The sizes of both ZnO and PdZn alloy crystallite size affects the catalytic performance.

Karim et al.[190] reported the effect of crystallite size and alloy formation on catalytic performance of PdZnO system containing about 15 wt % Pd prepared by coprecipitation method. They observed a lower selectivity for CO_2 in SRM when small crystallites were formed (1.5 nm). These small particles were thought to be metallic Pd that was eventually alloyed with Zn upon increasing reduction temperature, resulting in increased selectivity for CO_2. They also reported that larger PdZn particles did not adversely affect the catalytic activity. Dagle et al.[192] also recently reported that while the smaller PdZn crystallite size produce more CO, the large-sized PdZn crystallites dramatically suppress CO selectivity still exhibiting high activity for methanol steam reforming.

Conant et al.[193] recently reported that the activity of PdZn catalyst could be significantly improved by supporting them on alumina. They showed that the Pd/ZnO/Al_2O_3 catalysts have a better long-term stability compared with a commercial Cu–ZnO–Al_2O_3 catalyst, and that the former catalysts are stable under redox cycling. They also observed that reduction at higher temperatures (>420 °C) leads to Zn loss from the alloy nanoparticle surface resulting in a reduced catalytic activity. Lenarda et al.[191] prepared PdZn/Al_2O_3 catalyst by depositing Pd acetate on an organized mesoporous alumina, followed by prereduction at room temperature with a zinc borohydrate solution. By this procedure, it is possible to obtain intimately interacting Pd–Zn species that easily generates the 1:1 Pd/Zn alloy when reduced at 500 °C. Chen et al.[117] reported the use of a wall-coated, highly active PdZn/CuZnAl mixed catalyst for the OSRM in a microchannel reactor. High H_2 yields in the OSRM reaction was also observed in a microchannel reactor by Lyubovsky and Roychoudhury et al.[208] Suwa et al.[202] reported the performance of various supported Pd/ZnO-based catalysts. Although much more stable than Cu-based catalysts, deactivation of PdZn catalysts was reported. The deactivation could be reduced by supporting PdZn on a carbon support.

Researchers at the Pacific Northwest National Laboratory have explored the Pd/ZnO catalysts extensively and developed microchannel reactors for the reforming of methanol for H_2 production for portable fuel cell applications. Their research works have been published in recent reviews.[163,168,192]

Although the PdZnO catalysts are more stable compared with CuZn-based catalysts in methanol reforming reactions, the requirement of a high Pd loading (over 5 wt%) and the high cost of noble metals may prohibit their use in large-scale mobile applications. Cu-based catalysts with robust formulations such as those containing ZrO_2, CeO_2, La_2O_3, and Y_2O_3 that could improve the thermal stability of Cu-based catalysts may be potential catalyst candidates for large-scale applications. Loading a small amount of Pd (up to 1 wt%) on Cu-based catalysts can also reduce its pyrophoricity and improve thermal stability during onstream operation in SRM and OSRM reactions. The CuPd bimetallic catalysts are more stable than the Cu-based catalysts in oxygen-assisted WGS reaction.[209]

2.4.1.4 *Kinetics and Mechanism*

Two different reaction pathways for the SRM over Cu-based catalysts have been suggested in the literature. One is methanol decomposition followed by WGS reaction to form H_2 and CO_2 (pathway I), and the other is methanol dehydrogenation to formaldehydeyde/methylformate to formic acid followed by decomposition to H_2 and CO_2 (pathway II). These two reaction pathways are shown schematically in Figure 2.22. Since CO is the primary reaction product in pathway I, the CO concentrations in the product should be at least equal to or greater than the concentration of WGS equilibrium. However, experimental studies showed that the concentration of CO was much less than the equilibrium values. Spectroscopic and kinetic studies have shown that the reaction between methanol and water occurs directly to produce CO_2 and H_2, and consequently, the methyl formate reaction route (pathway II) has been suggested. Diffuse reflectance infrared Fourier transform (DRIFT) spectroscopic study of Frank et al.[188] recently showed that the catalyst surface is dominated by methoxy and formate species. Based on the observed results, they concluded that the dehydrogenation of methoxy groups in methanol is the RDS in SRM over various Cu-based catalysts.

The SRM reaction over Pd/ZnO catalysts also follows pathway II, producing formaldehyde and formate species followed by decomposition into H_2 and CO_2. This is especially true when Pd is alloyed with Zn. On the other hand, the formaldehyde species first formed from methanol by dehydrogenation undergoes decomposition to produce CO and H_2 when Pd is present in metallic state. This

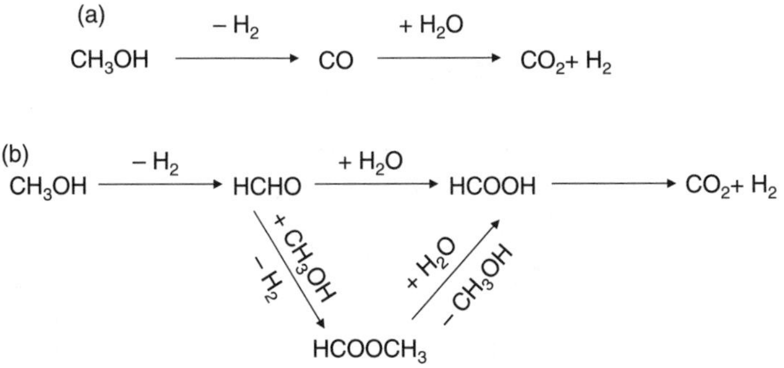

(a)
$$CH_3OH \xrightarrow{\;-H_2\;} CO \xrightarrow{\;+H_2O\;} CO_2 + H_2$$

(b)
$$CH_3OH \xrightarrow{\;-H_2\;} HCHO \xrightarrow{\;+H_2O\;} HCOOH \longrightarrow CO_2 + H_2$$

HCOOCH$_3$

Figure 2.22. Proposed reaction pathways for the steam reforming of methanol over Cu-based and Pd/ZnO-based catalysts. (a) Methanol decomposition followed by water-gas shift reaction. (b) Methanol dehydrogenation to formaldehyde/methylformate route.

difference in selectivity of Pd metal and PdZn alloy has been attributed to the difference in reactivity of the formaldehyde intermediate formed over these catalyst surfaces. Spectroscopic studies have shown that the structure of aldehyde species adsorbed on Cu surface is greatly different from that on group VIII metals such as Pd.[163] On Cu surface, the aldehyde species are adsorbed preferentially in an $\eta^1(O)$ structure (the oxygen in the carbonyl, C=O, is bonded to the Cu surface, retaining its double bond structure). On the other hand, the aldehyde species on group VIII metals are adsorbed as an $\eta^2(CO)$ structure (the carbon looses its double bond and adsorbs to the metal surface as does the oxygen). Thus, on Cu surfaces, the aldehyde species preserve its molecular identity and hence undergo reaction with water to produce formate species, which upon decomposition produces H_2 and CO_2, whereas on group VIII metals, the bonds are ruptured to produce CO and H_2. Hence, the Pd/ZnO catalysts mimic the performance of Cu-based catalysts when Pd and Zn are alloyed.

The rate of the SRM over a commercial $Cu/ZnO/Al_2O_3$ catalyst depends on the partial pressures of methanol and hydrogen. This indicates that the partial pressure terms of other components in the reaction mixture may not appear in the rate expression and that the adsorption of components other than methanol and hydrogen is insignificant. Based on these observations, and Langmuir adsorption isotherm, Frank et al.[188] showed the surface-adsorbed species involved in the SRM as illustrated in Figure 2.23. Accordingly, two kinds of active sites, namely S_A and S_B, exist on the catalyst surface. Adsorption of the methoxy species occurs on S_A while hydrogen occurs on S_B. Dehydrogenation of the adsorbed methoxy species to the adsorbed oxymethylene species is the RDS. The microkinetic rate equation for the SRM reaction is given Equation 2.78,

$$r_{MDH} = \frac{\left(k_{MDH} C_{SA}^T C_{SB}^T K^*_{CH3O(A)}\right)\left(p_{CH3OH}\big/p_{H2}^{0.5}\right)}{(\text{Methoxy} + \text{hydroxy} + \text{formate} + \text{carbon dioxide})\left(1 + K_{H(B)}^{0.5} p_{H2}^{0.5}\right)}, \quad (2.78)$$

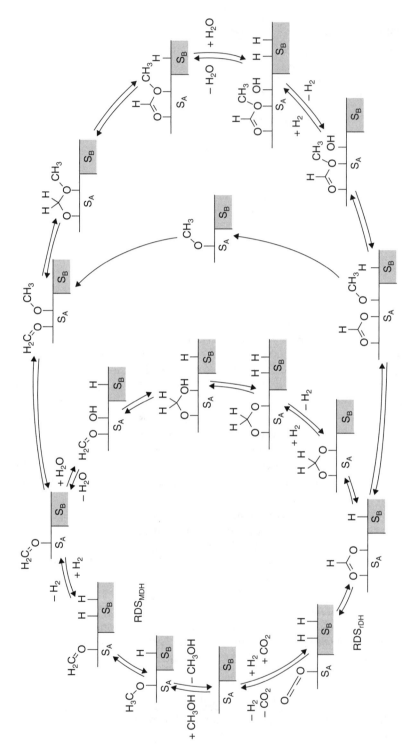

Figure 2.23. Catalysis cycle for the steam reforming of methanol involving different kinds of surface sites A and B. Adapted from Frank et al.[188]

where the terms, methoxy, hydroxy, formate, and carbon dioxide are given as follows:

- methoxy $= 1 + K^*_{CH3O(A)}\left(p_{CH3O(A)}/p_{H2}^{0.5}\right)$,
- hydroxy $= K^*_{OH(A)}\left(p_{H2O}/p_{H2}^{0.5}\right)$,
- formate $= K^*_{HCOO(A)}\,p_{CO2}\,p_{H2}^{0.5}$, and
- carbon dioxide $= K_{CO2(A)}p_{CO2}$.

K_{MDH} is the rate constant and C_{SA}^{T} and C_{SB}^{T} are the total concentrations of the type A and type B adsorption sites, respectively.

Equation 2.78 predicts that the rate of SRM reaction can be determined mainly by the partial pressure of methanol. For water, the rate shows a weak reverse dependence. The adsorption of carbon dioxide is competitive to that of methanol, water, and the oxygenate intermediates and thereby inhibiting the overall reaction. The apparent activation energy calculated based on these kinetic studies for various Cu-based catalysts are in the range between 70 and 90 kJ/mol. The rate expressions and activation energies for the SRM reaction over a few Cu-based catalysts reported in the recent literature are summarized in Table 2.25.[188,210–213]

Recently, Patel and Pant have developed a kinetic model for the OSRM reaction over Cu/ZnO/CeO$_2$/Al$_2$O$_3$ catalysts using the LH approach.[214] Their kinetic model for the OSRM reaction incorporates methanol partial oxidation, SRM, and RWGS reactions. Similar to that discussed above for the SRM reaction, they also considered the involvement of two different kinds of active sites on the catalyst surface: one for the adsorption of species containing carbon and oxygen and the other for the adsorption of hydrogen. The kinetic model based on the formation of formate species from oxymethylene as the RDS for methanol partial oxidation, dissociation of formic acid as the RDS for methanol reforming reaction, and the formation of adsorbed CO and surface hydroxyls from formate as an RDS of RWGS reaction has shown a good fit between experimental and predicted data.

2.4.2 Steam Reforming of Ethanol (SRE)

2.4.2.1 Chemistry

Similar to the methanol reforming methods discussed in Section 2.4.1, hydrogen can also be produced from ethanol by steam reforming (Eq. 2.79), partial oxidation (Eq. 2.80), and oxidative steam reforming or ATR (Eqs. 2.81 and 2.82). Among them, the SRE has been studied extensively.

1. SRE

$$CH_3CH_2OH(g) + 3H_2O(g) \rightarrow 6H_2(g) + 2CO_2(g)$$
$$\Delta H_{298}^{\circ} = +173.3 \text{ kJ/mol} \tag{2.79}$$

2. Partial oxidation of ethanol (POE)

$$CH_3CH_2OH(g) + 1.5O_2(g) \rightarrow 3H_2(g) + 2CO_2(g)$$
$$\Delta H_{298}^{\circ} = -552 \text{ kJ/mol} \tag{2.80}$$

Table 2.25. Kinetic Rate Expressions and Apparent Activation Energies for the Steam Reforming of Methanol over Various Cu-Based Catalysts Reported Recently

Rate Expression	Catalyst	Activation Energy (kJ/mol)	Reference
$-rM = kP_M^{0.6} P_W^{0.4}$	Cu/ZnO/Al$_2$O$_3$ (Süd-Chemie)	76	Purnama et al.[210]
$rMDH = \dfrac{k_{MDH} C_{SA}^T C_{SB}^T K^*_{CH3O(A)} \left(p_{CH3OH}/p_{H2}^{0.5}\right)}{\left(\text{Methoxy} + \text{hydroxy} + \text{formate} + \text{carbon dioxide}\right)\left(1 + K_{H(B)}^{0.5} p_{H2}^{0.5}\right)}$	CuO/ZnO/Al$_2$O$_3$ (C 18 HA)	76.9	Frank et al.[188]
	11.6% CuO/SiO$_2$	85.7	
$rM = \dfrac{kK_1\left(p_M/\sqrt{p_{H2}}\right)}{\left(1 + K_1\left(p_M/\sqrt{p_{H2}}\right)\right)\left(1 + \sqrt{K_2 p_{H2}}\right)}$	CuO/ZnO/Al$_2$O$_3$ (Synetix 33-5)	111	Lee et al.[211]
$-rM = kp_M^{0.63}\, p_W^{0.39}\, p_H^{-0.23}\, p_C^{-0.07}$	Cu/ZnO/Al$_2$O$_3$ (BASF K3-110)	74	Samms and Savinell[212]
$-rM = kp_M^{0.4}\left(1 - \left(p_C p_H^3 / K_E p_M p_W\right)\right)$	Cu/ZnO/Al$_2$O$_3$	83	Geissler et al.[213]

3. Oxidative steam reforming of ethanol (OSRE)
  • Using LHVs

$$CH_3CH_2OH(g) + 2.28H_2O(g) + 0.36O_2(g) \rightarrow 5.28H_2(g) + 2CO_2(g) \quad (2.81)$$
$$\Delta H^{\circ}_{298} \approx 0 \text{ kJ/mol}$$

  • Using HHVs

$$CH_3CH_2OH(l) + 1.78H_2O(l) + 0.61O_2(g) \rightarrow 4.78H_2(g) + 2CO_2(g) \quad (2.82)$$
$$\Delta H^{\circ}_{298} \approx 0.0 \text{ kJ/mol}$$

The SRE process involves the reaction between ethanol and water over a metal catalyst capable of breaking the C–C bond in ethanol to produce a mixture of H_2 and CO_2. The reaction is highly endothermic with a standard enthalpy, $\Delta H^{\circ}_{298} = +173.3 \text{ kJ/mol}$ of ethanol and occurs at relatively higher temperatures, typically between 300 and 800 °C. This reaction is considered as a combination of SRE to syngas (Eq. 2.83) followed by WGS (Eq. 2.2). Although a steam-to-ethanol (H_2O/EtOH) molar ratio of 3 is stoichiometrically required (see Eq. 2.79), higher H_2O/EtOH molar ratios, even up to 20, have been used because the actual bioethanol or the crude ethanol (fermentation broth) contains about 12 vol% of ethanol and about 86 vol% of water (H_2O/EtOH molar ratio about 18) together with traces of impurities such as lactic acid and glycerol.[167] Thus, the use of high H_2O/EtOH ratios has been considered advantageous, as it can minimize the cost of ethanol distillation. Also, the excess of water can contribute to WGS reaction to shift the CO formed by the steam reforming reaction and/or by ethanol decomposition (Eq. 2.84) into CO_2 and an additional mole of H_2. Although a high H_2O/EtOH molar ratio is advantageous, the use of higher H_2O/EtOH molar ratio in the SRE reaction will be limited by the energy cost of the system. A higher H_2O/EtOH molar ratio represents a higher energy cost because of the extra steam to be generated. Adding a small amount of oxygen (or air) in the steam reforming reaction can initiate the exothermic POE, and as a consequence, the overall reaction will either be thermally neutral, endothermic, or exothermic, depending on the amount of oxygen added in the reaction (Eqs. 2.81 and 2.82).

$$CH_3CH_2OH(g) + H_2O(g) \rightarrow 4H_2(g) + 2CO(g)$$
$$\Delta H^{\circ}_{298} = +255.5 \text{ kJ/mol} \tag{2.83}$$

Equation 2.2:

$$CO(g) + H_2O(g) \rightarrow CO_2(g) + H_2(g)$$

  • Decomposition of ethanol (DCE)

$$CH_3CH_2OH(g) \rightarrow H_2(g) + CO(g) + CH_4(g)$$
$$\Delta H^{\circ}_{298} = +49.7 \text{ kJ/mol} \tag{2.84}$$

2.4.2.2 Thermodynamics

The variation of free energy ΔG° values with respect to temperature for the SRE and accompanying reactions are shown in Figure 2.24. The values for the SRE reactions to produce syngas (line 1) and H_2 and CO_2 (line 6) become more negative above 200 °C, implying that these reactions are thermodynamically favorable with increasing temperature. However, several other competing reactions of ethanol, namely DCE to CH_4, CO, and H_2, (Eq. 2.84), dehydrogenation to acetaldehyde (Eq. 2.85), and dehydration to ethylene (Eq. 2.86) shown in lines 5, 3, and 4, respectively, in Figure 2.24 can also occur in parallel at lower temperatures, below 350 °C. DCE to CH_4, CO, and H_2 is a highly facile reaction and can occur even at room temperature (line 5).

$$CH_3CH_2OH(g) \rightarrow H_2(g) + CH_3CHO(g)$$
$$\Delta H^\circ_{298} = +68.4 \text{ kJ/mol} \tag{2.85}$$

$$CH_3CH_2OH(g) \rightarrow CH_2CH_2(g) + H_2O(g)$$
$$\Delta H^\circ_{298} = +45.4 \text{ kJ/mol} \tag{2.86}$$

The free energy changes for the reforming of acetaldehyde, ethylene, and CH_4, which could be formed as intermediates during the SRE reactions as discussed above, are shown in Figure 2.25. As can be seen, the decomposition of acetaldehyde to CH_4 and CO (line 2; Eq. 2.87 is favorable at room temperature, while a temperature of above 250 °C is required for the steam reforming of acetaldehyde (lines 3 and 5; Eqs. 2.88 and 2.89) and steam reforming of ethylene (line 4; Eq. 2.90). On

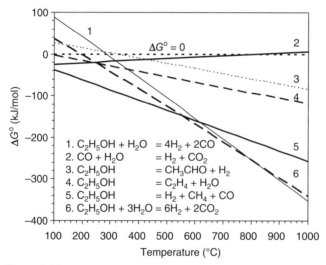

Figure 2.24. Free energy changes in the steam reforming, decomposition, dehydrogenation, and dehydration of ethanol. The data for water-gas shift reaction are also included. Adapted from Velu and Song.[167]

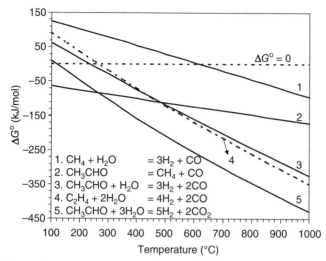

Figure 2.25. Free energy changes in the steam reforming of acetaldehyde, ethylene, and methane. Adapted from Velu and Song.[167]

the other hand, the SMR reaction (line 1) requires a high temperature, above 650 °C as discussed in Section 2.2. The WGS reaction to shift the CO formed in these reactions into CO_2 is favorable only at low temperatures, below 500 °C ($\Delta G°$ at 500 °C = $-10.4\,kJ/mol$) and will go to completion around 150 °C ($\Delta G°$ at 150 °C = $-23.5\,kJ/mol$). Above this temperature, the equilibrium will work against it and a significant amount of CO will remain unconverted.

$$CH_3CHO(g) \rightarrow CH_4(g) + CO(g)$$

$$\Delta H_{298}^{\circ} = -18.8 \text{ kJ/mol} \tag{2.87}$$

$$\Delta G_{298}^{\circ} = -54.5 \text{ kJ/mol}$$

$$CH_3CHO(g) + H_2O(g) \rightarrow 3H_2(g) + 2CO(g)$$

$$\Delta H_{298}^{\circ} = +187.1 \text{ kJ/mol} \tag{2.88}$$

$$\Delta G_{298}^{\circ} = +87.5 \text{ kJ/mol}$$

$$CH_3CHO(g) + 3H_2O(g) \rightarrow 5H_2(g) + 2CO_2(g)$$

$$\Delta H_{298}^{\circ} = +104.8 \text{ kJ/mol} \tag{2.89}$$

$$\Delta G_{298}^{\circ} = +30.3 \text{ kJ/mol}$$

$$CH_2CH_2(g) + H_2O(g) \rightarrow 4H_2(g) + 2CO(g)$$

$$\Delta H_{298}^{\circ} = +210.2 \text{ kJ/mol} \tag{2.90}$$

$$\Delta G_{298}^{\circ} = +114.5 \text{ kJ/mol}$$

Thus, the thermodynamic analysis suggests that the SRE into H_2 and CO_2 is thermodynamically favorable above 200 °C. A mixture of H_2, CO_2, CO, and CH_4 will be produced at lower temperatures, below 400 °C. The CH_4 formed will be

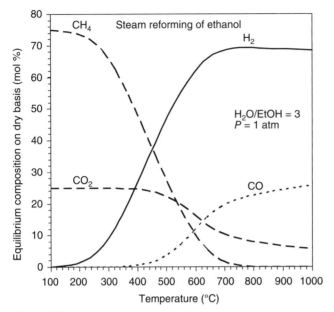

Figure 2.26. Thermodynamic equilibrium compositions on dry basis for the steam reforming of ethanol. All species are in gas phase. Initial concentrations of CO, CH_4, and CH_3CHO are taken as zero in the calculation. Adapted from Velu and Song.[167]

subsequently reformed into syngas at higher temperatures, above 650 °C. The thermodynamics become much more favorable under oxidative reforming conditions.

The equilibrium compositions (on the dry basis) of the SRE reaction calculated using $H_2O/EtOH$ molar ratio of 3 at 1 atm pressure is shown in Figure 2.26. Calculations were performed assuming only $EtOH(g)$ and $H_2O(g)$ as reactants and $H_2(g)$ and $CO_2(g)$ as products. The initial concentrations of other possible products, namely CO, CH_4, and CH_3CHO in gas phase were taken as zero, but they were allowed to form in the equilibrium calculations. The results indicate that the SRE reaction can produce H_2, CO, CO_2, and CH_4 as favorable products with CH_4 being predominant at lower temperatures, because of the decomposition reaction, and H_2 at higher temperatures. CH_4 is an undesirable by-product because it reduces the H_2 yield dramatically. As predicted from the change in free energy, a maximum H_2 yield of about 70% could be obtained above 650 °C because at this temperature, the CH_4 formed initially by the decomposition of ethanol and/or acetaldehyde can be subsequently steam reformed to produce syngas (Fig. 2.25). Thus, the exit composition of SRE at higher temperature should be close to that of the SMR reaction. It should be noted that, high temperature also favors the formation of a large amount of CO of over 20 mol %, while the yield of CO_2 decreases to about 10 mol % as the equilibrium limits the WGS reaction at high temperatures. However, the presence of even traces of CO is a poison to PEMFCs, and consequently, the high temperature steam reforming process would require multistage downstream CO cleanup processes consisting of WGS, Prox, methanation reactions and/or membrane/PSA, and so on. Thus, a highly selective catalyst that can reform ethanol directly into H_2 and CO_2 at relatively lower temperatures needs to be identified.

Thermodynamic equilibrium calculations for the SRE reaction have also been performed at different $H_2O/EtOH$ ratios and different pressures.[133,215–218] The analyses indicate that the formation of undesirable by-products such as CH_4 and CO could be minimized by performing the reaction with higher $H_2O/EtOH$ ratios and under low pressure, preferably at 1 atm. The increase in the total pressure leads to a decrease in the H_2 and CO_2 yields while the equilibrium composition of CH_4 increases. Thus, it is desirable to perform the reaction at atmospheric pressure, although high pressure reforming has the advantage of easy integration of the reformer with H_2 separation using membranes.[79]

Recently, Comas et al.[219] performed the thermodynamic analysis of the SRE reaction in the presence of CaO as a CO_2 sorbent. The equilibrium calculations indicate that the presence of CaO in the ethanol steam reforming reactor enhances the H_2 yield while reducing the CO concentrations in the outlet of the reformer. Furthermore, the temperature range at which maximum H_2 yield could be obtained also shifts from above 700 °C for the conventional steam reforming reaction without CaO to below 700 °C, typically around 500 °C in the presence of CaO. It appears that the presence of CaO along with ethanol reforming catalyst shift the WGS equilibrium in the forward direction and converts more CO into CO_2 that will be simultaneously removed by CaO by adsorption.

Figure 2.27 depicts the equilibrium compositions for the OSRE reaction calculated for $H_2O/EtOH$ and $O_2/EtOH$ stoichiometries shown in Equation 2.81. A

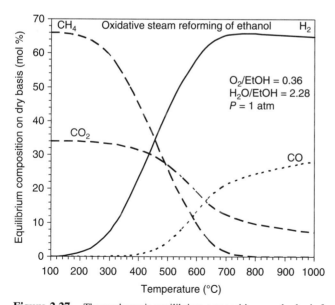

Figure 2.27. Thermodynamic equilibrium compositions on dry basis for the oxidative steam reforming of ethanol. All species are in gas phase. Initial concentrations of CO, CH_4, and CH_3CHO are taken as zero in the calculation. Adapted from Velu and Song.[167]

maximum yield of H_2 could be attained in the OSRE reaction above 600 °C because CH_4, if formed as an intermediate by ethanol decomposition, could be completely converted into syngas above this temperature.[167] The participation of RWGS reaction at high temperature leads to an increase in CO composition above 400 °C with a concomitant decrease in the composition of CO_2. Thus, at higher operating temperatures of above 600 °C, the system would produce mainly syngas rather than a mixture of H_2 and CO_2. A comparison of equilibrium compositions of SRE and OSRE reactions indicates that the maximum yield of H_2 (about 65%) is slightly less in OSRE compared with what could be obtained in the SRE (about 70%) due to the participation of steam in the overall reaction. Thus, thermodynamic analyses suggest that when a mixture of ethanol, steam, and O_2 with an appropriate $H_2O/EtOH$ and $O_2/EtOH$ ratios is passed through a catalyst bed in the temperature range between 200 and 900 °C, OSRE reaction occur to produce H_2 and CO/CO_2 as major products with maximum H_2 yield above 600 °C, if CH_4 is formed as an intermediate. A catalyst that has a high selectivity (fast kinetics) for ethanol reforming and poor selectivity (slow kinetics) for ethanol decomposition to CH_4 could produce H_2 and CO_2 in the OSRE reaction at relatively lower temperatures.

2.4.2.3 Catalysts

Compared with methanol, the simplest homologue in the alcohol series, ethanol has an additional carbon with a C–C bond energy of about 330 kJ/mol, making it more complicated to reform into H_2 and carbon oxides, because the reaction involves C–C bond rupture in addition to the cleavage of C–H bonds. Furthermore, as discussed above, several competing reactions, namely ethanol decomposition to CH_4, dehydrogenation to acetaldehyde, and dehydration to ethylene are also thermodynamically favorable to produce several undesirable by-products. An ideal catalyst should therefore favor only ethanol reforming to produce H_2 and carbon oxides rather than ethanol decomposition and/or dehydration reactions.

A wide variety of catalysts containing base metals, such as Cu, Ni, and Co, or noble metals, such as Rh, Pt, Ru, Pd, and Ir, have been investigated in the SRE. The type of catalysts used, reaction operating conditions employed, and the main results obtained are summarized in Tables 2.26 and 2.27.[220–264] A closer look at the literature reveals that among the catalysts explored, those containing Rh, Ni, or Co exhibit better performance, and these are the most extensively studied catalytic systems. Both Rh and Ni metals have been reported to be capable of breaking C–C and C–H bonds in ethanol. Addition of a second metal that has hydrogenation activity could help in recombining the atomic H to form a molecular H_2. Thus, a series of bimetallic catalysts such as RhPt, RhPd, and RhRu supported on CeO_2/ZrO_2 have been found to be more active than respective monometallic catalysts.[265,266]

The nature of support also plays a critical role in catalytic performance. Acidic supports such as Al_2O_3 favors the dehydration of ethanol to produce ethylene, which leads to carbon deposition on the catalyst surface.[242,252,254,259] The carbon deposition can be minimized over basic supports such as MgO.[235,241,258,259] However, these supports also favor condensation of alcohols to higher oxygenates.[260] Interestingly,

Table 2.26. Non-Noble Metal-Based Catalysts Employed in the Steam Reforming of Ethanol (SRE) for Hydrogen Production

Catalyst	Reaction Conditions	Description	Reference
15% Ni/Al_2O_3	320–520 °C; 1 g catalyst; H_2O/EtOH = crude ethanol; WHSV = 4.6–16.2 h^{-1}	Catalyst was prepared by coprecipitation technique. Work focused on reactor modeling and kinetic study of crude ethanol reforming. Simulation study demonstrated that plug flow and isothermal behavior could not be guaranteed. Suggests using a comprehensive model to verify whether or not plug flow behavior is attained.	Aboudheir et al.[220]
35 wt % Ni/Al_2O_3 derived from layered double hydroxide precursors	500 °C; 0.84 g catalyst; H_2O/EtOH = 1–6	Ethanol and methane steam reforming reactions were studied assuming that the exit composition of the ethanol reformer depends on the steam reforming of methane. The competition for the same active site for ethanol and methane reforming maximizes the H_2 and CO_2 production and minimizes the CO formation.	Comas et al.[221]
10–20 wt % Ni on ZnO, La_2O_3, MgO, and Al_2O_3	300–650 °C; 3 mL catalyst; H_2O/EtOH = 3–12; WHSV = 4–25 h^{-1}	Catalysts were prepared by incipient wet impregnation. Twenty weight percent Ni supported on ZnO exhibited better performance compared with that supported on La_2O_3, MgO, and Al_2O_3.	Yang et al.[222]
1–25 wt % Ni supported on ZnO–Al_2O_3	500–600 °C; 300 mg catalyst; H_2O/EtOH = 3.6–3.8	Catalysts containing 1–25 wt% Ni were synthesized by citrate sol-gel method. Catalyst with Ni loading between 18% and 25% exhibited better performance.	Barroso et al.[223]
3–15 wt % Ni on La_2O_3–Al_2O_3	500 °C; H_2O/EtOH = 3; WHSV = 24,487 h^{-1}	La_2O_3 (3, 6, and 15 wt %) was loaded on γ-Al_2O_3 by impregnation. About 20 wt% Ni was subsequently impregnated on La_2O_3-loaded Al_2O_3. Addition of La_2O_3 increased the selectivity of ethylene. Catalyst containing 15 wt % Ni exhibited a better and stable activity.	Idriss[265]

(Continued)

Table 2.26. *Continued*

Catalyst	Reaction Conditions	Description	Reference
12.5 and 21.5 wt % Co supported on ZnO and promoted by Na	500 °C; 0.6 g catalyst; H_2O/EtOH = 5–13; feed flow rate = 0.05 mL/min	Catalysts were prepared by coprecipitation. Reaction was performed in a membrane reactor. Higher Co loading and lower H_2O/EtOH ratio increased the ethanol conversion.	Sheng et al.[266]
15–20 wt % Ni loaded on Y_2O_3, La_2O_3, and Al_2O_3	250–350 °C; 4 g catalyst; H_2O/EtOH = 3; liquid flow rate = 0.05 mL/min	Catalysts were prepared by impregnation using nickel oxalate as an Ni precursor. The activity, stability, and H_2 selectivity decreased in the following order: $Ni/La_2O_3 > Ni/Y_2O_3 > Ni/Al_2O_3$	Sun et al.[225]
10–25 wt % Ni/Al_2O_3	320–520 °C; 1 g catalyst; H_2O/EtOH = crude ethanol; WHSV = 1.68–4.62 h^{-1}	Detailed catalyst synthesis and characterization and evaluation of performance and kinetics of the steam reforming of crude ethanol were investigated. Catalysts prepared by precipitation, coprecipitation, and impregnation methods were compared. Catalyst containing 15 wt % Ni synthesized by precipitation method produced smaller Ni crystallite sizes and better reducibility and hence a better catalytic performance.	Akande et al.[224] and Morgenstern and Fornango[226]
Cu-plated Raney Ni	250–300 °C; 70% EtOH in water	Cu was loaded on Raney Ni. The catalyst contained 68.9% Ni, 28.2% Cu, and 2.9% Al. Production of $H_2 + CH_4 + CO_2$ by coupling low-temperature ethanol decomposition and WGS reaction is more energy efficient and suitable for hybrid vehicles.	Bergamaschi et al.[227]
Cu and Ni supported on ZrO_2 microspheres	350–550 °C; 8 g catalyst; H_2O/EtOH = 0.5–3.0; flow rate = 1 mL/min	Cu and Ni were loaded by ion exchange technique. Catalyst containing 6 wt % Ni and 3 wt % Cu exhibits a high activity and H_2 selectivity.	Benito et al.[228]
Ni, Cu, or Co supported on ZrO_2	500–700 °C; H_2O/EtOH = 2.4; GHSV = 50,000–300,000 h^{-1}	Catalysts were prepared by proprietary methods. Among the catalysts, Co/ZrO_2 exhibits higher activity and stability. Some insights on the mechanism on the SRE reaction were also discussed.	Vargas et al.[229]

86

Catalyst	Conditions	Description	Reference
Ce–Zr–Co fluorite-type oxide catalysts	400–550 °C; 0.16 g catalyst; $H_2O/EtOH = 6$; $GHSV = 26,000\,h^{-1}$	Catalyst was synthesized by a pseudo sol-gel method, based on decomposition of propionate precursors. Nanoparticles of Co generated by partial reduction of $Ce_2Zr_{1.5}Co_{0.5}O_{8-\delta}$ exhibit higher conversion of ethanol or bioethanol containing methanol and higher alcohols and higher H_2 selectivity.	Nishiguchi et al.[230]
20 mol% CuO supported on CeO_2, Al_2O_3, and SiO_2	200–400 °C; $H_2O/EtOH = 5$; flow rate of ethanol and water were 1.2×10^{-4} and 6.0×10^{-4} mol/min, respectively	Two moles of H_2 was produced from 1 mol of ethanol. Reaction involves consecutive dehydrogenation and aldol condensation to produce significant amount of acetone.	Duan and Senkan[231]
42 metals from the periodic table supported on various oxides	300 °C; $H_2O/EtOH = 6$; $GHSV = 60,000\,h^{-1}$	Libraries of catalytic materials containing 0.5–5 wt% metals were prepared by impregnating the supports employing combinatorial approach. Among the catalysts tested, Pt/CeO_2 and Pt/TiO_2 exhibited better performance converting over 90% ethanol at 300 °C. However, the H_2 selectivities were low, around 30%.	Sun et al.[232]
Ni/Y_2O_3	250–350 °C; 4 g catalyst; $H_2O/EtOH = 3$; liquid flow rate = 0.05 mL/min	Catalysts were prepared in three different methods: impregnation, impregnation followed by treatment with $NaBH_4$, and precipitation of NiC_2O_4 on Y_2O_3. A high ethanol conversion of about 98% and H_2 selectivity of 55% were obtained at 380 °C. H_2 selectivity did not increase above 500 °C.	Batista et al.[233]
8–18 wt% Co/Al_2O_3 8–18 wt% Co/SiO_2	400 °C; 0.15 g catalyst; 0.15 g catalyst; $H_2O/EtOH = 3$	Catalysts were prepared by incipient wetness impregnation of commercial γ-Al_2O_3 and SiO_2. Catalytic activity increased with increasing Co content. Process produced H_2-rich stream with lower CO probably due to the involvement of subsequent WGS and CO methanation reactions.	Llorca et al.[262]

(Continued)

Table 2.26. *Continued*

Catalyst	Reaction Conditions	Description	Reference
1% Co/ZnO	400 °C; 0.1 g catalyst; H_2O/ EtOH = 13; GHSV = $5000 h^{-1}$	Catalyst was prepared by impregnation from n-hexane $CO_2(CO)_8$ solution on ZnO. The SRE reaction was studied using an in situ DRIFT–mass spectrometry study. The reaction proceeds through dehydrogenation to acetaldehyde followed by C–C scission.	Batista et al.[234]
8.6% Co/Al$_2$O$_3$, 7.8% Co/SiO$_2$, and 18% Co/MgO	400 °C; 0.15 g catalyst; H_2O/ EtOH = 3	Catalysts were prepared by incipient wetness impregnation of commercial supports using cobalt nitrate as a precursor. Metallic cobalt species were active centers in the ethanol steam reforming. Over 90% EtOH conversion was achieved. Nature of support influences the type of by-product formation. Ethylene, methane, and CO are formed over Co supported on Al$_2$O$_3$, SiO$_2$, and MgO, respectively.	Frusteri et al.[235]
Unpromoted and K promoted 19%–21% Ni/MgO	650 °C; 10–60 mg catalyst; 68% H$_2$O and 8% EtOH; GHSV = $40,000 h^{-1}$	Commercial MgO was impregnated with Ni acetate in toluene solution. Addition of 1–3 wt % K suppresses the Ni sintering and extends the catalyst stability. Catalysts were evaluated for 500 h of endurance test. Process was reported to be suitable for producing H$_2$ for MCFC.	Freni et al.[236] and Comas et al.[237]
35 wt % Ni/Al$_2$O$_3$	300–500 °C; 0.1–0.9 g catalyst; H$_2$O/EtOH = 1–6; residence time = 0–0.15 g·min/L	Catalyst developed at the Royal Military College, Canada. High activity and H$_2$ yield are obtained at higher temperatures, above 500 °C, but large amount of CO was also produced.	Casanovas et al.[267]
10 wt % Co/ZnO	300–400 °C; 0.1 g catalyst; H$_2$O/EtOH = 13; GHSV = $5000 h^{-1}$	Nitrate and carbonyl precursors of Co were used for the impregnation of ZnO support synthesized by the decomposition of 3ZnO·2ZnCO$_3$·3H$_2$O. Catalyst synthesized using carbonyl precursor was stable and selective for the production of CO-free H$_2$.	Llorca et al.[238]

Catalyst	Reaction conditions	Comments	References
1 wt% Co supported on MgO, Al_2O_3, SiO_2, TiO_2, V_2O_5, ZnO, La_2O_3, CeO_2, and Sm_2O_3	300–450 °C; H_2O/EtOH = 13; GHSV = 10,000 h^{-1}	Catalysts were prepared by impregnation of supports using $CO_2(CO)_8$ as a Co precursor. *In situ* magnetic characterization indicated that a mixed Co^0 and Co^{2+} contributes to catalytic activity in the SRE reaction.	Llorca et al.[239] and Freni et al.[240]
5–20 wt% Ni or Co supported on commercial MgO	650 °C; 15 mg catalyst; H_2O/EtOH = 8.4; GHSV = 10,000–80,000 h^{-1}	Catalysts were prepared by incipient wetness impregnation of MgO from two different commercial sources using nitrate salts and Ni acetylacetonate. Ni/MgO exhibited better performance than Co/MgO in the SRE reaction because of the lower tendency of Ni to oxidize during the reaction. The catalyst also exhibited a stable performance during 630 h of onstream operation.	Srinivas et al.[241]
NiO–CeO_2–ZrO_2 mixed oxides containing 1–40 wt% NiO	550 °C; 0.653 g catalyst; H_2O/EtOH = 8; LHSV = 6; 3 mL/h	Catalysts were synthesized by hydrothermal treatment and characterized by various spectroscopic methods. Bioethanol containing 5 ppm sulfur was used as a feedstock in the SRE reaction. Catalyst containing 40 wt% NiO exhibited a stable activity for over 500 h.	Fatsikostas et al.[242]
17 wt% Ni on La_2O_3, Al_2O_3, YSZ, and MgO	550–850 °C; 0.1 g catalyst; H_2O/EtOH = 2–3; gas flow rate = 160 cc/min	Catalysts were prepared by wet impregnation method. Among them, Ni/La_2O_3 exhibited high activity and selectivity for H_2 production in the SRE reaction. The enhanced activity has been attributed to scavenging of coke deposited on the Ni surface by lanthanum oxycarbonate species.	Fatsikostas et al.[243] and Mariño et al.[244]
Cu–Ni–K/Al_2O_3 with Cu or Ni 0–6 wt%	300 °C; 2.5 g catalyst; H_2O/EtOH = 2.5; LHSV = 1.8 h^{-1}	Catalysts were prepared by coimpregnation of γ-Al_2O_3. Addition of Ni enhances the gasification reaction and reduces the selectivity of acetaldehyde and acetic acid.	Mariño et al.[245] and Llorca et al.[246]

(Continued)

Table 2.26. *Continued*

Catalyst	Reaction Conditions	Description	Reference
Various transition metal oxides	300–450 °C; 0.1 g catalyst; $H_2O/EtOH = 13$; $GHSV = 5000 h^{-1}$	Among the oxides tested in the SRE reaction, ZnO was the most efficient, producing 5.1 mol of CO-free H_2 per mole of ethanol converted at 450 °C.	Freni et al.[247]
15 wt % Cu/SiO_2 and 0.5 wt % Ni/MgO two-layer catalysts	Two-layer catalyst bed, different temperature for each layer (300–650 °C); 0.12 g catalyst; $H_2O/EtOH = 8.2$; $GHSV = 109,000 h^{-1}$	Dehydrogenation of ethanol to acetaldehyde occurs on Cu/SiO_2 catalyst in the first layer at 370 °C, and steam reforming of acetaldehyde to syngas occurs in the second layer at 650 °C. This two-layer catalytic reactor prevents coke formation and produces H_2 approaching to the equilibrium.	Haga et al.[248]
7.4% Co supported on Al_2O_3, SiO_2, MgO, ZrO_2, and carbon	400 °C; 0.3 g catalyst; $W/F = 0.45$ g·s/cm^3	Catalysts were prepared by impregnation using cobalt (II) nitrate. Co/Al_2O_3 was the most active and selective catalyst. Suppresses the ethanol decomposition and CO methanation reactions.	Idriss[265]
Cu-, Ni-, and noble metal-supported catalysts	Fixed-bed reactor, 100–600 °C; $H_2O/EtOH = 6$–10; $LHSV = 1.6$–2.0 h^{-1}	Catalysts were prepared by impregnation and coprecipitation techniques. Among the catalysts tested in the SRE reaction, the $CuO/ZnO/Al_2O_3$ exhibited better performance.	Sheng et al.[266]
4% Ni–0.75% Cu–0.25% Cr on Al_2O_3	Fixed-bed reactor, 300–600 °C; $H_2O/EtOH = 0.4$–2.0; $LHSV = 2.5$–15.0 h^{-1}	Catalysts were prepared by chemically depositing Cu, Ni, and Cr on alumina. SRE reaction produces a mixture of H_2 and CO. Ni was thought to help in breaking the C–C bond while Cu and Cr for the oxidation of C1 species into CO and H_2.	Casanovas et al.[267]

WHSV, weight hourly space velocity; W/F, weight to flow rate ratio; LHSV, liquid hourly space velocity; MCFC, molten carbonate fuel cell.

Table 2.27. Noble Metal-Based Catalysts Employed in the Steam Reforming of Ethanol (SRE) for Hydrogen Production

Catalyst	Reaction Conditions	Results and Description	Reference
1–5 wt % Rh supported on γ-Al_2O_3, $MgAl_2O_4$, ZrO_2, and CeO_2–ZrO_2	450 °C; 50 mg catalyst; H_2O/EtOH = 2; WHSV = 133,333 mL gas/(g cata.h)	Rh was loaded by incipient wetness impregnation. SRE reaction over these catalysts revealed that ethanol hydration is favorable over acidic or basic catalysts, while dehydrogenation is favorable over redox catalysts. Among the catalysts, a 2% Rh/$Ce_{0.8}Zr_{0.2}O_2$ exhibited the best performance; this may be due to strong Rh-support interaction.	Zhang et al.[252]
2 wt % Ir, 15 wt % Co, and 15 wt % Ni supported on CeO_2	300–700 °C; 300 mg catalyst; H_2O/EtOH = 3; WHSV (gas phase) = 6000 mL/g/h	Catalysts were prepared by deposition–precipitation method over a commercial CeO_2. Among the catalysts, Ir/CeO_2 exhibited a stable activity for 300-h time onstream due to the prevention of metal sintering and coke resistance of highly dispersed Ir.	Aupretre et al.[253]
0.2–1 wt % Rh supported on $Mg_xNi_{1-x}Al_2O_3$	700 °C; 500 mg catalyst; H_2O/EtOH = 4; GHSV = 24,000 h^{-1}, 1 or 11 atm	Catalysts were prepared by wet impregnation of Rh salt on MgAl, prepared by solid–solid reaction and MgNiAl spinels supported on Al_2O_3 prepared by coimpregnation on Al_2O_3. Effect of acidic and basic properties of the support on the catalytic performance in the SRE has been studied by FTIR and DRIFT. Rh dispersed on MgNiAl/ Al_2O_3 support having higher basicity exhibited excellent performance at 1 atm. $RhCl_3$ is a preferred Rh precursor.	Wanat et al.[254]

(Continued)

Table 2.27. *Continued*

Catalyst	Reaction Conditions	Results and Description	Reference
Rh and Rh–Ce on washcoated fecralloy	800–900 °C; $H_2O/EtOH = 6$–8; residence time = 100–400 ms; used an autothermal flat plate catalytic wall reactor	Catalytic methane combustion and WGSs are coupled. Combined SRE and WGS reaction process produced syngas with H_2/CO of ≈ 30 and $H_2/EtOH$ ratio ≈ 5 at 99% EtOH conversion. The catalyst was stable during the 100 h onstream.	Diagne et al.[255]
2 wt % Rh supported on CeO_2, ZrO_2, and CeO_2–ZrO_2 synthesized by precipitation	300–500 °C; 0.1 g catalyst; $H_2O/EtOH = 8$; liquid flow rate = 0.77 mL/h	Rh was loaded on the support by impregnation. Among the catalysts tested in the SRE reaction, Rh/CeO_2–ZrO_2 was the best producing 5.8 mol of H_2/mole of ethanol injected and a high CO_2/CO ratio, around 35 at 450 °C.	Aupretre et al.[256]
0.2–0.3 wt % Rh on γ-Al_2O_3	700 °C; 0.25 g catalyst; $H_2O/EtOH = 4$; GHSV = 24,000 h^{-1}	Catalyst was prepared by wet impregnation using $Rh(NO_3)_3$ or $RhCl_3$ as an Rh precursor. Effect of Rh precursors on catalytic performance was investigated. Catalyst synthesized using $RhCl_3$ precursor was exhibited better and stable performance in the SRE reaction than $Rh(NO_3)_3$-based catalyst.	Frusteri et al.[257]
3 wt % Pd or Rh supported on a commercial MgO	650 °C; 10–60 mg catalyst; $H_2O/EtOH \approx 8$; GHSV = 40,000 h^{-1}	Catalysts were prepared by incipient wetness impregnation. Among them, the Rh/MgO showed a better and stable catalytic performance in the SRE reaction. Results were also compared with 21 wt % Ni- and Co-supported MgO catalysts.	Cavallaro et al.[258]
5 wt % Rh/Al_2O_3	550–650 °C; 0.01–0.06 g catalyst; $H_2O/EtOH = 4$–13. GHSV = 80,000 h^{-1}	Rh was loaded on a commercial γ-Al_2O_3 by incipient wetness impregnation. High temperature and low GHSV are required in the SRE reaction. Catalyst deactivation was observed due to carbon deposition. Better performance was observed under oxidative conditions.	Liguras et al.[263]

Catalyst	Conditions	Remarks	Reference
1–5 wt % Pt, Pd, Rh, and Ru on Al_2O_3, TiO_2, or MgO	600–900 °C; 0.1 g catalyst; H_2O/EtOH = 3	Noble metals were loaded on the supports by impregnation. At lower metal loadings, an Rh-based catalyst exhibited better performance in the SRE reaction. Catalytic performance increased with increasing metal loading on all the noble metal-based catalysts.	Breen et al.[264]
1 wt % Rh, Pt, or Pd supported on CeO_2–ZrO_2 and Al_2O_3	400–700 °C; 0.1 g catalyst; H_2O/EtOH = 3; total gas inlet flow rate = 152.4 cc/min	Catalysts were prepared by impregnation. Catalytic performance was compared with 5 wt % Ni on Al_2O_3 and CeO_2–ZrO_2. Dehydration of ethanol to ethylene occurs during the SRE reaction over catalysts supported on Al_2O_3 due to the acidity of the support. Ethylene was not formed over catalysts supported on CeO_2–ZrO_2. The activity was found to be in the following order: Pt≥Rh > Pd.	Diagne et al.[259]
2 wt % Rh supported on CeO_2, ZrO_2, and CeO_2–ZrO_2	300–500 °C; 0.1 g catalyst; H_2O/EtOH = 8; liquid flow rate = 0.77 mL/h	Rh was loaded on the homemade supports by impregnation. Among the catalysts, the Rh/CeO_2–ZrO_2 exhibited a better performance in the SRE reaction. The CO_2/CO ratio was found to be sensitive to Ce/Zr ratio of the support due to the difference in the CO_2 adsorption properties.	Galvita et al.[260]
2.5 wt % Rh supported on a commercial γ-Al_2O_3	650 °C; 0.016 g catalyst; H_2O/EtOH = 8.4. GHSV = 50,000–300,000 h^{-1}	Catalyst was prepared by incipient wetness impregnation using $RhCl_3$ as a precursor. SRE reaction produces mainly syngas. Consecutive reactions of ethanol to acetaldehyde and ethylene followed reforming of these intermediates occur. High temperature and high Rh loading (>5 wt %) reduces the coke formation.	Cavallaro[261]

redox supports such as CeO_2, ZrO_2, and CeO_2–ZrO_2 mixed oxides have been found to be highly favorable for ethanol reforming and WGS reactions to produce high yields of H_2.[240,242,252,255,256,259,260]

Some of these catalysts are capable of catalyzing SRE reaction at lower temperatures, below 450 °C, to directly produce H_2 and CO_2 with only traces of undesirable by-products such as CO, CH_4, and acetaldehyde. A few selected catalysts for the low-temperature SRE are listed in Table 2.28. As can be seen, cobalt metal supported on ZnO or Rh supported on ZrO_2 or CeO_2–ZrO_2 mixed oxides is highly favorable for SRE reaction to produce H_2 and CO_2 at low temperatures. Among them, Co/ZnO catalyst prepared using cobalt carbonyl complex as a precursor of cobalt exhibits high H_2 and CO_2 yields at very low temperature, around 350 °C. Such a catalytic process is highly attractive because it produces CO-free H_2 suitable for PEMFCs.

Table 2.28. Selected Best Performing Catalysts for the Low-Temperature Steam Reforming of Ethanol (SRE) Reaction[a]

Catalyst	Temperature (°C)	H_2/EtOH Converted[b]	Exit Composition (Dry Basis, %)[c]				Reference
			H_2	CO	CO_2	CH_4	
Co/ZnO	350	2.5	73.4	–	25	1.6	Llorca et al.[238]
Co/ZnO[d]	450	–	71.3	–	20.2	0.8	Freni et al.[240]
Co/ZnO[d]	400	–	70.3	–	19.8	0.2	Batista et al.[234]
Co/SiO₂[e]	400	–	70.0	<1	20.0	8.0	Frusteri et al.[235]
Rh/ZrO₂	450	–	71.1	2.1	20.3	ND	Aupretre et al.[256]
Rh/CeO₂ –ZrO₂	450	4.3	NA	2.8	16.0	6.3	Zhang et al.[252]
Rh/CeO₂ –ZrO₂	450	–	71.7	2.1	20.2	6.0	Galvita et al.[260]
Cu/CeO₂[f]	320	–	~63	–	16.2[h]	–	Duan and Senkan[231]
Ni/Al₂O₃[g]	400	–	70.5	6.0	12.2	11.3	Morgenstern and Fornango[226]

[a]Detailed experimental conditions are described in Table 2.26.
[b]Ratio of number of moles of H_2 formed per mole of ethanol converted.
[c]Approximate exit compositions close to 100% ethanol conversion as the data are extracted from respective figures.
[d]Traces of ethylene (below 2%) was also observed.
[e]Ethanol conversion over 70%.
[f]Acetaldehyde and acetone were also observed with a selectivity of 27.7% and 37.5%, respectively.
[g]Ethanol conversion about 85%.
[h]Selectivity defined based on carbon balance.
NA, not available; ND, not determined.

Table 2.29 lists a few selected catalysts for the SRE reaction in the middle temperature range, between 450 and 600 °C, and high temperature range, above 600 °C. Under these conditions, the listed catalysts produce relatively large amounts of CO, which is a poison to the PEMFCs. These catalytic processes would therefore require a multistage CO cleanup process in the downstream in order to produce pure H_2 suitable for fuel cells.

Figure 2.28 compares the exit compositions of H_2 and CO obtained in the SRE over a few selected low-temperature and high-temperature reforming catalysts with that of the equilibrium compositions. While the exit composition of H_2 from high-temperature reforming catalysts is close to that of the equilibrium, the composition of H_2 obtained from low-temperature reforming catalysts is significantly higher than that of the equilibrium composition, probably because SRE reaction over these

Table 2.29. Selected Best Performing Catalysts for the Middle- and High-Temperature Steam Reforming of Ethanol (SRE) Reaction[a]

Catalyst	Temperature (°C)	H_2/EtOH Converted[b]	Exit Composition (Dry Basis, %)[c]				Reference
			H_2	CO	CO_2	CH_4	
Ir/CeO$_2$	500	–	60	4.0	17.0	16.0	Aupretre et al.[253d]
Co/CeO$_2$	500	–	65	2.5	21.0	9.0	Aupretre et al.[253d]
Ni/CeO$_2$	500	–	65	2.0	23.0	6.0	Aupretre et al.[253d]
Rh/MgO/ NiO/Al$_2$O$_3$	700	4.68	70.2	13.4	13.6	2.7	Wanat et al.[254]
Co/ZrO$_2$	700	–	70	10.0	23.0	2.0	Vargas et al.[229]
Rh–Ce– monolith	900	4.7	75	12.5	12.5	<5	Diagne et al.[255]
Ni/Al$_2$O$_3$	500	–	57	9.0	27	7.0	Comas et al.[219]
Rh/Al$_2$O$_3$	800	–	71.7	6.0	18	1.5	Breen et al.[264]
Ni/MgO	650	5.3	–	7.5	17.0	<1.0	Srinivas et al.[241]
Ni/CeO$_2$– ZrO$_2$	550	–	69	3.7	21.3	6.0	Fatsikostas et al.[242]
Rh/Al$_2$O$_3$	700	–	70	18	9.0	<5.0	Diagne et al.[259]
Ni/La$_2$O$_3$	800	–	72	17	6.0	<5.0	Fatsikostas et al.[243]

[a]Detailed experimental conditions are described in Table 2.27.
[b]Ratio of number of moles of H_2 formed per mole of ethanol converted.
[c]Approximate exit compositions close to 100% ethanol conversion as the data are extracted from respective figures. The formation of acetaldehyde and ethylene is not reported.
[d]Traces of acetone was also observed.

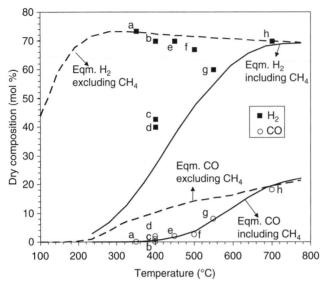

Figure 2.28. H_2 and CO selectivities in the steam reforming of ethanol over a few selected catalysts. Data reported at complete ethanol conversion and obtained using $H_2O/EtOH = 3$ and $P = 1$ atm are compared with that of equilibrium (Eqm.) data calculated by including CH_4 (solid lines) and excluding CH_4 (broken lines) in the calculation. Catalysts: (a) Co(CO)/ZnO, (b) Co/SiO$_2$, (c) Ni/Y$_2$O$_3$, (d) Ni/Al$_2$O$_3$, (e) Ni/CeO$_2$, (f) Co/CeO$_2$, (g) 6% Ni–3% Cu/ZrO$_2$, and (h) Rh/CeO$_2$–ZrO$_2$. Adapted from Velu and Song.[167]

catalysts does not involve CH_4 as an intermediate. The equilibrium compositions are therefore recalculated excluding CH_4 in the calculation, and this shows a maximum H_2 of over 70% at low temperatures, above 250 °C itself. Interestingly, the exit H_2 concentration obtained from low-temperature reforming catalysts is close to the equilibrium composition of H_2 calculated by excluding CH_4 in the calculation, thus supporting the assumption that the SRE reaction over low-temperature reforming catalysts proceeds without involving CH_4 as an intermediate, and the exit H_2 concentration does not depend on the CH_4 composition.

Since Ni-based catalysts have been successfully employed in the SMR process, the use of similar Ni-based catalysts in the SRE reaction has been explored extensively and found that Ni supported on Y$_2$O$_3$, La$_2$O$_3$, CeO$_2$, ZrO$_2$, and CeO$_2$–ZrO$_2$ exhibit better performance. Fatsikostas et al.[243] observed a stable activity for over 100 h of onstream operation over La$_2$O$_3$-supported Ni catalyst. The enhanced stability of the catalyst has been attributed to the formation of a thin over layer of lanthana on top of the Ni particles under the reaction conditions.

The type of catalysts used and reaction operating conditions employed in the oxidative stream reforming (OSR) and ATR of ethanol for H_2 production are summarized in Table 2.30.[173,267–278] In principle, the catalysts that are active in the steam reforming and POE should also be active in the OSR reactions. As can be seen from

Table 2.30. Selected Catalysts for the Oxidative Steam Reforming (OSR) and Autothermal Reforming of Ethanol for Hydrogen Production

Catalyst	Reaction Conditions	Results and Description	Reference
2.8 wt % Pd/ZnO	275–450 °C; 0.1 g catalyst; $H_2O/EtOH = 13$; $O_2/EtOH = 0.5$; $GHSV = 5200\,h^{-1}$	Catalysts were prepared by the incipient wetness impregnation. PdZn alloy formation favors the oxidative dehydrogenation of ethanol to acetaldehyde rather than CH_4, thereby producing H_2 with high yield in the OSR at low temperatures.	Fierro et al.[270]
5 wt % Ni–1 wt % Rh supported on three different CeO_2 supports	300–600 °C; 0.1 g catalyst; $H_2O/EtOH = 4$; $O_2/EtOH = 0.4$; $GHSV = 244,000\,h^{-1}$	Ni and Rh were loaded by impregnation. The higher the Rh dispersion, the better is the catalytic activity, H_2 selectivity, and stability in the OSR of ethanol. Higher Rh dispersion was obtained when Rh was supported on CeO_2 with smaller particle sizes and higher BET surface area. The mechanism of the OSR reaction on these catalysts was investigated.	Liguras et al.[271] and Fierro et al.[272]
CuNi/ZnO–Al_2O_3 and CoNi/ZnO–Al_2O_3 catalysts derived from layered double hydroxides	200–300 °C; 0.1 g catalyst; $H_2O/EtOH = 3$; $O_2/EtOH = 0.4$; $WHSV = 107\,mol/h/kg$	Catalysts containing various Cu : Ni : Zn : Al and Co : Ni : Zn : Al molar ratios were synthesized by coprecipitation. Cu-rich catalysts favor the formation of acetaldehyde by dehydrogenation, while Ni-rich catalysts favor the formation of CH_4. The CoNi-based catalysts exhibit better catalytic performance with lower selectivity to undesirable acetaldehyde, CO, and CH_4. The Cu-based catalysts were characterized by XPS.	Therdthianwong et al.[279] and Shamsi et al.[281]

(Continued)

Table 2.30. *Continued*

Catalyst	Reaction Conditions	Results and Description	Reference
2.5 wt% Rh, Ru, Pt, or Pd supported on Al_2O_3 ceramic foams with various promoters	700 °C; contact time ≈5–10 ms; H_2O/EtOH = 2.5–9; O_2/EtOH = 0.5–1.5	Catalysts were prepared by washcoating. CeO_2-promoted Rh catalyst exhibited the highest activity and selectivity to syngas. Addition of steam in the reaction feed increased the H_2 selectivity and decreased CO selectivity. The process produced H_2-rich gas containing large amount of CO.	Akande[280]
Commercial Ni/Al_2O_3 catalysts containing 11 and 20 wt% Ni/Al_2O_3 doped with Cu, Zn, Fe, and Cr	H_2O/EtOH = 1.6; O_2/EtOH = 0.68; 600–800 °C; gas flow rate = 80 cc/min	Metal promoters were loaded on the commercial Ni/Al_2O_3 catalyst by impregnation. The addition of a small amount of promoter metal to 20 wt% Ni/Al_2O_3 catalyst increased the H_2 production only at higher temperatures, around 800 °C, and the H_2 production was in the order Ni–Zn > Ni–Fe > Ni–Cr > unpromoted > Ni–Cu.	Cavallaro et al.[273]
0.75 wt% Pt supported on Al_2O_3, Al_2O_3–La_2O_3, Al_2O_3–CeO_2, and Al_2O_3–La_2O_3–CeO_2	H_2O/EtOH = 2.28; O_2/EtOH = 0.36; 575–725 °C; GHSV = 19,492 h⁻¹	Pt was loaded by impregnation of the supports prepared from a commercial Al_2O_3. Pt/Al_2O_3–CeO_2 performs better in the OSR of ethanol because of the higher Pt dispersion and Pt–Ce interaction. The catalyst exhibited a higher EtOH conversion and H_2 selectivity around 700 °C.	Reitz et al.[173]
5 wt% Rh containing Ce additive supported on ceramic foam	Feed sprayed at 140 °C; catalyst exit temperature 700 °C; contact time ≈5–100 ms; GHSV ~ 10⁵ h⁻¹, O_2/EtOH = 1	Rh–Ce catalyst exhibited a high ethanol conversion over 95% and high selectivity to H_2 and CO in a very short residence time of <10 ms under autothermal condition. Increasing steam/carbon ratio increased the H_2 yield due to the participation of WGS reaction. The POE and WGS reactions were also performed in a two-stage reactor.	Deluga et al.[166]

Catalyst	Conditions	Remarks	Reference
5 wt% Ru/γ-Al₂O₃ on cordierite monolith/ceramic foams and pellets	$H_2O/EtOH = 2\text{–}4$; $O_2/EtOH = 0.4\text{–}1.0$; 600–750 °C; $GHSV = 7245\,h^{-1}$	Ru impregnated γ-Al₂O₃ catalyst was washcoated on cordierite monolith/ceramic forms and pellets. Among them, the Ru supported on ceramic foams exhibited better performance in the autothermal reforming of ethanol probably due to smaller pore size and higher tortuosity of the support.	Fierro et al.[274]
20 wt% Ni/SiO₂, NiCu/SiO₂, and 5% Pd, Pt, Ru, and Rh on Al₂O₃	50 mg catalyst; $H_2O/EtOH = 1\text{–}2$; $O_2/EtOH = 0.5\text{–}1.0$; 400–800 °C; contact time = 0.2 min/kg/mol	Among the catalysts tested in the OSR of ethanol, the Ni–Cu/SiO₂ and Rh/Al₂O₃ showed better activity and selectivity for H₂. Addition of Cu to Ni/SiO₂ decreased the coke formation due to the formation of Ni–Cu alloy. A large amount of CO was however present in the reformed gas stream.	Klouz et al.[275]
5 wt% Rh supported on a commercial γ-Al₂O₃	16 mg catalyst; $H_2O/EtOH = 2\text{–}3$; $O_2/EtOH = 0.2\text{–}1.1$; 650 °C; $GHSV = 150{,}000\,h^{-1}$	Rh was loaded by impregnation. Addition of a small amount of O₂ in the autothermal reforming reaction increases the overall H₂ yield, decreases the operating temperature, and reduces the coke and CH₄ formation. Detailed studies on the effect of H₂O/EtOH and O₂/EtOH on the catalytic performance were evaluated.	Velu et al.[276]
18.4 wt% Ni-Cu bimetallic system supported on SiO₂	15–50 mg catalyst; $H_2O/EtOH = 1.5\text{–}5.5$; $O_2/EtOH = 0\text{–}1.7$; 300–600 °C; total flow rate = 50 mL/min	Catalysts were prepared by impregnation. OSR of ethanol over these catalysts was tested off-board under dilute condition as well as on-board conditions. The addition of O₂ along with bioethanol suppressed the CH₄ and coke formation and improved the stability of the catalyst during onstream operation.	Salge et al.[277] and Velu et al.[278]

XPS, X-ray photoelectron spectroscopy.

Table 2.30, a wide range of catalysts such as Pd/ZnO, noble metals supported on Al_2O_3, or ceramic monolith/foams with or without additives such as CeO_2 or La_2O_3, as well as bimetallic catalysts such as Ni–Cu, Ni–Co, and Ni–Rh, supported on $ZnO–Al_2O_3$, Al_2O_3, and CeO_2 have been employed in the OSRE.

The catalysts for low-temperature and high-temperature reforming of ethanol under OSR/autothermal conditions are listed in Tables 2.31 and 2.32, respectively. In the absence of added O_2, each mole of ethanol converted should produce 6 mol of H_2 and 2 mol of CO_2, which corresponds to an exit H_2 composition of 75%. Depending upon the O_2/EtOH ratio used in the OSRE reaction, the composition of H_2 will vary between 50% for O_2/EtOH ratio of 1.5 and H_2O/EtOH ratio of 0, and 75% for O_2/EtOH ratio of 0 and H_2O/EtOH ratio of 3.

The H_2 yield for the low-temperature reforming catalysts such as Pd/ZnO and NiRh/CeO_2 shown in Table 2.31 is close to that of the equilibrium composition determined by excluding CH_4 in the calculation rather than that derived by including CH_4 in the calculation (see Fig. 2.29), suggesting that the OSRE reaction over these catalysts proceeds without involving CH_4 as an intermediate.[267–269] The observed CO concentration of <1% over these catalysts is well below the equilibrium composition. This could be due to slightly different H_2O/EtOH and O_2/EtOH ratios used in testing these catalytic systems (see Table 2.31) than those used in determining the equilibrium compositions. The formation of Pd–Zn alloy in the Pd/ZnO catalyst has been reported to be more selective toward ethanol reforming via acetaldehyde intermediate rather than ethanol decomposition to produce CH_4 as an intermediate, thus favoring the OSR reaction at relatively lower temperatures to directly produce H_2 and CO_2 with only traces of CO and CH_4.[267] The exit H_2 composition of high-temperature reforming catalysts listed in Table 2.32 is close to that of equilibrium compositions, except for Rh supported on CeO_2-modified ceramic foam, which exhibit lower H_2

Table 2.31. Selected Best Performing Catalysts for the Low-Temperature Oxidative Steam Reforming (OSR) of Ethanol

Catalyst	Temperature (°C)	H_2/EtOH Converted[a]	Exit Composition (Dry Basis, %)[b]					Reference
			H_2	CO	CO_2	CH_4	Acetaldehyde	
Pd/ZnO[c]	450	–	61	<1	22.0	3.1	1.1	Fierro et al.[270]
NiRh/ CeO_2	450	–	52	5.0	27.7	15.0	0.3	Liguras et al.[271] and Fierro et al.[272]
Ni–Co/ ZnO– Al_2O_3	300	3.0	~50	20.6	25.9	35.6	17.9	Shamsi et al.[281]

[a] Moles of H_2 produced per mole of ethanol converted.

[b] Approximate exit compositions close to 100% ethanol conversion as the data are extracted from respective figures.

[c] Traces of ethylene was also detected.

Table 2.32. Selected Catalysts for the High-Temperature Oxidative Steam Reforming/Autothermal Reforming of Ethanol

Catalyst	Temperature (°C)	H_2/EtOH Converted	Exit Composition (Dry Basis, %)[a]						Reference
			H_2	CO	CO_2	CH_4	Acetaldehyde	Ethylene	
Rh–CeO$_2$–Al$_2$O$_3$ foam	700	3.3[b]	47.0[c]	15.7	12.8	~2	Trace	Trace	Deluga et al.[166]
Rh–CeO$_2$–Al$_2$O$_3$ foam	700; 400[d]	3.9	55.6	~9	14.3	–	–	–	Deluga et al.[166]
Rh–CeO$_2$–Al$_2$O$_3$ foam	700	3.6	51.3	11.4	13.2	~4	Trace	Trace	Akande[280]
Ni–Cu/Al$_2$O$_3$	800	5.3	72.5	22	5.5	Trace	Trace	Trace	Cavallaro et al.[273]
Rh/Al$_2$O$_3$	700	–	70.2	15.7	11.1	Trace	Trace	Trace	Salge et al.[277]
Rh/Al$_2$O$_3$	800	–	71.5	19.3	5.5	Trace	Trace	Trace	Klouz et al.[275]
Rh/Al$_2$O$_3$	650	5.5	–	6.1	21.3	<1	–	–	Velu et al.[276]
Ru/Al$_2$O$_3$	600	–	70.0	10.5	13.8	2.8	Trace	Trace	Fierro et al.[274]

[a] Approximate exit compositions close to 100% ethanol conversion as the data are extracted from respective figures.

[b] Reported H_2 selectivity is about 110% H_2. One hundred percent H_2 = 3 mol of H_2/mole of ethanol converted.

[c] Stoichiometrically, 167% H_2 corresponds to 5 mol of H_2/mole of ethanol converted, which corresponds to an exit composition of 71.4% H_2 and 28.6% CO/CO$_2$. The observed H_2 selectivity of 110% would therefore correspond to an exit composition of about 47% H_2.

[d] Autothermal reforming coupled with WGS reaction ~400 °C over Pt-based catalyst in the downstream.

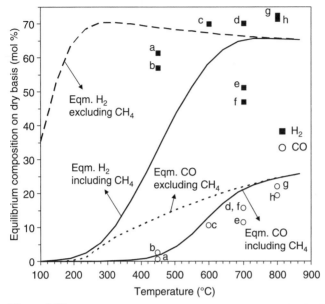

Figure 2.29. H_2 and CO selectivities in the oxidative steam reforming/autothermal reforming of ethanol over selected catalysts. Data reported at complete conversion of ethanol and obtained using $O_2/EtOH = 0.36$, $H_2O/EtOH = 2.28$, and $P = 1$ atm and calculated by including CH_4 (solid lines) and excluding CH_4 (broken lines) in the calculations. Catalysts: (a) Pd/ZnO, (b) NiRh/CeO$_2$, (c) Ru/Al$_2$O$_3$, (d) Rh/Al$_2$O$_3$, (e) Rh–CeO$_2$–Al$_2$O$_3$ foam, (f) Rh–CeO$_2$–Al$_2$O$_3$ foam, (g) NiCu/Al$_2$O$_3$, and (h) Rh/Al$_2$O$_3$. Adapted from Velu and Song.[167] Eqm., equilibrium.

yield. This could be due to poor WGS reaction under the operating conditions of high temperature and millisecond contact times. The H_2 yield increased with increasing contact time or coupling the WGS reaction in the downstream at lower temperature over a Pt-based catalyst along with this autothermal reaction.[166,277]

2.4.2.4 Kinetics

Although the SRE reaction has been studied over various noble metal- and Ni-based catalysts, kinetic studies have been reported only for a few Ni-based catalysts. For Ni/Al$_2$O$_3$, Ni/Y$_2$O$_3$, and Ni/La$_2$O$_3$, Sun et al.[225] observed that the reaction was first order with respect to ethanol and employed the following rate equation,

$$r = k[\text{ethanol}][\text{H}_2\text{O}]^n, \tag{2.91}$$

where k is the rate constant and n is the order of the reaction with respect to the partial pressure of water. On the other hand, for the similar Ni/Al$_2$O$_3$ catalyst, Therdthianwong et al.[279] observed that the order of the ethanol steam reforming at 400 °C and atmospheric pressure was 2.52 with respect to ethanol and 7 with respect to steam, and proposed the following rate expression:

$$-r_{\text{EtOH}} = 77.8 \left(P_{\text{EtOH}} \right)^{2.52} \left(P_{\text{H}_2\text{O}} \right)^7. \tag{2.92}$$

Akande,[280] in the reforming of crude ethanol containing $H_2O/EtOH$ ratio of about 18, employed a power law model and expressed the rate equation as shown in Equation 2.93,

$$-r_{\text{A}} = k_0 e^{-E/RT} C_{\text{A}}^n, \tag{2.93}$$

where r_{A} is the rate of crude ethanol conversion, k_0 is the collision frequency, E is the energy of activation, R is the gas constant, and T is the absolute temperature in kelvin. C_{A} is the crude ethanol concentration, and the order of reaction n with respect to ethanol was determined to be 0.43. Using the ER-type kinetic model, Aboundheir et al.[220] modified the above rate equation as

$$-r_{\text{A}} = \frac{k_0 e^{-E/RT} \left(N_{\text{A}} - N_{\text{C}}^2 N_{\text{D}}^6 / K p N_{\text{B}}^3 \right)}{\left(1 + K_{\text{A}} N_{\text{A}} \right)^2}, \tag{2.94}$$

where N_{A}, N_{B}, N_{C}, and N_{D} are molar flow rates (kmol/s) of species such as ethanol, water, CO_2, and H_2, respectively, Kp is the equilibrium constant, and K_{A} is the adsorption constant. The authors also determined these parameters for a hypothetical industrial crude ethanol reformer and compared with that of the lab-scale crude ethanol reformer.

The type of catalyst, the rate constant for ethanol conversion, and energy of activation over various Ni-based catalysts are summarized in Table 2.33. The activation energy for ethanol steam reforming reaction over similar Ni-based supported catalysts varied between 2 and 17 kJ/mol. A very high activation energy of 149 kJ/mol reported by Morgenstern and Fornango[226] could be the activation energy for the DCE to CH_4, CO, and H_2 rather than ethanol reforming into H_2-rich syngas.

Table 2.33. Kinetic Data of the Steam Reforming of Ethanol over Various Supported Ni Catalysts

Catalyst	Temperature (°C)	Rate Constant	Order w.r.t. Ethanol	Energy of Activation (E = kJ/mol)	Reference
Ni/Al_2O_3	400	77.8 kmol/kg$_{\text{cat}}$/s/atm$^{9.52}$	2.52	NA	279
Ni/Y_2O_3	250–350	2.95×10^{-3} m^3/kg$_{\text{cat}}$/s	1	7.04	Sun et al.[225]
Ni/Al_2O_3	250–350	2.32×10^{-3} m^3/kg$_{\text{cat}}$/s	1	16.88	Sun et al.[225]
Ni/La_2O_3	250–350	19.1×10^{-3} m^3/kg$_{\text{cat}}$/s	1	1.87	Sun et al.[225]
Ni/Al_2O_3	320–520	NA	0.43	4.41	280
Cu-plated Raney nickel	250–300	NA	1	149[a]	Bergamaschi et al.[227]

[a]Could be the activation energy for the decomposition of ethanol to CH_4, CO, and H_2 rather than ethanol reforming into H_2-rich syngas.

NA, not available.

2.4.2.5 Reaction Pathway

Experimental results on the SRE reaction over a wide range of catalysts produced both acetaldehyde and CH_4 as intermediates. Depending on the type of catalyst used and reaction operating conditions employed, the selectivity to these intermediates varied. Diagne et al.[259] observed a selectivity about 75% toward acetaldehyde at 300 °C over Ni/La_2O_3 catalyst, and this selectivity was found to decrease with an increase in temperature with a value close to zero at around 550 °C. However, CH_4 selectivity peaked to about 20% at around 600 °C. On the other hand, Cavallaro et al.[258,261] observed more CH_4 than acetaldehyde over 5% Rh/Al_2O_3 catalyst at 650 °C. It appears in general that the primary reactions in the SRE are dehydrogenation (Eq. 2.85) and decomposition (Eq. 2.84) to produce acetaldehyde and CH_4, respectively. Dehydration to ethylene (Eq. 2.86) could also occur over Al_2O_3-supported catalysts. The extent of these reactions and selectivity of acetaldehyde, CH_4, and ethylene depend on the nature of catalysts and reaction operating conditions employed. These intermediates could subsequently undergo steam reforming to produce H_2 and CO followed by WGS reaction to convert CO into CO_2 and additional H_2. Taking all these elementary reactions into account, the overall reaction of ethanol steam reforming can be schematically represented as shown in Figure 2.30.

According to this scheme (Fig. 2.30), a catalyst with high selectivity toward ethanol dehydrogenation to acetaldehyde followed by the steam reforming of acetaldehyde (SRE pathway-1) could produce high yields of H_2 and CO_2 at lower temperatures compared with ethanol decomposition to a mixture of CO and CH_4 followed by steam reforming (SRE pathway-2) or ethanol dehydration to ethylene followed by steam reforming (SRE pathway-3). Spectroscopic studies revealed that most of the low-temperature reforming catalysts listed in Table 2.28, especially those con-

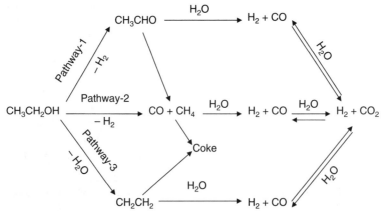

Figure 2.30. Proposed reaction pathway for the steam reforming of ethanol. Adapted from Velu and Song.[167]

taining Co, Rh, and Cu, supported on ZnO, CeO_2, ZrO_2, or CeO_2–ZrO_2 mixed oxides, proceed through the formation of acetaldehyde, which subsequently undergo steam reforming and WGS reaction to produce H_2 and CO_2 with only small amounts of CO at relatively lower temperatures. The exit composition of SRE reaction in this pathway could be expected to be close to that of the steam reforming of acetaldehyde (Eqs. 2.88 and 2.89). In some cases, the formation of other oxygenates such as acetone has also been observed,[228,231,234–236,253,262] probably by a network of reactions involving aldol-type condensation of acetaldehyde and oxidation of intermediate aldehyde followed by decarboxylation and dehyderogenation.

On the other hand, catalysts that are more selective toward ethanol decomposition to CH_4 would require higher temperatures to subsequently reform CH_4 into syngas.[229,254,255,259] It appears that the high-temperature reforming catalysts listed in Table 2.29 proceed through the formation of CH_4 as an intermediate, thereby requiring a higher reaction temperature in order to completely reform CH_4 into syngas. Since, WGS reaction is limited by equilibrium at higher temperatures (see Fig. 2.24), the SRE reaction performed at higher temperatures produce large amounts of CO in the exit gas (see Table 2.29 and Fig. 2.28).

The reaction pathway involved in the OSR of ethanol is a little different from that of the SRE reaction and includes steps involved in the partial oxidation. Velu et al.[278] and Kugai et al.[268] performed a systematic study over Ni–Cu/ZnO–Al$_2$O$_3$ and Ni–Rh/CeO$_2$ catalysts, respectively. As reported for the POE,[167] the reaction in the OSR was found to proceed through acetaldehyde intermediate formed by the oxidative dehydrogenation of ethanol (Eq. 2.95).[268,269,278] Acetaldehyde could further undergo an oxidative steam reforming (Eq. 2.96) to produce H_2 and CO_2. As described above, the OSR of acetaldehyde is thermodynamically a highly favorable reaction with a large negative free energy change.[167] In fact, a comparison of experimental results on the steam reforming and OSR of acetaldehyde over NiRh/CeO$_2$ catalyst revealed that the later reaction is much more efficient, with acetaldehyde conversion close to 100%, producing H_2 and CO_2 selectively (see Table 2.34).[268] On the other hand, very low conversion of acetaldehyde (below 10%) has been observed in the steam reforming of acetaldehyde without added O_2. Thus, the overall reaction

Table 2.34. Steam Reforming and Oxidative Steam Reforming of Acetaldehyde over Ni–Rh/CeO$_2$ Catalyst[268]

Reaction	CH$_3$CHO Conversion (mol%)	Product Distribution (mol%)					
		H$_2$	CO	CO$_2$	CH$_4$	EtOH	Acetone
Steam reforming	10.6	52.0	20.7	12.6	9.5	1.8	3.4
Oxidative steam reforming	94.6	52.0	3.3	35.0	9.6	0.0	0.1

Reaction conditions: temperature = 350 °C; GHSV = 24,379 h^{-1}; H$_2$O/CH$_3$CHO = 4 in steam reforming, while in oxidative steam reforming, H$_2$O/CH$_3$CHO = 4 and O$_2$/CH$_3$CHO = 0.4.

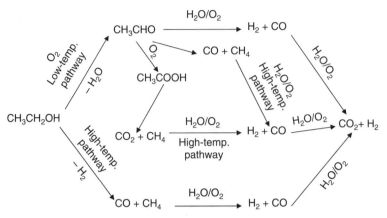

Figure 2.31.   Proposed reaction pathway for the oxidative steam reforming of ethanol. Adapted from Velu and Song.[167]

pathway for the OSR of ethanol shown in Figure 2.31 could be similar to those proposed for steam reforming and partial oxidation reactions, except that the intermediates acetaldehyde and CH_4 could undergo OSR/ATR rather than steam reforming/partial oxidation to produce H_2 and CO_2. Furthermore, catalysts such as Pd/ZnO and NiRh/CeO$_2$ would proceed through acetaldehyde as an intermediate to produce H_2 and CO_2 directly in a low-temperature pathway rather than ethanol decomposition to CH_4, which would follow a high-temperature pathway.[267,268]

$$CH_3CH_2OH(g) + 0.5O_2(g) \rightarrow CH_3CHO(g) + H_2O(g)$$
$$\Delta H^{\circ}_{298} = -153.0 \text{ kJ/mol} \tag{2.95}$$

$$CH_3CHO(g) + H_2O(g) + O_2(g) \rightarrow 3H_2(g) + 2CO_2(g)$$
$$\Delta H^{\circ}_{298} = -379.0 \text{ kJ/mol} \tag{2.96}$$

2.5 CARBON FORMATION AND CATALYST DEACTIVATION

One of the most critical issues in developing catalytic reformers, especially for the reforming of hydrocarbon fuels, is the risk of carbon deposition on the catalyst surface and consequent catalyst deactivation. Carbon formation can occur in several regions of the steam reformer where hot fuel gas is present. Natural gas for example will decompose when heated in the absence of air or steam at temperatures above 650 °C via pyrolysis reaction as shown in Equation 2.4.

Similar pyrolysis reaction can also occur during reforming of higher hydrocarbons. In fact, higher hydrocarbons tend to decompose more easily than methane and therefore the risk of carbon formation is even higher with vaporized liquid petroleum fuels than with natural gas. Another source of carbon formation is from the disproportionation of carbon monoxide, also known as Boudouard reaction shown in Equation 2.3. These reactions are catalyzed by metals such as nickel, and therefore, there is a high risk of occurring on steam reforming catalysts that contain nickel and nickel-containing stainless steel used for fabrication of the reactor.

Carbon formation on steam reforming catalysts has been the subject of intense study over the years and is relatively well understood.[5,8–10,13,16,54–57,281] Carbon formed via pyrolysis and Boudouard reaction adopts different forms such as amorphous carbon and filamental carbon. The nature of the carbons formed can be identified by scanning or TEM or temperature-programmed oxidation (TPO) methods. The filamental carbon is the most damaging form of carbon, and it grows while attached to the nickel crystallites within the catalyst. Such carbon formation can be very fast. If the steam flow to the reformer reactor is shut down for some reason, the consequences can be disastrous; carbon formation occurs within seconds, leading to permanent breakdown and fouling of the catalyst and plugging of the reactor.

There are several ways to reduce the risk of carbon formation in steam reforming reactions. Some of the approaches include:

- use of excess steam than stoichometrically required,
- use of a catalyst with smaller crystal sizes,
- addition of promoters to the Ni-based catalysts,
- use of noble metals instead of Ni or other base metals,
- alloying of Ni with other base metal or noble metal,
- passivation of some of the Ni surfaces that favors carbon formation by sulfur,
- use of a prereformer, and
- conducting reaction under oxidative conditions or oxidative steam reforming.

For instance, the thermodynamic minimum S/C ratio required for the reforming of diesel, jet fuel, gasoline, and methane are shown in Figure 2.32. In the steam reforming of methane, an S/C ratio of 2.5 to 3.0 has been used although only 1.0 is stoichiometrically required. Even higher S/C ratio of about 5 is used in the reforming of higher hydrocarbons such as gasoline, diesel, or jet fuel. The added steam promotes the shift reaction, thereby reducing the partial pressure of carbon monoxide in the fuel gas stream. The excess steam also leads to carbon gasification reaction (Eq. 2.5).

The conventional nickel-based catalysts could be modified by adding oxide promoters such as potassium, lanthanam, cerium, and molybdenum in the catalyst formulations. It is believed that the added promoters improve the dispersion of nickel metal on the catalyst surface, thereby reducing the chance of carbon accumulation. Noble metals such as Pd, Pt, Ru, and Ir have been found to be more carbon tolerant as the solubility of carbon is less in these metals.[54–57] However, they are more expensive than nickel-based catalyst, and as a consequence, they are less attractive for large-scale commercial applications. Alloying of nickel with other base metals such as Cu, Co, or noble metals such as Au, Pt, and Re has also been found to decrease

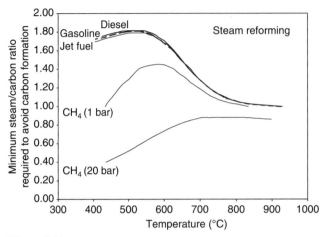

Figure 2.32. Thermodynamic minimum S/C ratios required to prevent carbon formation in diesel, jet fuel, gasoline, and methane steam reforming.

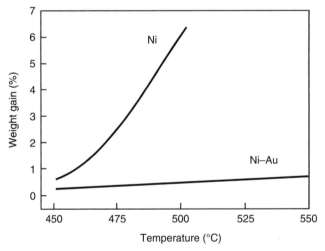

Figure 2.33. TGA data showing the effect of alloying of nickel with gold on the amount of carbon deposited on the catalyst surface during steam reforming of n-butane.[10]

the rate of carbon formation. As an example, Figure 2.33 shows that the carbon formation during steam reforming of n-butane can be dramatically suppressed by alloying of Ni with 1.85 wt% of Au. The nickel and gold do not mix in the bulk but may form stable ally-like structures in the outermost layer. As a result, it was possible to control the surface coverage and to eliminate carbon formation.

As has been discussed in Section 2.3.6, the use of an adiabatic prereformer also helps in alleviating the risk of carbon formation in the steam reforming of higher hydrocarbons. The prereforming is carried out before the main reforming at relatively low temperature, between 350 and 550 °C over nickel-based catalyst or the catalyst similarly used in the main reformer. The reformer operates adiabatically; that is, heat is neither supplied to it nor removed from it.

Carbon formation and catalyst deactivation have also been observed in the SRE. Approaches similar to those discussed above for the steam reforming of hydrocarbons are also employed to suppress the carbon formation in ethanol reforming as well.[167]

2.6 RECENT DEVELOPMENTS IN REFORMING TECHNOLOGIES

The CSR technology using conventional packed-bed reactors has been highly successful in the large-scale production of hydrogen. In order to employ this technology for small-scale portable and mobile applications, where compactness, easy start-up of the fuel processor, and good temperature control are critical, new reactor designs such as microreactors, plate reformers, and membrane reactors have been developed in recent years.[163,168] The recent developments in reactor design for the SMR reaction has been discussed in Section 2.2.1.7. In this section, the reactor designs developed for the reforming of other hydrocarbon and alcohol fuels are briefly discussed.

2.6.1 Microreactor Reformer

Microreactors have channel gaps in the order of microscale, typically $<1000 \mu m$ or mesoscale ($1000 \mu m$ to a few centimeters). Microreactors are much more suitable for the distributed production of hydrogen compared with conventional systems as they offer numerous advantages. High heat and mass transfer rates, for example, enable reactions to be performed under more aggressive conditions that favor overall kinetics or space time yields. These high rates are due to high surface-to-volume ratios and short transfer distances in the reactors. For reactions operating in mass transfer-limited regimes, microreaction devices could be considerably smaller than their conventional counterparts at the same throughput. These small reactors are being designed to provide hydrogen for varied power requirements (subwatt power to hundreds of watts), primarily using methanol as a fuel.

Researchers at Pacific Northwest National Laboratory, Battelle, in the U.S. are recently developing a methanol reformer to provide subwatt to over 100 W power from methanol fuel using microreactor technology (Fig. 2.34).[168] The fuel processors consist of fuel vaporizer, preheater, steam reforming reactant, and recuperators. The reactors were assembled using a combination of welding, brazing, and diffusion bonding techniques. The reactor was operated with methanol–water mixture at a gas hourly space velocity (GHSV) of 36,000 h^{-1} continuously over 1000 h without any significant activity loss over a nonpyrophoric CO_2 selective methanol reforming catalyst such as Pd/ZnO. The water-to-methanol ratio in the mixture varied between 3.0:1 and 1.2:1 without any significant difference in catalyst lifetime or CO_2 selectiv-

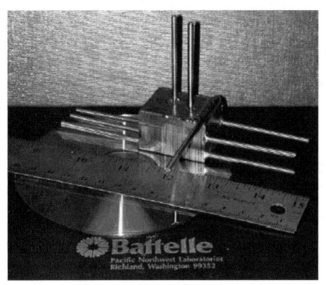

Figure 2.34. A 50- to 100-W integrated fuel processor design developed by the Pacific Northwest National Laboratory.[168]

ity. Recently, the catalyst for the downstream CO cleanup has also been included in the above design, thereby improving the efficiency of the fuel processing system.

2.6.2   Plate Reformer

Plate reformers are also under development as a compact heat exchange reformer for the steam reforming of methane.[131] In a plate reformer, a stack of alternate combustion and reforming chambers are separated by plates. The chambers are filled with suitable catalysts to promote the combustion and reforming reactions, respectively. These reactions take place in more intimate contact on either side of the heat exchanging plate, allowing higher heat transfer rates. The heat released from the combustion reaction is used to drive the reforming reaction. The largest promise appears to be shown by designs utilizing thin catalyst layers coated on compact heat exchangers. Potential heat transfer rates of more than $20 \, kW/m^2$ of heat exchange surface have been demonstrated in the laboratory tests. Such reactors are potentially capable of achieving a power density of up to 13 kWe/I when running on natural gas. This kind of reactor is currently being developed by Advantica and BG Technology Ltd. in U.K. and Pacific Northwest National Laboratory in the U.S.

2.6.3   Membrane Reformer

Hydrogen is able to permeate selectively through palladium or palladium alloy membranes. This has led to the demonstration of membrane reformers in the laboratory. In these reformers, H_2 is selectively removed from the reformer as it is produced. This simultaneous removal of H_2 increases the methane conversion at lower temperatures due to shifting of thermodynamic equilibrium. The H_2 produced in this method is pure, thereby avoiding downstream CO cleanup to use in the PEMFCs.

Table 2.35. Some of the Organizations Worldwide Involved in Developing Fuel Processors for H_2 Production for Fuel Cells[6]

Organization	Focus Fuel	Focus Stack Type	Unit Size (kW)	Application	Product Status
Ansaldo Fuel Cells	NG, diesel, landfill gas	MCFC	200	Stationary	Commercial
Aspen Products, Inc.	Diesel	SOFC	5–10	Transportation	Prototype
Ballard Power	NG, methanol	PEMFC	1.0	Stationary	Commercial
Clean Fuel Generation, LLC	NG, LPG, propane	PEMFC	1–10	Transportation	Prototype
FuelCell Energy	NG, anaerobic digester gas	MCFC	1.5–300	Stationary	Commercial
General Electric	NG	SOFC	–	Transportation	Development
HydrogenSource	NG, methanol, anaerobic digester gas	PEMFC	–	Stationary/ transportation	Commercial
IdaTech	Methanol, diesel, biodiesel	PEMFC		Stationary	Commercial
Innovatek	Gasoline, diesel			1–50	Transportation	Prototype
Membrane Reactor Technologies	NG	Steam reforming hydrogen generator	–	Transportation	Commercial
Osaka Gas Company	NG	PEMFC	1,500	Stationary	Commercial
Plug Power	NG, liquid propane	PEMFC	5	Stationary	Commercial
Haldor Topsoe	NG	PAFC	–	On-site H_2	Near commercial

NG, natural gas; MCFC, molten carbonate fuel cell; PAFC, phosphoric acid fuel cell.

2.6.4 Plasma Reforming (PR)

PR technology is being developed as an alternative to the conventional steam reforming technology for H_2 production. The PR technology utilizes the enhanced reactivity of chemical species in the excited states that are present in plasmas. Plasma can be generated by a variety of methods including electric arc, pulsed microwave, and induction methods. A wide variety of fuels such as alcohols, diesel, biomass wastes, and natural gas could be processed by using PR technology. A key factor that affects the attractiveness of PR is the requirement of electric power in order to generate the plasma.

High-temperature PR has been investigated in a variety of modes including steam reforming, partial oxidation, and pyrolysis. Massachusetts Institute of Technology (MIT) in the U.S. has been working on the PR technology.[131] Some of the organizations worldwide involved in developing fuel processors using steam reforming and other reforming technologies to produce H_2 for various fuel cell stacks are summarized in Table 2.35.

2.7 SUMMARY

Hydrogen and syngas can be produced from various gaseous and liquid hydrocarbon fuels as well as alcohol fuels by CSR, partial oxidation, and ATR or oxidative steam reforming methods. Among them, the steam reforming method has been studied extensively. It is a mature technology and being employed for large-scale production of H_2 from natural gas. This technology has been extended to produce H_2 on a small scale from hydrocarbon and alcohol fuels, using fuel cells for stationary and transportation applications. Steam reforming of higher hydrocarbon fuels poses several challenges, including carbon deposition and catalyst deactivation. The chemistry, thermodynamics, catalysts used, kinetics, reaction pathways, and process developments for the steam reforming of various gaseous and liquid hydrocarbon fuels, as well as alcohol fuels such as methanol and ethanol, have been analyzed and discussed in this chapter. Significant progress has been made in recent years, in catalyst development and reactor design, to improve the efficiency of these processes.

REFERENCES

1. ZITTEL, W. and WURSTER, R. *Hydrogen in the Energy Sector*. Ottobrunn: Ludwig-Bolkow-Systemtechnik GmbH, **1996**.
2. MAURA, D.G. Hydrogen market growth—No end in sight. Refinery hydrogen leads the way. *Cryogas International*, **2008**, 46 (2), 22.
3. BROWN, F.L. A comparative study of fuels for on-board hydrogen production for fuel-cell-powered automobiles. *International Journal of Hydrogen Energy*, **2001**, 26 (4), 381.
4. MCINTOSH, S. and GORTE, R.J. Direct hydrocarbon solid oxide fuel cells. *Chemical Reviews*, **2004**, 104 (10), 4845.
5. ROSTRUP-NIELSEN, J.R. and ROSTRUP-NIELSEN, T. Large-scale hydrogen production. *Cattech*, **2002**, 6, 150.
6. FERREIRA-APARICIO, P., BENITO, M.J., and SANZ, J.L. New trends in reforming technologies: From hydrogen industrial plants to multifuel microreformers. *Catalysis Reviews*, **2005**, 47, 491.
7. BRUNGS, A.J., YORK, A.P.E., CLARIDGE, J.B., MÁRQUEZ-ALVAREZ, C., and GREEN, M.L.H. Dry reforming of methane to synthesis gas over supported molybdenum carbide catalysts. *Catalysis Letters*, **2000**, 70 (3), 117.

8. ARMOR, J.N. The multiple roles for catalysis in the production of H_2. *Applied Catalysis. A, General,* **1999**, 176 (2), 159.

9. ROSTRUP-NIELSEN, T. Manufacture of hydrogen. *Catalysis Today,* **2005**, 106 (1–4), 293.

10. ROSTRUP-NIELSEN, J.R., SEHESTED, J., and NØRSKOV, J.K. Hydrogen and syngas by steam reforming. *Advances in Catalysis,* **2002**, 47, 65.

11. SPATH, P.L. and DAYTON, D.C. Preliminary screening—Technical and economic assessment of synthesis gas to fuels and chemicals with emphasis on the potential for biomass-derived syngas. *Technical Report of the U.S. Department of Energy, National Renewable Energy Laboratory, NREL/TP-510-34929,* December **2003**.

12. TWIGG, M.V. *Catalyst Handbook.* London: Wolfe Publishing, **1989**.

13. TRIMM, D.L. The formation and removal of coke from nickel catalyst. *Catalysis Reviews: Science and Engineering,* **1977**, 16 (1), 155.

14. BARTHOLOMEW, C.H. Carbon deposition in steam reforming and methanation. *Catalysis Reviews: Science and Engineering,* **1982**, 24 (1), 67.

15. ROSTRUP-NIELSEN, J.R. In *Catalytic Steam Reforming Catalysis, Science & Technology,* Vol. 5 (eds. J.R. Anderson, M. Boudart). Berlin: Springer, pp. 1–118, **1984**.

16. TRIMM, D.L. Coke formation and minimisation during steam reforming reactions. *Catalysis Today,* **1997**, 37, 233.

17. SEHESTED, J. Four challenges for nickel steam-reforming catalysts. *Catalysis Today,* **2006**, 111 (1–2), 103.

18. HOU, Z. and YASHIMA, T. Meso-porous Ni/Mg/Al catalysts for methane reforming with CO_2. *Applied Catalysis. A, General,* **2004**, 261 (2), 205.

19. KIM, J.-H., SUH, D.J., PARK, T.-J., and KIM, K.-L. Effect of metal particle size on coking during CO_2 reforming of CH_4 over Ni-alumina aerogel catalysts. *Applied Catalysis. A, General,* **2000**, 197 (2), 191.

20. LIU, B.S. and AU, C.T. Carbon deposition and catalyst stability over $La_2NiO_4/[gamma]-Al_2O_3$ during CO_2 reforming of methane to syngas. *Applied Catalysis. A, General,* **2003**, 244 (1), 181.

21. FONSECA, A. and ASSAF, E.M. Production of the hydrogen by methane steam reforming over nickel catalysts prepared from hydrotalcite precursors. *Journal of Power Sources,* **2005**, 142 (1–2), 154.

22. ZHANG, L., WANG, X., TAN, B., and OZKAN, U.S. Effect of preparation method on structural characteristics and propane steam reforming performance of Ni-Al_2O_3 catalysts. *Journal of Molecular Catalysis A: Chemical,* **2009**, 297 (1), 26.

23. ZHANG, Q.-H., LI, Y., and XU, B.-Q. Reforming of methane and coalbed methane over nanocomposite Ni/ZrO_2 catalyst. *Catalysis Today,* **2004**, 98 (4), 601.

24. TAKEHIRA, K., KAWABATA, T., SHISHIDO, T., MURAKAMI, K., OHI, T., SHORO, D., HONDA, M., and TAKAKI, K. Mechanism of reconstitution of hydrotalcite leading to eggshell-type Ni loading on MgAl mixed oxide. *Journal of Catalysis,* **2005**, 231 (1), 92.

25. KIKUCHI, R., IWASA, Y., TAKEGUCHI, T., and EGUCHI, K. Partial oxidation of CH_4 and C_3H_8 over hexaaluminate-type oxides. *Applied Catalysis. A, General,* **2005**, 281 (1–2), 61.

26. SATRIO, J.A., SHANKS, B.H., and WHEELOCK, T.D. Development of a novel combined catalyst and sorbent for hydrocarbon reforming. *Industrial & Engineering Chemistry Research,* **2005**, 44 (11), 3901.

27. LISBOA, J.D.S., SANTOS, D.C.R.M., PASSOS, F.B., and NORONHA, F.B. Influence of the addition of promoters to steam reforming catalysts. *Catalysis Today,* **2005**, 101 (1), 15.

28. BOROWIECKI, T. and GOŁCEBIOWSKI, A. Influence of molybdenum and tungsten additives on the properties of nickel steam reforming catalysts. *Catalysis Letters,* **1994**, 25 (3), 309.

29. BOROWIECKI, T., GIECKO, G., and PANCZYK, M. Effects of small MoO_3 additions on the properties of nickel catalysts for the steam reforming of hydrocarbons: II. Ni-Mo/Al_2O_3 catalysts in reforming, hydrogenolysis and cracking of n-butane. *Applied Catalysis. A, General,* **2002**, 230 (1–2), 85.

30. BOROWIECKI, T., GOLEBIOWSKI, A., and STASINSKA, B. Effects of small MoO_3 additions on the properties of nickel catalysts for the steam reforming of hydrocarbons. *Applied Catalysis. A, General,* **1997**, 153 (1–2), 141.

31. NATESAKHAWAT, S., OKTAR, O., and OZKAN, U.S. Effect of lanthanide promotion on catalytic performance of sol-gel Ni/Al_2O_3 catalysts in steam reforming of propane. *Journal of Molecular Catalysis A: Chemical,* **2005**, 241 (1–2), 133.

32. WANG, S. and LU, G.Q. Role of CeO_2 in Ni/CeO_2-Al_2O_3 catalysts for carbon dioxide reforming of methane. *Applied Catalysis. B, Environmental*, **1998**, 19 (3–4), 267.

33. ZHUANG, Q., QIN, Y., and CHANG, L. Promoting effect of cerium oxide in supported nickel catalyst for hydrocarbon steam-reforming. *Applied Catalysis*, **1991**, 70 (1), 1.

34. HUANG, T.-J. and JHAO, S.-Y. Ni-Cu/samaria-doped ceria catalysts for steam reforming of methane in the presence of carbon dioxide. *Applied Catalysis. A, General*, **2006**, 302 (2), 325.

35. NURUNNABI, M., FUJIMOTO, K.-I., SUZUKI, K., LI, B., KADO, S., KUNIMORI, K., and TOMISHIGE, K. Promoting effect of noble metals addition on activity and resistance to carbon deposition in oxidative steam reforming of methane over NiO-MgO solid solution. *Catalysis Communications*, **2006**, 7 (2), 73.

36. NURUNNABI, M., KADO, S., SUZUKI, K., FUJIMOTO, K.-I., KUNIMORI, K., and TOMISHIGE, K. Synergistic effect of Pd and Ni on resistance to carbon deposition over NiO-MgO solid solution supported Pd catalysts in oxidative steam reforming of methane under pressurized conditions. *Catalysis Communications*, **2006**, 7 (7), 488.

37. NURUNNABI, M., MUKAINAKANO, Y., KADO, S., LI, B., KUNIMORI, K., SUZUKI, K., FUJIMOTO, K.-I., and TOMISHIGE, K. Additive effect of noble metals on NiO-MgO solid solution in oxidative steam reforming of methane under atmospheric and pressurized conditions. *Applied Catalysis. A, General*, **2006**, 299, 145.

38. NURUNNABI, M., MUKAINAKANO, Y., KADO, S., MIYAZAWA, T., OKUMURA, K., MIYAO, T., NAITO, S., SUZUKI, K., FUJIMOTO, K.-I., KUNIMORI, K., and TOMISHIGE, K. Oxidative steam reforming of methane under atmospheric and pressurized conditions over Pd/NiO-MgO solid solution catalysts. *Applied Catalysis. A, General*, **2006**, 308, 1.

39. WANG, L., MURATA, K., and INABA, M. Development of novel highly active and sulphur-tolerant catalysts for steam reforming of liquid hydrocarbons to produce hydrogen. *Applied Catalysis. A, General*, **2004**, 257 (1), 43.

40. ROSTRUP-NIELSEN, J.R. and ALSTRUP, I. Innovation and science in the process industry: Steam reforming and hydrogenolysis. *Catalysis Today*, **1999**, 53 (3), 311.

41. MATSUMURA, Y. and NAKAMORI, T. Steam reforming of methane over nickel catalysts at low reaction temperature. *Applied Catalysis. A, General*, **2004**, 258 (1), 107.

42. SIDJABAT, O. and TRIMM, D.L. Nickel-magnesia catalysts for the steam reforming of light hydrocarbons. *Topics in Catalysis*, **2000**, 11–12 (1), 279.

43. CHOUDHARY, V.R., UPHADE, B.S., and MAMMAN, A.S. Simultaneous steam and CO_2 reforming of methane to syngas over NiO/MgO/SA-5205 in presence and absence of oxygen. *Applied Catalysis. A, General*, **1998**, 168 (1), 33.

44. SUETSUNA, T., SUENAGA, S., and FUKASAWA, T. Monolithic Cu-Ni-based catalyst for reforming hydrocarbon fuel sources. *Applied Catalysis. A, General*, **2004**, 276 (1–2), 275.

45. LAOSIRIPOJANA, N. and ASSABUMRUNGRAT, S. Catalytic dry reforming of methane over high surface area ceria. *Applied Catalysis. B, Environmental*, **2005**, 60 (1–2), 107.

46. URASAKI, K., SEKINE, Y., KAWABE, S., KIKUCHI, E., and MATSUKATA, M. Catalytic activities and coking resistance of Ni/perovskites in steam reforming of methane. *Applied Catalysis. A, General*, **2005**, 286 (1), 23.

47. ROSS, J.R.H. Natural gas reforming and CO_2 mitigation. *Catalysis Today*, **2005**, 100 (1–2), 151.

48. ROSTRUP-NIELSEN, J.R. and BAK-HANSEN, J.H. Carbon dioxide reforming of methane over transition metals. *Journal of Catalysis*, **1993**, 144, 38.

49. GOKON, N., OSAWA, Y., NAKAZAWA, D., and KODAMA, T. Kinetics of CO_2 reforming of methane by catalytically activated metallic foam absorber for solar receiver-reactors. *International Journal of Hydrogen Energy*, **2009**, 34 (2), 1787.

50. MARK, M.F., MAIER, W.F., and MARK, F. Reaction kinetics of the CO_2 reforming of methane. *Chemical Engineering & Technology*, **1997**, 20 (6), 361.

51. RICHARDSON, J.T. and PARIPATYADAR, S.A. Carbon dioxide reforming of methane with supported rhodium. *Applied Catalysis*, **1990**, 61 (1), 293.

52. TRIMM, D.L. and ÖNSAN, Z.I. Onboard fuel conversion for hydrogen-fuel-cell-driven vehicles. *Catalysis Reviews: Science and Engineering*, **2001**, 43, 31.

53. Xu, J. and FROMENT, G.F. Methane steam reforming, methanation and water-gas shift. *AIChE Journal*, **1989**, 35, 88.

54. WEI, J. and IGLESIA, E. Isotopic and kinetic assessment of the mechanism of reactions of CH_4 with CO_2 or H_2O to form synthesis gas and carbon on nickel catalysts. *Journal of Catalysis*, **2004**, 224 (2), 370.

55. WEI, J. and IGLESIA, E. Structural and mechanistic requirements for methane activation and chemical conversion on supported iridium clusters. *Angewandte Chemie*, **2004**, 43 (28), 3685.

56. WEI, J. and IGLESIA, E. Mechanism and site requirements for activation and chemical conversion of methane on supported Pt clusters and turnover rate comparisons among noble metals. *The Journal of Physical Chemistry. B*, **2004**, 108, 4094.

57. WEI, J. and IGLESIA, E. Isotopic and kinetic assessment of the mechanism of methane reforming and decomposition reactions on supported iridium catalysts. *Physical Chemistry Chemical Physics*, **2004**, 6, 3754.

58. CHORKENDORFF, I. and NIEMANTSVERDRIET, J.W. Chapter 8. Heterogeneous catalysis in practice: Hydrogen. In *Concepts of Modern Catalysis and Kinetics*. Weinheim: Wiley-VCH Verlag GmbH & Co. KGaA, **2003**, pp. 301–307.

59. LEE, D.K., BAEK, I.H., and YOON, W.L. A simulation study for the hybrid reaction of methane steam reforming and *in situ* CO_2 removal in a moving bed reactor of a catalyst admixed with a CaO-based CO_2 acceptor for H_2 production. *International Journal of Hydrogen Energy*, **2006**, 31 (5), 649.

60. LEE, D.K., BAEK, I.H., and YOON, W.L. Modeling and simulation for the methane steam reforming enhanced by in situ CO_2 removal utilizing the CaO carbonation for H_2 production. *Chemical Engineering Science*, **2004**, 59 (4), 931.

61. OCHOA-FERNÁNDEZ, E., RUSTEN, H.K., JAKOBSEN, H.A., RØNNING, M., HOLMEN, A., and CHEN, D. Sorption enhanced hydrogen production by steam methane reforming using Li_2ZrO_3 as sorbent: Sorption kinetics and reactor simulation. *Catalysis Today*, **2005**, 106 (1–4), 41.

62. CAO, C., WANG, Y., and ROZMIAREK, R.T. Heterogeneous reactor model for steam reforming of methane in a microchannel reactor with microstructured catalysts. *Catalysis Today*, **2005**, 110 (1–2), 92.

63. YU, W., OHMORI, T., YAMAMOTO, T., ENDO, A., NAKAIWA, M., HAYAKAWA, T., and ITOH, N. Simulation of a porous ceramic membrane reactor for hydrogen production. *International Journal of Hydrogen Energy*, **2005**, 30 (10), 1071.

64. RAKIB, M.A. and ALHUMAIZI, K.I. Modeling of a fluidized bed membrane reactor for the steam reforming of methane: Advantages of oxygen addition for favorable hydrogen production. *Energy & Fuels*, **2005**, 19, 2129.

65. WANG, Y.-N. and RODRIGUES, A.E. Hydrogen production from steam methane reforming coupled with in situ CO2 capture: Conceptual parametric study. *Fuel*, **2005**, 84 (14–15), 1778.

66. POSADA, A. and MANOUSIOUTHAKIS, V. Heat and power integration of methane reforming based hydrogen production. *Industrial & Engineering Chemistry Research*, **2005**, 44 (24), 9113.

67. TAKEUCHI, T., AIHARA, M., and HABUKA, H. CFD-simulation of membrane reactor for methane steam reforming. *AIChE Fall 2004 Meeting*, Abstract 392e.

68. ZANFIR, M. and GAVRIILIDIS, A. Influence of flow arrangement in catalytic plate reactors for methane steam reforming. *Chemical Engineering Research & Design*, **2004**, 82 (2), 252.

69. GALLUCCI, F., PATURZO, L., and BASILE, A. A simulation study of the steam reforming of methane in a dense tubular membrane reactor. *International Journal of Hydrogen Energy*, **2004**, 29 (6), 611.

70. AYTURK, M.E., ENGWALL, E.E., and MA, Y.H. Challenges in the formation of composite Pd-Ag-alloy/porous stainless steel (PSS) membranes for high temperature H_2 separation. *Proceedings of the International Hydrogen Energy Congress and Exhibition*. Istanbul, Turkey, **2005**.

71. PATIL, C.S., van SINT ANNALAND, M., and KUIPERS, J.A.M. Design of a novel autothermal membrane-assisted fluidized-bed reactor for the production of ultrapure hydrogen from methane. *Industrial & Engineering Chemistry Research*, **2005**, 44 (25), 9502.

72. TONG, J. and MATSUMURA, Y. Pure hydrogen production by methane steam reforming with hydrogen-permeable membrane reactor. *Catalysis Today*, **2006**, 111 (3–4), 147.

73. TONG, J., MATSUMURA, Y., SUDA, H., and HARAYA, K. Thin and dense Pd/CeO₂/MPSS composite membrane for hydrogen separation and steam reforming of methane. *Separation and Purification Technology*, **2005**, 46 (1–2), 1.

74. TONG, J., MATSUMURA, Y., SUDA, H., and HARAYA, K., Experimental study of steam reforming of methane in a thin Pd-based membrane reactor. *Industrial & Engineering Chemistry Research*, **2005**, 44 (5), 1454.

75. TONG, J., SU, L., KASHIMA, Y., SHIRAI, R., SUDA, H., and MATSUMURA, Y. Simultaneously depositing Pd-Ag thin membrane on asymmetric porous stainless steel tube and application to produce hydrogen from steam reforming of methane. *Industrial & Engineering Chemistry Research*, **2006**, 45 (2), 648.

76. TSURU, T., YAMAGUCHI, K., YOSHIOKA, T., and ASAEDA, M. Methane steam reforming by microporous catalytic membrane reactors. *AIChE Journal*, **2004**, 50 (11), 2794.

77. NOMURA, M., SESHIMO, M., AIDA, H., NAKATANI, K., GOPALAKRISHNAN, S., SUGAWARA, T., ISHIKAWA, T., KAWAMURA, M., and NAKAO, S. Preparation of a catalyst composite silica membrane reactor for steam reforming reaction by using a counter-diffusion CVD method. *Industrial & Engineering Chemistry Research*, **2006**, 45 (11), 3950.

78. KWANG, B.Y. and HARRISON, D.P. Low-pressure sorption-enhanced hydrogen production. *Industrial & Engineering Chemistry Research*, **2005**, 44 (6), 1665.

79. JOHNSEN, K., RYU, H.J., GRACE, J.R., and LIM, C.J. Sorption-enhanced steam reforming of methane in a fluidized bed reactor with dolomite as CO₂-acceptor. *Chemical Engineering Science*, **2006**, 61 (4), 1195.

80. LIN, S., HARADA, M., SUZUKI, Y., and HATANO, H. CO₂ separation during hydrocarbon gasification. *Energy*, **2005**, 30 (11–12), 2186.

81. WU, S., BEUM, T.H., YANG, J.I., and KIM, J.N. The characteristics of a sorption-enhanced steam-methane reaction for the production of hydrogen using CO₂ Sorbent. *Chinese Journal of Chemical Engineering*, **2005**, 13 (1), 43.

82. KATO, M., NAKAGAWA, K., ESSAKI, K., MAEZAWA, Y., TAKEDA, S., KOGO, R., and HAGIWARA, Y. Novel CO₂ absorbents using lithium-containing oxide. *International Journal of Applied Ceramic Technology*, **2005**, 2 (6), 467.

83. KUSAKABE, K., FUMIO, S., EDA, T., ODA, M., and SOTOWA, K.-I. Hydrogen production in zirconia membrane reactors for use in PEM fuel cells. *International Journal of Hydrogen Energy*, **2005**, 30 (9), 989.

84. XIU, G.H., LI, P., and RODRIGUES, A.E. Subsection-controlling strategy for improving sorption-enhanced reaction process. *Chemical Engineering Research & Design*, **2004**, 82 (2), 192.

85. ROYCHOUDHURY, S., CASTALDI, M., LYUBOVSKY, M., LAPIERRE, R., and AHMED, S. Microlith catalytic reactors for reforming iso-octane-based fuels into hydrogen. *Journal of Power Sources*, **2005**, 152, 75.

86. CHIKAZAWA, Y., KISOHARA, N., USUI, S., KONOMURA, M., SAWA, N., SATO, M., and TANAKA, T. Feasibility study of a compact loop type fast reactor without refueling for a remote place power source. *Proceedings of GLOBAL*. Tsukuba, Japan, October 9–13, **2005**.

87. CHIKAZAWA, Y., OKANO, Y., HORI, T., OHKUBO, Y., SHIMAKAWA, Y., and TANAKA, T. A feasibility study on a small sodium cooled reactor as a diversified power source. *Journal of Nuclear Science and Technology*, **2006**, 43 (8), 829.

88. GLÖCKLER, B., GRITSCH, A., MORILLO, A., KOLIOS, G., and EIGENBERGER, G. Autothermal reactor concepts for endothermic fixed-bed reactions. *Chemical Engineering Research & Design*, **2004**, 82 (2), 148.

89. KOLIOS, G., GRITSCH, A., MORILLO, A., TUTTLIES, U., BERNNAT, J., OPFERKUCH, F., and EIGENBERGER, G. Heat-integrated reactor concepts for catalytic reforming and automotive exhaust purification. *Applied Catalysis. B, Environmental*, **2007**, 70 (1–4), 16.

90. HOGLEN, W. and VALENTINE, J. Coriolis flowmeters improve hydrogen production: Accurate steam-to-carbon ratio control provided efficient operation. *Hydrocarbon Processing*, **2007**, 86 (8), 71.

91. TONKOVICH, A.Y., PERRY, S., WANG, Y., QIU, D., LAPLANTE, T., and ROGERS, W.A. Microchannel process technology for compact methane steam reforming. *Chemical Engineering Science*, **2004**, 59 (22–23), 4819.

92. GALVITA, V. and SUNDMACHER, K. Hydrogen production from methane by steam reforming in a periodically operated two-layer catalytic reactor. *Applied Catalysis. A, General*, **2005**, 289 (2), 121.

93. NOZAKI, T., MUTO, N., KADO, S., and OKAZAKI, K. Dissociation of vibrationally excited methane on Ni catalyst: Part 1. Application to methane steam reforming. *Catalysis Today*, **2004**, 89 (1–2), 57.

94. NOZAKI, T., MUTO, N., KADIO, S., and OKAZAKI, K. Dissociation of vibrationally excited methane on Ni catalyst: Part 2. Process diagnostics by emission spectroscopy. *Catalysis Today*, **2004**, 89 (1–2), 67.

95. MOLLER, S., KAUCIC, D., and SATTLER, C. Hydrogen production by solar reforming of natural gas: A comparison study of two possible process configurations. *Journal of Solar Energy Engineering*, **2006**, 128 (1), 16.

96. FUKUSHIMA, K. and OGAWA, T. Conceptual design of low-temperature hydrogen production and high-efficiency nuclear reactor technology. *JSME International Journal. Series B, Fluids and Thermal Engineering*, **2004**, 47 (2), 340.

97. CHEEKATAMARLA, P.K. and FINNERTY, C.M. Reforming catalysts for hydrogen generation in fuel cell applications. *Journal of Power Sources*, **2006**, 160 (1), 490.

98. RECUPERO, V., PINO, L., VITA, A., CIPITI, F., CORDARO, M., and LAGANÀ, M. Development of a LPG fuel processor for PEFC systems: Laboratory scale evaluation of autothermal reforming and preferential oxidation subunits. *International Journal of Hydrogen Energy*, **2005**, 30 (9), 963.

99. GÖKALILER, F., Selen Çaglayan, B., Ilsen Önsan, Z., and Erhan Aksoylu, A. Hydrogen production by autothermal reforming of LPG for PEM fuel cell applications. *International Journal of Hydrogen Energy*, **2008**, 33 (4), 1383.

100. MOON, D.J. Hydrogen production by catalytic reforming of gaseous hydrocarbons (Methane & LPG). *Catalysis Surveys from Asia*, **2008**, 12 (3), 188.

101. CIPITÌ, F., RECUPERO, V., PINO, L., VITA, A., and LAGANÀ, M. Experimental analysis of a 2 kWe LPG-based fuel processor for polymer electrolyte fuel cells. *Journal of Power Sources*, **2006**, 157 (2), 914.

102. RESINI, C., HERRERA DELGADO, M.C., ARRIGHI, L., ALEMANY, L.J., MARAZZA, R., and BUSCA, G. Propene versus propane steam reforming for hydrogen production over Pd-based and Ni-based catalysts. *Catalysis Communications*, **2005**, 6 (7), 441.

103. SHEKHAWAT, D., BERRY, D.A., GARDNER, T.H., and SPIVEY, J.J. Catalytic reforming of liquid hydrocarbon fuels for fuel cell applications. *Catalysis*, **2006**, 19, 184.

104. KOLB, G., ZAPF, R., HESSEL, V., and LÖWE, H. Propane steam reforming in micro-channels—Results from catalyst screening and optimisation. *Applied Catalysis. A, General*, **2004**, 277 (1–2), 155.

105. BESENBACHER, F., CHORKENDORFF, I., CLAUSEN, B.S., HAMMER, B., MOLENBROEK, A.M., NORSKOV, J.K., and STENSGAARD, I. Design of a surface alloy catalyst for steam reforming. *Science*, **1998**, 279 (5358), 1913.

106. CHIN, Y.-H., KING, D.L., ROH, H.-S., WANG, Y., and HEALD, S.M. Structure and reactivity investigations on supported bimetallic AuNi catalysts used for hydrocarbon steam reforming. *Journal of Catalysis*, **2006**, 244 (2), 153.

107. AVCI, A.K., TRIMM, D.L., AKSOYLU, A.E., and ÖNSAN, Z.I. Hydrogen production by steam reforming of n-butane over supported Ni and Pt-Ni catalysts. *Applied Catalysis. A, General*, **2004**, 258 (2), 235.

108. GRAF, P.O., MOJET, B.L., van OMMEN, J.G., and LEFFERTS, L. Comparative study of steam reforming of methane, ethane and ethylene on Pt, Rh and Pd supported on yttrium-stabilized zirconia. *Applied Catalysis. A, General*, **2007**, 332 (2), 310.

109. LAOSIRIPOJANA, N. and ASSABUMRUNGRAT, S. Hydrogen production from steam and autothermal reforming of LPG over high surface area ceria. *Journal of Power Sources*, **2006**, 158 (2), 1348.

110. WANG, X. and GORTE, R.J. Steam reforming of n-butane on Pd/ceria. *Catalysis Letters*, **2001**, 73 (1), 15.

111. LAOSIRIPOJANA, N., SANGTONGKITCHAROEN, W., and ASSABUMRUNGRAT, S. Catalytic steam reforming of ethane and propane over CeO₂-doped Ni/Al₂O₃ at SOFC temperature: Improvement

of resistance toward carbon formation by the redox property of doping CeO_2. *Fuel*, **2006**, 85 (3), 323.

112. TAKENAKA, S., ORITA, Y., UMEBAYASHI, H., MATSUNE, H., and KISHIDA, M. High resistance to carbon deposition of silica-coated Ni catalysts in propane stream reforming. *Applied Catalysis. A, General*, **2008**, 351 (2), 189.

113. ÇAĞLAYAN, B.S., İLSEN ÖNSAN, Z., and ERHAN AKSOYLU, A. Production of hydrogen over bimetallic Pt–Ni/δ-Al$_2$O$_3$: II. Indirect partial oxidation of LPG. *Catalysis Letters*, **2005**, 102 (1), 63.

114. IGARASHI, A., OHTAKA, T., and MOTOKI, S. Low-temperature steam reforming of n-butane over Rh and Ru catalysts supported on ZrO_2. *Catalysis Letters*, **1992**, 13 (3), 189.

115. MODAFFERI, V., PANZERA, G., BAGLIO, V., FRUSTERI, F., and ANTONUCCI, P.L. Propane reforming on Ni-Ru/GDC catalyst: H_2 production for IT-SOFCs under SR and ATR conditions. *Applied Catalysis. A, General*, **2008**, 334 (1–2), 1.

116. LI, D., NISHIDA, K., ZHAN, Y., SHISHIDO, T., OUMI, Y., SANO, T., and TAKEHIRA, K. Sustainable Ru-doped Ni catalyst derived from hydrotalcite in propane reforming. *Applied Clay Science*, **2009**, 43 (1), 49.

117. CHEN, G., LI, S., and YUAN, Q. Pd-Zn/Cu-Zn-Al catalysts prepared for methanol oxidation reforming in microchannel reactors. *Catalysis Today*, **2007**, 120 (1), 63.

118. JENSEN, M.B., RÅBERG, L.B., OLAFSEN SJÅSTAD, A., and OLSBYE, U. Mechanistic study of the dry reforming of propane to synthesis gas over a Ni/Mg(Al)O catalyst. *Catalysis Today*, **2009**, 47 (2), 340.

119. SOLYMOSI, F., TOLMACSOV, P., and ZAKAR, T.S. Dry reforming of propane over supported Re catalyst. *Journal of Catalysis*, **2005**, 233 (1), 51.

120. ROSTRUP-NIELSEN, J.R., CHRISTENSEN, T.S., and DYBKJAER, I. Steam reforming of liquid hydrocarbons. *Studies in Surface Science and Catalysis*, **1998**, 113, 81.

121. OGDEN, J.M., STEINBUGLER, M.M., and KREUTZ, T.G. A comparison of hydrogen, methanol and gasoline as fuels for fuel cell vehicles: Implications for vehicle design and infrastructure development. *Journal of Power Sources*, **1999**, 79 (2), 143.

122. KRUMPELT, M., KRAUSE, T.R., CARTER, J.D., KOPASZ, J.P., and AHMED, S. Fuel processing for fuel cell systems in transportation and portable power applications. *Catalysis Today*, **2002**, 77 (1–2), 3.

123. Zymax. Forensic geochemistry. http://www.dpra.com/index.cfm/m/158 (accessed October 3, 2009). View topics on: The origin and chemistry of petroleum and identifying hydrocarbons.

124. VELU, S., MA, X., and SONG, C. Selective adsorption for removing sulfur from jet fuel over zeolite-based adsorbents. *Industrial & Engineering Chemistry Research*, **2003**, 42, 5293.

125. MELO, F. and MORLANÉS, N. Synthesis, characterization and catalytic behaviour of NiMgAl mixed oxides as catalysts for hydrogen production by naphtha steam reforming. *Catalysis Today*, **2008**, 133–135, 383.

126. DREYER, B.J., LEE, I.C., KRUMMENACHER, J.J., and SCHMIDT, L.D. Autothermal steam reforming of higher hydrocarbons: n-Decane, n-hexadecane, and JP-8. *Applied Catalysis. A, General*, **2006**, 307 (2), 184.

127. ALVAREZ-GALVAN, M.C., NAVARRO, R.M., ROSA, F., BRICEÑO, Y., RIDAO, M.A., and FIERRO, J.L.G. Hydrogen production for fuel cell by oxidative reforming of diesel surrogate: Influence of ceria and/or lanthana over the activity of Pt/Al$_2$O$_3$ catalysts. *Fuel*, **2008**, 87 (12), 2502.

128. SHI, L. and BAYLESS, D.J. Analysis of jet fuel reforming for solid oxide fuel cell applications in auxiliary power units. *International Journal of Hydrogen Energy*, **2008**, 33 (3), 1067.

129. KOPASZ, J.P., LIU, D.-J., LOTTES, S., AHLUWALIA, R., NOVICK, V., and AHMED, S. Hydrogen, fuel cells, and infrastructure technologies. *Progress Report* (submitted to US Department of Energy), **2003**.

130. MING, Q., HEALEY, T., ALLEN, L., and IRVING, P. Steam reforming of hydrocarbon fuels. *Catalysis Today*, **2002**, 77, 51.

131. SIDDLE, A., POINTON, K.D., JUDD, R.W., and JONES, S.L. Fuel processing for fuel cells-A status review and assessment of prospects. A Report of Advantica Ltd. *ETSU F/03/00252/REP. URN 031644*, **2003**.

132. KOPASZ, J.P., MILLER, L.E., and APPLEGATE, D.V. Society of automotive engineers. *SP-1790*, p. 59, **2003**.

133. SASAKI, K. and TERAOKA, Y. Equilibria in fuel cell gases. I. Equilibrium compositions and reforming conditions. *Journal of the Electrochemical Society*, **2003**, 150 (7), A878.

134. FUKUNAGA, T., KATSUNO, H., MATSUMOTO, H., TAKAHASHI, O., and AKAI, Y. Development of kerosene fuel processing system for PEFC. *Catalysis Today*, **2003**, 84 (3–4), 197.
135. LU, Y., CHEN, J., LIU, Y., XUE, Q., and HE, M. Highly sulfur-tolerant $Pt/Ce_{0.8}Gd_{0.2}O_{1.9}$ catalyst for steam reforming of liquid hydrocarbons in fuel cell applications. *Journal of Catalysis*, **2008**, 254 (1), 39.
136. KIM, D.H., KANG, J.S., LEE, Y.J., PARK, N.K., KIM, Y.C., HONG, S.I., and MOON, D.J. Steam reforming of n-hexadecane over noble metal-modified Ni-based catalysts. *Catalysis Today*, **2008**, 136 (3–4), 228.
137. STROHM, J.J., ZHENG, J., and SONG, C. Low-temperature steam reforming of jet fuel in the absence and presence of sulfur over Rh and Rh-Ni catalysts for fuel cells. *Journal of Catalysis*, **2006**, 238 (2), 309.
138. CHEEKATAMARLA, P.K. and THOMSON, W.J. Catalytic activity of molybdenum carbide for hydrogen generation via diesel reforming. *Journal of Power Sources*, **2006**, 158 (1), 477.
139. CHEEKATAMARLA, P.K. and THOMSON, W.J. Hydrogen generation from 2,2,4-trimethyl pentane reforming over molybdenum carbide at low steam-to-carbon ratios. *Journal of Power Sources*, **2006**, 156 (2), 520.
140. MARIN FLORES, O.G. and HA, S. Study of the performance of Mo_2C for iso-octane steam reforming. *Catalysis Today*, **2008**, 136 (3–4), 235.
141. GOUD, S.K., WHITTENBERGER, W.A., CHATTOPADHYAY, S., and ABRAHAM, M.A. Steam reforming of n-hexadecane using a Pd/ZrO_2 catalyst: Kinetics of catalyst deactivation. *International Journal of Hydrogen Energy*, **2007**, 32 (14), 2868.
142. MURATA, K., WANG, L., SAITO, M., INABA, M., TAKAHARA, I., and MINURA, N. Hydrogen production from steam reforming of hydrocarbon oven alkaline-earth metal modified Fe- or Ni-based catalysts. *Energy & Fuels*, **2004**, 18, 122.
143. FLYTZANI-STEPHANOPOULOS, M. and VOECK, G.E. Autothermal reforming of n-tetradecane and benzene solutions of naphthalene on pellet catalysts, and steam reforming of n-hexane on Peppet and monolithic catalyst beds. *U.S. DOE Report DE-AI03-78ET-11326*, October **1980**.
144. WANG, X. and GORTE, R.J. A study of steam reforming of hydrocarbon fuels on Pd/ceria. *Applied Catalysis. A, General*, **2002**, 224 (1–2), 209.
145. SUZUKI, T., IWANAMI, H.-I., and YOSHINARI, T. Steam reforming of kerosene on Ru/Al_2O_3 catalyst to yield hydrogen. *International Journal of Hydrogen Energy*, **2000**, 25 (2), 119.
146. PATT, J., MOON, D.J., PHILLIPS, C., and THOMPSON, L.T. Molybdenum carbide catalysts for water–gas shift. *Catalysis Letters*, **2000**, 65 (4), 193.
147. BEJ, S.K., BENNETT, C.A., and THOMPSON, L.T. Acid and base characteristics of molybdenum carbide catalysts. *Applied Catalysis. A, General*, **2003**, 250 (2), 197.
148. YORK, A.P.E., CLARIDGE, J.B., BRUNGS, A.J., TSANG, S.C., and GREEN, M.L.H. Molybdenum and tungsten carbides as catalysts for the conversion of methane to synthesis gas using stoichiometric feedstocks. *Chemical Communications*, **1997**, 39.
149. SEHESTED, J., JACOBSEN, C.J.H., ROKNI, S., and ROSTRUP-NIELSEN, J.R. Activity and stability of molybdenum carbide as a catalyst for CO_2 reforming. *Journal of Catalysis*, **2001**, 201 (2), 206.
150. THOMPSON, L.T. High surface area carbides and nitrides. *Abstracts of Papers, 227th ACS National Meeting*. Anaheim, CA, March 28–April 1, **2004**.
151. TØTTRUP, P.B. Evaluation of intrinsic steam reforming kinetic parameters from rate measurements on full particle size. *Applied Catalysis*, **1982**, 4 (4), 377.
152. ROSTRUP-NIELSEN, J.R. Activity of nickel catalysts for steam reforming of hydrocarbons. *Journal of Catalysis*, **1973**, 31 (2), 173.
153. PRAHARSO, X., ADESINA, A.A., TRIMM, D.L., and CANT, N.W. Kinetic study of iso-octane steam reforming over a nickel-based catalyst. *Chemical Engineering Journal*, **2004**, 99 (2), 131.
154. SHI, L., BAYLESS, D.J., and PRUDICH, M. A model of steam reforming of iso-octane: The effect of thermal boundary conditions on hydrogen production and reactor temperature. *International Journal of Hydrogen Energy*, **2008**, 33 (17), 4577.
155. PACHECO, M., SIRA, J., and KOPASZ, J. Reaction kinetics and reactor modeling for fuel processing of liquid hydrocarbons to produce hydrogen: Isooctane reforming. *Applied Catalysis. A, General*, **2003**, 250 (1), 161.

156. CHRISTENSEN, T.S. Adiabatic prereforming of hydrocarbons—An important step in syngas production. *Applied Catalysis. A, General*, **1996**, 138 (2), 285.

157. SUZUKI, T., IWANAMI, H.-I., IWAMOTO, O., and KITAHARA, T. Pre-reforming of liquefied petroleum gas on supported ruthenium catalyst. *International Journal of Hydrogen Energy*, **2001**, 26 (9), 935.

158. ZHENG, J., STROHM, J.J., and SONG, C. Steam reforming of liquid hydrocarbon fuels for micro-fuel cells. Pre-reforming of model jet fuels over supported metal catalysts. *Fuel Processing Technology*, **2008**, 89 (4), 440.

159. GHENCIU, A.F. Review of fuel processing catalysts for hydrogen production in PEM fuel cell systems. *Current Opinion in Solid State & Materials Science*, **2002**, 1, 389.

160. VELU, S., SUZUKI, K., and OSAKI, T. Oxidative steam reforming of methanol over CuZnAl(Zr)-oxide catalysts; a new and efficient method for the production of CO-free hydrogen for fuel cells. *Chemical Communications*, **1999**, 2341.

161. OLAH, G.A., GOEPPERT, A., and PRAKASH, G.K.S. Chapter 10. In *Beyond Oil and Gas: The Methanol Economy*. Weinheim: Wiley-VCH Verlag GmbH & Co. kGaA., **2006**, pp. 168–172.

162. SONG, C. Fuel processing for low-temperature and high-temperature fuel cells: Challenges, and opportunities for sustainable development in the 21st century. *Catalysis Today*, **2002**, 77 (1–2), 17.

163. PALO, D.R., DAGLE, R.A., and HOLLADAY, J.D. Methanol steam reforming for hydrogen production. *Chemical Reviews*, **2007**, 107 (10), 3992.

164. AGRELL, J., LINDSTROM, B., PETTERSSON, L., and JARAS, S. Catalytic hydrogen generation from methanol. *Catalysis*, **2002**, 16, 67.

165. JOENSEN, F. and ROSTRUP-NIELSEN, J.R. Conversion of hydrocarbons and alcohols for fuel cells. *Journal of Power Sources*, **2002**, 105 (2), 195.

166. DELUGA, G.A., SALGE, J.R., SCHMIDT, L.D., and VERYKIOS, X.E. Renewable hydrogen from ethanol by autothermal reforming. *Science*, **2004**, 303, 993.

167. VELU, S. and SONG, C. Advances in catalysis and processes for hydrogen production from ethanol reforming. *Catalysis*, **2007**, 20, 65.

168. HOLLADAY, J.D., WANG, Y., and JONES, E. Review of developments in portable hydrogen production using microreactor technology. *Chemical Reviews*, **2004**, 104, 4767.

169. NAVARRO, R.M., PENA, M.A., and FIERRO, J.L.G. Hydrogen production reactions from carbon feedstocks: Fossil fuels and biomass. *Chemical Reviews*, **2007**, 107 (10), 3952.

170. LIU, S., TAKAHASHI, K., FUCHIGAMI, K., and UEMATSU, K. Hydrogen production by oxidative methanol reforming on Pd/ZnO: Catalyst deactivation. *Applied Catalysis. A, General*, **2006**, 299, 58.

171. LIU, S., TAKAHASHI, K., UEMATSU, K., and AYABE, M. Hydrogen production by oxidative methanol reforming on Pd/ZnO catalyst: Effects of the addition of a third metal component. *Applied Catalysis. A, General*, **2004**, 277 (1–2), 265.

172. MURCIA-MASCARÓS, S., NAVARRO, R.M., GÓMEZ-SAINERO, L., COSTANTINO, U., NOCCHETTI, M., and FIERRO, J.L.G. Oxidative methanol reforming reactions on CuZnAl catalysts derived from hydrotalcite-like precursors. *Journal of Catalysis*, **2001**, 198 (2), 338.

173. REITZ, T.L., LEE, P.L., CZAPLEWSKI, K.F., LANG, J.C., POPP, K.E., and KUNG, H.H. Time-resolved XANES investigation of CuO/ZnO in the oxidative methanol reforming reaction. *Journal of Catalysis*, **2001**, 199 (2), 193.

174. SHAN, W., FENG, Z., LI, Z., ZHANG, J., SHEN, W., and LI, C. Oxidative steam reforming of methanol on $Ce_{0.9}Cu_{0.1}O_Y$ catalysts prepared by deposition-precipitation, coprecipitation, and complexation-combustion methods. *Journal of Catalysis*, **2004**, 228 (1), 206.

175. LIU, S., TAKAHASHI, K., UEMATSU, K., and AYABE, M. Hydrogen production by oxidative methanol reforming on Pd/ZnO. *Applied Catalysis. A, General*, **2005**, 283 (1–2), 125.

176. UMEGAKI, T., MASUDA, A., OMATA, K., and YAMADA, M. Development of a high performance Cu-based ternary oxide catalyst for oxidative steam reforming of methanol using an artificial neural network. *Applied Catalysis. A, General*, **2008**, 351 (2), 210.

177. VELU, S., SUZUKI, K., OKAZAKI, M., KAPOOR, M.P., OSAKI, T., and OHASHI, F. Oxidative steam reforming of methanol over CuZnAl(Zr)-oxide catalysts for the selective production of hydrogen

for fuel cells: Catalyst characterization and performance evaluation. *Journal of Catalysis*, **2000**, 194 (2), 373.

178. VELU, S., SUZUKI, K., and OSAKI, T. Selective production of hydrogen by partial oxidation of methanol over catalysts derived from CuZnAl-layered double hydroxides. *Catalysis Letters*, **1999**, 62 (2), 159.

179. SHISHIDO, T., YAMAMOTO, Y., MORIOKA, H., TAKAKI, K., and TAKEHIRA, K. Active Cu/ZnO and Cu/ZnO/Al$_2$O$_3$ catalysts prepared by homogeneous precipitation method in steam reforming of methanol. *Applied Catalysis. A, General*, **2004**, 263 (2), 249.

180. LIU, Y., HAYAKAWA, T., SUZUKI, K., and HAMAKAWA, S. Production of hydrogen by steam reforming of methanol over Cu/CeO$_2$ catalysts derived from Ce$_{1-x}$Cu$_x$O$_{2-x}$ precursors. *Catalysis Communications*, **2001**, 2 (6–7), 195.

181. CLANCY, P., BREEN, J.P., and ROSS, J.R.H. The preparation and properties of coprecipitated Cu-Zr-Y and Cu-Zr-La catalysts used for the steam reforming of methanol. *Catalysis Today*, **2007**, 127 (1–4), 291.

182. OGUCHI, H., NISHIGUCHI, T., MATSUMOTO, T., KANAI, H., UTANI, K., MATSUMURA, Y., and IMAMURA, S. Steam reforming of methanol over Cu/CeO$_2$/ZrO$_2$ catalysts. *Applied Catalysis. A, General*, **2005**, 281 (1–2), 69.

183. AGRELL, J., BIRGERSSON, H., BOUTONNET, M., MELIÁN-CABRERA, I., NAVARRO, R. M., and FIERRO, J.L.G. Production of hydrogen from methanol over Cu/ZnO catalysts promoted by ZrO$_2$ and Al$_2$O$_3$. *Journal of Catalysis*, **2003**, 219 (2), 389.

184. VELU, S., SUZUKI, K., KAPOOR, M.P., OHASHI, F., and OSAKI, T. Selective production of hydrogen for fuel cells via oxidative steam reforming of methanol over CuZnAl(Zr)-oxide catalysts. *Applied Catalysis. A, General*, **2001**, 213 (1), 47.

185. VELU, S. and SUZUKI, K. Selective production of hydrogen for fuel cells via oxidative steam reforming of methanol over CuZnAl oxide catalysts: Effect of substitution of zirconium and cerium on the catalytic performance. *Topics in Catalysis*, **2003**, 22, 235.

186. MATTER, P.H., BRADEN, D.J., and OZKAN, U.S. Steam reforming of methanol to H$_2$ over nonreduced Zr-containing CuO/ZnO catalysts. *Journal of Catalysis*, **2004**, 223 (2), 340.

187. BREEN, J.P. and ROSS, J.R.H. Methanol reforming for fuel-cell applications: Development of zirconia-containing Cu-Zn-Al catalysts. *Catalysis Today*, **1999**, 51 (3–4), 521.

188. FRANK, B., JENTOFT, F.C., SOERIJANTO, H., KRÖHNERT, J., SCHLÖGL, R., and SCHOMÄCKER, R. Steam reforming of methanol over copper-containing catalysts: Influence of support material on microkinetics. *Journal of Catalysis*, **2007**, 246 (1), 177.

189. VELU, S., SUZUKI, K., GOPINATH, C.S., YOSHIDA, H., and HATTORI, T. XPS, XANES and EXAFS investigations of CuO/ZnO/Al$_2$O$_3$/ZrO$_2$ mixed oxide catalysts. *Physical Chemistry Chemical Physics*, **2002**, 4, 1990.

190. KARIM, A., CONANT, T., and DATYE, A. The role of PdZn alloy formation and particle size on the selectivity for steam reforming of methanol. *Journal of Catalysis*, **2006**, 243 (2), 420.

191. LENARDA, M., MORETTI, E., STORARO, L., PATRONO, P., PINZARI, F., RODRÍGUEZ-CASTELLÓN, E., JIMÉNEZ-LÓPEZ, A., BUSCA, G., FINOCCHIO, E., MONTANARI, T., and Frattini, R. Finely dispersed Pd-Zn catalyst supported on an organized mesoporous alumina for hydrogen production by methanol steam reforming. *Applied Catalysis. A, General*, **2006**, 312, 220.

192. DAGLE, R.A., CHIN, Y.-H., and WANG, Y. The effects of PdZn crystallite size on methanol steam reforming. *Topics in Catalysis*, **2007**, 46 (3), 358.

193. CONANT, T., KARIM, A.M., LEBARBIER, V., WANG, Y., GIRGSDIES, F., SCHLÖGL, R., and DATYE, A. Stability of bimetallic Pd-Zn catalysts for the steam reforming of methanol. *Journal of Catalysis*, **2008**, 257 (1), 64.

194. IWASA, N., MASUDA, S., OGAWA, N., and TAKEZAWA, N. Steam reforming of methanol over Pd/ZnO: Effect of the formation of PdZn alloys upon the reaction. *Applied Catalysis. A, General*, **1995**, 125 (1), 145.

195. IWASA, N., MAYANAGI, T., NOMURA, W., ARAI, M., and TAKEZAWA, N. Effect of Zn addition to supported Pd catalysts in the steam reforming of methanol. *Applied Catalysis. A, General*, **2003**, 248 (1–2), 153.

196. IWASA, N. and TAKEZAWA, N. New supported Pd and Pt alloy catalysts for steam reforming and dehydrogenation of methanol. *Topics in Catalysis*, **2003**, 22 (3), 215.
197. TAKEZAWA, N. and IWASA, N. Steam reforming and dehydrogenation of methanol: Difference in the catalytic functions of copper and group VIII metals. *Catalysis Today*, **1997**, 36 (1), 45.
198. CHIN, Y.-H., DAGLE, R., HU, J., DOHNALKOVA, A.C., and WANG, Y. Steam reforming of methanol over highly active Pd/ZnO catalyst. *Catalysis Today*, **2002**, 77 (1–2), 79.
199. CHIN, Y.-H., WANG, Y., DAGLE, R.A., and LI, X.S. Methanol steam reforming over Pd/ZnO: Catalyst preparation and pretreatment studies. *Fuel Processing Technology*, **2003**, 83 (1–3), 193.
200. XIA, G., HOLLADAY, J.D., DAGLE, R.A., JONES, E.O., and WANG, Y. Development of highly active Pd-ZnOAl₂O₃ catalysts for microscale fuel processor applications. *Chemical Engineering and Technology*, **2005**, 28, 515.
201. RANGANATHAN, E.S., BEJ, S.K., and THOMPSON, L.T. Methanol steam reforming over Pd/ZnO and Pd/CeO₂ catalysts. *Applied Catalysis. A, General*, **2005**, 289 (2), 153.
202. SUWA, Y., ITO, S.-I., KAMEOKA, S., TOMISHIGE, K., and KUNIMORI, K. Comparative study between Zn-Pd/C and Pd/ZnO catalysts for steam reforming of methanol. *Applied Catalysis. A, General*, **2004**, 267 (1–2), 9.
203. AGRELL, J., GERMANI, G., JÄRÅS, S.G., and BOUTONNET, M. Production of hydrogen by partial oxidation of methanol over ZnO-supported palladium catalysts prepared by microemulsion technique. *Applied Catalysis. A, General*, **2003**, 242 (2), 233.
204. PFEIFER, P., SCHUBERT, K., LIAUW, M.A., and EMIG, G. PdZn catalysts prepared by washcoating microstructured reactors. *Applied Catalysis. A, General*, **2004**, 270 (1–2), 165.
205. PENNER, S., JENEWEIN, B., GABASCH, H., KLÖTZER, B., WANG, D., KNOP-GERICKE, A., SCHLÖGL, R., and HAYEK, K. Growth and structural stability of well-ordered PdZn alloy nanoparticles. *Journal of Catalysis*, **2006**, 241 (1), 14.
206. PFEIFER, P., KÖLBL, A., and SCHUBERT, K. Kinetic investigations on methanol steam reforming on PdZn catalysts in microchannel reactors and model transfer into the pressure gap region. *Catalysis Today*, **2005**, 110 (1–2), 76.
207. CUBEIRO, M.L. and FIERRO, J.L.G. Partial oxidation of methanol over supported palladium catalysts. *Applied Catalysis. A, General*, **1998**, 168 (2), 307.
208. LYUBOVSKY, M. and ROYCHOUDHURY, S. Novel catalytic reactor for oxidative reforming of methanol. *Applied Catalysis. B, Environmental*, **2004**, 54 (4), 203.
209. FOX, E.B., VELU, S., ENGELHARD, M.H., CHIN, Y.-H., MILLER, J.T., KROPF, J., and SONG, C. Characterization of CeO₂-supported Cu-Pd bimetallic catalyst for the oxygen-assisted water-gas shift reaction. *Journal of Catalysis*, **2008**, 260 (2), 358.
210. PURNAMA, H., RESSLER, T., JENTOFT, R.E., SOERIJANTO, H., SCHLÖGL, R., and SCHOMÄCKER, R. CO formation/selectivity for steam reforming of methanol with a commercial CuO/ZnO/Al2O3 catalyst. *Applied Catalysis. A, General*, **2004**, 259 (1), 83.
211. LEE, J.K., KO, J.B., and KIM, D.H. Methanol steam reforming over Cu/ZnO/Al2O3 catalyst: Kinetics and effectiveness factor. *Applied Catalysis. A, General*, **2004**, 278 (1), 25.
212. SAMMS, S.R. and SAVINELL, R.F. Kinetics of methanol-steam reformation in an internal reforming fuel cell. *Journal of Power Sources*, **2002**, 112 (1), 13.
213. GEISSLER, K., NEWSON, E., VOGEL, F., TRUONG, T.-B., HOTTINGER, P., and WOKAUN, A. Autothermal methanol reforming for hydrogen production in fuel cell applications. *Physical Chemistry Chemical Physics*, **2001**, 3, 289.
214. PATEL, S. and PANT, K.K. Kinetic modeling of oxidative steam reforming of methanol over Cu/ZnO/CeO₂/Al₂O₃ catalyst. *Applied Catalysis. A, General*, **2009**, 356 (2), 189.
215. MAS, V., KIPREOS, R., AMADEO, N., and LABORDE, M. Thermodynamic analysis of ethanol/water system with the stoichiometric method. *International Journal of Hydrogen Energy*, **2006**, 31 (1), 21.
216. GARCÍA, E.Y. and LABORDE, M.A. Hydrogen production by the steam reforming of ethanol: Thermodynamic analysis. *International Journal of Hydrogen Energy*, **1991**, 16 (5), 307.
217. FISHTIK, I., ALEXANDER, A., DATTA, R., and GEANA, D. A thermodynamic analysis of hydrogen production by steam reforming of ethanol via response reactions. *International Journal of Hydrogen Energy*, **2000**, 25 (1), 31.

218. VASUDEVA, K., MITRA, N., UMASANKAR, P., and DHINGRA, S.C. Steam reforming of ethanol for hydrogen production: Thermodynamic analysis. *International Journal of Hydrogen Energy*, **1996**, 21 (1), 13.

219. COMAS, J., LABORDE, M., and AMADEO, N. Thermodynamic analysis of hydrogen production from ethanol using CaO as a CO_2 sorbent. *Journal of Power Sources*, **2004**, 138 (1–2), 61.

220. ABOUDHEIR, A., AKANDE, A., IDEM, R., and DALAI, A. Experimental studies and comprehensive reactor modeling of hydrogen production by the catalytic reforming of crude ethanol in a packed bed tubular reactor over a Ni/Al_2O_3 catalyst. *International Journal of Hydrogen Energy*, **2006**, 31 (6), 752.

221. COMAS, J., DIEUZEIDE, M.L., BARONETTI, G., LABORDE, M., and AMADEO, N. Methane steam reforming and ethanol steam reforming using a Ni(II)-Al(III) catalyst prepared from lamellar double hydroxides. *Chemical Engineering Journal*, **2006**, 118 (1–2), 11.

222. YANG, Y., MA, J., and WU, F. Production of hydrogen by steam reforming of ethanol over a Ni/ZnO catalyst. *International Journal of Hydrogen Energy*, **2006**, 31 (7), 877.

223. BARROSO, M.N., GOMEZ, M.F., ARRÚA, L.A., and ABELLO, M.C. Hydrogen production by ethanol reforming over NiZnAl catalysts. *Applied Catalysis. A, General*, **2006**, 304, 116.

224. AKANDE, A.J., IDEM, R.O., and DALAI, A.K. Synthesis, characterization and performance evaluation of Ni/Al_2O_3 catalysts for reforming of crude ethanol for hydrogen production. *Applied Catalysis. A, General*, **2005**, 287 (2), 159.

225. SUN, J., QIU, X.-P., WU, F., and ZHU, W.-T. H_2 from steam reforming of ethanol at low temperature over Ni/Y_2O_3, Ni/La_2O_3 and Ni/Al_2O_3 catalysts for fuel-cell application. *International Journal of Hydrogen Energy*, **2005**, 30 (4), 437.

226. MORGENSTERN, D.A. and FORNANGO, J.P. Low-temperature reforming of ethanol over copper-plated Raney nickel: A new route to sustainable hydrogen for transportation. *Energy & Fuels*, **2005**, 19 (4), 1708.

227. BERGAMASCHI, V.S., CARVALHO, F.M.S., RODRIGUES, C., and FERNANDES, D.B. Preparation and evaluation of zirconia microspheres as inorganic exchanger in adsorption of copper and nickel ions and as catalyst in hydrogen production from bioethanol. *Chemical Engineering Journal*, **2005**, 112 (1–3), 153.

228. BENITO, M., SANZ, J.L., ISABEL, R., PADILLA, R., ARJONA, R., and DAZA, L. Bio-ethanol steam reforming: Insights on the mechanism for hydrogen production. *Journal of Power Sources*, **2005**, 151, 11.

229. VARGAS, J.C., LIBS, S., ROGER, A.-C., and KIENNEMANN, A. Study of Ce-Zr-Co fluorite-type oxide as catalysts for hydrogen production by steam reforming of bioethanol. *Catalysis Today*, **2005**, 107–108, 417.

230. NISHIGUCHI, T., MATSUMOTO, T., KANAI, H., UTANI, K., MATSUMURA, Y., SHEN, W.-J., and IMAMURA, S. Catalytic steam reforming of ethanol to produce hydrogen and acetone. *Applied Catalysis. A, General*, **2005**, 279 (1–2), 273.

231. DUAN, S. and SENKAN, S. Catalytic conversion of ethanol to hydrogen using combinatorial methods. *Industrial & Engineering Chemistry Research*, **2005**, 44 (16), 6381.

232. SUN, J., QIU, X., WU, F., ZHU, W., WANG, W., and HAO, S. Hydrogen from steam reforming of ethanol in low and middle temperature range for fuel cell application. *International Journal of Hydrogen Energy*, **2004**, 29 (10), 1075.

233. BATISTA, M.S., SANTOS, R.K.S., ASSAF, E.M., ASSAF, J.M., and TICIANELLI, E.A. High efficiency steam reforming of ethanol by cobalt-based catalysts. *Journal of Power Sources*, **2004**, 134 (1), 27.

234. BATISTA, M.S., SANTOS, R.K.S., ASSAF, E.M., ASSAF, J.M., and TICIANELLI, E.A. Characterization of the activity and stability of supported cobalt catalysts for the steam reforming of ethanol. *Journal of Power Sources*, **2003**, 124 (1), 99.

235. FRUSTERI, F., FRENI, S., CHIODO, V., SPADARO, L., BONURA, G., and CAVALLARO, S. Potassium improved stability of Ni/MgO in the steam reforming of ethanol for the production of hydrogen for MCFC. *Journal of Power Sources*, **2004**, 132 (1–2), 139.

236. FRENI, S., CAVALLARO, S., MONDELLO, N., SPADARO, L., and FRUSTERI, F. Steam reforming of ethanol on Ni/MgO catalysts: H_2 production for MCFC. *Journal of Power Sources*, **2002**, 108 (1–2), 53.

237. COMAS, J., MARIÑO, F., LABORDE, M., and AMADEO, N. Bio-ethanol steam reforming on Ni/Al₂O₃ catalyst. *Chemical Engineering Journal*, **2004**, 98 (1–2), 61.
238. LLORCA, J., DALMON, J.-A., RAMÍREZ DE LA PISCINA, P., and HOMS, N. In situ magnetic characterisation of supported cobalt catalysts under steam-reforming of ethanol. *Applied Catalysis. A, General*, **2003**, 243 (2), 261.
239. LLORCA, J., HOMS, N., SALES, J., and de la PISCINA, P.R. Efficient production of hydrogen over supported cobalt catalysts from ethanol steam reforming. *Journal of Catalysis*, **2002**, 209 (2), 306.
240. FRENI, S., CAVALLARO, S., MONDELLO, N., SPADARO, L., and FRUSTERI, F. Production of hydrogen for MC fuel cell by steam reforming of ethanol over MgO supported Ni and Co catalysts. *Catalysis Communications*, **2003**, 4 (6), 259.
241. SRINIVAS, D., SATYANARAYANA, C.V.V., POTDAR, H.S., and RATNASAMY, P. Structural studies on NiO-CeO₂-ZrO₂ catalysts for steam reforming of ethanol. *Applied Catalysis. A, General*, **2003**, 246 (2), 323.
242. FATSIKOSTAS, A.N., KONDARIDES, D.I., and VERYKIOS, X.E. Production of hydrogen for fuel cells by reformation of biomass-derived ethanol. *Catalysis Today*, **2002**, 75 (1–4), 145.
243. FATSIKOSTAS, A.N., KONDARIDES, D.I., and VERYKIOS, X.E. Steam reforming of biomass-derived ethanol for the production of hydrogen for fuel cell applications. *Chemical Communications*, **2001**, 851.
244. MARIÑO, F., BOVERI, M., BARONETTI, G., and LABORDE, M. Hydrogen production from steam reforming of bioethanol using Cu/Ni/K/[gamma]-Al₂O₃ catalysts. Effect of Ni. *International Journal of Hydrogen Energy*, **2001**, 26 (7), 665.
245. MARIÑO, F.J., CERRELLA, E.G., DUHALDE, S., JOBBAGY, M., and LABORDE, M.A. Hydrogen from steam reforming of ethanol. Characterization and performance of copper-nickel supported catalysts. *International Journal of Hydrogen Energy*, **1998**, 23 (12), 1095.
246. LLORCA, J., de la PISCINA, P.R., SALES, J., and HOMS, N. Direct production of hydrogen from ethanolic aqueous solutions over oxide catalysts. *Chemical Communications*, **2001**, 641.
247. FRENI, S., MONDELLO, N., CAVALLARO, S., CACCIOLA, G., PARMON, V.N., and SOBYANIN, V.A. Hydrogen production by steam reforming of ethanol: A two step process. *Reaction Kinetics and Catalysis Letters*, **2000**, 71 (1), 143.
248. HAGA, F., NAKAJIMA, T., MIYA, H., and MISHIMA, S. Catalytic properties of supported cobalt catalysts for steam reforming of ethanol. *Catalysis Letters*, **1997**, 48 (3), 223.
249. LUENGO, C.A., CIAMPI, G., CENCIG, M.O., STECKELBERG, C., and LABORDE, M.A. A novel catalyst system for ethanol gasification. *International Journal of Hydrogen Energy*, **1992**, 17 (9), 677.
250. CAVALLARO, S. and FRENI, S. Ethanol steam reforming in a molten carbonate fuel cell. A preliminary kinetic investigation. *International Journal of Hydrogen Energy*, **1996**, 21 (6), 465.
251. ROH, H.-S., WANG, Y., KING, D., PLATON, A., and CHIN, Y.-H. Low temperature and H₂ selective catalysts for ethanol steam reforming. *Catalysis Letters*, **2006**, 108 (1), 15.
252. ZHANG, B., TANG, X., LI, Y., CAI, W., XU, Y., and SHEN, W. Steam reforming of bio-ethanol for the production of hydrogen over ceria-supported Co, Ir and Ni catalysts. *Catalysis Communications*, **2006**, 7 (6), 367.
253. AUPRETRE, F., DESCORME, C., DUPREZ, D., CASANAVE, D., and UZIO, D. Ethanol steam reforming over MgₓNi₁₋ₓAl₂O₃ spinel oxide-supported Rh catalysts. *Journal of Catalysis*, **2005**, 233 (2), 464.
254. WANAT, E.C., VENKATARAMAN, K., and SCHMIDT, L.D. Steam reforming and water-gas shift of ethanol on Rh and Rh-Ce catalysts in a catalytic wall reactor. *Applied Catalysis. A, General*, **2004**, 276 (1–2), 155.
255. DIAGNE, C., IDRISS, H., PEARSON, K., GÓMEZ-GARCÍA, M.A., and KIENNEMANN, A. Efficient hydrogen production by ethanol reforming over Rh catalysts. Effect of addition of Zr on CeO₂ for the oxidation of CO to CO₂. *Comptes Rendus Chimie*, **2004**, 7, 617.
256. AUPRETRE, F., DESCORME, C., and DUPREZ, D. Hydrogen production for fuel cells the catalytic ethanol steam reforming. *Topics in Catalysis*, **2004**, 30–31 (1), 487.
257. FRUSTERI, F., FRENI, S., SPADARO, L., CHIODO, V., BONURA, G., DONATO, S., and CAVALLARO, S. H₂ production for MC fuel cell by steam reforming of ethanol over MgO supported Pd, Rh, Ni and Co catalysts. *Catalysis Communications*, **2004**, 5 (10), 611.

258. CAVALLARO, S., CHIODO, V., FRENI, S., MONDELLO, N., and FRUSTERI, F. Performance of Rh/Al$_2$O$_3$ catalyst in the steam reforming of ethanol: H$_2$ production for MCFC. *Applied Catalysis. A, General*, **2003**, 249 (1), 119.

259. DIAGNE, C., IDRISS, H., and KIENNEMANN, A. Hydrogen production by ethanol reforming over Rh/CeO$_2$-ZrO$_2$ catalysts. *Catalysis Communications*, **2002**, 3 (12), 565.

260. GALVITA, V.V., SEMIN, G.L., BELYAEV, V.D., SEMIKOLENOV, V.A., TSIAKARAS, P., and SOBYANIN, V.A. Synthesis gas production by steam reforming of ethanol. *Applied Catalysis. A, General*, **2001**, 220 (1–2), 123.

261. CAVALLARO, S. Ethanol steam reforming on Rh/Al$_2$O$_3$ catalysts. *Energy & Fuels*, **2000**, 14 (6), 1195.

262. LLORCA, J., HOMS, N., and RAMÍREZ DE LA PISCINA, P. In situ DRIFT-mass spectrometry study of the ethanol steam-reforming reaction over carbonyl-derived Co/ZnO catalysts. *Journal of Catalysis*, **2004**, 227 (2), 556.

263. LIGURAS, D.K., KONDARIDES, D.I., and VERYKIOS, X.E. Production of hydrogen for fuel cells by steam reforming of ethanol over supported noble metal catalysts. *Applied Catalysis. B, Environmental*, **2003**, 43 (4), 345.

264. BREEN, J.P., BURCH, R., and COLEMAN, H.M. Metal-catalysed steam reforming of ethanol in the production of hydrogen for fuel cell applications. *Applied Catalysis. B, Environmental*, **2002**, 39 (1), 65.

265. IDRISS, H. Ethanol reactions over the surfaces of noble metal/cerium oxide catalysts. *Platinum Metals Review*, **2004**, 48 (3), 105.

266. SHENG, P.Y., YEE, A., BOWMAKER, G.A., and IDRISS, H. H$_2$ production from ethanol over Rh-Pt/CeO$_2$ catalysts: The role of Rh for the efficient dissociation of the carbon-carbon bond. *Journal of Catalysis*, **2002**, 208 (2), 393.

267. CASANOVAS, A., LLORCA, J., HOMS, N., FIERRO, J.L.G., and RAMÍREZ DE LA PISCINA, P. Ethanol reforming processes over ZnO-supported palladium catalysts: Effect of alloy formation. *Journal of Molecular Catalysis A: Chemical*, **2006**, 250 (1–2), 44.

268. KUGAI, J., VELU, S., and SONG, C. Low-temperature reforming of ethanol over CeO$_2$-supported Ni-Rh bimetallic catalysts for hydrogen production. *Catalysis Letters*, **2005**, 101 (3), 255.

269. KUGAI, J., SUBRAMANI, V., SONG, C., ENGELHARD, M.H., and CHIN, Y.-H. Effects of nanocrystalline CeO$_2$ supports on the properties and performance of Ni-Rh bimetallic catalyst for oxidative steam reforming of ethanol. *Journal of Catalysis*, **2006**, 238 (2), 430.

270. FIERRO, V., AKDIM, O., PROVENDIER, H., and MIRODATOS, C. Ethanol oxidative steam reforming over Ni-based catalysts. *Journal of Power Sources*, **2005**, 145 (2), 659.

271. LIGURAS, D.K., GOUNDANI, K., and VERYKIOS, X.E. Production of hydrogen for fuel cells by catalytic partial oxidation of ethanol over structured Ru catalysts. *International Journal of Hydrogen Energy*, **2004**, 29 (4), 419.

272. FIERRO, V., AKDIM, O., and MIRODATOS, C. On-board hydrogen production in a hybrid electric vehicle by bio-ethanol oxidative steam reforming over Ni and noble metal based catalysts. *Green Chemistry*, **2003**, 5, 20.

273. CAVALLARO, S., CHIODO, V., VITA, A., and FRENI, S. Hydrogen production by auto-thermal reforming of ethanol on Rh/Al$_2$O$_3$ catalyst. *Journal of Power Sources*, **2003**, 123 (1), 10.

274. FIERRO, V., KLOUZ, V., AKDIM, O., and MIRODATOS, C. Oxidative reforming of biomass derived ethanol for hydrogen production in fuel cell applications. *Catalysis Today*, **2002**, 75 (1–4), 141.

275. KLOUZ, V., FIERRO, V., DENTON, P., KATZ, H., LISSE, J.P., BOUVOT-MAUDUIT, S., and MIRODATOS, C. Ethanol reforming for hydrogen production in a hybrid electric vehicle: Process optimisation. *Journal of Power Sources*, **2002**, 105 (1), 26.

276. VELU, S., SUZUKI, K., VIJAYARAJ, M., BARMAN, S., and GOPINATH, C.S. In situ XPS investigations of Cu 1-x Ni x ZnAl-mixed metal oxide catalysts used in the oxidative steam reforming of bio-ethanol. *Applied Catalysis. B, Environmental*, **2005**, 55, 287.

277. SALGE, J.R., DELUGA, G.A., and SCHMIDT, L.D. Catalytic partial oxidation of ethanol over noble metal catalyst. *Journal of Catalysis*, **2005**, 235, 69.

278. VELU, S., SATOH, N., GOPINATH, C.S., and SUZUKI, K. Oxidative reforming of bio-ethanol over CuNiZnAl mixed oxide catalysts for hydrogen production. *Catalysis Letters*, **2002**, 82, 145.
279. THERDTHIANWONG, A., SAKULKOAKIET, T., and THERDTHIANWONG, S. Hydrogen production by catalytic ethanol steam reforming. *Science Asia*, **2001** 27, 193.
280. AKANDE, A.J. Production of hydrogen by reforming of crude ethanol. Master of Science Thesis, University of Saskatchewan, Canada, February **2005**.
281. SHAMSI, A., BALTRUS, J.P., and SPIVEY, J.J. Characterization of coke deposited on Pt/alumina catalyst during reforming of liquid hydrocarbons. *Applied Catalysis. A: General*, **2005**, 293, 145.

Chapter 3

Catalytic Partial Oxidation and Autothermal Reforming

KE LIU,[1] GREGG D. DELUGA,[1] ANDERS BITSCH-LARSEN,[2] LANNY D. SCHMIDT,[2] AND LINGZHI ZHANG[1]

[1]Energy & Propulsion Technologies, GE Global Research Center
[2]Department of Chemical Engineering & Materials Science, University of Minnesota

3.1 INTRODUCTION

The production of hydrogen for use as a clean fuel source is not a new technology. Felice Fontana discovered in 1780 that "blue water gas" could be produced by passing steam over red-hot coal; the name "blue water gas" was coined due to the flame color obtained when burning the gas, which was a mixture of carbon monoxide and hydrogen.[1] In 1812, the streets of London were illuminated by "town gas" obtained from coal. However, the process at that time suffered from incomplete conversion of coal. In the middle of the 19th century, a cyclic process was introduced, with higher conversion achieved.[2] In 1850, Sir William Siemens developed the water-gas process in which hot coke was exposed to alternating blasts of air and steam for the production of CO and H_2:

$$C + H_2O \rightarrow CO + H_2, \Delta H_r^\circ = +130 \text{ kJ/mol}. \quad (3.1)$$

The puff of air allowed some of the coke to combust:

$$C + \frac{1}{2}O_2 \rightarrow CO_2, \Delta H_r^\circ = -343 \text{ kJ/mol}. \quad (3.2)$$

The alternating blasts of steam and air maintained the bed temperature at ~1000 °C. Thus when steam was added, thermodynamic equilibrium favored the production of CO and H_2, not CO_2. Although natural gas has been widely used in most of Europe and America with the discovery of reliable supplies and construction

Hydrogen and Syngas Production and Purification Technologies, Edited by Ke Liu, Chunshan Song and Velu Subramani

127

of infrastructure during the post-World War II, town gas derived from gasification remains as a heating or cooking source and is still popular in parts of China.

In recent years, H_2 and CO have become important starting materials for the chemical industry. The mixture has come to be known as synthesis gas, or syngas. Currently, production of syngas is dominated by steam reforming:[3]

$$CH_4 + H_2O \rightarrow CO + 3H_2, \Delta H_r^\circ = +206 \text{ kJ/mol}. \tag{3.3}$$

This highly endothermic process is typically catalyzed by a nickel-based supported catalyst placed in a large furnace. A clear and complete discussion of the steam reforming process can be found in Twigg[4] and in Chapter 2 of this book. The largest use of hydrogen in the world today is for the synthesis of ammonia for use as fertilizer; however, many believe the future of energy is a return to the "days of old," production of hydrogen for use as a fuel source for fuel cells, gas turbines, and so on.

Due to the increasing demand of transportation fuel and growing environmental concerns, hydrogen used for fuel upgrading via hydrogenation is becoming more and more crucial as crude oil gets heavier and dirtier. In fact, if one can produce H_2 cost-effectively, the U.S. oil imports can be reduced by one-third today.[5] The reason is simple: since clean transportation fuels (e.g., gasoline or diesel) have an approximate stoichiometry of CH_2, but crude oil is $CH_{-1.4}$; additional H_2 is needed to upgrade the bottom barrel to clean transportation fuel. Furthermore, stricter environmental regulations require fuels to undergo hydrogen desulfurization (HDS), hydrogen denitrogenation (HDN), and hydrogen demetallization (HDM), processes that would all require additional hydrogen input. Natural gas (mainly CH_4) has the highest H/C ratio among all available molecules. Therefore, it is straightforward that natural gas is used to produce hydrogen to enable refineries to squeeze more fuels from crude oil.

An alternative to the use of hydrogen in fuel upgrading is as a replacement to natural gas for electric power generation. This scenario could become reality if natural gas prices continue to be near historic highs, as shown in Figure 3.1, forcing U.S. power plants to shift to Integrated Gasification Combined Cycle (IGCC) or nuclear power. Under this scenario, more natural gas will be available for H_2 production, which leads to economical and clean liquid fuels production from U.S. domestic oil sources. Furthermore, gas-to-liquids (GTL) plants are likely to be built in geographic regions where natural gas is abundant, such as Qatar and other areas of the Middle East. Stricter environmental regulations may force existing Natural Gas Combined Cycle (NGCC) plants, such as those located in Europe, to reform all or part of the natural gas and capture the CO_2 while utilizing the H_2-rich syngas in a highly efficient gas turbine to avoid carbon taxes in certain countries. All these need the state of the art natural gas reforming technology to produce H_2 from natural gas for different applications.

Finally, fuel cells—the highly efficient electrochemical devices that convert chemical energy to electrical energy directly—may be used for distributed power generation or in some vehicles in the future. The most efficient and durable fuel cells use hydrogen as their fuel source. All of the technologies that will be discussed in

(a)

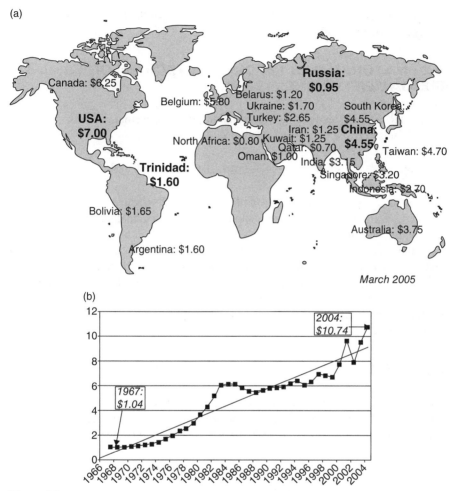

March 2005

Figure 3.1. (a) Geographic variability of natural gas prices in $U.S./million British Thermal Units (MMBTU) and (b) residential natural gas prices in the U.S. in 1967–2004 ($/thousand cubic feet). *Source:* EIA.

this chapter can be used, with cost reduction, as the hydrogen generator for a fuel cell-powered car, backup generator, or regional power plant. It is clear that hydrogen production is already an important chemical process. However, with the world facing limited fossil reserves and an ever-increasing demand for energy and environmental protection, the production of H_2 is likely to increase in the future. These facts lead to a demand for cleaner, simpler, and more efficient technologies for hydrogen production.

In summary, H_2 and syngas production will become increasingly important in future energy industry. This chapter reviews an important H_2 and syngas production technology: the catalytic partial oxidation (CPO) including autothermal reforming (ATR). The CPO and ATR technologies have current applications and important technical features, which makes them an integral part of the H_2 and syngas production technology portfolio.

3.2 NATURAL GAS REFORMING TECHNOLOGIES: FUNDAMENTAL CHEMISTRY

CPO, ATR, and noncatalytic partial oxidation (POX) of methane all share the same POX chemistry:

$$CH_4 + \frac{1}{2}O_2 \rightarrow CO + 2H_2, \Delta H_r^\circ = -36 \text{ kJ/mol}. \tag{3.4}$$

Figure 3.2 shows that the thermodynamic syngas production is maximized at ~1100 °C at atmospheric pressure, yielding near complete methane conversion and minimizing CO_2 production. The technological differentiation of reforming comes from the method by which the heat is generated and provided. In conventional endothermic steam reforming technologies, heat is supplied by burning fuel outside reactor tubes and the steam reforming catalyst is packed inside. For catalytic POX or ATR, a portion of the fuel is oxidized within the reactor to generate the heat required to drive the endothermic steam reforming reaction occurring over the same catalyst bed.

3.2.1 ATR

ATR is the reaction of natural gas or liquid hydrocarbons with steam and oxygen at high temperature and high pressure to produce syngas. The reaction is exothermic and catalysts are used to improve hydrogen yield. The ATR feed generally has a

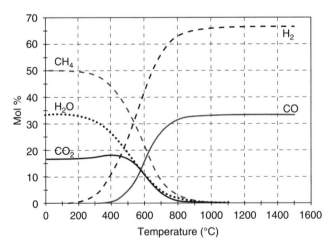

Figure 3.2. Equilibrium composition of Equations 3.3 and 3.4 as a function of temperature at 1 bar absolute pressure. Syngas selectivity is maximized at ~1200 °C reaction temperature.

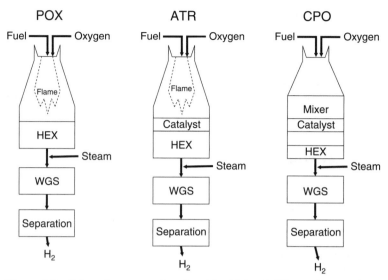

Figure 3.3.   Schematic representation of partial oxidation (POX), autothermal reforming (ATR), and catalytic partial oxidation (CPO) reformers followed by H_2 purification steps. HEX, heat exchanger.

higher steam-to-carbon ratio than POX in order to facilitate the steam reforming and water-gas shift (WGS) reactions to produce more hydrogen. The oxidant can be air, enriched air, or oxygen, depending on the demands of the downstream purification processes. A plant includes a reformer, a shift reactor, and hydrogen purification equipment as depicted in Figure 3.3.

ATR reformer temperatures are in the range of 900–1150 °C, and the reformers with pressure between 1 and 80 bar have been designed and built so far. The ATR reactor is more compact than a steam reformer but larger than a POX unit. The main advantage of the ATR process is that it will produce a syngas with very favorable H_2-to-CO ratio for downstream usage in chemical synthesis. The ratio should be ~2 for Fischer–Tropsch liquids or methanol synthesis.

Since ATRs combine some of the best features of steam reforming and partial or full oxidation, some groups have developed compact catalyst systems to eliminate the need for a robust burner and mixer design. The catalyst system also reduces the formation of carbon and soot. Farrauto et al.[6] and Giroux et al.[7] discussed ATR-based systems for fuel cell applications in detail.

With the right mixture of input fuel, air, and steam, the POX reaction supplies all the heat needed to drive the endothermic catalytic steam reforming reaction. Unlike the steam methane reformer, the autothermal reformer requires no external heat source and no indirect heat exchangers. These features make autothermal reformers simpler and more compact than steam reformers and thus can be built for a relatively low capital cost.

The catalysts in these systems are traditionally two-stage catalysts. A combustion catalyst dominates the first stage to provide heat for the next stage. The second stage is optimized for endothermic steam reforming reactions. The kinetics and space velocities of these two reactions must be delicately balanced for catalytic ATR reactors to work efficiently. Recent developments in catalysis have led to single-stage catalysts integrating both steam reforming and combustion moieties within a single catalyst bed or a monolith. These catalysts and ATR systems are robust and much more compact than a steam methane reforming (SMR) or a homogeneous POX system.

3.2.2   Homogeneous POX

Homogeneous POX is the uncatalyzed reaction of natural gas or liquid hydrocarbons with oxygen at high temperature and high pressure to produce syngas. Some processes use a small amount of steam cofeed. The reaction is exothermic; catalysts are not required because the high temperature produced in the diffusion flame area of the reactor can drive the homogeneous chemistry in the gas phase. The oxidant can be air, enriched air, or oxygen; however, large systems generally incorporate an oxygen plant to reduce not only the size of the POX reactors, but more importantly to reduce the downstream syngas cleanup unit. A hydrogen production plant includes a POX reactor, followed by a shift reactor and hydrogen purification equipment (Fig. 3.3). Large-scale POX systems are typically used to produce hydrogen from hydrocarbons such as residual oil for general refinery operations. The technological feature that is most important to the plant's design and operation is the burner/feed injector. The burner allows a safe mixing of oxygen and fuel to avoid hot spots, coke formation, and explosions. Small-scale POX systems have a fast response time, making them attractive for subsequent rapidly varying loads, and can handle a variety of fuels, including methane, ethanol, methanol, and gasoline. These systems have recently become commercially available, but are still undergoing intensive R&D.[8]

Large-scale reactors typically have temperatures in the range of 1150–1500 °C and pressures between 25 and 80 bar. The H_2 production efficiency of the POX unit is typically in the range of 70%~80%. The POX reactor is more compact than a steam reformer but is typically less efficient, because the higher temperatures require burning additional fuel and the heat is not efficiently recycled. The main advantage of the POX process is that it will operate on virtually any hydrocarbon feedstock from natural gas to petroleum residue and petroleum coke or to coal (gasification). Another advantage is that the process is unlikely to produce NOx or SOx due to lower flame temperature and reducing environment compared with combustion processes in steam reformers. This process can also be considered a fuel-rich burner producing incomplete combustion products.

The capital cost of POX can be high because of the need for post treatment of the raw syngas to remove carbon and acid gases. There are also issues of coke and soot formation if the oxidation temperature becomes too low or the mixing of the feed components is incomplete. The addition of steam to the process allows for greater flame temperature control and suppression of carbon; however, the hydrogen production efficiency is reduced due to more fuel being consumed in the combustion process. The increased operating cost is primarily due to the cost of high pressure and pure oxygen required by the economics of the downstream processes.[9]

3.2.3 CPO

CPO is the heterogeneous reaction of natural gas or liquid hydrocarbons with oxygen and steam at high space velocity over a solid catalyst to produce syngas. The catalyst, support, and optimized operation conditions are the key technological aspects of the process. The oxidant used in the reactor can be air, enriched air, or oxygen.

The pioneering work of Hickman and Schmidt employed ceramic foam coated with rhodium to demonstrate the feasibility of CPO in a small compact system with high gas hour space velocities.[10,11] More than a decade later, active research is still being conducted in order to understand and develop better catalyst systems for these processes. Recently, Lyubovsky et al.[12] discussed the application of microlith screens to elucidate the reaction mechanism within a POX catalyst bed. In this system, metal screens are coated with alumina and catalyst. The use of metal screens allowed the scientists to extract both the temperatures and gas components at various locations along the catalyst bed. A representation of the data collected is shown in Figure 3.4.

The main feature of CPO is the complete conversion of oxygen and near complete conversion of methane within the catalyst bed. Figure 3.5 shows a picture of a working CPO catalyst from a quartz reactor.[8] The temperature of the top portion (inlet) of the CPO catalyst is much hotter than the bottom portion of the catalyst (outlet).

It has been speculated that oxygen is consumed in the first few millimeters of the catalyst bed, increasing the temperature of the bed to 1000~1250 °C (depending

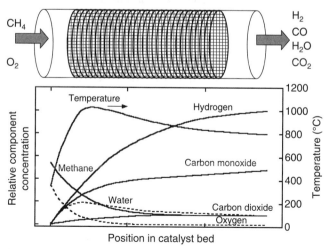

Figure 3.4. Typical temperature profile of a CPO reactor with concentration profiles of fuel and O_2 along a CPO catalyst bed.

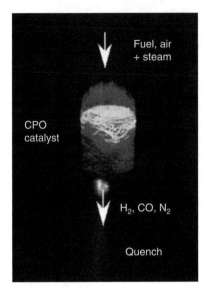

Figure 3.5.   Picture of a working CPO catalyst. See color insert.

on the O_2/C ratio), then the endothermic SMR reaction proceeds, lowering the temperature down to ~800 °C as illustrated in Figure 3.6. Therefore, it is suggested that the very fast exothermic oxidation or POX reactions dominate in the top ~20% of a catalyst bed until most of the O_2 is consumed. The WGS and the endothermic SMR reactions dominate the bottom portion of a CPO catalyst. However, there was no experimental proof of such a theory.

In order to understand the mechanism of CPO, Prof. Lanny Schmidt's group together with a team from GE Global Research developed a powerful technique that allows sampling and measurement of temperature and concentration profiles inside the catalyst with a spatial resolution of ~300 μm, which is on the order of the characteristic length of the CPO catalyst bed support.[13] Figure 3.7 shows the experimental setup. A capillary tube was inserted inside the CPO catalyst bed. Sampling of the gases at different positions of the catalyst bed was achieved by moving the capillary tube through an electronically controlled stepper motor. This sampling method introduced minimal disturbance in flow. Gas samples were sent to a mass spectroscopy (MS), allowing *in situ* quantitative analysis of concentration profiles of the gas components along the CPO catalyst bed. The temperature was monitored using both optical pyrometer and a thermocouple inserted into the catalyst bed. Such a system allowed *in situ* sampling in a system with very high temperature and concentration gradients. More details on the technique can be found in references.[13,14]

Figure 3.8 shows the temperature and concentration profiles along a CPO catalyst bed. The front face of the catalyst bed was at 0 mm. From these results, clearly two zones in CPO catalyst bed can be identified: first a short ~2-mm oxidation zone, where all O_2 is consumed and second, a longer endothermic steam reforming zone. These measured results confirm the CPO mechanism shown in Figure 3.6. In Lyuborsky and colleagues' work,[15] it was demonstrated that oxygen is consumed in

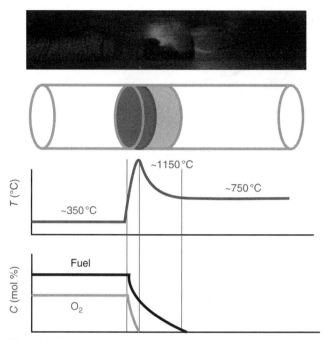

Figure 3.6.   Speculated temperature profile of a CPO catalyst bed based on the observation from Figure 3.5. See color insert.

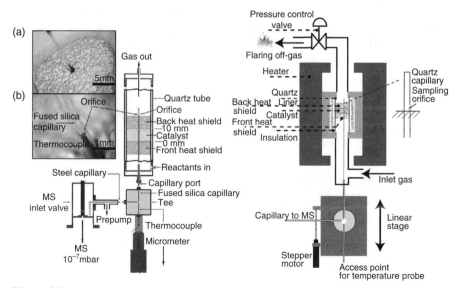

Figure 3.7.   High-pressure CPO system with an integrated mass spectrometer.

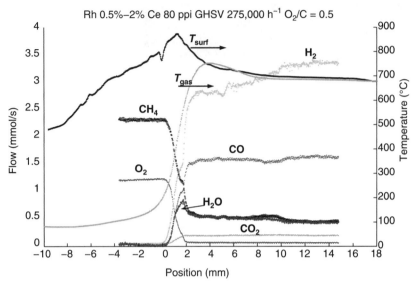

Figure 3.8. Measured temperature and concentration profile of a typical CPO catalyst bed. 0 mm indicates the start of the catalyst, whereas 10 mm denotes the end. Prior to and after the 0–10 mm range are two blank monoliths that act as heat shields.

the first 20% of the catalyst, accompanied by a rapid drop in methane concentration. Another observation is a small water formation peak at the end of this zone. Some believe that the consumption of oxygen in this small zone is caused by combustion; however, one would then expect a peak in CO_2 concentration in the same location as the water. This is not the case. The CO_2 concentration rises slowly along the catalyst bed. This calls for more work to elucidate the details of the CPO reaction mechanism.

Another indication from Figure 3.8 is that temperature rapidly rises in the beginning of the catalyst bed, and then smoothly decreases along the catalyst bed. This allows the reaction to reach near equilibrium, before cooling down and producing a favorable product distribution.

3.3 DEVELOPMENT/COMMERCIALIZATION STATUS OF ATR, POX, AND CPO REFORMERS

Autothermal reformers and CPO are being developed by a number of groups, mostly for fuel processors of gasoline, diesel, and JP-8 fuels and for natural gas-fueled proton exchange membrane fuel cell (PEMFC) cogeneration systems. A few examples are the following:

- Shell and UTC jointly developed the CPO technology for an on-board gasoline reformer together with a 150-kW reforming system based on the state of the art natural gas CPO technology.[8] In this program, the natural gas CPO was demonstrated with over 6000 h of continuous run without significant activity decay. The methane slip through the CPO catalyst was <2000 ppm after 6000 h of operation. In an earlier program, UTC Fuel Cells (FC) designed an ATR that ran on JP-8 fuel. This technology was used by BWX and McDermott Technology, Inc. (MTI) to design a reformer system starting from naval distillate for shipboard fuel cells.

- Johnson Matthey developed a Hot Spot autothermal reformer capable of reforming methanol and methane.[16]

- MTI and Catalytica are developing a small autothermal reformer for diesel and logistics fuels on ships. A regenerable desulfurization stage is important for navy diesel fuel with up to 1% sulfur. Partners in this program are MTI, Catalytica Advanced Technologies, Ballard, BWX Technologies, and Gibbs and Cox.

- The Idaho National Energy and Environment Laboratory (INEEL), with MTI and Pacific Gas and Electric, has recently begun working on developing a 10-kW ATR system for hydrogen refueling station applications.

A number of companies are involved in developing small-scale POX systems that are either homogeneous or catalytic. For example:

- UTC Fuel Cells has partnered with Shell Hydrogen to develop a variety of fuel processors for natural gas, gasoline, and diesel feed for PEMFC, phosphoric acid fuel cell (PAFC), and distributed H_2 production applications.[8]

- Argonne National Laboratory has developed a POX reformer suitable for use in vehicles. The U.S. Department of Energy (DOE) supports work on POX systems for on-board fuel processors for fuel cell vehicles through the Office of Transportation Technologies Fuel Cell Program.[17]

In addition, a number of automotive companies are in joint ventures to develop gasoline fuel processors based on POX or CPO technology. These include the following:

- General Motors has collaborated with ExxonMobil to develop an on-board gasoline fuel processor in the late 1990s and early 2000s.

- UTC Fuel Cells has partnered with Nissan and Shell Hydrogen to develop a 50-kW on-board gasoline fuel processor with a volume of 78 L and weight of 65 kg. The start-up time is 3.5 min.[8]

Projects employing POX systems in stationary fuel cells include:

- Tokyo Gas Company has demonstrated a partial oxidation system for 1 kW fuel cell cogeneration system.

- MTI and Catalytica are working together to develop compact fuel processors for use with PEMFCs and solid oxide fuel cells (SOFCs). This system is designed to reform gasoline and Naval Distillate for PEMFCs.

3.4 CPO CATALYSTS

CPO uses a fixed-bed catalyst to produce syngas. The role of catalysts in POX includes the following:

- To enhance reaction rates and allow high gas velocities while still achieving thermodynamic equilibrium;
- To scavenge radicals formed and thus prevent the formation of soot;
- To prevent high temperatures and avoid the formation of HCN and NH_3. These compounds are not desired in the product stream as they are not only toxic, but also consume hydrogen.

The desired characteristics of catalysts are the following:

- The CPO process must be able to achieve a high conversion of the hydrocarbon feedstock.
- It should be operable at extreme high gas hourly space velocities (GHSVs), while the pressure drop over the catalyst bed should be low.
- The selectivity of the process to the desired products of carbon monoxide and hydrogen must be high.
- It should have high stability to allow for long periods of continuous operation without significant changes in the performance.

The above factors must be met using process equipment that is economical to construct and operate. First-row transition metals (Ni, Co, and Fe) and noble metals (Ru, Rh, Pd, Pt, and Ir) have been reported as active catalysts for the POX of methane. The Ni-based catalysts are the most studied due to the low cost. However, rapid deactivation due to carbon deposition, carbon dissolution in the nickel, or metal loss at high temperatures has been reported. Claridge et al.[18] showed that methane decomposition is the primary route for carbon formation over supported nickel catalysts at a typical methane POX temperature of 780 °C. Noble metals, on the other hand, inhibit dissolution of carbon and therefore have the potential to be stable CPO catalysts.

Another function of the CPO catalysts is to avoid the formation of carbon when there is rapid temperature change in the system.[19] For proper operation of catalysts, a uniform mixing between the fuel and oxidant is required before reaching the catalyst. Therefore, to ensure fast response time required by variable loads, small-scale CPO systems are more favorable.

Lab-scale experiments have demonstrated the feasibility of a wide range of fuels for CPO reformers. Some of the fuels examined are methane, ethanol, methanol, JP-8, diesel, and bio-oil or biodiesel. Products from different fuels can be tuned toward either syngas or olefins by changing the O_2/C, steam/C ratio, and the catalyst.[20,21]

3.4.1 Nickel-Based CPO Catalysts

Due to the widespread use of nickel-based catalysts for steam reforming, they have received great interest for the POX of methane. The earliest study of Ni as a POX catalyst was reported by Prettre et al.[22] Ni was attributed to be the active component for syngas formation. A steep temperature gradient was observed with a maximum located close to the reactor inlet. Dissanayake et al.[23] examined Ni–Al_2O_3 catalysts for the POX and detected different forms of Ni phase with varying catalytic ability along different regions of the catalyst bed.

A major problem with Ni catalyst systems is carbon deposition. Carbon, once formed on Ni metal surface, can dissolve and diffuse through the metal. Carbon accumulated at the rear side of the Ni particles can form whisker-like fibers and lift the active metals away from the support, leading to marked activity loss and severe pressure buildup in the reactor.[24] Therefore, extensive efforts have been made to eliminate carbon formation on the catalyst surface. This can be achieved by different methods from literature studies. One way is the addition of a second or third metal component to reduce the carbon deposition either by promoting surface gasification of the deposited carbon, or selectively poisoning acidic sites that may contribute to coking, and/or inhibiting carbon solubility in the metal particle. Bi- or trimetallic nickel-based mixtures can be found in a significant body of research. Similar initial activity was obtained for those catalysts, with reduced carbon deposition. For example, Nakagawa et al.[25] studied a supported iridium–nickel bimetallic catalyst system. Enhanced performance was observed for Ir (0.25 wt %)–Ni (0.5 wt %)/La_2O_3 catalyst, achieving 36% methane conversion ($CH_4/O_2 = 5$) and CO and H_2 selectivity of above 90% at 800 °C, and 25.3% methane conversion and >80% H_2 and CO selectivity at 600 °C. Monometallic of either Ir or Ni alone on La_2O_3 was evaluated, but they did not show significant syngas formation at 600 °C. Rare earth metals were also examined as a second component for an Ni-based catalyst system. Zhu and Flytzani-Stephanopoulos[26] conducted a research with Ni–Ce catalysts for methane POX. The 5 wt % Ni–Ce (La)O_x catalyst, comprising highly dispersed nickel oxide in ceria, showed excellent resistance to carbon deposition and stable performance during 100 h onstream at 650 °C. When the Ni content was increased to >10 wt %, both dispersed nickel and bulk nickel oxide particles were formed, which is believed to be the cause for unstable time-on-stream performance, with marked coking on the catalyst surface.

Wang et al.[27] also studied the effect of promoters such as Ce, La, and Ca and the loading thereof on the catalytic performance of Ni catalyst in a continuous fixed-bed reactor for the CPO of methane. However, loading large amounts of promoters tended to decrease the catalytic activity due to coverage of active sites. It was found that the effect of promoters was greater for α-Al_2O_3 support than γ-Al_2O_3, with a Ce loading of 1% giving the best results with respect to selectivity and conversion. The reason for the better performance at low loading was ascribed to high dispersion of Ce and formation of bulk CeO_2 with increasing loading. In addition to improving stability, rare earth metals are found to facilitate nickel oxide reduction, which is beneficial for CPO performance as well. This was reported in Chu and colleagues'[28] work on NiO/La_2O_3/γ-Al_2O_3.

Boron was found to be an effective promoter from Chen and colleagues'[29] studies. The two low Ni-loading catalysts (1 wt % Ni and 1 wt % NiB) showed high activity/selectivity for the POX, as good as those of 1 wt % Rh and 10 wt % Ni catalysts supported on commercial γ-Al_2O_3. The coking resistance was remarkably better than that of 10 wt % Ni/Al_2O_3, with coke formation rate in the following order:

10 wt % $Ni/Al_2O_3 \ggg 1$ wt % Ni/Ca–AlO > 1 wt % NiB/Ca–AlO ~ 1 wt % Rh/Al_2O_3. This was ascribed to much smaller metallic Ni particle size induced by lowering Ni loading combined with the addition of B promoter. The presence of B promoter enhanced Ni particle dispersion and thermal stability, which greatly prevented the formation of crystalline carbon during CPO.

Catalyst preparation is another important factor influencing CPO stability. Zhang et al.[30] prepared $NiO/\gamma\text{-}Al_2O_3$ catalysts using both impregnation and sol-gel methods. Catalysts were evaluated at 850 °C and a GHSV as high as 1.8×10^5 L/(kg·h). It was shown that the sol-gel method yielded less carbon deposition than the catalyst prepared by impregnation method. For the sol-gel, an 80-h test did not show apparent activity loss caused by carbon deposition, the loss or/and sintering of nickel. However, a gradual decline in activity was observed, which was attributed to the transition of the support from γ to α phase. Therefore, a complex agent-assisted sol-gel method was employed to stabilize the support phase and enhance performance as well. Xu and Wang[31] employed coprecipitation method to prepare $CeO_2\text{-}ZrO_2$ solid solution as support for Ni. A strong interaction between Ni and $CeO_2\text{-}ZrO_2$ was confirmed by the H_2-temperature programmed reduction (TPR) analysis. High activity and stability were observed, with $Ni/Ce_{0.25}Zr_{0.75}O_2$ formulation exhibiting the best activity and coking resistance. The support prepared by coprecipitation was found to play a large role in stabilizing the nickel catalyst into the desired oxidation state.

Kim et al.[32] reported the preparation, characterization, and catalytic performance of a finely dispersed and thermally stable nickel catalyst incorporated into mesoporous alumina. Mesoporous alumina catalysts that incorporate Ni (Ni–alumina) with different Ni/Al molar ratios were synthesized by a one-step sol-gel method using lauric acid as a template. The prepared Ni–alumina catalysts showed a relatively high surface area with a narrow pore size distribution after calcination at 700 °C; these effects were independent of the Ni/Al molar ratio. The Ni–alumina catalysts were found to be highly active in the POX of methane. The deactivation of catalysts examined in this work was not due to catalyst sintering, but mainly to carbon deposition.

Not only the preparation method but also the support materials play a major role in producing a highly active and stable CPO catalyst. Instead of conventional alumina support, Sun et al.[33] studied 10 wt % Ni/SiC and detected tubular and amorphous carbon after time-on-stream. However, the deactivated catalyst was easily regenerated through burning carbon in air, and the activity completely recovered to the fresh condition. Another advantage is that the catalyst can be directly used in the reaction without prereduction.

Calcium aluminate supports obtained with varying CaO/Al_2O_3 ratios were examined in a work by Goula et al.[34] Metallic nickel (Ni^0) was shown to be the active species for the methane POX reaction. The supports led to varying surface nickel species reducibility, which ultimately influenced catalyst stability and activity.

Yang et al.[35] studied various supported metal catalysts (Ni, Co, Pt, Pd, Ru) prepared using a novel nanoporous material $12CaO:Al_2O_3$ (C12A7) as support. With an increase in space velocity, activity values for the catalysts were 1% Pt/C12A7 > 5% Co/C12A7 > 5% Ni/C12A7 > 1% Ru/C12A7 > 1% Pd/C12A7. The activity and selectivity of Ni/Cl2A7 and Pt/Cl2A7 increased with metal loading, and the activity of 10% Ni/C12A7 was comparable to that of 1% Pt/C12A7. Ni/C12A7 exhibited a low coke formation rate and the material was more active than nickel supported on CaO, γ-Al$_2$O$_3$, 3CaO–Al$_2$O$_3$ (C3A), or CaO–Al$_2$O$_3$ (CA), owing to the good dispersion of NiO on C12A7 and the existence of active oxygen ions incorporated in its nanocages.

Since the metallic nickel seems to favor carbon formation so strongly, studies have been performed to find methods to incorporate nickel into other structures such as perovskites or hexa-aluminates. Chu et al. suggested the special structure of the nickel containing barium hexa-aluminate to prevent its lattice from being destroyed with the Ni loading as high as 20 wt %.[36] The "layered aluminate type" structure of the barium hexa-aluminate support can accommodate Ni ions in the structure so that a strong interaction exists between Ni ions and the support, which explains excellent stability and capability to suppress carbon deposition. Over all compositions of study, the reduced Ni-modified hexa-aluminates BaNi$_y$Al$_{12-y}$O$_{19-\delta}$ exhibit significant catalytic activity and stability for the POX of methane to syngas, with 92% CH$_4$ conversion and 95% CO selectivity at 850 °C. After 100 h of time-on-stream, no loss of nickel was detected and carbon deposition is nearly negligible, although Ni loss was reported as a concern.[37]

Ni catalysts on perovskite-type oxides (CaTiO$_3$, SrTiO$_3$, and BaTiO$_3$) prepared by a solid-phase crystallization method have been reported to be resistant to coke formation.[38] Perovskite-type oxides of La$_{1-x}$Ca$_x$Ru$_{1-x}$Ni$_x$O$_3$ were synthesized by the citrate sol-gel method. All perovskites showed a well-defined perovskite structure with surface areas between 2 and 17 m^2/g. Among the calcium series, La$_{0.8}$Ca$_{0.2}$Ru$_{0.8}$Ni$_{0.2}$O$_3$ proved to be the most active precursor catalyst with the highest selectivity to syngas. Gou et al.[39] studied the performances of BaTi$_{1-x}$Ni$_x$O$_3$ perovskites, prepared using the sol-gel method. Experimental studies showed that the calcination temperature and Ni content exhibit a significant influence on catalytic activity. Among the catalysts they tested, the BaTi$_{0.8}$Ni$_{0.2}$O$_3$ catalyst exhibited the best activity and stability.

3.4.1.1 Summary of Ni-Based CPO Catalysts

Nickel-based catalysts are viable for the POX of methane. The major problem associated with these catalysts is the formation of carbon and consequent loss of catalytic activity. Catalyst stability can be improved by different approaches as reviewed in this section. The addition of a second or third promoter can facilitate surface carbon gasification, or neutralizing acidic sites that may contribute to coking, and/or inhibiting carbon solubility in the metal particle. Ni metal particle size and dispersion are found to be closely dependent on support materials and preparation methods, which strongly influences coke formation reaction. Highly dispersed Ni on the surface tends to inhibit coke deposition. Another alternative is to place Ni into a structure that can stabilize Ni oxidation state, thus reducing carbon formation. Perovskites and hexaluminates have been shown to produce syngas under mild conditions; however, these materials lack thermal stability and further improvement is needed.

3.4.2 Precious Metal CPO Catalysts

It should be clear from the discussion of nickel catalysts that the use of inexpensive nickel comes with a price because of problems with deactivation from carbon formation. One way to solve this problem is to avoid the use of nickel and replace it with elements in which carbon does not dissolve.[40] This is the main reason that research has focused on precious metal catalysts for CPO. The added cost of metal is to be justified by better performance in terms of stability and yields. It is also clear that the high space velocities obtainable in CPO will help reduce the total amount of catalyst and thus alleviate one of the inherent problems with the cost of the precious metals.

Some of the earliest work on precious metal catalysts for CPO of methane was carried out by Ashcroft et al.[41] In their work, $Pr_2Ru_2O_7$ was evaluated for CPO performance. A methane conversion of >90% and selectivity of 94%–99% were achieved. Pure Ru metal was detected on the catalyst surface and believed to be the active component during the reaction.

Following the pioneering work by Ashcroft and his coworkers, other researchers also examined Ru as active metal for CPO. Torniainen et al.[42] investigated α-Al_2O_3 foam monolith-supported Ru catalysts for natural gas autothermal reaction and found that Ru would not sustain the reaction. Rabe et al.[43] evaluated Ru catalysts for dry CPO performance by coupling with thermogravimetric and infrared (IR) spectroscopic analysis. It was found that at lower temperatures (<450 °C), catalysts are in an oxidized state, which yields a high CO_2 selectivity. CO selectivity rises with reduction of the catalysts. With S/C at 0.8 and O/C at 1.0, 5Ru/5Ce-γ-Al_2O_3 catalyst showed a stable performance during an 80-h period, with methane conversion constantly about 65%. Kikuchi et al.[44] prepared hexa-aluminate catalysts based on $BaRu_xAl_{12-x}O_{19-y}$ ($x = 0.25$, 0.5, 1.0, and 1.5) via the alkoxide method and investigated the POX of CH_4 and C_3H_8. The catalyst exhibited excellent activities for CH_4 and C_3H_8 POX under a high space velocity of 120,000 L/(kg·h). It was also found that the strong interaction between Ru and the base oxide in the Ru-substituted hexa-aluminate greatly inhibited Ru evaporation during the reaction. Similarly, in Perkas and colleagues'[45] work on Ru CPO catalysts, to improve stability, a one-step ultrasound assisted polyol reduction procedure was employed to disperse Ru nanoparticles on mesoporous TiO_2. The Ru^{3+} ions were created in a narrow size distribution within the mesoporous TiO_2 support. High activity and selectivity toward CO and H_2 up to 95% and 96% at 800 °C, respectively, have been achieved during POX of methane at a GHSV of 14,400 h^{-1}.

In addition to Ru, other precious metals investigated for CPO include Ag, Pt, Re, Ir, Pd, and Rh.[42–44,46–51] Bharadwaj et al.[46] examined the catalytic oxidation of alkanes on supported Ag catalysts at high temperatures (>800 °C) in both fluidized bed and monolithic reactors at contact times of 50–200 and 5–10 ms, respectively. Oxidation of CH_4, C_2H_6, and i-C_4H_{10} were examined on 0.25–0.5 wt % Ag/α-Al_2O_3 in fluidized beds, while CH_4 and C_2H_6 oxidations and ammoxidations were examined on 4–35 wt % Ag monoliths supported on α-Al_2O_3. However, results showed that it was not possible to maintain reaction with only CH_4 and O_2 in the feed on Ag in either the fluidized bed or on the monoliths, which indicates that Ag is not very active in catalyzing the POX of methane. Rh, Pt, Ir, Ru, Pd, and Re catalysts on α-Al_2O_3 foam monoliths were studied in Torniainen and colleagues'[42] work. Catalysts were evaluated for methane CPO at a GHSV of 100,000 h^{-1}. Rh, Pt, and Ir displayed similar performance, with Rh showing better selectivity. All these catalysts showed stable performance over time, although carbon was found downstream on the tube wall for Pt. Ir catalyst was found to sinter into large crystals with time, but performance was not reported to be affected. Ru catalyst was found not able to sustain the operation. Pd catalysts suffered from serious coking, whereas hot spots were detected on Re catalysts (1430 °C), causing severe active metal loss. Claridge et al.[47] reported similar findings for Rh, Pt, Ir, and Pd catalysts. However, in their studies, instead of Re vaporization in Torniainen and colleagues'[42] work, Re catalysts demonstrated comparable activity as Rh, Pt, and Ir at high temperatures (780 °C). Otsuka et al.[52] studied Pt supported on CeO_2 catalysts and attributed performance enhancement through Fourier transform IR spectroscopy (FT-IR) and temperature programmed desorption (TPD) analysis to Pt accelerating desorption of H_2 and CO from the surface, which was identified as the rate-determining step.

Through literature studies, Pt and Rh were identified to be active and stable CPO catalysts, with Pt having lower hydrogen yield and activity than Rh, as attributed to a higher hot spot temperature by Rabe et al.[43] Hickman and Schmidt[10,11] investigated Al_2O_3 monolith-supported Rh and Pt catalysts for the POX of methane and found that Rh has a higher H_2 selectivity (85%) than Pt (60% H_2 selectivity) with a feed of CH_4/O_2 at 1.6.

Arpentinier et al.[53] reported nanostructured catalysts for the POX of methane using Rh as the active metal. The hydrotalcite (HT)-type precursors, which contain carbonates or silicates as interlayer anions, help create mixed oxide or silicate phases after calcination. A strong interaction was found between the metal and support, which explains better catalyst stability at high temperatures compared with those obtained by deposition techniques. A 50-h test was performed, and X-ray diffraction (XRD) and Bruauer Emmett Teller (BET) surface area measurements showed no significant sign of changing with time-on-stream, indicating a stable catalyst.

The influence of precursors was also examined by Ruckenstein and Wang[54] in Rh/MgO catalysts. Large differences were found in the dispersion and surface area of Rh depending on the precursor; however, no difference in the conversion or selectivity was detected. This was attributed to diffusion-limited condition or insufficient amount of catalyst to reach thermodynamic equilibrium. The resistance to methane decomposition was found to be a function of the precursor, with $Mg(NO_3)_2 \cdot 6H_2O$ deactivating the fastest and $4MgCO_3 \cdot Mg(OH)_2 \cdot 4H_2O$ being the most resistant. Stability tests over 90-h on all supports did not show any sign of deactivation.

In another work, Ruckenstein and Wang[54] examined the effects of support on the performance of Rh catalysts for the POX of methane. Two kinds of metal oxide supports, relatively easy reducible (CeO_2, Nb_2O_5, Ta_2O_5, TiO_2, ZrO_2) and not easy reducible (γ-Al_2O_3, La_2O_3, MgO, SiO_2, Y_2O_3), were studied. It was found that reducible oxide-supported Rh catalyst with the exception of Ta_2O_5, which rapidly deactivated, gave much lower activities and selectivity to syngas. One possible reason is that suboxide generated via the reduction of reducible oxides could migrate onto the surface of the metal particles and decrease the number of active Rh sites. Among the irreducible oxides, MgO provided the highest catalytic activity with high product selectivity and stability. Strong interactions between rhodium and magnesium oxides (especially the formation of $MgRh_2O_4$) are responsible for the high stability of MgO-supported Rh catalyst. Metallic rhodium, which is believed to be the active sites for methane POX, can be generated via the reaction between methane and rhodium oxide (or Rh-containing compound) even in the presence of oxygen and induces the POX of methane.

Bruno et al.[55] investigated the activity of Rh/γ-Al$_2$O$_3$ and Rh/ZrO$_2$ in CH$_4$ POX in an annular reactor, operating at high space velocity, under kinetically controlled conditions and minimum temperature gradients along the catalyst bed. Effects of temperature, dilution, space velocity, and CH$_4$/O$_2$ ratio were explored. Rh/γ-Al$_2$O$_3$ provided higher H$_2$ yields than Rh/ZrO$_2$ at all conditions. Concerning the process kinetics, an indirect-consecutive kinetic scheme for synthesis gas formation prevailed over Rh/γ-Al$_2$O$_3$; the observed trends were in fact in line with the combined presence of methane total oxidation, methane reforming reactions, and consecutive oxidations of H$_2$ and CO. Over Rh/ZrO$_2$, the additional contribution of a direct route to synthesis gas could not be excluded. The complexity of the reaction scheme seems associated with the existence of different active sites, whose concentration could be affected by metal-support interactions and Rh reconstruction during conditioning. Indeed, once exposed to the reaction atmosphere, the fresh catalysts showed an initial conditioning during which a specific activation of the high temperature routes (responsible for H$_2$ production) was observed.

Eriksson et al.[56] investigated CPO of methane to syngas, over supported Rh catalysts at atmospheric pressure at a GHSV of 99,000–110,000 h^{-1}. The influence of support material, Rh loading, and the presence of steam on the methane conversion efficiency and the product gas composition were studied. The catalysts containing ceria in the support material showed the highest activity and formation of H$_2$ and CO, which was ascribed to higher activity to WGS reaction. By increasing the Rh loading, a decrease of the ignition temperature was obtained. Addition of steam to the reactant gas mixture was found to increase the ignition temperature and hydrogen formation, which is favorable for combustion applications where the CPO stage is followed by H$_2$-stabilized homogeneous combustion.

Hohn and Schmidt studied the effect of space velocity on the POX of methane using different support geometries with Rh as a catalyst.[57] Syngas selectivity drops as space velocity is increased to above 4×10^5 h^{-1} on foam alumina monolith support, while nonporous alumina spheres allow high reactant conversions and syngas selectivity even at high space velocities of 1.8×10^6 h^{-1}. The different results between monoliths and spheres are explained by the major role played by the differences in heat transfer within the two support geometries. A convective heat transfer model was used to show that higher rates of convection in a monolith will lead to lower front temperatures than a sphere bed, which is important at high space velocities, leading to blowout and lower syngas selectivity. The performance of Rh-coated spheres is relatively independent of sphere size, although the largest sphere size ($3200\,\mu$m) gave significantly worse results. Internal surface area has no effect on performance as indicated by the nearly identical results of sintered and unsintered spheres. Pressure drop is higher in spheres than in a monolith due to smaller void fraction in the bed. A foam monolith has higher convective heat transfer but lower conductive transfer, while spheres have higher conductive heat transfer maintaining a uniform bed temperature at high gas velocities.

Leclerc et al.[58] in Schmidt's group showed that the POX of CH_4 to produce H_2 and CO can be accomplished within ~5 s from room temperature by spark igniting a stoichiometric feed of CH_4 and air. The ignition rapidly heats Rh on α-alumina catalyst to a temperature where it is possible to switch the feed to the syngas-producing regime. This configuration should be useful for fast light-off in hydrogen generation in transportation and stationary applications of fuel cells for electricity generation. It was shown that the selectivity and conversion did not vary with varying GHSV from 160,000 to 430,000 h^{-1}, making the system suitable for variable loads.

Natural gas is a widely used feedstock for syngas production. The presence of parts per million levels of sulfur compounds poses a challenge for CPO catalyst development. There are not many studies devoted to this research topic. GE Global Research Center, in collaboration with Prof. Schmidt, evaluated Rh- and Pt-based CPO catalysts in the presence of sulfur. CPO performance was negatively impacted by parts per million level of sulfur, as indicated by a decrease in CH_4 conversion and H_2 selectivity. However, upon removal of sulfur in natural gas, the CPO performance was recovered. This was demonstrated in a 300-h long-term run by GE.[59] In the test, Rh-based catalyst was evaluated at a 200,000 h^{-1} GHSV and 8 ppm sulfur in natural gas. A sharp increase in catalyst back-face temperature was observed immediately upon sulfur doping. However, this temperature returned to normal when the sulfur was removed. Fundamental studies were performed to understand the influence of sulfur on CPO catalysts. Spatially resolved temperature and concentration gradients in the presence of sulfur during CPO were collected by Schmidt's group using the profiling technique described earlier in this chapter. The poisoning effect is shown to be due to severe hindrance of steam reforming by adsorbed sulfur, whereas the POX occurring in the top portion of the catalyst bed was much less affected. It was suggested that the presence of sulfur could limit surface diffusion of $H(s)$ and/or $O(s)$, causing a decrease in CPO performance.[60]

3.4.2.1 *Summary of Precious Metal CPO Catalysts*

The presented review on precious metal catalysts for CPO has shown that it is possible to obtain high conversion of methane with good selectivity to syngas at high GHSV. Rhodium has been shown to be a better catalyst over Pt, Ru, and Pd. Much of the current research is therefore focused on increasing the performance and stability of the rhodium catalyst by investigating different supports. It is still not clear what the best support is; however, the use of foam monoliths and sphere beds are very promising, and in those systems, it is not a question of increasing the catalytic activity but improving mass and heat transfer.

One of the key challenges remaining in the development of precious metal catalysts for CPO is attaining high yields at high pressures. The desire to operate the reactors at pressure stems from the fact that the downstream processes utilizing syngas operate at high pressures and that it will allow maintaining compact systems while increasing the yields. Increasing the pressure while maintaining the volume of catalyst will increase the temperature; this in turn increases the potential for nonselective gas-phase reactions during mixing/preheating or in the reactor. Overcoming these challenges require the design of specialized mixers that avoid gas-phase ignition and development of very effective catalysts that permit the desired surface-catalyzed reactions to proceed much faster than the competing gas-phase reactions.

Another aspect that needs more research efforts is sulfur tolerant CPO catalyst design and understanding of sulfur deactivation mechanism to predict catalyst lifetime, which is critical for industrial application of the CPO technology.

3.5   CPO MECHANISM AND KINETICS

To meet the challenges for effective CPO catalyst design, it is necessary to have a fundamental understanding of what is happening on the catalyst surface. Different detail levels of kinetic models can be found from literature studies. The most fundamental ones are based on elementary reactions with a large number of steps and detailed accounting regarding the surface coverage, while the least fundamental ones are based on one global reaction rate, which is normally dependent on the partial pressure of the species. It should be clear that the choice of detail largely depends on the purpose of the study. For catalyst design, the computational costs of solving large models may be defended, whereas in other systems when catalyst is not the focus area, simple rate laws may fulfill the needs. The development of kinetic modeling allows a safe investigation of the operational settings that potentially could be dangerous experimentally. Insights gained from the modeling can guide catalyst development, reactor scale-up, and process operation.

3.5.1   Ni Catalyst Mechanism and Reactor Kinetics Modeling

There has been a large amount of literature examining the steam reforming mechanism over Ni catalysts, which greatly contributes to the understanding of CPO mechanism. A detailed summary of Ni catalyst steam reforming mechanism and kinetics can be found in the steam reforming chapter of this book. In De Groote and Froment's studies,[61] the overall CPO reaction mechanism was simulated, incorporating different reactions: catalytic combustion, steam reforming, and WGS. Methane decomposition and the Boudouard reaction were also included to account for carbon formation. Kinetic modeling was conducted with an adiabatic fixed-bed reactor, with total combustion occurring first followed by the steam reforming and WGS reactions. Effectiveness factors were employed to take intraparticle transport into consideration. In another study by De Groote and Froment,[62] the POX of CH_4/O_2 and CH_4–air mixtures into synthesis gas was modeled on a $Ni/MgO/Al_2O_3$ catalyst in an adiabatic fixed-bed reactor to further investigate carbon formation. Diffusion limitations were accounted for through effectiveness factors, averaged over the operating conditions, and calculated from a number of off-line pellet simulations. Coke deposition as a function of time was included in the simulation to examine whether or not a steady-state reactor operation was possible. The influence of the feed composition (addition of steam and carbon dioxide) and the operating conditions on the coke content was also investigated.

3.5.2 Precious Metal Catalyst Mechanism and Reactor Kinetics Modeling

The reaction mechanism is an important and open question in CPO research. Direct and indirect mechanisms were proposed, as summarized in a review paper by York et al. in 2003.[63] Some authors support the direct mechanism whereas some find evidence for the indirect mechanism. Others advocate for a mechanism in between. Professor Schmidt's group reported extensive fundamental studies on CPO reaction mechanism.[64] For CPO reaction, the direct mechanism assumes that H_2 and CO are primary reaction products formed by POX in the presence of gas phase O_2. This can be shown by the following equation, with competitive formation of H_2O and CO_2 included ($0 \leq x \leq 2$, $0 \leq y \leq 1$):

$$CH_4 + \left(2 - \frac{x}{2} - \frac{y}{2}\right)O_2 \rightarrow xH_2 + yCO + (2 - x)H_2O + (1 - y)CO_2. \qquad (3.5)$$

The indirect mechanism postulates a two-zone model with strongly exothermic CH_4 combustion to H_2O and CO_2 (Eq. 3.6) at the catalyst entrance, followed by strongly endothermic steam (see Eq. 3.3) and CO_2 reforming downstream (Eq. 3.7):

$$CH_4 + 2O_2 \rightarrow CO_2 + 2H_2O, \Delta H_r^\circ = -803 \text{ kJ/mol}, \qquad (3.6)$$

$$CH_4 + CO_2 \rightarrow 2CO + 2H_2, \Delta H_r^\circ = +247 \text{ kJ/mol}. \qquad (3.7)$$

Two approaches are typically adopted in literature for mechanistic studies.[64] One is to compare reactor exit data obtained from realistic methane CPO experiments with numerical simulations derived from reaction mechanism assumption. For the other approach, well-defined isothermal low-pressure or diluted conditions different from technical applications are employed. Mechanistic conclusions are drawn based on experimental product distribution under those conditions.

In the mechanistic studies conducted by Horn et al.,[64] methane CPO reactions were run under as close to industrial conditions as possible. A high-resolution ($\approx 300 \mu m$) spatial profiling technique was developed to measure the temperature and species profiles along the centerline of the catalyst bed. The unprecedented high-resolution data make it possible to accurately speculate reactions occurring along the catalyst bed (10 mm) in a very short contact time. It was revealed that for Rh-supported CPO catalysts, H_2 and CO are formed in both the initial 0~2-mm

POX and the subsequent 2~10-mm steam reforming zone. No CO_2 reforming was observed. Higher H_2 and CO yields were detected at high catalyst temperatures, which were attributed to increased CH_4 conversion and H_2 and CO selectivities in the oxidation zone and accelerated steam reforming reactions downstream. For Pt catalysts, a longer oxidation zone and lower steam reforming activity were measured, yielding much less syngas in comparison with Rh.

Based on experimental results and mechanistic studies, extensive kinetics and reactor modeling work has been conducted. Hickman and Schmidt combined a 19-step elementary reaction mechanism and reactor modeling to simulate methane CPO on Rh and Pt. All reactions were assumed independent of coverage due to low pressures used for the reactions.[65] Reactor models were varied depending on the mass transfer rates. Syngas selectivities and yields predicted from the model were found in good agreement with experimental results. This reactor modeling study was revised by Deutschmann and Schmidt[66,67] to include surface coverage-dependent reaction rates and gas-phase chemistry. A two-dimensional model (1D spatial and 1D temporal) for methane CPO on Rh in monolithic reactors was employed. Therefore, it offers a detailed description of the complex interaction between mass and heat transfer as well as the reaction chemistry. An extremely rapid variation of temperature, velocity, and transport coefficients was seen at the reactor entrance, accompanied by a large amount of CO_2 and H_2O from complete methane oxidation. The competition between complete and partial methane oxidation was ascribed to different extents of surface coverage caused by reaction pressures. Methane conversions as well as H_2 and CO selectivity are found to increase with increasing temperature. To better predict the reaction at pressures, Deutschmann et al.[68] simulated the reaction using 38 elementary steps and a three-dimensional reactor model, taking reaction–diffusion into account. This model provides a more accurate estimation of gas-phase reactions, inlet rapid variations, and the steam reforming reaction.

Bizzi et al.[69] examined the physicochemical features of short contact time CPO of methane on Rh with experimental and modeling activities. Instead of using an elementary reaction mechanism, three global reactions were considered: total oxidation, POX, and WGS. The theoretical analysis was developed by coupling transport phenomena and chemical kinetics in a diffusion–reaction system. Oxygen transport from the bulk gas phase to the catalyst was found to be the rate-limiting step. In another study by Bizzi et al.,[70] the effect of thermal conductivity in the support was highlighted. In their work, reactions could only take place when oxygen was present, which implies that steam reforming, total, and POX reactions all happen in parallel. It was shown that increasing the thermal conductivity does not change the reactor performance significantly unless the conductivity is changed independently in the preheat and the reaction zone. It should be noted that this simplified mechanism was incapable of predicting blowout. In order to predict blowout, Bizzi et al.[71] incorporated the surface mechanism by Deutschmann and Schmidt[66] and Deutschmann et al.[68] to examine the effects of the support geometry, feed composition, and reactor scale-up. Biesheuvel et al.[72] studied the catalytic conversion of a gaseous hydrocarbon fuel with air and steam to synthesis gas by ATR and CPO using a 1D stationary reactor model. This model relies on effective rate equations to reflect mass transfer to the catalyst surface. The reactor was simulated with two sections: an upstream oxidation section and a downstream reforming section. In the oxidation section, oxygen was completely converted. Kinetics was not considered in this section, and an empirical fuel utilization ratio was used to quantify fuel conversion. In the reforming section, the reactor was modeled as plug flow, and methane conversion is determined based on kinetics and how close the local composition is to equilibrium. The gas temperature rapidly increased in the oxidation section and peaked at the intersection with the reforming section.

3.6 START-UP AND SHUTDOWN PROCEDURE OF CPO

The most hazardous part of the reactor operation for CPO is in the transient mode of start-up and shutdown, where improper control of the carbon-to-oxygen ratio can lead to catastrophic results. Safe start-up of CPO reactors can be approached in several ways. One way is to pass preheated inert gas over the catalyst to reach a certain minimum temperature required for "light-off." When the temperature is sufficient (~350 °C for methane CPO), fuel and oxygen can be admitted in a controlled manner to start the reaction.[46] If a high enough O_2/C ratio is initially fed to the reactor, the reaction will rapidly increase the temperature of the catalyst, perhaps to a level that is undesirable. Ideally, a fraction of the desired O_2/C ratio will be initially fed, followed by a systematic increase to the final desired value while continuously monitoring the catalyst temperature. Tight control of the O_2/C ratio is necessary to stay well below the stoichiometric ratio needed for combustion of the fuel. Ratios approaching this stoichiometric limit run high risk of exceeding temperatures of 1000 °C and damaging the catalyst and possibly reactor components. These risks often require safety controls of catalysts inlet and outlet temperatures as well as fuel and air (O_2) flow rates.

Alternatively, one can introduce either NH_3 or H_2 with air and either decrease or eliminate the need for preheat. These reactions will proceed at much lower temperatures and provide the heat necessary to warm the catalyst before switching to the desired fuel. When the catalyst temperature is high enough, NH_3 or H_2 can be incrementally replaced with the desired fuel. It should be stressed that the feed composition should always be outside the flammable range in order to avoid upstream flames. In fact, in practical design, flashback protection and temperature alarms are common in the upstream portion of CPO reactors.

Yet another approach to initiate a CPO reaction is by the use of spark ignition. This technique uses an electric spark situated slightly upstream of the catalyst to ignite a stoichiometric mixture of methane and air for combustion. This feed composition is then altered to the desired composition. This technique has been used for rapid-start application and has been reported to achieve production of syngas in ~5 s.[58]

Because CPO reactors are normally operated in the fuel-rich regime, the proper way to shut them down is to remove the oxidant and thus avoid feed compositions that are within the flammable range that could potentially lead to explosions.

Another potential concern of proper shutdown is the potential formation of coke on the catalyst or in the reactor itself. Coke formation could occur if the O_2/C ratio is maintained too low in the absence of steam for an extended period of time. In order to ensure a safe shutdown, the oxidant is removed from the system while fuel and steam continue to flow for a short period to ensure all oxidizants have been removed from the system. Subsequently, the reactor can be blanketed with an inert gas for cooldown.

CPO reactor design and operation is still a field with ongoing investigation. However, this review of the scientific literature has shown that a theoretical framework exists. The framework has identified important process parameters, and with this knowledge, it should be possible to design reliable systems for commercial applications. It should be noted that steady-state operation might not be the only way to operate these systems since interesting results have been obtained using a cyclic or reverse flow mode.

3.7 CPO OF RENEWABLE FUELS

In view of the growing concern over greenhouse gas emissions, fuels from renewable sources have received increasing interest in recent years.[73-76] Recent work on CPO has therefore focused on using feedstocks that are renewable.

The work by Deluga et al.[76] showed that it is possible to obtain complete conversion of ethanol and high selectivity to hydrogen (80%) over Rh–Ce catalysts on Al_2O_3 foams at millisecond contact time (GHSV $1.5 \times 10^5 h^{-1}$). This selectivity could be further increased by the addition of steam not only to convert all the hydrogen in the ethanol but also some hydrogen from the steam as well. Operation with steam is furthermore favorable since one of the main costs of ethanol is closely linked to the cost of water vaporization. By adding a catalyst bed of Pt–Ce, it was possible to drive the WGS reaction toward CO_2 and H_2 thus lowering the CO content. This work was featured by almost complete absence of by-products like acetaldehyde and CH_4. There were reported problems with catalyst stability in the work by Salge et al.,[21] where pure Rh catalyst on Al_2O_3 would disintegrate with time-on-stream. The reason for this was not described in the work but was speculated to be relevant with the formation of hot spots in the catalyst. It was furthermore shown that the selectivity to CH_4 increased with increasing GHSV, indicating that it is first formed and then reformed inside the catalyst.

While the use of ethanol opens the door to using renewable fuels in a CPO reactor, it still requires a volatile fuel. However, most of the available biomass is nonvolatile. Thus, there is great interest in extending the CPO technology to non-volatile renewable fuels. This was demonstrated by Duane and Schmidt[74] where soy oil was successfully partially oxidized at millisecond contact times. This was achieved by spraying the soy oil directly on the hot catalyst surface. The extreme heat flux prevented the formation of soot by rapidly pyrolyzing the oil into smaller hydrocarbons that were subsequently partially oxidized over the catalyst. This resulted in almost complete conversion of the fuel to syngas with only a small amount of olefins produced.

3.8 SUMMARY

This review has shown the current state of the art in connection with CPO and ATR of methane to syngas. The focus incorporated both commercial and academic areas of interest.

There is a vast amount of scientific literature that describes various aspects of the catalyst systems used in CPO. The initial choice stands between using an inexpensive Ni catalyst and choosing a precious metal catalyst, such as Rh. The former has been investigated for 60+ years in a variety of configurations, while precious metals have been researched for over 15 years. As in all catalyst development, the focus is on maintaining the activity for long-term operation. The main problem with lifetime is the formation of carbon on nickel. However, for both types of catalysts, current research focuses on the support used to form stronger support–metal interactions to avoid sintering and to maintain the catalyst in its active oxidation state. Further research on the support includes optimizing mass and heat transfer, which would allow for even more compact designs.

Through results from surface science and direct measurement of rate constants, it has been possible to develop detailed surface mechanism for Rh and Pt, which gives a thorough understanding of the elementary steps and allows one to make predictions of operating conditions that are not easily obtained safely in a laboratory, such as high pressure. The main focus of the mechanistic work is to establish whether the formation of syngas proceeds via a direct mechanism or through deep oxidation followed by reforming. The models suggest that both mechanisms are important in accurately predicting experimental results. The level of computational cost incurred from detailed models may not always be warranted from the specific application. For this reason, simplified models have been proposed. They do not rely on elementary steps, but on overall reactions such as total oxidation, reforming, WGS, and POX. Within the realm of use, these models have been shown to provide adequate accuracy at much lower computational cost.

The CPO technology has, in recent years, started to migrate from research laboratories into commercial application, exemplified by the growing number of patents, where it is seen as an alternative to steam reformers for hydrogen production. The reason for this is the effectiveness of catalysts, which would give high selectivity to hydrogen and yield themselves to a compact design, which allows for rapid response and relatively low capital cost. This has been achieved through fundamental knowledge of the process through years of research. The commercial application relying on CPO and ATR have thus far focused primarily on production of hydrogen for fuel cells, where the compact design makes CPO an obvious choice since many of these systems are designed to be portable. However, the future will certainly expand the areas where this technology will find its application.

ACKNOWLEDGMENTS

The authors would like to thank Dr. Rick Watson for his contribution in reviewing this chapter, Dr. Lakshmi Tulasi for her contribution on the literature search and review.

REFERENCES

1. SHADLE, L.J., BERRY, D.A., and SYAMLAL, M. Coal conversion processes. In *Kirk-Othmer Encyclopedia of Chemical Technology*, 3rd edition. New York: John Wiley & Sons, pp. 224–377, **2002**.

2. PROBSTEIN, R.F. and HICKS, R.E. *Synthetic Fuels*. New York: McGraw-Hill, p. 1, **1979**.

3. BAADE, W.I., UDAY, N.P., and VENKAT, S.R. Hydrogen. In *Kirk-Othmer Encyclopedia of Chemical Technology*. New York: John Wiley & Sons, **2001**.

4. TWIGG, M.V. *Catalyst Handbook*. London: Wolfe Publishing, **1989**.

5. SHINNAR, R. The hydrogen economy: Clarifications on solar cell technologies. *Chemical Engineering Progress*, **2005**, 101, 5.

6. FARRAUTO, R., HWANG, S., SHORE, L., RUETTINGER, W., LAMPERT, J., GIROUX, T., LIU, Y., and ILINICH, O. New material needs for hydrocarbon fuel processing: Generating hydrogen for the PEM fuel cell. *Annual Review of Materials Research*, **2003**, 33, 1.

7. GIROUX, T., HWANG, S., LIU, Y., RUETTINGER, W., and SHORE, L. Monolithic structures as alternatives to particulate catalysts for the reforming of hydrocarbons for hydrogen generation. *Applied Catalysis. B, Environmental*, **2005**, 56, 185.

8. LIU, K., BUGLUSS, J., and KOCUM, F. Overview of fuel processing technologies for fuel cells. *AIChE Spring National Meeting*. New Orleans, LA, April **2004**.

9. OGDEN, J.M., KREUTZ, T.G., STEINBUGLER, M., COX, A.B., and WHITE, J.W. Hydrogen energy systems studies. *Proceedings of the U.S. DOE Hydrogen Program Review*. Miami, FL, p. 125, May **1996**.

10. HICKMAN, D.A. and SCHMIDT, L.D. Production of syngas by direct catalytic oxidation of methane. *Science*, **1992**, 259, 343.

11. HICKMAN, D.A. and SCHMIDT, L.D. Synthesis gas formation by direct oxidation of methane over Pt monoliths. *Journal of Catalysis*, **1992**, 138, 267.

12. LYUBOVSKY, M., ROYCHOUDHURY, S., and LAPIERRE, R. Catalytic partial oxidation of methane to syngas at elevated pressures. *Catalysis Letters*, **2005**, 99, 113.

13. LIU, K., DELUGA, G.A., MOOREFIELD, C., WEI, W., HAYNES, J., and GILLETTE, G. Integrated short contact time hydrogen generator. *DOE Quarterly Progress Report, DE-FG36-05GO15023, Q4*, **2006**.

14. HORN, R., WILLIAMS, K.A., DEGENSTEIN, N.J., and SCHMIDT, L.D. Syngas by catalytic partial oxidation of methane on rhodium: Mechanistic conclusions from spatially resolved measurements and numerical simulations. *Journal of Catalysis*, **2006**, 242, 92.

15. LYUBOVSKY, M., KARIM, H., MENACHERRY, P., BOORSE, S., LAPIERRE, R., PFEFFERLE, W.C., and ROYCHOUDHURY, S. Complete and partial catalytic oxidation of methane over substrates with enhanced transport properties. *Catalysis Today*, **2003**, 83, 183.

16. EDWARDS, N., ELLIS, S.R., FROST, J.C., GOLUNSKI, S.E., VAN KEULEN, A.N.J., LINDEWALD, N.G., and REINKINGH, J.G. On-board hydrogen generation for transport applications: The hot spot methanol processor. *Journal of Power Sources*, **1998**, 71, 123.

17. CHALK, S.G., MILLER, J.F., and WAGNER, F.W. Challenges for fuel cells in transport applications. *Journal of Power Sources*, **2000**, 86, 40.

18. CLARIDGE, J.B., GREEN, M.L.H., TSANG, S.C., YORK, A.P.E., ASHCROFT, A.T., and BATTLE, P.D. A study of carbon deposition during partial oxidation of methane to synthesis gas. *Catalysis Letters*, **1993**, 22, 299.

19. BURKE, N. and DAVID, T. Coke formation during high pressure catalytic partial oxidation of methane to syngas. *Reaction Kinetics and Catalysis Letters*, **2005**, 84, 137.

20. O'CONNOR, R.P., KLEIN, E.J., HENNING, D., and SCHMIDT, L.D. Tuning millisecond chemical reactors for the catalytic partial oxidation of cyclohexane. *Applied Catalysis. A, General*, **2002**, 238, 29.

21. SALGE, J.R., DELUGA, G.A., and SCHMIDT, L.D. Catalytic partial oxidation of ethanol over noble metal catalyst. *Journal of Catalysis*, **2005**, 235, 69.

22. PRETTRE, M., EICHNER, C., and PERRIN, M. The catalytic oxidation of methane to carbon monoxide and hydrogen. *Transactions of the Faraday Society*, **1946**, 42, 335.

23. DISSANAYAKE, D., ROSYNEK, M.P., KHARAS, K.C.C., and LUNSFORD, J.H. Partial oxidation of methane to carbon monoxide and hydrogen over a nickel/alumina catalyst. *Journal of Catalysis*, **1991**, 132, 117.

24. ROSTRUP-NIELSEN, J.R. and BAK-HANSEN, J.H. Carbon dioxide reforming of methane over transition metals. *Journal of Catalysis*, **1993**, 144, 38.

25. NAKAGAWA, K., IKENAGA, N., TENG, Y., KOBAYASHI, T., and SUZUKI, T. Partial oxidation of methane to synthesis gas over iridium-nickel bimetallic catalysts. *Applied Catalysis. A, General*, **1999**, 180, 183.

26. ZHU, T. and FLYTZANI-STEPHANOPOULOS, M. Catalytic partial oxidation of methane to synthesis gas over Ni-CeO$_2$. *Applied Catalysis. A, General*, **2001**, 208, 403.

27. WANG, H.-T., LI, Z.-H., and TIAN, S.-X. Effect of promoters on the catalytic performance of Ni/Al$_2$O$_3$ catalyst for partial oxidation of methane to syngas. *Reaction Kinetics and Catalysis Letters*, **2004**, 83, 245.

28. CHU, Y., LI, S., LIN, J., GU, J., and YANG, Y. Partial oxidation of methane to carbon monoxide and hydrogen over NiO/La$_2$O$_3$ and γ-Al$_2$O$_3$ catalysts. *Applied Catalysis. A, General*, **1996**, 134, 67.

29. CHEN, L., LU, Y., HONG, Q., LIN, J., and DAUTZENBERG, F.M. Catalytic partial oxidation of methane to syngas over Ca-decorated-Al$_2$O$_3$-supported Ni and NiB catalysts. *Applied Catalysis. A, General*, **2005**, 292, 295.

30. ZHANG, Y., XIONG, G., SHENG, S., and YANG, W. Deactivation studies over NiO/γ-Al$_2$O$_3$ catalysts for partial oxidation of methane to syngas. *Catalysis Today*, **2000**, 63, 517.

31. XU, S. and WANG, X. Highly active and coking resistant Ni/CeO$_2$-ZrO$_2$ catalyst for partial oxidation of methane. *Fuel Processing Technology*, **2005**, 84, 563.

32. KIM, P., KIM, Y., KIM, H., SONG, I.K., and YI, J. Synthesis and characterization of mesoporous alumina with nickel incorporated for use in the partial oxidation of methane into synthesis gas. *Applied Catalysis. A, General*, **2004**, 272, 157.

33. SUN, W.-Z., JIN, G.-Q., and GUO, X.-Y. Partial oxidation of methane to syngas over Ni/SiC catalysts. *Catalysis Communications*, **2005**, 6, 135.

34. GOULA, M.A., LEMONIDOU, A.A., GRUNERT, W., and BAERNS, M. Methane partial to synthesis gas using nickel on calcium aluminate catalysts. *Catalysis Today*, **1996**, 32, 149.

35. YANG, S., KONDO, J.N., HAYASHI, K., HIRANO, M., DOMEN, K., and HOSONO, H. Partial oxidation of methane to syngas over promoted C12A7. *Applied Catalysis. A, General*, **2004**, 277, 239.

36. CHU, W., YANG, W., and LIN, L. Selective oxidation of methane to syngas over NiO/barium hexaaluminate. *Catalysis Letters*, **2001**, 74, 139.

37. CHU, W., YANG, W., and LIN, L. The partial oxidation of methane to syngas over the nickel-modified hexaaluminate catalysts BaNi$_y$Al$_{12-y}$O$_{19-\delta}$. *Applied Catalysis. A, General*, **2002**, 235, 39.

38. GOLDWASSER, M.R., RIVAS, M.E., PIETRI, E., PEREZ-ZURITA, M.J., CUBEIRO, M.L., GRIVOBAL-CONSTANT, A., and LECLERCQ, G. Perovskites as catalysts precursors: Synthesis and characterization. *Journal of Molecular Catalysis. A, Chemical*, **2005**, 228, 325.

39. GUO, C., ZHANG, J., LI, W., ZHANG, P., and WANG, Y. Partial oxidation of methane to syngas over BaTi$_{1-x}$Ni$_x$O$_3$ catalysts. *Catalysis Today*, **2004**, 98, 583.

40. ROSTRUP-NIELSEN, J.R., SEHESTED, J., and NØRSKOV, J.K. Hydrogen and syngas by steam reforming. *Advances in Catalysis*, **2002**, 47, 65.

41. ASHCROFT, A.T., CHEETHOM, A.K., FOORD, J.S., GREEN, M.L.H., GREY, C.P., MURREL, C.P., and VERNON, P.D.F. Selective oxidation of methane to syntheis gas using transition metal catalysts. *Nature*, **1990**, 344, 319.

42. TORNIAINEN, P.M., CHU, X., and SCHMIDT, L.D. Comparison of monolith-supported metals for the direct oxidation of methane to syngas. *Journal of Catalysis*, **1994**, 146, 1.

43. RABE, S., TRUONG, T.-B., and VOGEL, F. Low temperature catalytic partial oxidation of methane for gas-to-liquids applications. *Applied Catalysis. A, General*, **2005**, 292, 177.

44. KIKUCHI, R., IWASA, Y., TAKEGUCHI, T., and EGUCHI, K. Partial oxidation of CH$_4$ and C$_3$H$_8$ over hexaaluminate-type oxides. *Applied Catalysis. A, General*, **2005**, 281 (1–2), 61.

45. PERKAS, N., ZHONG, Z., CHEN, L., BESSON, M., and GEDANKEN, A. Sonochemically prepared high dispersed Ru/TiO$_2$ mesoporous catalyst for partial oxidation of methane to syngas. *Catalysis Letters*, **2005**, 103, 9.

46. BHARADWAJ, S.S., YOKOYAMA, C., and SCHMIDT, L.D. Catalytic partial oxidation of alkanes on silver in fluidized bed and monolith reactors. *Applied Catalysis. A, General*, **1996**, 140, 73.

47. CLARIDGE, J.B., GREEN, M.L.H., and TSANG, S.C. Methane conversion to synthesis gas by partial oxidation and dry reforming over rhenium catalysts. *Catalysis Today*, **1994**, 21, 455.

48. SANTOS, A.C.S.F., DAMYANOVA, S., TEIXEIRA, G.N.R., MATTOS, L.V., NORONHA, F.B., PASSOS, F.B., and BUENO, J.M.C. The effect of ceria content on the performance of Pt/CeO₂/Al₂O₃ catalysts in the partial oxidation of methane. *Applied Catalysis. A, General*, **2005**, 290, 123.

49. ALBERTAZZI, S., ARPENTINIER, P., BASILE, F., Del GALLO, P., FORNASARI, G., GARY, D., and VACCARI, A. Deactivation of a Pt/γ-Al₂O₃ catalyst in the partial oxidation of methane to synthesis gas. *Applied Catalysis. A, General*, **2003**, 247, 1.

50. SILBEROVA, B., VENVIK, H.J., WALMSLEY, J.C., and HOLMEN, A. Small-scale hydrogen production from propane. *Catalysis Today*, **2005**, 100, 457.

51. REYES, S.C., SINFELT, J.H., and FEELEY, J.S. Evolution of processes for synthesis gas production: Recent developments in an old technology. *Industrial & Engineering Chemistry Research*, **2003**, 42, 1588.

52. OTSUKA, K., WANG, Y., SUNADA, E., and YAMANAKA, I. Direct partial oxidation of methane to synthesis gas by cerium oxide. *Journal of Catalysis*, **1998**, 175, 152.

53. ARPENTINIER, P., BASILE, F., DEL GALLO, P., FORNASARI, G., GARY, D., ROSETTI, V., and VACCARI, A. Role of the hydrotalcite-type precursor on the properties of CPO catalysts. *Catalysis Today*, **2005**, 9, 99.

54. RUCKENSTEIN, E. and WANG, Y.H. Partial oxidation of methane to synthesis gas over MgO-supported Rh catalysts: The effect of precursor of MgO. *Applied Catalysis. A, General*, **2000**, 198, 33.

55. BRUNO, T., BERETTA, A., GROPPI, G., RODERI, M., and FORZATTI, P. A study of methane partial oxidation in annular reactor: Activity of Rh/α-Al₂O₃ and Rh/ZrO₂ catalysts. *Catalysis Today*, **2005**, 99, 89.

56. ERIKSSON, S., NILSSON, M., BOUTONNET, M., and JÄRÅS, S. Partial oxidation of methane over rhodium catalysts for power generation applications. *Catalysis Today*, **2005**, 100, 447.

57. HOHN, K.L. and SCHMIDT, L.D. Partial oxidation of methane to syngas at high space velocities over Rh-coated spheres. *Applied Catalysis. A, General*, **2001**, 211, 53.

58. LECLERC, C.A., REDENIUS, J.M., and SCHMIDT, L.D. Fast light off of millisecond reactors. *Catalysis Letters*, **2002**, 79, 39.

59. LIU, K., HONG, J.K., WEI, W., ZHANG, L., CUI, Z., SINGH, S., and SUBRAMANIAN, R. Integrated short contact time hydrogen generator. *DOE Final Report DE-FG36-05GO15023*, **2009**.

60. BITSCH-LARSEN, A., DEGENSTEIN, N.J., and SCHMIDT, L.D. Effect of sulfur in catalytic partial oxidation of methane over Rh-Ce coated foam monoliths. *Applied Catalysis. B, Environmental*, **2008**, 78 (3–4), 364.

61. DE GROOTE, A.M. and FROMENT, G.F. Synthesis gas production from natural gas in a fixed-bed reactor with reversed flow. *Canadian Journal of Chemical Engineering*, **1996**, 74, 735.

62. DE GROOTE, A.M. and FROMENT, G.F. The role of coke formation in catalytic partial oxidation for synthesis gas production. *Catalysis Today*, **1997**, 37, 309.

63. YORK, A.P.E., XIAO, T., and GREEN, M.L.H. Brief overview of the partial oxidation of methane to synthesis gas. *Topics in Catalysis*, **2003**, 22 (3), 345.

64. HORN, R., WILLIAMS, K.A., DEGENSTEIN, N.J., BITSCH-LARSEN, A., DALLE NOGARE, D., TUPY, S.A., and SCHMIDT, L.D. Methane catalytic partial oxidation on autothermal Rh and Pt foam catalysts: Oxidation and reforming zones, transport effects, and approach to thermodynamic equilibrium. *Journal of Catalysis*, **2007**, 249 (2), 380.

65. HICKMAN, D.A. and SCHMIDT, L.D. Reactors, kinetics and catalysis—Steps in CH₄ oxidation on Pt and Rh surfaces: High-temperature reactor simulations. *AIChE Journal*, **1993**, 39 (7), 1164.

66. DEUTSCHMANN, O. and SCHMIDT, L.D. Two-dimensional modeling of partial oxidation of methane on rhodium in a short contact time reactor. *Proceedings of the International Symposium on Combustion*, **1998**, 2, 2283.

67. DEUTSCHMANN, O. and SCHMIDT, L.D. Modeling the partial oxidation of methane in a short-contact-time reactor. *AIChE Journal*, **1998**, 44 (11), 2465.

68. DEUTSCHMANN, O., SCHWIEDERNOCH, R., MAIER, L.I., and CHATTERJEE, D. Natural gas conversion in monolithic catalysts: Interaction of chemical reactions and transport phenomena. *Symposium on Natural Gas Conversion VI. Studies in Surface Science and Catalysis*, **2001**, 136, 251.

69. BIZZI, M., BASINI, L., SARACCO, G., and SPECCHIA, V. Short contact time catalytic partial oxidation of methane: Analysis of transport phenomena effects. *Chemical Engineering Journal*, **2002**, 90, 97.

70. BIZZI, M., BASINI, L., SARACCO, G., and SPECCHIA, V. Modeling a transport phenomena limited reactivity in short contact time catalytic partial oxidation reactors. *Industrial & Engineering Chemistry Research*, **2003**, 42, 62.

71. BIZZI, M., SARACCO, G., SCHWIEDERNOCH, R., and DEUTSCHMANN, O. Modeling the partial oxidation of methane in a fixed bed with detailed chemistry. *AIChE Journal*, **2004**, 50, 1289.

72. BIESHEUVEL, P.M. and KRAMER, G.J. Two-section reactor model for autothermal reforming of methane to synthesis gas. *AIChE Journal*, **2003**, 49, 1827.

73. REITZ, T.L., AHMED, S., KRUMPELT, M., KUMAR, R., and KUNG, H.H. Characterization of CuO/ZnO under oxidizing conditions for the oxidative methanol reforming reaction. *Journal of Molecular Catalysis A: Chemical*, **2000**, 162, 275.

74. DUANE, A.G. and SCHMIDT, L.D. Microsecond catalytic partial oxidation of alkanes. *Science*, **1996**, 271, 1560.

75. VELU, S. and SONG, C. Advances in catalysis and processes for hydrogen production from ethanol reforming. *Catalysis*, **2007**, 20, 65.

76. DELUGA, G.A., SALGE, J.R., SCHMIDT, L.D., and VERYKIOS, X.E. Renewable hydrogen from ethanol by autothermal reforming. *Science*, **2004**, 303, 993.

Chapter 4

Coal Gasification

KE LIU,[1] ZHE CUI,[1] AND THOMAS H. FLETCHER[2]

[1]*Energy & Propulsion Technologies, GE Global Research Center*
[2]*Department of Chemical Engineering, Brigham Young University*

4.1 INTRODUCTION TO GASIFICATION

Coal gasification is a well-established technology to convert coal partially or completely to syngas, which generally consists of CO, H_2, CO_2, CH_4, and impurities such as H_2S and NH_3. Coal gasifiers use either O_2 or air to provide heat by combustion of some of the coal, and steam or CO_2 is added for gasification reactions. After purification, syngas is used in one of three ways: (a) combustion in a gas turbine (GT) to produce electricity; (b) raw material for chemical syntheses, such as ammonia, Fischer–Tropsch for liquid fuel, and methanol production; or (c) methanation for synthetic natural gas production. In certain gasification processes, water is used to make a coal–water slurry (CWS) that can be pumped into a high-pressure gasifier. In this case, the water in the slurry acts as a gasifying agent.

Gasification is a booming activity all over the world. Currently, 128 gasification plants are in operation, with a total of 366 gasifiers producing about 42,700 MW of syngas.[1] Also, gasification projects under development or construction will provide another 24,500 MW of syngas. In the current syngas capacity listed above, about 27,000 MW are based on coal gasification.[2]

The term coal gasification excludes coal combustion, because the product flue gases from combustion are CO_2 and H_2O, which have no residual heating value. Gasification includes the technologies of pyrolysis, partial oxidation, and hydrogenation.[3] In early years, pyrolysis was the dominant technology. However, in recent years, gasification technology dominates.

Coal gasification has a wide range of applications as shown schematically in Figure 4.1. Coal is shown here as a feedstock, but the technology is applicable to

Hydrogen and Syngas Production and Purification Technologies, Edited by Ke Liu, Chunshan Song and Velu Subramani
Copyright © 2010 American Institute of Chemical Engineers

156

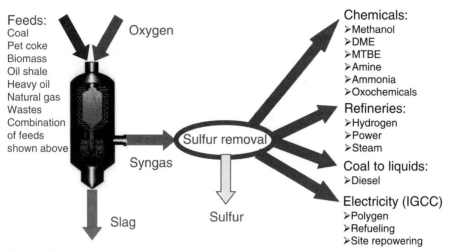

Figure 4.1.   Applications of coal gasification. MTBE, methyl tertiary butyl ether. See color insert.

any hydrocarbon feedstock, including natural gas, biomass, and heavy refinery residues (including petroleum coke). Hydrogasification and catalytic gasification processes can produce large quantities of synthetic natural gas from coal and yield clean-burning fuel. Steam-oxygen and steam-air gasification processes produce a carbon monoxide- and hydrogen-rich syngas. Gas produced in coal gasification can be used as fuel for power generation (Integrated Gasification Combined Cycle [IGCC]), for synthesis, as a source of hydrogen for the manufacture of ammonia or hydrogenation applications in refineries, or for the production of liquid fuels. Many gasification technologies have been developed by petroleum companies for coal-to-chemical applications in the last several decades.[4] For methane production, an H_2/CO ratio of 3:1 is desired, while for Fischer–Tropsch gasoline production, an H_2/CO ratio of 2:1 is desired. To achieve the desired product H_2/CO ratio, an additional process called the "water-gas shift" is sometimes required, which uses steam to convert CO to CO_2 and H_2.

One of the main current interests in coal gasification is the IGCC process for electricity generation. The typical IGCC process is briefly shown in Figure 4.2. Fuel gas is produced from a gasification process and is cleaned and burned with compressed air in a GT to produce high-pressure hot gas. The high-pressure hot gas is expanded in the GT and generates electrical power. The clean low-pressure hot flue gas is subsequently used to raise steam in a boiler. The steam is expanded through a steam turbine and generates more electrical power. For power generation, IGCC is recognized as a cleaner and more efficient method compared with traditional power generation via coal combustion. Therefore, electrical power generation through coal gasification could be an important aspect of coal utilization in the future.

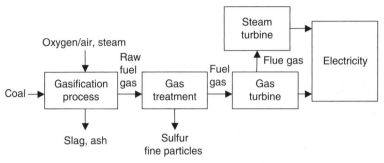

Figure 4.2. Typical plant components of an IGCC system.

4.2 COAL GASIFICATION HISTORY

Although coal has been used as a fuel for a long time, converting coal to combustible gas was not practiced commercially until the end of the 18th century. In 1792, the Scottish engineer William Murdock heated coal in a retort in the absence of air. He partially converted coal to gas, with a residue of coke. In 1812, gas production by pyrolysis finally became a commercial process with the foundation of the London Gas, Light and Coke Company. In 1816, the first coal gasification company in the U.S., Baltimore Gas Company, was established based on the same technology. The process of heating coal to produce coke and gas is still used in the metallurgical industry.

In the 1850s, the gas producer was invented, and the water-gas process was discovered in Europe. In the gas producer, coal and coke were completely converted to gas by reacting coal with air and steam continuously in a downward moving bed at atmospheric pressure. The gas obtained in this method, called producer gas, had a low heating value ($3500–6000\,kJ/m^3$).

To enhance the heating value of the gas product, a cyclic steam-air process was developed in 1873. This process produced water gas, composed chiefly of carbon monoxide and hydrogen, which had a higher heat value ($12,000–13,000\,kJ/m^3$) compared with the producer gas. Furthermore, by adding oil to the reactor, the heating value was enhanced (to $19,000–20,500\,kJ/m^3$). This type of fuel gas, carbureted water gas, became the standard for gas distributed to residences and industry in the U.S. until the 1940s.

In the 1920s, Carl von Linde commercialized the cryogenic separation of air. The availability of pure O_2 led to the development of some of the important fully continuous oxygen-based processes for the production of syngas in coal gasification industry. The forerunners of many modern gasification units were soon initiated in the U.S. and other countries, including the Winkler fluidized-bed process (1926), the Lurgi moving-bed pressurized gasification process (1931), and the Koppers–Totzek entrained-flow process (1940s). In 1950, South Africa Synthetic Oil Limited (SASOL) was formed. Coal gasification and Fischer–Tropsch synthesis are still used

Table 4.1.   Composition of Coal Syngas from Different Processes

Composition (mol %)	Oven Gas	Producer Gas	Water Gas	Carbureted Water Gas	Synthetic Coal Gas
CO	6.9	27.1	42.8	33.4	15.8
H_2	47.3	13.9	49.9	34.6	40.6
CH_4	33.8	3.0	0.5	10.4	10.9
CO_2	2.2	4.5	3.0	3.9	31.3
N_2	6.0	50.9	3.3	7.9	N/A

N/A, no significant concentration of N_2 is observed.

Table 4.2.   Typical Composition of Gasification Products from Modern Gasifiers

Composition (mol %)	BGL (Fixed Dry)	Shell (Dry Feeding)	Texaco (Slurry)
CO	57	65	49
H_2	28	29	34
CH_4	6	<0.1	0.2
CO_2	4	2	10
N_2	2	2	1

in this plant. With the extensions made in the late 1970s, SASOL is the largest coal gasification center in the world.

The composition of the syngas generated through different gasification methods can be totally different. Also, the gas products depend on the coal used, the oxygen purity, and the gasification conditions (e.g., temperature, pressure, reactants, and coal/oxygen ratio). Compositions of some typical coal gas products from different processes are listed in Tables 4.1 and 4.2.

After the oil crises in the 1970s, a renewed interest in coal gasification was triggered as a means of replacing or supplementing oil or natural gas. Much of the effort focused on coal hydrogasification. In hydrogasification, coal is hydrogenated directly to methane as a substitute natural gas (SNG). However, due to the high pressure required, few hydropressure processes were commercially successful.

The older processes were also further developed in the last several decades. A slagging version of the long-established Lurgi dry ash gasifier was developed and demonstrated by the British Gas PLC (BGL) in collaboration with Lurgi Energies und Umwet GmbH in the early 1990s. Koppers and Shell produced a pressurized version of the Koppers–Totzek gasifier jointly. Texaco developed a coal slurry-fed gasification process, and Rheinbraun developed the high-temperature Winkler (HTW) fluidized-bed process. Also, many other processes have been developed to

Table 4.3. Different Gasification Processes

Gasifier Type	Coal Feed	Ash Handling	Gasification Technology
Moving bed	Dry	Dry	Lurgi
		Slag	BGL
Fluidized bed	Dry	Agglomerating	Kellogg–Rust–Westinghouse
			Utility gas
		Nonagglomerating	High-temperature Winkler
Entrained-flow reactor	Slurry	Slag	Texaco (now General Electric)
			Dow (Destec)
	Dry		Shell
			Koppers–Totzek
			Pressurized entrained flow (PRENFLO)

a pilot stage but not fully commercialized, such as the Hygas process, the Cogas process, and the U-gas process.[5] Table 4.3 shows the different types of gasifiers currently being used worldwide.

4.3 COAL GASIFICATION CHEMISTRY

Coal gasification is a complex chemical process. During coal gasification, both homogeneous and heterogeneous reactions occur. Homogeneous reactions in the gas phase can be described by relatively simple equations. Many of the gas-phase reactions achieve chemical equilibrium at the temperatures and pressures used in modern gasifiers. However, the heterogeneous reactions between gas and solid phases are much more complicated, due to heat and mass transfer effects. The kinetics of coal gasification have been extensively studied over the past few decades. However, the understanding of gasification kinetics is not yet as clear as that of gasification thermodynamics.

Coal gasification is actually comprised of several processes: (a) evaporation of moisture; (b) coal pyrolysis, releasing volatile matter (tar, CO, CH_4, H_2, CO_2, etc.); (c) reaction of volatiles in the gas phase; (d) heterogeneous reaction of char with gas-phase species (such as H_2O and CO_2); and (e) mineral matter release and transformation. Coal moisture is rank dependent; low-rank coals such as lignites and subbituminous coals may have up to 35% moisture by weight, whereas bituminous coals generally contain less than 5% moisture by weight. The heat to evaporate this moisture must come from the combustion of volatiles and reduces the efficiency of the process. However, the evaporated moisture acts as a char gasification agent later in the process, decreasing the steam requirement. Moisture evaporation occurs in the early stages of coal heating, at temperatures from 100 to 150 °C. The weight percent ash in the coal varies depending on the quality of the coal, and the composition of the ash also varies. The ash composition greatly affects the slagging behavior of deposits in the gasifier. The processes of coal pyrolysis, volatile combustion, heterogeneous gasification, and ash–slag transformations are discussed in the following sections.

4.3.1 Pyrolysis Process

Pyrolysis is the process of chemical decomposition of organic materials by heating to an appropriate temperature in the absence of oxygen. The words pyrolysis and devolatilization are often used interchangeably.

Coal can be thought of as a polymer network of aromatic clusters connected by aliphatic bridges, as illustrated in Figure 4.3. Side chains of different functional groups may also be attached to the clusters, such as methyl, ethyl, or carboxylic groups. As the coal particle temperature increases, the bridges between clusters can rupture, eventually breaking up the coal polymer into smaller fragments. As these fragments become smaller, their vapor pressure becomes high enough to cause the fragments to evaporate as coal tar. Coal tar is defined as volatile matter that con-

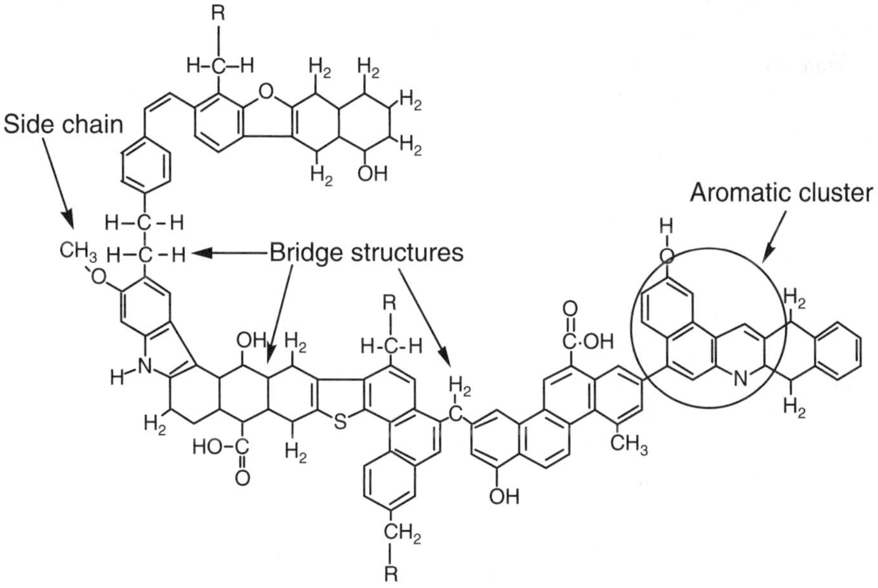

Figure 4.3. Hypothetical coal structure model, illustrating aromatic clusters, bridge structures, and side chains (adapted from Perry[6]).

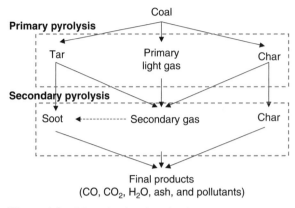

Figure 4.4.   Schematic of coal combustion process.

denses at room temperature and pressure. In addition, at increased temperatures, the side chains may also detach from the polymer or the tar to form light hydrocarbon gases. A large amount of cross-linking also occurs in the partially reacted coal during pyrolysis, resulting in a stable char. A schematic of the pyrolysis process is shown in Figure 4.4.

The pyrolysis process is dependent on the properties of the coal and the operating conditions (temperature, pressure, heating rate, particle size, and residence time). Volatile yields as high as 70% (on a dry ash-free basis) have been achieved for some coals, although 40%–50% is more typical of many industrially relevant coals. High-rank coals, such as low-volatile bituminous or anthracite coals, have much lower volatile yields. The pyrolysis conditions also greatly affect the structure and the composition of the char obtained,[7] which is the starting point for heterogeneous gasification reactions.

The temperature at which pyrolysis reactions occur is very dependent on the heating rate: as heating rate increases, the temperature at which pyrolysis reactions occur increases. For example, the temperature where the main volatile release occurs was shown to increase from 400 to 525 °C when the heating rate was increased from 1K/s to 1000K/s. The volatile yield also increases slightly for most coals as heating rate increases. At particle heating rates typical of entrained flow gasification (~10^6K/s), the main volatile release occurs at approximately 700 °C.

The average molecular weight of coal tar from atmospheric pyrolysis is about 350 amu. The percentage of volatiles that becomes tar is dependent on coal rank; lignites and subbituminous coals generally have much lower fractions of tar than high-volatile bituminous coals. At elevated pressure, the higher molecular weight compounds do not have enough vapor pressure to evaporate, hence the tar yield decreases and the average molecular weight also decreases. Hence, since low-rank coals do not have high tar yields, their total volatile yields do not change much as a function of pressure. For example, Anthony and Howard[8] showed that the volatile

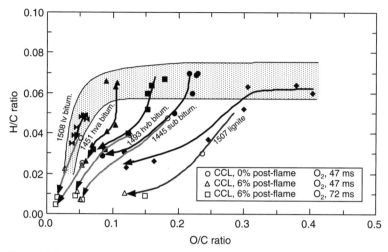

Figure 4.5. Coalification diagram showing the elemental compositions of coals of different ranks versus the mass ratio of O/C and H/C, along with compositions of chars from those coals. The shaded region represents the typical range of unreacted coals. The dark solid lines are reaction pathways from a drop tube reactor, and the gray lines are from a flat-flame burner system at higher temperatures and heating rates (from Fletcher[14]). CCL, carbon per cluster.

yield from a high-volatile bituminous coal changes from 54% (as received) to 37% as the pressure was changed from 0.001 to 69 atm at a heating rate of 1000K/s in He, while Suuberg[9] showed that volatile yields from a lignite only changed from 44% to 40% under the same conditions. More recent data on the effects of pressure on coal pyrolysis are available.[7,10,11]

As the temperature increases, in fuel-rich environments, tar forms soot particles. Increased pressure enhances the formation of soot. The soot acts as a carrier for the nitrogen that was in the coal[12] and as a heat transfer agent by radiating heat away from the flame.[13]

The elemental compositions of different coals are often shown on a coalification diagram, similar to that shown in the shaded region of Figure 4.5.[14] Also shown in Figure 4.5 is the pathway of elemental composition during pyrolysis in N_2 at 1200K and combustion in post-flame gases in a flat-flame burner at about 1500K. Note that the resulting chars are not pure carbon but still retain some oxygen and hydrogen.

4.3.2 Combustion of Volatiles

When released from the coal in the pyrolysis process, volatiles (mainly CO, CH_4, H_2, hydrocarbon liquids, and tars) combust in the gas phase with the oxidant surrounding the coal particles. These reactions are very exothermic, raising the temperature of the product gases significantly. However, in a gasification environment, especially when the coal has a high volatile content, volatile combustion may be far from complete due to the low oxygen/coal ratio. This may lead to extra soot formation. Recirculation of syngas is observed in many gasification reactors. Recirculated combustion flue gas consists mainly of carbon dioxide, nitrogen (in air-blown gasification), and water vapor. The temperature is then reduced due to the moderating effect of the carbon dioxide and water vapor. In some oxygen-blown gasification cases, the recycled gas can contain significant quantities of carbon monoxide and hydrogen. When these gases come into contact with oxygen, the local temperature will increase significantly.

Volatile combustion is a gas-phase homogeneous reaction, so there is no mass transport limitation between phases. There is, however, a mass transfer resistance due to gas-phase mixing of combustible gases and oxidizers. In general, volatile combustion is much faster than the pyrolysis process, which in turn is much faster than the char gasification process.

4.3.3   Char Gasification Reactions

Coal char gasification is a complicated physicochemical process. Char gasification is slow compared with other reaction processes in the gasifier (i.e., pyrolysis, volatile combustion, or even char oxidation); therefore char gasification is the rate-controlling process in the gasifier. In developing char gasification rates and mechanisms, coal is always assumed to be pure carbon and reacts with other reactant components in the gasifier. Also, most gasification processes take place at high temperatures in the range of 800–1800 °C. The gas-phase reaction rates are sufficiently high in the upper regions of this temperature range, so that the main gaseous components achieve thermodynamic equilibrium. Using equilibrium gas-phase compositions instead of solving detailed kinetic equations gives reasonably close results in most commercial reactor designs.[3,15]

The main reactions that take place in a gasification system are discussed below, emphasizing the heterogeneous gas–char reactions.

4.3.3.1   Reactions between Carbon and Oxygen

The main reactions of carbon and O_2 in gasifiers are

$$C(s) + O_2 \rightarrow CO_2, \tag{R-4.1}$$
$$2C(s) + O_2 \rightarrow 2CO, \tag{R-4.2}$$
$$2CO + O_2 \rightarrow 2CO_2. \tag{R-4.3}$$

All of these reactions are highly exothermic, providing heat for the subsequent endothermic gasification reactions. Because gasifiers operate under fuel-rich conditions, the O_2 is completely consumed before these particular reactions have consumed all of the carbon. Reaction R-4.3 achieves thermodynamic equilibrium quickly enough at the high temperature conditions in entrained flow gasifiers so that chemical equilibrium can be assumed.

4.3.3.2   Reactions between Carbon and CO_2

The reaction of carbon and CO_2 in gasifiers is

$$C(s) + CO_2 \leftrightarrow 2CO. \tag{R-4.4}$$

This reaction is called the reverse Boudouard reaction and is an endothermic reaction. The reaction rate of the reverse Boudouard reaction is several orders of magnitude lower than reaction rates of R-4.1 and R-4.2 (at the same temperature in the absence of a catalyst). Also, the reverse Boudouard reaction is inhibited by the presence of products, namely carbon monoxide.

4.3.3.3   Reactions between Carbon and Steam

The reactions of carbon and H_2O in gasifiers are

$$C(s) + H_2O \leftrightarrow CO + H_2, \tag{R-4.5}$$
$$C(s) + 2H_2O \leftrightarrow CO_2 + 2H_2, \tag{R-4.6}$$
$$CO + H_2O \leftrightarrow CO_2 + H_2. \tag{R-4.7}$$

Reactions R-4.5 and R-4.6 are main reactions to generate H_2 and CO. Both reactions are endothermic and have high activation energies. This means that they are slow compared with reactions R-4.1 and R-4.2, and slow down as they proceed since the temperature decreases with extent of reaction. The reaction rates are proportional to the partial pressure of the steam in the system.

Reaction R-4.7, the water-gas shift reaction, is an exothermic reaction. The water-gas shift reaction has influence on the CO/H_2 ratio in the gasification product, which is very important when the gas is used for synthesis purpose. Therefore, the shift process can be found in almost all the ammonia plants and hydrogen generation process in gas plants. The shift reaction can generally be taken into account using thermodynamic chemical equilibrium, since gas-phase temperatures are high.

4.3.3.4   Methanation Reactions

The reaction pathways to form CH_4 from carbon in gasifiers are

$$C(s) + 2H_2 \leftrightarrow CH_4, \tag{R-4.8}$$
$$CO + 3H_2 \leftrightarrow CH_4 + H_2O, \tag{R-4.9}$$
$$2CO + 2H_2 \leftrightarrow CH_4 + CO_2, \tag{R-4.10}$$
$$CO_2 + 4H_2 \leftrightarrow CH_4 + 2H_2O. \tag{R-4.11}$$

These reactions increase the heating value of the gas product, since methane has a high heat of combustion. However, these reactions are very slow except under high pressure and in the presence of a catalyst. Another source of the methane in the syngas is the pyrolysis process. Reaction R-4.11 is the reverse steam methane reforming reaction. All reactions that produce methane are exothermic reactions.

Due to the high temperatures (800–1800 °C), no hydrocarbons other than methane are present in any appreciable quantity in the outlet of gasifiers.

4.3.3.5 Other Reactions in Gasifiers

At high temperatures, reactions between nitrogen, sulfur, and reactant gas (O_2, H_2O, H_2, and some components in product, e.g., CO, CO_2) may occur, some of which are shown below:

$$S + O_2 \rightarrow SO_2, \tag{R-4.13}$$
$$SO_2 + 3H_2 \leftrightarrow H_2S + 2H_2O, \tag{R-4.14}$$
$$SO_2 + 2CO \leftrightarrow S + 2CO_2, \tag{R-4.15}$$
$$2H_2S + SO_2 \rightarrow 3S + 2H_2O, \tag{R-4.16}$$
$$C(s) + 2S \leftrightarrow CS_2, \tag{R-4.17}$$
$$CO + S \leftrightarrow COS, \tag{R-4.18}$$
$$N_2 + 3H_2 \leftrightarrow 2NH_3, \tag{R-4.19}$$
$$2N_2 + 2H_2O + 4CO \leftrightarrow 4HCN + 3O_2, \tag{R-4.20}$$
$$N_2 + xO_2 \leftrightarrow 2NO_x. \tag{R-4.21}$$

The quantities of these products from the impurities have a negligible effect on the main syngas components of hydrogen and carbon monoxide. However, it is necessary to consider these reactions in the design of gasification systems because the products of these reactions have a significant effect on the resulting compounds downstream of the gas production, for example, environmental emissions and catalyst poisons. Studies on these reactions indicate that the main sulfur compound is H_2S and the main nitrogen compound is NH_3, as predicted by equilibrium. However, equilibrium considerations are not always adequate to predict the exact quantities of these pollutant species (especially NH_3), hence reaction modeling becomes necessary.

Heats of reaction (ΔH_r) of some gasification reactions are listed in Table 4.4. Some of the reactions are endothermic (i.e., require heat in order to continuously proceed), as indicated by the positive values of ΔH_r in Table 4.4. The heat required for these endothermic reactions is supplied by the complete or partial oxidation of a small proportion of the coal volatiles and/or char in O_2 or in air.

4.3.4 Ash–Slag Chemistry

The fate of the inorganic matter in gasifiers is perhaps the dominant design variable and greatest challenge for continuous operation. Because the minerals have accumulated in the coal seams over geologic timescales, most elements known to man occur to some extent in coal. Typical coals of industrial interest contain less than 15% mineral matter. The composition of the minerals in a given coal varies with rank and geographic location. Some of the minerals occur as organically associated species, present as salts of organic acid groups and chelates, often associated with

Table 4.4. Reaction Enthalpies of Some Gasification Reactions

T (K)	ΔH_r (kJ/mol)							
	$C + O_2 =$ CO_2	$C + 0.5O_2 =$ CO	$C + CO_2 =$ $2CO$	$C + H_2O =$ $CO + H_2$	$C + 2H_2O =$ $CO_2 + 2H_2$	$CO + H_2O =$ $CO_2 + H_2$	$C + 2H_2 =$ CH_4	$CO + 3H_2 =$ $CH_4 + H_2O$
298	−394	−111	173	131	90	−41	−75	−206
500	−394	−110	174	134	94	−40	−81	−215
1000	−395	−112	171	136	101	−35	−90	−226
1100	−395	−113	170	136	102	−34	−91	−227
1200	−395	−113	169	136	103	−33	−91	−227
1300	−396	−114	168	136	104	−32	−92	−228
1400	−396	−115	167	135	104	−31	−92	−228
1500	−396	−115	165	135	105	−30	−93	−228
1600	−396	−116	164	135	105	−29	−93	−227
1700	−397	−117	163	134	106	−29	−93	−227
1800	−397	−118	162	134	106	−28	−93	−226

elements such as Na, Mg, Ca, K, and Sr. Low-rank coals have a greater abundance of these organically associated groups than high-rank coals, due to the higher abundance of organic acid groups in low-rank coals.[16] Discrete mineral grains of inorganic material are also a major component of coals, including quartz, kaolinite, illite, pyrite, and calcite as major species. In pulverized coal, particles that are pure mineral grains are termed "excluded" minerals, while mineral grains contained in an organic particle are termed "included" minerals. In the U.S., eastern province coals are usually high in aluminosilicate minerals (Si, Al, K) and low in Mg, Ca, and Mn.[17] Interior province coals are noted for high contents of S, Fe, Cu, and Zn, and are low in the aluminosilicate minerals. Western province coals are noted for high contents of Ca, Mg, and Sr, depending on rank.

The inorganic matter in the coal transforms both physically and chemically according to the chemical composition of the inorganic matter and to the reaction conditions (temperature, heating rate, and local gas composition). Figure 4.6 shows the physical transformations that can occur during pulverized coal combustion, depending on if the minerals are included or liberated (including excluded minerals). Particle swelling is influenced by coal type as well as heating rate.[18] Fly-ash distributions often have a bimodal size distribution, as indicated in Figure 4.6, with submicron particles formed by condensation of flame-volatilized species and larger particles ($\sim 10 \mu$m) formed from single or combined discrete mineral grains.[19]

Ash composition and size affect the deposition characteristics, along with reactor geometry, gas and wall temperature, and velocity. Fickian diffusion is important for the deposition of vapors and small particles ($<1 \mu$m), thermophoresis is important for the deposition of intermediate-sized particles ($1–10 \mu$m), and inertial

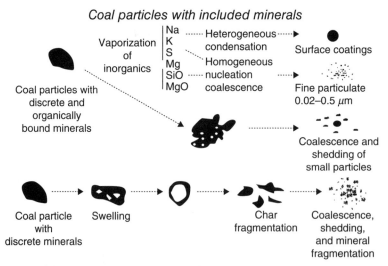

Figure 4.6. Schematic diagram of the transformations of inorganic constituents during coal combustion (from Benson et al.[19]).

impaction is important for larger particles ($>10\,\mu m$). As the deposit forms, the surface of the deposit may become sticky or molten, since the deposit surface is at a higher temperature than the wall–deposit interface. A sticky or molten deposit surface will enhance additional deposition, capturing particles that would have bounced off a nonsticky deposit. Since different sizes of ash particles have different compositions, the composition of the initial deposit differs from the outer layers of the deposit.

The primary bonding or sintering mechanisms for slagging deposits are due to silicate liquid phases.[19] However, other species can act as fluxing agents to increase slag viscosity or to enhance the strength of the deposit (making it hard to remove). In particular, calcium and iron species can often serve as fluxing agents, lowering the critical temperature at which deposits form a molten slag. Some correlations of this critical temperature for pulverized coal boilers are reviewed by Raask.[20] For example, a correlation published by Hoy et al.[21] shows the dependency on several species:

$$T_c\,(\text{in K}) = 3263 - 1470\,(SiO_2/Al_2O_3) + 360\,(SiO_2/Al_2O_3)^2 -$$
$$14.7\,(Fe_2O_3 + CaO + MgO) + 0.15\,(Fe_2O_3 + CaO + MgO)^2 \quad (4.1)$$

and

$$SiO_2 + Al_2O_3 + \text{equiv. } Fe_2O_3 + CaO + MgO = 100. \quad (4.2)$$

This correlation shows quantitatively the effects of the fluxing agents Fe_2O_3, CaO, and MgO. However, the local gas environment in a gasifier has a significant effect on the slagging behavior. Minerals form metal oxides (SiO_2, Al_2O_3, Fe_2O_3, CaO, etc.) in areas of the gasifier that are oxygen rich, depending on the temperature and residence time. These oxides do not form or may be transformed in regions of the gasifier that are fuel rich. Since the oxide forms generally exhibit the highest melting temperatures, the slagging temperatures decrease in fuel-rich regions. In addition, the corrosion potential increases in fuel-rich regions, since the minerals are not in the oxide form and attack refractory and metal tubes. The chemistry and nature of slagging and fouling deposits in boilers are much more quantified than in gasifiers, and more research is needed on mineral transformations in gasifiers.

4.4 GASIFICATION THERMODYNAMICS

Chemical equilibrium dictates the extent of reaction possible for a given temperature and pressure. For single simple reactions, an equilibrium constant approach can be used to determine the equilibrium concentration of gases for a given reaction. At equilibrium, the forward and the reverse reaction rates are equal. The equilibrium constant is calculated from the Gibbs free energy, as follows:

$$\Delta G^\circ_{rxn} = -RT \ln K, \quad (4.3)$$

where the ΔG°_{rxn} is calculated from the Gibbs free energy of formation ($\Delta G^\circ_{f,i}$) at 1 bar pressure for products minus reactants. For example, in a CO shift reaction

(reaction R-4.7), the forward reaction rate r_1 is proportional to the partial pressure of CO and H_2O in the system, or

$$r_1 = k_1 p_{CO} p_{H2O}. \tag{4.4}$$

The reverse reaction rate r_2 can be calculated as

$$r_2 = k_2 p_{CO2} p_{H2}, \tag{4.5}$$

where the constant of proportionality k_1 and k_2 are temperature dependent. At equilibrium, the two reaction rates are equal, and the equilibrium constant is defined in terms of partial pressures:

$$K_p = \frac{k_1}{k_2} = \frac{p_{CO2} p_{H2}}{p_{CO} p_{H2O}}, \tag{4.6}$$

where K_p is the temperature-dependent equilibrium constant for the CO shift reaction.

Similarly, the equilibrium constants for other reactions can be expressed as

$$K_p = \frac{p_{CO}^2}{p_{CO2}} \tag{4.7}$$

for the Boudouard reaction (reaction R-4.4),

$$K_p = \frac{p_{CO} p_{H2}}{p_{H2O}} \tag{4.8}$$

for the water-gas reaction (reaction R-4.5), and

$$K_p = \frac{p_{CO} p_{H2}^3}{p_{CH4} p_{H2O}} \tag{4.9}$$

for the reforming reaction (reaction R-4.9), where p is the partial pressure of the gas component in the gasifier.

Many researchers have studied the thermodynamic equilibrium of the reactions occurring in the gasification process. The reaction enthalpies and equilibrium constants of some reactions at different gasification temperature are listed in Table 4.5.

Table 4.5. Equilibrium Constants of Some Reactions in Gasification Process

	Temperature (K)				
	298	500	800	1000	1500
$CO + H_2O = CO_2 + H_2$	9.9×10^4	131.5	4.038	1.378	0.375
$C + 2H_2 = CH_4$	7.9×10^8	2688	1.412	0.098	0.00256
$CO + 3H_2 = CH_4 + H_2O$	7.87×10^{24}	1.15×10^{10}	32.06	0.0376	0.000421
$CO_2 + 4H_2 = CH_4 + 2H_2O$	8.58×10^5	9333	5.246	0.0273	3.76×10^{-8}
$2C + 2H_2O = CH_4 + CO_2$	0.00785	0.0817	0.2506	0.3545	0.5819

The effects of pressure on equilibrium mole fractions can be illustrated with Equation 4.9, changed into variables of mole fraction and total pressure:

$$K_p = \frac{y_{CO} P_{tot}\ y_{H2}{}^3 P_{tot}^3}{y_{CH4} P_{tot}\ y_{H2O} P_{tot}} = \frac{y_{CO} y_{H2}{}^3}{y_{CH4} y_{H2O}} P_{tot}^2. \tag{4.10}$$

Note that the value of the equilibrium constant K_p does not change with pressure, since it is calculated from ΔG^o_{rxn} at 1 bar standard state pressure, but that the mole fractions change with pressure if the number of moles of reactants is different from the number of moles of products. In this case (Eq. 4.10), as pressure rises, less CO and H_2 will be produced since K_p is a constant with pressure.

The equilibrium constant approach works well when single simple reactions occur, but not when there are competing reactions. The formal definition of chemical equilibrium is that the total Gibbs free energy is at a minimum:

$$dG = 0, \tag{4.11}$$

where G is the Gibbs free energy of the system ($G = \sum_i G_i$ for all species i) and is defined as

$$G = H - TS, \tag{4.12}$$

where H is the enthalpy, T is the temperature, and S is the entropy. Chemical equilibrium programs have been developed to perform the Gibbs free energy minimization in combination with atomic balances on each element in the system.[22,23] Species are added or removed as the minimization takes place. An example using this technique is shown in Figure 4.7 for a system initially consisting of a 1:1 mole ratio of steam and carbon at 1 bar total pressure. The C(s) is included in the reported mole fractions, even though it is a solid. It can be seen that at equilibrium, C(s) and CH_4 are present at temperatures as high as 1200K. At higher temperatures, C(s) is converted to CO and H_2O is converted to H_2. At a pressure of 50 atm, C(s) (and CH_4,

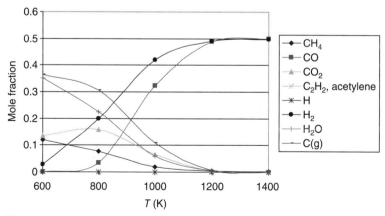

Figure 4.7.   Equilibrium mole fractions from a 1:1 molar ratio of steam and carbon at 1 bar total pressure.

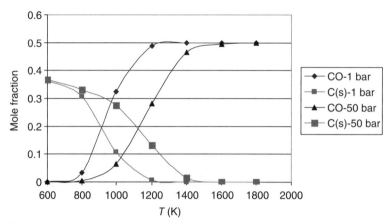

Figure 4.8. Equilibrium mole fractions of C(s) and CO from a 1:1 molar ratio of steam and carbon at 1 and 50 bar total pressure.

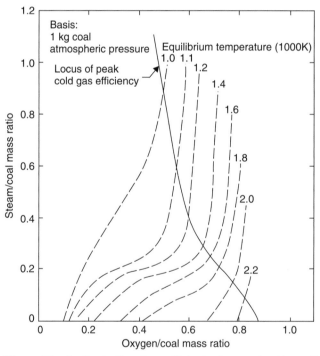

Figure 4.9. Predicted effect of steam/O_2/coal ratio on peak cold gas efficiency and equilibrium temperature (adapted from Smoot and Smith[15]).

which is not shown) persists at higher temperatures, as illustrated in Figure 4.8. Similar trends occur in a C–CO_2 system. This means that at elevated pressures, higher temperatures are needed if 100% conversion of the coal char to syngas is desired.

In general, both O_2 and steam are added to a coal gasifier to achieve the optimal heating value of the syngas produced. One method of assessing the value of the syngas is to calculate the heating value at room temperature of all the species produced in the gasifier. This is called the cold gas heating value. The cold gas efficiency can then be calculated as the ratio of the cold gas heating value to the heating value of the unreacted coal. The cold gas efficiency can be calculated from the equilibrium concentrations of species, depending on the steam/coal and O_2/coal ratios in the gasifier. Figure 4.9 shows the locus of maximum coal gas efficiencies as a function of the steam/coal and O_2/coal ratios, as well as the adiabatic flame temperatures for these same ratios (using a typical heating value bituminous coal).

4.5 GASIFICATION KINETICS

In gasification processes, oxidation of part of the coal with oxygen or air generates the heat required for the gasification processes. In general, the gasification temperature increases with the increase of the oxygen/coal ratio. However, with a higher oxygen/coal ratio, the content of CO_2 and H_2O in the gas product is also higher. By lowering the oxygen/coal ratio, a higher CO and H_2 level in the gas product can be achieved. However, coal conversion will be incomplete below a certain oxygen/coal ratio because the temperature will decrease. Therefore, a good understanding of coal reactivity and the kinetics of the gasification reactions are important for the optimization of gasification processes. Coal gasification kinetics has been extensively studied, particularly in the last three decades. Despite this, the mechanism of coal gasification is still unclear.

Generally, gasification kinetics research includes the study of rates and mechanisms of gasification reactions and the effect of factors on the reaction rates. Reactions in coal gasification are mainly heterogeneous. Besides the chemical reaction (heterogeneous and homogeneous), some physical processes such as adsorption, diffusion, heat transfer, and fluid mechanics also play important roles in this complicated process.

In a gas–solid reaction, generally there are seven steps:

1. diffusion of reactant gas from gas phase to solid surface,
2. diffusion of reactant gas from solid surface to solid internal surface through pores,
3. adsorption of reactant gas on the surface of solid,
4. reaction on the surface,
5. desorption of the product from the surface of solid,
6. diffusion of product from the internal surface to solid surface,
7. diffusion of product from solid surface to gas phase.

Each step has its own resistance. The overall reaction rate is limited by the step with the highest resistance (the rate-limiting step). At a lower temperature range ($T < 1000\,°C$), surface reaction (step 4) is the rate-limiting step in the gasification process. As temperature increases, the diffusion of the reactants through the pores (step 2) and through the boundary layer (step 1) has more and more effect on the apparent reaction rate.

It is possible that some O_2 is still present in the early stages of char gasification. Some methods for calculating char–O_2 reactions have been presented in the literature;[7,24] no further discussion is presented here. The slowest reactions that govern the overall reaction rate in gasification processes are the heterogeneous reactions with carbon, including Boudouard, water–gas, and hydrogenation reactions. The Boudouard and water-gas reaction are comparable and several orders of magnitude faster than the hydrogenation reaction.[16]

4.5.1 Reaction Mechanisms and the Kinetics of the Boudouard Reaction

The rate-limiting step in the Boudouard reaction is the chemical reaction at lower temperature ($T < 1000\,°C$) for small particle sizes ($d_p < 300\,\mu m$). Under these conditions, the reaction takes place on the internal surface of the coal particle.

There are several mechanisms describing the Boudouard reaction. Ergun[25] proposed a two-step process model to describe the Boudouard reaction as

$$C_{fas} + CO_2 \underset{k_2}{\overset{k_1}{\Longleftrightarrow}} C(O) + CO, \tag{R-4.22}$$

$$C(O) \overset{k_3}{\rightarrow} CO + C_{fas}. \tag{R-4.23}$$

In this model, the first step is the dissociation of CO_2 at a carbon free active site (C_{fas}), releasing CO and forming an oxidized surface complex [C(O)]. In the second step, the carbon–oxygen complex subsequently produces CO and a new free active site. The reverse reaction is relatively slow compared with the forward reaction, so the second reaction can be treated as an irreversible reaction. In this model, desorption of the carbon–oxygen surface complex is the rate-limiting step. The rate for this mechanism can be described by the Langmuir–Hinshelwood rate equation. Furthermore, the C/CO_2 reaction rate is dependent on the CO and CO_2 partial pressures and is inhibited by the presence of carbon monoxide. A widely utilized reaction rate equation based on this mechanism is

$$R = \frac{k_1 p_{CO2}}{1 + \dfrac{k_1}{k_2} p_{CO2} + \dfrac{k_2}{k_3} p_{CO}}, \tag{R-4.24}$$

where p is the partial pressure for each component, and k_1, k_2, and k_3 are the rate constants for reaction R-4.22, reverse of reactions R-4.22 and R-4.23, respectively.

Another mechanism was proposed by Gadsby et al.[26] In this model, besides the two steps described in Ergun's model, there is a third step:

$$C_{fas} + CO \Leftrightarrow C(CO). \qquad \text{(R-4.25)}$$

4.5.2 Reaction Mechanisms and the Kinetics of the Water-Gas Reaction

The models for the water-gas reaction are

$$C_{fas} + H_2O \underset{k_5}{\overset{k_4}{\Leftrightarrow}} C(O) + H_2, \qquad \text{(R-4.26)}$$

$$C(O) \overset{k_6}{\rightarrow} CO + C_{fas}. \qquad \text{(R-4.27)}$$

In the first step, a water molecule dissociates at a carbon free active site to form H_2 and an oxidized surface complex. In the second step, the carbon–oxygen complex produces carbon monoxide and a new free active site. Based on this oxygen exchange mechanism, the reaction rate is obtained:

$$R = \frac{k_4 p_{H2O}}{1 + \dfrac{k_4}{k_6} p_{H2O} + \dfrac{k_5}{k_6} p_{H2}}. \qquad \text{(4.13)}$$

In this model, only hydrogen inhibits the water-gas reaction. The rate-limiting step is the desorption of the carbon–oxygen surface complex, which has an activation energy of 250–300 kJ/mol.

By adding a third reaction, a hydrogen-inhibition model is proposed:

$$C_{fas} + H_2 \underset{k_8}{\overset{k_7}{\Leftrightarrow}} C(H)_2. \qquad \text{(R-4.28)}$$

In this case, the first reaction (reaction R-4.26) is not reversible. Hydrogen inhibition is attributed to the formation of the $C(H)_2$ complex. The reaction rate becomes

$$R = \frac{k_4 p_{H2O}}{1 + \dfrac{k_4}{k_6} p_{H2O} + \dfrac{k_7}{k_8} p_{H2}}. \qquad \text{(4.14)}$$

This approach was used successfully to describe the gasification of several Australian coal chars by Roberts and Harris.[27,28] In the second hydrogen-inhibition model, the third reaction (reaction R-4.28) is replaced by another reaction (reaction R-4.29) as

$$C_{fas} + \frac{1}{2} H_2 \underset{k_{10}}{\overset{k_9}{\Leftrightarrow}} C(H). \qquad \text{(R-4.29)}$$

Reaction R-4.29 describes dissociative chemisorption of hydrogen. The reaction rate derived from this model is then

$$R = \frac{k_4 p_{H2O}}{1 + \dfrac{k_4}{k_6} p_{H2O} + \dfrac{k_9}{k_{10}} p_{H2}^{1/2}}. \tag{4.15}$$

From all the reaction rates shown above, it can be shown that the reaction rate decreases linearly with increasing carbon conversion. This means that the reaction rate is almost first order in carbon and that the reactivity of char is almost constant during gasification. The presence of hydrogen in the system reduces the initial reaction rate of the gasification and results in a gradual decrease in the reactivity of the char with increasing carbon conversion.

In reaction R-4.27, carbon monoxide is one of the products and may inhibit the water-gas reaction. The carbon monoxide reaction is

$$C(O) + CO \underset{k_{12}}{\overset{k_{11}}{\rightleftharpoons}} CO_2 + C_{fas}. \tag{R-4.30}$$

The reaction rate then becomes

$$R = \frac{k_4 p_{H2O} + k_{11} p_{CO2}}{1 + \dfrac{k_4}{k_6} p_{H2O} + \dfrac{k_5}{k_6} p_{H2} + \dfrac{k_{12}}{k_{11}} p_{CO}}, \tag{4.16}$$

where both carbon monoxide and hydrogen inhibit the reaction.

Based on experimental results, Wen and Lee[29] developed an experimental correlation for the water-gas reaction,

$$\frac{dx}{dt} = k_v \left(p_{H2O} - \frac{p_{H2} p_{CO} RT}{K_p} \right)(1 - x), \tag{4.17}$$

where x is the carbon conversion of the char, k_v is the rate constant, and K_p is the equilibrium constant. More complicated expressions for the steam–char gasification rate were developed by Mühlen et al.[30] and Liu and Niksa[31].

For smaller particle size $(d_p < 500 \mu m)$ and lower temperature $(1000\,°C < T < 1200\,°C)$, char–$H_2O$ reaction is chemical reaction controlled. When partial pressure of the steam is low, the reaction order is 1. With increasing steam partial pressure, the reaction order decreases gradually to 0.

4.6 CLASSIFICATION OF DIFFERENT GASIFIERS

The coal gasification process converts coal into a syngas composed mainly of carbon monoxide and hydrogen, which can be used as a fuel to generate electricity or as a basic chemical building block for a large number of chemicals. Dry or slurried feedstock reacts with steam and oxygen (or air) in the gasifier at high temperature and pressure in a reducing atmosphere. Clean syngas can be obtained after particles, sulfur, and other impurities are removed in a subsequent purification process.

According to the feedstock and the oxidant flow through the gasifier, coal gasification technologies can be classified into three basic types: moving- or fixed-bed, fluidized-bed, and entrained-bed technology. Among all these processes, the entrained-flow gasification process has the largest treatment ability and the smallest environmental impact. Hence, entrained-flow processes occupy most of the commercial market. At present, General Electric Company (GE), Shell, ConocoPhillips, and GSP (Gas Schwarze Pumpe) are the four leading entrained-bed coal gasification technology suppliers.

Moving-bed gasifiers (fixed-bed gasifiers) operate on lump coal. Coal is loaded from the top of the gasifier, while the reactant agent is introduced from the bottom. The coal moves slowly downward under gravity as it is gasified by the countercurrent blast. In a moving-bed gasifier, coal is preheated and pyrolyzed by the rising hot syngas from the gasification zone. Compared with other gasification processes, the oxygen consumption in this process is low but pyrolysis products are present in the syngas product. The temperature of the syngas at the outlet is generally low. Moving-bed gasifiers are simple to design and operate, and have high reliability. Furthermore, the countercurrent contact between coal and reactant gas results in a high coal conversion, thus high heat efficiency.

Fluidized-bed gasifiers use small coal particles or pulverized coal, as well as a bed of sand or some other inert particle. Coal particles are fluidized by the oxidant stream in the gasifer. Fluidized-bed gasifiers offer a homogeneous temperature environment and promote the heat and mass transfer between gas and solid phases. As with other fluidized bed applications, the sizing of the coal particles in the feed is critical. Also, the fluidized bed reactor can only operate at the temperatures below the softening point of the ash. The lower operating temperature also means that they are suited for gasifying reactive feedstocks such as lower-rank coals. However, as a continuous stirred-tank reactor (CSTR), the carbon conversion in the fluidized bed is generally lower compared with other gasification methods.

Entrained-flow reactors operate in cocurrent flow. Pulverized coal is entrained in the oxidant (oxygen and steam) and introduced into the gasifier (sometimes using a coal water slurry [CWS] and pumped into the gasifier as in GE gasifiers). The reaction takes place at a high temperature (1500–1900 °C) and generates CO, H_2, CO_2, and other gases. Ash is removed as molten slag from the bottom of the reactor. Due to the short residence time in the reactor, high temperatures are required to ensure a good coal conversion, and therefore the operating temperature is high. The oxygen consumption is also high in the entrained-flow gasifier to maintain a high operating temperature. Due to the high operating temperature, entrained-flow gasifiers do not have any specific technical limitations on the type of feedstock used. However, coals with a high moisture or ash content require higher oxygen consumption and it reduces the heat efficiency compared with alternate processes.

Recently, some novel gasification processes have also been developed, including *in situ* gasification of coal in the underground seam as well as molten bath processes. The different gasification processes are described in the following sections.

4.7 GE (TEXACO) GASIFICATION TECHNOLOGY WITH CWS FEEDING

4.7.1 Introduction to GE Gasification Technology

The Texaco Gasification Process (TGP) was developed in the late 1940s using natural gas as feedstock. The process was commercialized with gas feed in 1950 and later with liquid feed in 1956. During the oil shortage period, Texaco successfully developed its coal gasification process.

Texaco developed their CWS feeding technology in 1948. Based on this technology, Texaco developed the TGP (GE acquired it from Texaco in 2004) and built a 15 t/day pilot-scale gasifier in Montebello, CA, in the U.S. In the original TGP, coal was dry pulverized to fine particles, then mixed with water and other additives to make a CWS. The coal concentration in the CWS was as high as 50 wt %, depending on the coal properties. To minimize the energy penalty in the gasifier, the CWS was preheated, and water was separated from the slurry before it was injected into the gasifier in the original TGP. In 1958, the Montebello tests were stopped due to technical difficulties.

Many disadvantages limited the application of the original TGP. These disadvantages included the following: (a) slurry additives had not been developed at that time, resulting in a low coal concentration in the CWS; (b) the coal was dry pulverized, which made the process complicated and not environmentally friendly; (c) the slurry was abrasive during the evaporation, causing a myriad of problems; and (d) some pulverized coal was lost with the vapor during the prevaporizing process.

Due to the worldwide oil shortage, Texaco resumed work on coal gasification at the beginning of the 1970s, recommissioning the Montebello facility and building a new 2.5 MPa gasifier in 1975. With the development of new additives and CWS preparation techniques, CWS could be injected into the gasifier directly without the prevaporizing step. This simplified the process and enhanced the system reliability significantly. In 1975, Texaco built its first demonstration-scale gasifier (165 t/day) in Oberhausen, Germany, in conjunction with Ruhrchemie AG (RCH) and Ruhrkohle AG (RAG). Over 20 kinds of feedstock were gasified in this demonstration gasifier. Some operating condition and parameters from these demonstration tests are shown in Table 4.6. Two 8.5 MPa high-pressure gasifiers were built in 1978 and 1981 for

Table 4.6. RCH/RAG Demonstration Gasifier Data

	T (°C)	P (MPa)	Coal (t/day)	Syngas Yield (m³/h)	Slurry Concentration (%)	Dry Syngas Composition (%)				C Conversion (%)
						CO	H_2	CO_2	CH_4	
Designed parameter	1500	4	6	10,000	55–60	45–55	30–40	15–20	1	95
Operating parameter	1200–1600	4	2.9–8.2	15,200	Up to 71	54	34	11	<0.1	99

Table 4.7. Commercial Plants Based on TGP

Project	Capacity (t/day)	Purpose
Eastman	1100	Coal to methanol/acetic anhydride
Cool Water	1000	Coal to power (120 MW)
Ube, Japan	1650	Coal and petroleum coke to ammonia
SAR, W. Germany	800	Coal to oxo-chem/H_2
Lunan, China	550	Coal to ammonia
Shanghai, China	1800	Coal to town gas/methanol
Weihe, China	1650	Coal to ammonia/urea
TECO	2000	Coal to power (250 MW)
Huainan, China	900	Coal to ammonia/urea

the evaluation of coals. Since then, over 20 various coals from all over the world have been evaluated on these pilot-scale facilities.

Commercial-scale gasification plants based on Texaco gasification technology were then started in Ube, Japan (1650 t/day) and Kingsport, TN (1100 t/day, Eastman).

Since 1990, several additional commercial coal-based plants have been brought into services with Texaco gasification technologies, as summarized in Table 4.7.

In June 2004, GE acquired the assets of Chevron Texaco's gasification technology business. To date, more than 65 commercial gasification facilities using GE technology are in operation or advanced development stages.

4.7.2 GE Gasification Process

The GE gasification process includes a CWS preparation, a high-pressure entrained-flow slagging gasifier, and a gray water treatment. In the gasification step, GE gasification technology offers both a radiant boiler and a total quench as syngas cooling concepts. In the gray water treatment step, pressurized flash and vacuum flash are normally utilized. Since the CWS preparation and the gray water treatment methods are simple and more standard, gasification processes with two different syngas cooling configurations will be further discussed.

At the heart of the GE coal gasification process is a top-fired slurry-fed slagging entrained-flow gasifier. Schematics of the typical Texaco gasification system are shown in Figures 4.10 and 4.11.

The coal is wet-milled with additives to create a stable slurry with approximately 60% solid particles. The CWS is introduced with the oxidant stream (oxygen or air) into the gasifier through a multihead, water-cooled feed injector that is located at the top of the gasifier. As the slurry enters the gasifier, the water evaporates and pyrolysis of the coal particles occurs. The volatiles released react with oxygen and generate large amounts of heat. Char is then gasified by the steam and other

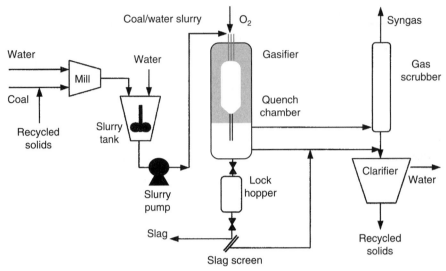

Figure 4.10.   Schematic of a typical GE coal gasification system with quench chamber.

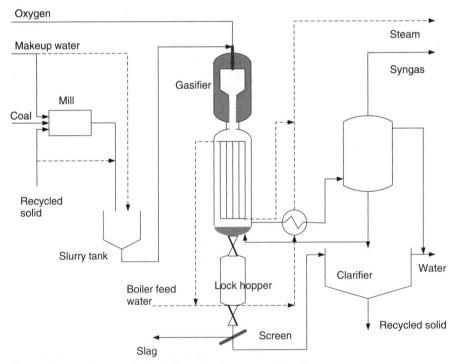

Figure 4.11.   Schematic of typical GE gasification system with radiant syngas cooler.

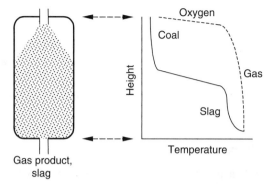

Figure 4.12. Typical temperature profile in a GE coal gasifier.

gasification agents at high temperature, producing carbon monoxide and hydrogen. The typical temperature profile in a GE gasifier is shown in Figure 4.12.

Following gasification, the syngas and slag are cooled in either of two configurations, the choice of which depends on the subsequent application of the syngas.

4.7.2.1 Quench Configurations

The quench configuration is shown in Figure 4.10. The hot syngas leaves the gasifier, together with the liquid slag, and is quenched in water. The water-saturated gas product leaves the quench chamber at a temperature between 200 and 300 °C. The solidified slag is removed from the quench chamber via a lock hopper. The water is then separated and can be recycled for slurry preparation. The raw syngas is further cleaned in a scrubber to remove particulates entrained in the gas phase. In the quench configuration, the gasifier is connected to the quench chamber directly, making a compact configuration. Energy in the high-temperature raw syngas is recovered by the quench water/steam directly. Slag is solidified and separated from the gasifier. The quench configuration is simple and easily maintained. Furthermore, the syngas is saturated with steam, making it ideal for the downstream water-gas shift process.

As such, the quench configuration is ideally suited for the ammonia industry or other industries that require hydrogen concentrated in the syngas.

4.7.2.2 Radiant Cooler Configurations

In the radiant cooler configuration shown in Figure 4.11, hot syngas and slag enter a tall heat exchanger. The molten slag falls to a quench bath, solidifies, and is removed via a lock hopper. The heat in the syngas is exchanged into water-fed cooling pipes. All coolers are used to raise high temperature steam that is sent to a steam turbine for further energy recovery. The moisture content in the raw syngas generated from this configuration is low. Therefore, this configuration is suitable for CO production, industrial fuel gas generation, or IGCC or other process that do not require a high H_2/CO ratio, but high cold gas efficiency. The radiant cooler configuration is more energy efficient than the quench configuration; however, this is achieved at additional capital expense (CAPEX). In both configurations, a final hot-gas scrubber is installed to remove particulates and hydrogen chloride from the gas product.

The most important operating parameters are coal concentration in the CWS, gasification temperature, gasification pressure, residence time, and oxygen/carbon ratios. These effects are discussed below.

4.7.2.3 Effects of Coal Slurry Properties

Coal concentration in the slurry is a unique parameter in GE gasification technology. Coal slurries are non-Newtonian fluids and can be treated as Bingham plastics. The slurry rheology is described with an apparent viscosity. The apparent viscosity of coal slurry increases with shear time and decreases with shear rate. Generally, the apparent viscosity of coal slurry used in real application is around 1 Pa·s.[32] The rheological properties of coal slurry depend on the following: type of coal, particle size distribution (PSD), solid concentration, and type of additive and concentration. Generally, viscosity increases with increasing coal concentration and decreasing particle size. Surfactants are used as additives to permit increased coal concentration at a certain slurry viscosity. The rheological properties are important parameters for piping design and pump selection.

Figure 4.13 shows the effects of temperature and coal concentration on gasification efficiency. The gasification efficiency increases with coal concentration in the slurry. Generally, gasification efficiency (i.e., carbon conversion) increases with decreasing particle size. However, decreased particle size will also lead to a higher apparent viscosity and additional grinding costs. For a certain coal, the optimum PSD and coal concentration is typically determined experimentally.

Figure 4.14 shows the effect of coal concentration in the slurry on the syngas composition. It is shown that increasing coal concentration is helpful for a higher

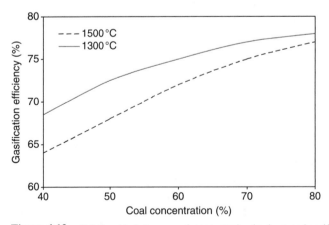

Figure 4.13.   Relationship between coal concentration in slurry and gasification efficiency.

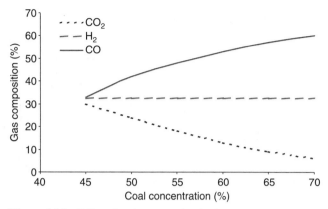

Figure 4.14.   Effect of coal concentration in slurry on syngas composition.

Table 4.8.   Coal Slurries with Different Agents

Coal Slurry	Description	Coal in Slurry (%)	Status
Coal–oil slurry	Fine coal particles slurried with heavy oil and additives	50	Commercialized in Japan
Ultrafine oil slurry	Ultrafine coal particle slurried with heavy oil and additives	40	Commercialized in the U.S.
Coal–water slurry	Coal particles slurried with water	50	Commercialized
High concentration coal–water slurry	Wide PSD coal particles slurried with water and additives	~75	Commercialized
High concentration ultrafine coal–water slurry	Ultrafine coal particle slurried with water and additives	~70	Under development
Coal methanol slurry	Coal particles slurried with methanol	~80	Under development
Coal–CO_2 slurry	Coal particle slurried with liquid carbon dioxide	~75	Under development

$CO + H_2$ yield in the product. However, it is also shown that the concentration of $CO + CO_2$ is constant with increased coal concentration in the slurry. This is because the additional water in the slurry vaporizes and reacts with CO, generating more H_2 and CO_2.

In the last several decades, for different purposes, several other agents have been used to prepare coal slurry. These coal slurries are summarized in Table 4.8.

4.7.2.4   Effects of Pressure and Temperature

The gasification temperature and gasification pressure play very important roles in coal gasification. At high temperatures, the gasification efficiency is improved and thus the required residence time decreases. Also, as in other entrained-flow gasifiers, slag is discharged from the bottom in a liquid phase. Therefore, the operating temperature should be higher than the ash fusion temperature (AFT). To increase the lifetime of the refractory, gasification temperatures range from 1350 to 1500 °C. For coals with AFTs higher than 1500 °C, additives are required to lower the AFT.

Normally, gasifier capacity is proportional to $\sqrt{p}$.[32] The pressure may be as high as 10 MPa in a GE gasification system. Also, the operating pressure is selected based on the purpose of the gas product; in the ammonia industry, the operating pressure ranges from 8.5 to 10 MPa, while in methanol industry, the pressure is 6–7 MPa.

4.7.2.5   Effects of Residence Time

As discussed in the previous section, the main step of the gasification process is the heterogeneous reaction of char. The reaction rate of char gasification is much lower than that of oil gasification. Therefore, a longer residence time is necessary for reasonable coal conversion. In the GE gasification process, the residence time is typically 4–10 s. Residence time is determined by reactor size, coal particle size, char reactivity, and operating conditions (T and P).

4.7.2.6   Effect of O/C Ratio

The oxygen/carbon ratio is an important parameter in the operation of the GE gasifier. The O/C ratio has a significant effect on the gasification temperature, coal conversion, and cold gas efficiency. Normally, at a higher O/C ratio, the gasifier temperature and coal conversion increase as shown in Figure 4.15. However, if the O/C ratio is too high, the overall cold gas efficiency will decrease (Fig. 4.15) because some of the CO/H_2 in the product will react with oxygen. Also, to reach a higher O/C ratio, more oxygen will be consumed, which necessitates a larger air separation unit (ASU) in the system. In addition, increased O_2 leads to increased CO_2 in the product gas.

4.7.3   Coal Requirements of the GE Gasifier

In principle, almost all kinds of coal can be gasified in the GE gasification process, including lignite, bituminous, and anthracite. However, to optimize the performance of the gasifier, bituminous coal with high volatile and high reactivity is recommended.

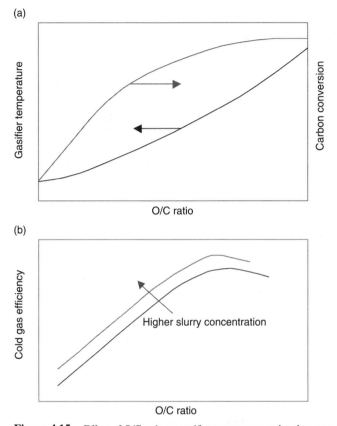

Figure 4.15. Effect of O/C ratio on gasifier temperature and carbon conversion. (a) Effect of O/C ratio on GE gasifier parameters. (b) Effect of O/C ratio on cold gas efficiency.

4.7.3.1 Moisture Content

There are two kinds of moisture that exist in the coal: surface moisture and inherent moisture. Since the CWS is basically a mixture of coal particle and water, the surface moisture of coal does not have too much impact in the GE gasification process. However, a higher surface moisture content means a higher transportation cost. Inherent moisture includes all the moisture that is physically or chemically held within coal. In the CWS preparation step, inherent moisture of the coal plays an important role. With an increase in the inherent moisture, it has been found that the maximum coal concentration possible in a CWS decreases.

4.7.3.2 Ash

Ash is the inert material contained in the coal. Although ash is not directly involved in the gasification, some of the heat generated in the oxidation reaction is consumed

Table 4.9. Performance of the GE Gasifier with Different Coals

Coal	Coal Composition (%)								Syngas Composition (%)			
	C	H	N	S	O	Ash	CO	H_2	CO_2	Steam	CH_4	H_2S/COS
Pittsburgh #8	74.2	5.2	1.2	3.3	6.7	9.5	40.0	30.1	11.4	16.4	0.0	0.9
Utah coal	68.2	4.8	1.2	0.4	15.7	9.7	30.9	26.7	15.9	25.7	0.2	0.1
French coal	78.1	5.3	0.9	0.5	8.2	7.1	37.4	29.3	13.3	19.4	0.2	0.1
German coal	73.9	4.7	1.5	1.1	5.9	13.0	39.5	29.3	12.6	17.5	0.3	0.3
Pet coke	87.7	3.0	1.3	6.9	1.4	0.5	46.7	26.5	11.9	12.5	0.1	1.8

by heating the ash. Also, as discussed above, slag is abrasive to the refractory in the gasifier, and hence reduces the lifetime of the refractory. Therefore, coal with lower ash content is more favorable in the GE gasification process.

In the GE gasification process, slag is discharged in the liquid phase. The viscosity of the slag decreases with increases in the operating temperature. The AFT is one of the most important properties for a certain coal. Although most of ash in coal is mainly SiO_2, Al_2O_3, CaO, and Fe_2O_3, the AFT varies from coal to coal. Generally, ash with a higher Fe_2O_3, CaO, and MgO content has a lower AFT; and ash with a higher SiO_2 and Al_2O_3 content has a higher AFT.

When coal with a comparatively low AFT is gasified, clinkers form in the static fuel beds. Furthermore, slag deposits on gasifier walls and reduce the lifetime of the refractory. Therefore, coal with too low of an AFT should be avoided in the GE gasification system. When some coals with higher AFT (>1400 °C) are gasified in the GE gasifier, additives are needed to reduce the AFT. Normally, CaO/$CaCO_3$ and/ or Fe_2O_3 are used as additives.

To reduce the operating temperature and lengthen the lifetime of the refractory, coals with lower AFTs are recommended in the GE gasification process.

Technically, GE gasifiers can be used to gasify all kinds of coals. However, coals with low inherent moisture (<8 wt %), low ash content (<13 wt %), low AFT (<1300 °C), high heating value (>25 MJ/kg), and low impurities content (Cl, As, S, etc.) are strongly recommended.

For a better understanding of the performance of the GE gasification system, results from some of the demonstration plant are shown in Table 4.9.

4.7.4 Summary of GE Slurry Feeding Gasification Technology

The coal–water feeding technology is the key feature of GE gasifiers. Texaco developed the CWS feeding gasification technology in the 1950s. After 20 years, the technology was successfully demonstrated in the 1970s when a new slurry preparation technique was available. Since then, coal–water feeding technology (GE gasification technology) has become one of the most widely used gasification technologies.

Compared with other existing gasification technologies, the GE gasification process has some unique advantages: (a) it can gasify varieties of coals including lignite, bituminous, and anthracite; (b) it is one of the most inexpensive technologies with lowest CAPEX; (c) due to its simple design, the GE gasifier is robust and thus has the highest reliability; and (d) GE gasifiers are operated at a high temperature (~1400 °C), so a high coal conversion can be achieved.

Disadvantages of the GE gasification process include the following: (a) more O_2 is needed to maintain a high operating temperature; (b) compared with dry feeding technology, the gas product of the GE process has a higher CO_2 content (~17 mol %), resulting in a lower heating value; (c) due to the high operating temperature, the lifetime of the injector and the refractory is short compared with other gasification technologies.

4.8 SHELL GASIFICATION TECHNOLOGY WITH DRY FEEDING

4.8.1 Introduction to Dry-Feeding Coal Gasification

The first dry-feeding gasification was the low pressure Koppers–Totzek (K-T) gasifier developed by Krupp-Koppers in 1938. This technology was mainly used in the ammonia industry; approximately 90% of the ammonia plants used the K-T gasifier in that era. The coal conversion and the efficiency of the K-T gasifier were relatively low, and thus high-pressure dry-feeding gasifiers gradually replaced the low-pressure K-T gasifiers. In 1972, Shell collaborated with Krupp-Koppers and developed a high-pressure dry-feeding gasification technology. The target of their gasification technology was defined as follows:

1. ability to gasify almost every type of coal in the world,
2. environmentally friendly,
3. high T gasification to increase the coal conversion and to prevent the formation of tar oil and phenol type of organic by-products,
4. good reliability,
5. high efficiency, and
6. high syngas productivity from a single gasifier.

Shell built a 6 t/day lab/pilot-scale gasifier to test different types of coal in 1976. It also conducted tests to collect the fundamental data on cleanup of the syngas produced by this gasifier and to assess the technologies to meet environmental regulations.

At the same time, Shell also built a 150 t/day pilot-scale gasifier in one of Shell's refineries at Hamburg, Germany, in 1978. The major task of this pilot plant was to validate the model developed in the Amsterdam lab.

Shell conducted tests in this 150 t/day demonstration plant to determine the design principles of key gasifier parts, including the nozzle, feeding system, slagging system, and the gasifier. The experience gained from the demonstration plant was used to construct a plant named SCGP-1, located in an existing petrochemicals complex at Deer Park in Houston, TX. This plant operated at about 2–4 MPa, produced 16 t of steam per hour, and was sized to gasify 220 t/day of bituminous coal or 365 t/day of high-moisture, high-ash lignite. Shell tested a total of 21 types of coal in this bench-scale gasifier between 1978 and 1983.[32]

By 1983, Shell had conducted over 6000 h of tests in its pilot plant, including 1000 h of continuous testing. Based on the pilot-plant test results, Shell found many areas in which to improve their gasifier and radiant syngas cooler (RSC) design. It started to design the 250–400 t/day demonstration plant in Shell's Deer Park refinery in 1983 and completed the construction and started operation in 1987. This unit had operated over 15,000 h by 1991 and tested a total of 18 types of coal including low-rank coal and petroleum coke. The testing results from this demonstration plant were better than expected. The Deer Park gasifier proved the ability of the Shell Coal Gasification Process (SCGP) to gasify a wide range of coals.[33] By 1991, Shell was ready to commercialize its gasification process.

In 1988, the Holland Government decided to build a 250 MW IGCC plant in Buggeum, Holland, using Shell's gasification technology. The process incorporated Shell's Sulfinol-M technology, a Siemens' V94.2 GT for power generation, and an integrated feed of compressed air to the ASU. This IGCC plant was designed to deliver a total of 253 MW of power with an efficiency of 43% (lower heating value [LHV]). This IGCC plant started to feed coal in April 1994. However, severe problems were encountered with the GT at the beginning of the project, which were finally resolved in September 1996. The availability of the plant reached 85% in late 1997. The availability of the gasifier and syngas cleanup system reached 95% in the same period. The nozzle life exceeded 7500 hours, and 14 kinds of different coals were gasified. The IGCC plant was transferred to the local utility company for operation on January 1, 1998. By this time, the SCGP finally reached commercial success and became an efficient technology for producing syngas from different types of coals. This demonstrated that Shell's gasifier was ready for commercial applications.[34,35]

On November 2, 2001, Shell and the Sinopec signed a cooperation deal, totaling $136 million. According to the deal, the two parties were to establish a coal gasification plant in Yueyang, Central China's Hunan Province. The coal consumption of the plant will reach 2000 t/day. Coal will be transformed to gas and then serve as the raw materials for the Dongting Fertilizer Plant. In addition, the two parties agreed to build coal gasification plants using Shell's technology in China's Hubei, Anhui, and other provinces. Table 4.10 shows the recent gasification projects in China that use Shell's technology.

Table 4.10.   Shell Coal Gasification Projects in China

Name	Scale (t/day)	End Product	Subscription Date
Sinopec-Shell Yueyang Coal Gasification Co.	2000	Ammonia	2001
Hubei Shuanghuan Chemical Group Co.	900	Ammonia	2001
Liuzhou Chemical Ltd.	900	Ammonia	2001
Sinopec Hubei Chemical Fertilizer Ltd.	2000	Ammonia	2003
Sinopec Anqing Co.	2000	Ammonia	2003
Yunnan Tian'an Chemical Co.	2000	Ammonia	2003
Yunnan Guhua Co.	2000	Ammonia	2003
Dahua Group Co.	2000	Methanol	2004
Yongcheng Meidian Group Co.	2000	Methanol	2004
China Shenhua Coal to Liquid Co.	2000*2	Hydrogen	2004

4.8.2   Shell Gasification Process

Figure 4.16 shows a schematic of the typical Shell gasification process. The coal is ground into fine particles, dried to <2% water, and then carried by N_2 or CO_2 to the storage tank and lock hopper. High-pressure N_2 or CO_2 is used to transport the coal particles from the lock hopper to the two symmetric feed injectors of the gasifier. Steam and O_2 are also injected into the gasifier via the same nozzles. The temperature of the gasifier is controlled to be in the range of 1400–1700 °C by adjusting the flow rates of coal, oxygen, and steam. Most of the ash forms slag inside the gasifier at these high temperatures and then flows down and exits from the bottom of gasifier. Water quench is used at the bottom of the gasifier, and the cooled slag is then ground before being sent to the slag lock hopper system. Finally, the pressure is reduced in the lock hopper, and the slag is sent to the slag storage tank. Figure 4.17 shows the structure of the SCGP. The gasifier vessel consists of a carbon steel pressure shell, within which is a gasification chamber enclosed by a refractory-lined membrane wall. The temperature in the reaction center ranges between 1500 and 2200 °C, depending on the coal type. Shell gasifiers typically operate at pressure ranges of 2–4 MPa.[32] Water circulated through the membrane wall is used to control the temperature of the gasifier wall and raises saturated steam. High-pressure steam is produced in the gasification and heat recovery section and can then be used, increasing the efficiency of the whole process. Dried pulverized coal, O_2, and steam are fed

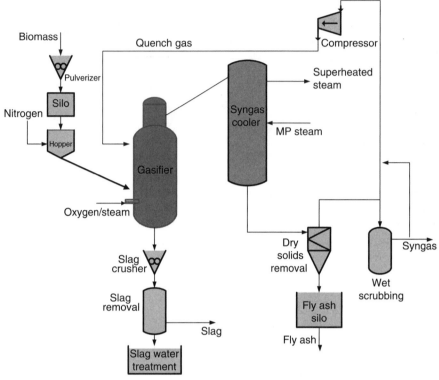

Figure 4.16.   Schematic of typical Shell gasification process. MP, medium pressure.

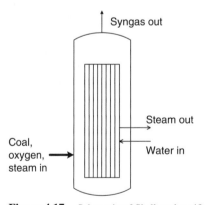

Figure 4.17.   Schematic of Shell coal gasifier.

through opposed burners at the bottom of the gasifier, which operates at 2–4 MPa. Gasification occurs at temperatures of >1500 °C, which ensures that the ash in the coal melts and forms a molten slag. The slag runs down the inner surface of the gasifier wall and is quenched in a water bath at the bottom of the gasifier. A portion of the slag adheres to the wall of the gasifier and cools, forming a protective layer.

The membrane wall's life was estimated to be over 25 years. The process can handle a wide variety of solid feedstocks, including lignite, brown coal, subbituminous coal, bituminous coal, anthracite, and petroleum coke. This gasifier can handle different coal types and is relatively insensitive to the size, condition, or other physical properties of the raw coal.

The hot raw syngas with small slag droplets is quenched with cold recycled syngas, thus the slag droplets are cooled to form solid particles and fly ash, which will not stick onto the wall of the syngas cooler. The syngas cooler is a shell-tube type of heat exchanger with water and steam flows inside the tube and hot syngas in the shell side. The high-pressure steam generated from the syngas cooler is used to drive the steam turbine in an IGCC plant. The cooled raw syngas is then sent to the fly-ash removal system (ceramic filters), followed by a wash system to remove all the fly ash, and then sent to the acid gas removal (AGR) system. A part of the cooled syngas is recompressed and sent back to the gasifier to quench the hot syngas.

The SCGP has the following technical features:[36]

- dry feed of pulverized coal, coal/nitrogen (or CO_2) ratio between 50 and 500 kg/m^3, normally 400 kg/m^3;
- compact gasifier and other equipment due to the pressurized entrained flow and oxygen blown design. Coal conversion is over 98%;
- a slagging, membrane wall that allows high temperatures because of insulation and protection of wall by solid inert slag layer; and
- multiple, opposed burners resulting in good mixing of coal and blast, large turndown, and scale-up potential.

Figure 4.18 shows a typical heat balance in an SCGP plant. Gasification of the coal forms a raw fuel gas that is predominantly 25%–30% H_2 and 60%–65% CO,

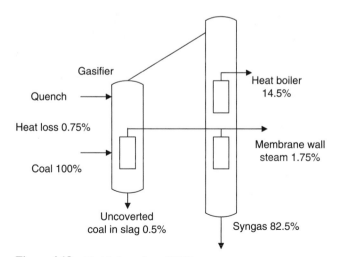

Figure 4.18.   Heat balance in an SCGP system.

Table 4.11.   SCGP-1 Performance

Item	Design Value	Change Scope	Demonstration Value
Capacity (t/day)	229	115–235	229
Pressure (MPa)	2.51	2.44–2.51	2.44
Syngas components (%)			
H_2	29.9	22.7–34.6	27.7–29.8
CO	62.9	54.8–69.0	66.5–69.0
CO_2	5.8	2.2–10.8	2.2–2.4
CH_4	0.04	0.001–0.004	0.001–0.004
H_2S + COS	1	1.1–1.8	1.1–1.2
C conversion rate (%)	98.5	96–99	97–98

with a little CO_2 and some entrained slag particles. The total cold syngas efficiency can reach over 83% and another 15% can be recovered by high-pressure steam.[37] At the gasifier outlet, the raw gas is quenched with recycled, cooled fuel gas to lower the temperature to ~900 °C. This cooling "freezes" the slag particles, rendering them less sticky and less prone to fouling surfaces.

The fuel gas is then cooled to ~300 °C in the syngas cooler, raising high and medium-pressure steam. In contrast to the syngas cooler for Shell's oil gasification process, the SCGP syngas cooler has the gas on the shell side. The syngas cooler thus has a complex tube bundle comprising various economizers, medium- and high-pressure evaporators, and some superheaters.

The cooled syngas is filtered using ceramic filters. About 50% of the cooled syngas is then recycled to the top of the gasifier to act as the quenching medium for the gas. The remainder is washed to remove halides and NH_3 and then sent to the downstream AGR system.

Typical process data from the Shell gasifier are shown in Table 4.11.[38]

The major benefits of Shell's gasification technology are the following:

1. The ability to gasify different types of coal including low-rank coal and petroleum coke. The carbon conversion is over 98%. The amount of sulfur, oxygen, and ash in the coal does not significantly impact the Shell gasification process.

2. The CO + H_2 concentration is >90% in the product gas stream of a Shell gasifier. The CO_2 concentration in the syngas stream is rather low, thus reducing the cost of AGR for ammonia or other coal-derived liquids or chemicals. In addition, there is no tar oil in the product gas stream.

3. Due to the dry feeding of the Shell gasifier, the O_2 consumption is reduced, and the efficiency is increased by ~2% compared with the CWS feeding system. However, the maximum gasification pressure is limited to about 5 MPa due to the dry-feeding system, as opposed to 8 MPa for some CWS feeding systems.

4. The gasifier can be scaled-up to a 3000 t/day unit. One can relatively easily remove the sulfur and ammonia in the syngas stream.

5. The active cooling membrane wall can last much longer than the bricks utilized in the GE (Texaco) gasifier.

Typical temperatures of Shell gasifiers are in the range of 1440–1600 °C. The typical O_2 consumption rate of Shell gasifiers is 0.9 kg O_2/kg of coal (on a dry, ash-free basis). The ratio of steam consumption to O_2 consumption ratio is 0.14. The cold syngas efficiency is ~82% and the carbon conversion can reach >99%. The wet syngas compositions (mol %) from the gasification of a Cerrejon coal from Columbia are

- CO: 62.1%,
- H_2: 31.0%,
- N_2: 3.1%,
- $H_2S + COS$: 0.23%,
- Ar: 0.8%, and
- steam: 1.7%.

4.8.3 Coal Requirements of Shell Gasification Process

Shell gasifiers can gasify almost every type of coal from anthracite to brown coal (lignite). However, they still have the following requirements for coal.

4.8.3.1 Moisture Content

The dry-feeding system of the Shell gasifier requires the water content in the bituminous coal to be less than ~2 wt %. For high-moisture, low-rank coal, the required water content could be less than ~10 wt%.

4.8.3.2 Ash Content and AFT

Ash is the inert material inside the coal and does not significantly impact the gasification reactions. However, ash characteristics impact coal transportation, coal feeding, and the ash processing system. Shell's gasifier utilizes an active-cooling membrane wall and uses a solid slag layer to protect the gasifier. If the ash content is too low, the solid slag protection layer may not be formed properly, increasing heat loss and impacting the durability of the gasifier.

The Shell gasifier is an entrained bed with slag formation. In order for the slag to flow out of the gasifier, the gasification temperature has to be 100–150 °C higher than the AFT. If the AFT is too high, one has to increase the gasification temperature, which can impact the operation, durability, and economics of the gasifier. Thus, low or medium AFT coals are preferred for the Shell gasifier for easy removal of the slag. For a high AFT coal, one can add an ash-fluxing agent to lower the AFT in order to ensure smooth gasifier operation.

4.8.3.3 Particle Size, Volatiles, and Reactivity of Coal

Shell gasifiers operate at high temperature, thus the residence time of gas and coal are relatively short. The gas–solid diffusion is the rate-limiting step of coal conversion. Thus, relatively small coal particles (<0.1 mm) are required compared with the fluid-bed or fixed-bed gasifiers; however, the requirements for the volatiles and reactivity of coal are not as strict as those of a fixed-bed gasifier. A lot of energy (electricity) is required to grind the coal into very fine particles; thus, for coal with higher volatiles and reactivity, the requirement for small coal particle sizes can be relaxed. However, for low volatile or low reactivity coal, the smaller coal particles are better from a gasification kinetics point of view. The optimum size of coal particles for dry feeding gasification is a function of many system variables.

4.8.4 Summary of Dry-Feeding Shell Gasifier

The dry-feeding technology is the key feature of the Shell gasifier. In the early 1950s, initial development of the continuous coal dry-feeding system started, but significant progress was not made until the Shell–Koppers gasification process was developed in 1978. In this process, inert gas is used to transport the coal particles into the ambient pressure feed hopper, and the inert gas is vented via the particulate filter. The coal particles are then dropped into the lock hopper and sealed. The pressure of the coal in the lock hopper is increased gradually using N_2, and coal particles are conveyed into the pressurized gasifier by high pressure N_2 or CO_2. The pressurized entrained flow (PRENFLO), SCGP, Thermal and Power Research Institute (TPRI), and GSP gasification technologies developed later are all based on this type of dry-feeding system.

The advantages of the process include the following:

1. It can gasify varieties of coals including lignite and anthracite.
2. It uses 15%–25% less oxygen consumption compared with slurry feed gasification processes. The trend for dried pulverized coal gasification is to replace N_2 with CO_2 and increase the gasification pressure, which can further reduce the O_2 consumption.[37]
3. The use of a membrane wall leads to a higher reliability.

The disadvantages of the SCGP include the following:

1. The dried pulverized coal system is quite complex; the nitrogen from dense entrained flow may affect the subsequent chemical process. Thus, the CAPEX is much higher compared with a CWS feeding gasifier for methanol or other coal-to-chemicals plants.

2. The low gasification pressure requires extra syngas compression for the methanol synthesis process. The bulk density of the released dust is much higher than that of CWS gasification.

3. The reliability of the dry-feeding gasifier is not as good as the matured CWS feeding system and the initial investment is much higher.[39]

4.9 OTHER GASIFICATION TECHNOLOGIES

4.9.1 GSP Gasification Technology

The GSP coal gasification process was developed by the German Fuel Institute of former East Germany in the 1970s. The early-stage purpose of the process was to produce town gas, and a 3 MW pilot-scale factory was set up in Freiburg to verify the technology. In 1982, a 130 MW commercial plant with 30 t/h coal processing capability was built in Scharwze Pumpe, Germany, to produce electricity. This plant operated for 5 years without changing nozzles and refractory bricks. Since 1989, the key technology development was changed to use biomass and different kinds of waste including city sludge, industrial residue, and petroleum coke as feedstock. Table 4.12 shows the gas composition with different feedstocks of the GSP gasification process.[40]

Table 4.12.   GSP Gas Composition with Different Feedstocks

Item	Lignite	Sludge	Waste Residue	Petrol Coke
Feedstock				
Moisture	0.5–16.2	3.7–8.6	4.3	1
Ash	7.5–40.2	38.1–57.3	36.8	0.6
S	0.6–6.2	0.5–1.2	1.6	5.3
C		25.0–34.3	36.8	87
H	54.1–83.1	3.5–5.0	4.9	3.9
N	0.1–5.3	1.6–3.2	7.4	2
O	0.0	10.1–19.9	7.2	0.2
Cl	0.0–5.0	0.0–0.2	0.7	0.0
Gas composition				
H_2	31	24–32	46	31.2
CO	55	45–48	43	47.5
CO_2	9	16–23	6	7.8
CH_4	<0.10	<0.1	<0.1	<0.1
N_2	4.7	4.7	4.9	13.5
H_2S/COS	0.27	0.2–1.0	0.1	19.7 g/m³

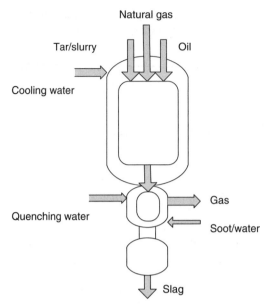

Figure 4.19.   Schematic of a typical GSP gasifier.

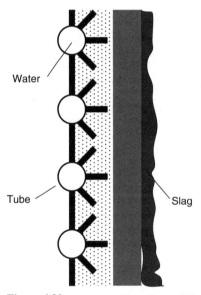

Figure 4.20.   Water-cooling wall of a GSP gasifier.

Figure 4.19 shows a schematic of the GSP process, including the nozzle, reaction chamber, the water wall, and the quench chamber. Figure 4.20 shows the structure of the water-cooling wall, which is a gastight membrane wall structure.[41] In this gasification process, slag is trapped by the internal surface of the water-

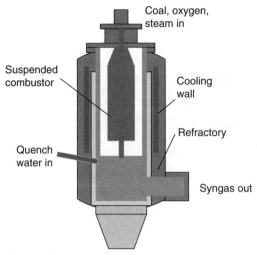

Figure 4.21.   Schematic of a typical GSP plus gasifier. See color insert.

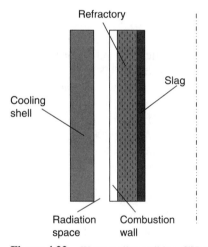

Figure 4.22.   Water-cooling wall in a GSP plus gasifier.

cooling wall in the high-temperature reaction condition, forming a solid layer and a flowing layer that varies in thickness according to the heat flux and the composition of the slag.

Figure 4.21 shows a schematic of the GSP Plus process. The distinct difference from the former GSP gasifier is that the water wall is changed from a pressure-bearing structure to a nonpressure-bearing structure, which is suspended in the gasifier. The structure can be seen in Figure 4.22.

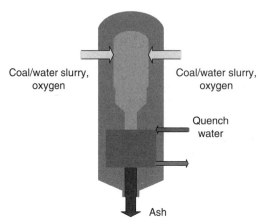

Figure 4.23. Schematic of an ECUST gasifier.

4.9.2 East China University of Science and Technology (ECUST) Gasifier

The Institute of Clean Coal Technology (ICCT) at the ECUST started to develop coal gasification technology with the YanKuang Coal Mine Group in the early 1990s.[42] ECUST incorporates both CWS feeding gasification and N_2 or CO_2 dry-feeding technology. Figure 4.23 shows a schematic of the ECUST gasifier.[42] This gasifier consists of four opposed-jet nozzles, with a down-flow configuration and slagging equipment.

With the support of the Ministry of Science and Technology of China, a 22 t/day CWS pilot gasifier was successfully started in August 2000, at the Lunan Chemical Fertilizer Plant in the Shandong province. Four years later, the first commercialized four-opposed-jet CWS gasifier with 750 t/day capacity was started up at Hualu Hengsheng Chemical Co. for ammonia production. In July 2005, a 1150 t/day gasifier went into operation in Lunan, Shandong Province. After having succeeded in developing CWS gasification technology, ICCT is now focusing on dry-feed pulverized coal gasification technology. In July 2004, it successfully demonstrated a pulverized coal gasifier with N_2 to convey the dry particles. The pilot gasifier's capacity is between 15 and 45 t of coal per day, the pressure is 2.0–2.5 MPa, and the temperature is 1300–1400 °C. The $CO + H_2$ ratio in the syngas is 89%–93%, the carbon conversion rate is 98%–99%, the oxygen consumption is 300–320 Nm^3 O_2/1000 Nm^3 ($CO + H_2$), the coal consumption is 530–540 kg coal/1000 Nm^3 ($CO + H_2$), and cold gas efficiency is over 84%. In June 2005, the gasification performance obtained by replacing nitrogen with carbon dioxide as the pulverized coal carrier was studied at the pulverized coal gasification pilot. The operating pressure was 1.0–3.0 MPa and the working temperature was about 1350 °C. As a result, the dry mole fraction of nitrogen in the syngas obtained was less then 1% and the dry mole fraction of ($CO + H_2$) was in the range of 90%–95%.

4.9.3 TPRI Gasifier

TPRI's gasifier is a two-stage dry-feed, water-cooled gasification process.[43] Approximately 80%–85% of the pulverized coal is fed into the first stage of the gasifier and reacts with pure oxygen and steam. The steam and the remaining 15%–20% of the pulverized coal are fed into the second stage from different nozzle channels and react with the high-temperature syngas from the first stage at about 1400–1500 °C. The temperature of the outlet syngas is decreased to ~900 °C after the second stage's endothermic reaction, not only freezing the slag particles but also increasing the gasifier's thermal efficiency.

In 2004, a 0.5 t/day pilot gasifier was completed to verify this gasifier concept. The result showed that the O_2 consumption was lower, the thermal efficiency was 1% higher, and the outlet syngas temperature was 200 °C lower than that of a one-stage gasifier.

A new 24–36 t/day pilot-scale demonstration gasifier aimed at verifying the whole process (reliability, availability, and coal compatibility) was built and tested recently. TPRI is building a 1000 t/day commercial-scale gasifier based on the data collected in this pilot plant.

4.9.4 Fluidized-Bed Gasifiers

The application of fluidized-bed reactors in gasification processes started in the 1920s. The atmospheric fluidized-bed process was patented in 1922. The first gasification plant was built by Winkler in 1925. Since then, fluidized beds have been widely used in gasification processes.

Generally, to avoid bed agglomeration, fluidized-bed gasifiers are operated at a maximum temperature that is lower than the AFT. The reactant gases (steam, air/O_2) are normally used as the fluidizing medium. To achieve a stable fluidization, raw coal must be ground to <10 mm in size before injection into the gasifier. As coal particles are consumed and fragmented during the reaction, smaller particles are generated and entrained with the gas product and then leave the gasifier. In general, these char particles are recovered by cyclones and recycled into the reaction system. In a fluidized bed, due to the extensive mixing and strong turbulence, the temperature distribution is much more uniform than those in moving-bed gasifiers and entrained-bed gasifiers. A typical temperature profile in a fluidized-bed gasifer is shown in Figure 4.24.

In the last several decades, many gasification processes have been developed using all three regimes of fluidization: bubbling fluidized bed, circulating fluidized bed, and transport reactor. Each gasification process exploits the particular characteristics of a fluidization regime.

The first application of the fluidized-bed gasifier in the gasification process is the original Winkler process. In the Winkler process, steam and air (or O_2) are used at atmospheric pressure. The schematic of a typical Winkler gasifier is shown in Figure 4.25. The coal (brown coal coke, subbituminous, or bituminous coals) is

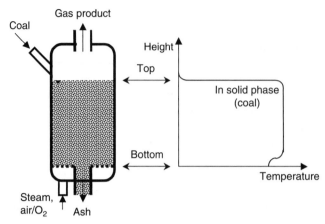

Figure 4.24.   Typical temperature profile in a fluidized-bed gasifier.

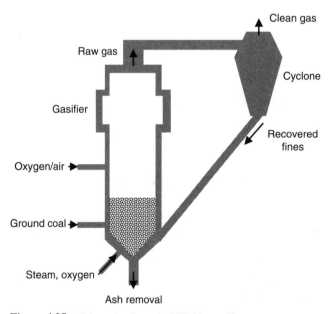

Figure 4.25.   Schematic of a typical Winkler gasifier.

milled to a particle size below 10 mm for best fluidization. The coal particles are fed
to the gasifier by a screw conveyor. The gas blast (steam and oxygen) is introduced
from the bottom of the gasifier and fluidizes the coal particles. The operating temperature is maintained below the ash melting point (950–1050 °C for most commercial plants). The gas velocity in the gasifier can be as high as 5 m/s. To minimize
the small particles entrained in the gas product, a secondary blast (oxygen) is often
fed in above the fluidized bed.

The Winkler gasification process has many advantages such as low operating temperature, high reliability, and low tar in the gas product. However, due to the low operating pressure and temperature, the carbon conversion is also low (20%–30% carbon in ash). Most plants using the Winkler process have been shut down for almost entirely economic reasons.

To improve the gas yield and increase the carbon conversion, many gasification processes (HTW, Kellogg Brown and Root [KBR], Kellogg–Rust–Westinghouse [KRW], U-Gas, etc.) have been developed with increased operating temperatures and pressures, and modified gas distributors.

The HTW technique, developed by Rheinbraun, utilizes increased pressure (up to 3.0 MPa) and operating temperature. In this process, a lock hopper and screw feeder are used to feed the coal particles into the reactor from a high-pressure charge bin. Due to the increases in both temperature and pressure, coal conversion in the HTW process is much higher than in the Winkler process.

The KBR transport gasifier was developed to operate in the high-velocity fluidization regime. To achieve higher throughput and better mass and heat transfer rates in the gasifier, the KBR gasifier operates at higher circulation rates, gas velocities, and riser densities. The superficial gas velocity in the riser can be as high as 11~18 m/s.[44] Limestone is used as the sorbent for sulfur removal and fed to the gasifier together with ground coal. In the mixing zone, solids are mixed with oxidant and steam. The sorbent reacts with sulfur and forms CaS. CaS then leaves the reactor through a screw cooler together with a char–ash mixture. These solids and fines are combusted in an atmospheric fluidized-bed combustor. The gasification takes place at 900–1000 °C under 1.0–2.0 MPa, yielding carbon conversion rates as high as 98%.

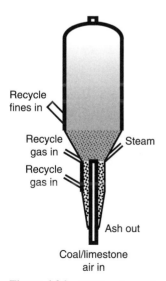

Figure 4.26.   KRW agglomerating fluidized-bed gasifier.

To minimize the carbon loss in the ash and increase the carbon conversion, agglomerating fluidized-bed gasification processes have been developed. In the agglomerating fluidized-bed gasifier, char and ash that are entrained in the product gas are recycled to a hot agglomerating and jetting zone. The temperature in the agglomerating and jetting zone is high enough to soften the ash particles. The ash particles agglomerate and fall to the bottom of the gasifier, where they are cooled and removed from the gasifier. Agglomerating gasifiers can provide a higher carbon conversion compared with conventional fluidized gasifiers.

Figure 4.26 shows an agglomerating fluidized-bed gasifier developed by KRW. In the KRW process, limestone can be injected with the coal for sulfur capture. The KRW process is employed in a 100 MW IGCC Piñon Pine Plant near Reno, NV. Due to some difficulties in the hot gas filtering section, the plant could not be started up successfully.[1]

4.9.5   ConocoPhillips Gasifier

The ConocoPhillips process is a CWS-fed, pressurized, two-stage process. The process was originally developed by Dow Chemical in the 1970s. In 1989, the technology was named Destec. The technology has been chosen for a repowering IGCC at Wabash River, IN. The gasifier shown in Figure 4.27 consists of a high-pressure vessel lined with refractory bricks.[45]

In the lower part of the gasifier, CWS is injected with O_2 via two opposed nozzles. The reaction temperature in this first stage is normally 1350–1400 °C and the pressure is about 3 MPa. The molten slag runs down the gasifier and is discharged

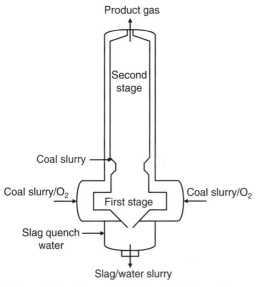

Figure 4.27.   Schematic of a ConocoPhillips gasifier.

Table 4.13. ConocoPhillips Gasifier Operating Conditions

Temperature (°C)	1370–1540 (1st stage)		
	980–1020 (2nd stage)		
Pressure (MPa)	2.0–4.0		
Carbon conversion	>98%		
Cold gas efficiency	70%–80%		
Solid fuel	Subbituminous	Bituminous	Pet Coke
Unit capacity (t/day)	2750	2750	2750
Sulfur content (%)	0.4	3.7	6.2
Heating value dry (HHV BTU/lb)	11,700	12,500	14,100
Oxygen consumption (t/day)	2220	2330	2540
Syngas			
CO (%)	42.9	46.5	60.4
H_2 (%)	32.6	32.2	21.7
CO_2 (%)	20.2	15.1	10.2
CH_4 (%)	1.7	3.2	4.7
Inert (%)	2.6	3	3
Heat content (HHV BTU/h)	2030	2200	2620
Heating value (HHV BTU/SCF)	260	285	311
Steam from syngas cooler			
Flow (kg/h)	23,600	22,700	199,800
Potential H_2 (MMSCFD)	113	116	132

HHV, high heat value.

into the quench chamber. The high-temperature syngas from the first stage flows to the upper second stage of the gasifier, where the remaining 20% of the coal slurry is injected and reacts, cooling the gas to about 1050 °C. The crude syngas is then cooled in a fire tube syngas cooler, followed by the cleanup using filters that remove large ash and char particles. The char obtained can be recycled to the gasifier. Typical operating conditions and syngas components are shown in Table 4.13.[45]

The only operational ConocoPhillips gasifier is at the Wabash River IGCC and runs on bituminous coal. Extensive tests have been carried out over the years on subbituminous coals and petroleum coke.

4.9.6 Moving-Bed and Fixed-Bed Gasifiers: Lurgi's Gasification Technology

Gasification processes based on moving beds (fixed beds) have been used for more than 100 years for making producer gas or water gas using air gasification.

In the producer gas process, coal is fed from the top and moves slowly downwards while humidified air is blown upward through the bed of coal or coke. Ash is removed from the bottom of the gasifier through a rotating grate. The gas product contains approximately 50 mol % N_2 because air is used as the gasification agent. Therefore, the heating value of the product gas is very low. Currently, this process is only used in some parts of China.

In the water-gas process, hydrogen and carbon monoxide are generated from the reaction between steam and high-temperature coke in a two-step process. In the first step, the coal bed is heated to about 1300 °C with upward blown air. The reactant gas is then switched to steam, creating the syngas and cooling the coal bed. To make optimum use of the heat in the system, steam is blown first upward then downward. When the bed temperature drops to about 900 °C, the steam is stopped and the next cycle is started. The product from the water-gas process can be used for ammonia or methanol synthesis.

In 1927, a patent was granted for the Lurgi dry ash process (coal pressure gasification). In 1931, Lurgi upgraded the existing atmospheric producer gas technology by pressurizing it and using an oxygen blast. The Lurgi dry ash process, together with its continuous developments, had been the only pressure gasification system for many years. Figure 4.28 shows a typical Lurgi dry ash gasification system. In this process, coal is first loaded from an overhead bunker into a lock hopper. The lock hopper is then pressurized with syngas and opened to the reactor, in which coal and oxygen flow countercurrently. The coal is fed into the reactor through a distributor and then slowly moves down through the gasifier. The gasification agent (O_2 and steam) is introduced from the bottom of the gasifier through a grate. Four reaction zones are distinguishable from top to bottom: (a) the coal-drying zone, (b) the coal devolatilization zone, (c) the gasification zone, and (d) the combustion zone. The cold inlet stream of gasification agent (steam/O_2) is preheated by the hot ash at the bottom, while hot raw gas exchanges heat with the cold incoming coal. This process results in a relatively low ash and raw gas product temperature at the exit of the

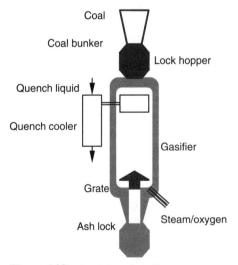

Figure 4.28.   Lurgi dry ash gasifier.

gasifier, which improves the thermal efficiency and lowers the steam consumption.[46] Gasification takes place in the gasification zone and unconverted coal is combusted in the combustion zone. The Lurgi process has a low O_2 consumption and high steam demand. Also, in the gas product, there is a high content of methane. Lurgi-type moving-bed dry ash gasifiers have been widely used around the world. In the U.S., Lurgi dry ash gasifiers are used for SNG production. In China, the Lurgi process is used to produce town gas, hydrogen, and ammonia.

4.9.7 Summary of Different Gasification Technologies

The major differences between various types of gasification processes and the technical trends are briefly summarized in Table 4.14.

Table 4.14. Comparison of Different Gasification Processes

	Reactor	Feed System	Gasification Agent	P (MPa)	T (°C)	O/Coal Ratio (kg/kg)	Carbon Conversion (%)
GE	Entrained flow	Coal–water slurry	O_2	3.0–6.5	1260–1540	~0.9	97.2
Shell	Entrained flow	Dry feeding	O_2	~3	1500–2000	~0.86	99
BGL	Fixed/moving bed	Dry feeding	O_2	~2.5	>2000	~0.52	99.9
HTW	Fluidized bed	Dry feeding	O_2/air	~1	<1000	0.54	96
U-Gas	Fluidized bed	Dry feeding	O_2/air	0.4–3.2	950–1090	~0.6	95.3
KRW	Fluidized bed	Dry feeding	O_2/air	2.1	870–1040	0.68	95
ConocoPhillips	2-step entrained flow	Coal–water slurry	O_2	2.0–4.0	1350–1400		>98%
E-Gas	Entrained flow	Slurry	O_2	2.2–3.0	<1040	0.89	98.9
PRENFLO	Entrained flow	Dry feeding	O_2	3	1500–2000	1.03	99.3
GSP	Entrained flow	Dry feeding	O_2	3	1600–2000	0.8	98
ECUST	Entrained flow	Coal–water slurry	O_2	2–6.5	1300–1400		98–99
TPRI	2-step entrained flow	Dry feeding	O_2		1400–1500		

4.10 CHALLENGES IN GASIFICATION TECHNOLOGY: SOME EXAMPLES

4.10.1 High AFT Coals

The AFT is a very important parameter when designing different kinds of gasifiers. For entrained-bed gasifiers, the reaction temperature must be above the fluid temperature (FT) of the ash to keep the ash flowing continuously. The AFT can be obtained by experiments specified in the ASTM D1857, "Fusibility of Coal and Coke Ash," or similar specifications, such as ISO 540.[47] In these methods, the temperatures measured relate to the behavior of an ash sample under specified conditions and are reported as initial deformation temperature (DT), softening temperature (ST), hemispherical temperature (HT), and FT. For gasifier applications, the ash-melting characteristics should be determined under reducing conditions, since these data may differ considerably from data in oxidizing conditions. It is thought that pressure does not significantly affect these properties.

An additional property required for slagging gasifiers is the slag viscosity–temperature relationship. It is generally accepted that for reliable, continuous slag tapping, a viscosity of less than 25 Pa·s (250 poise) is required. The temperature required to achieve this viscosity (T_{250}) is therefore sometimes used in the literature. Some slags are characterized by a typical exponential relationship between viscosity and temperature over a wide temperature range. For other slags, the behavior changes at a critical temperature (T_{cv}), at which temperature the viscosity increases very rapidly with decreasing temperature. For a slagging gasifier to operate at a reasonable temperature, it is necessary for the slag to have a $T_{cv} < 1400\,°C$.

The relationship between ash-melting characteristics and composition is very complicated, and depends largely on the quaternary mixture of SiO_2–Al_2O_3–CaO–FeO. In general, slags that are high in silica and/or alumina will have high ash-melting points, but this is reduced by the presence of both iron and calcium, hence the use of limestone as an ash-fluxing agent. The SiO_2/Al_2O_3 ratio is also important, and where the calcium content is already high (as in some German lignites), there can be some advantage to lowering the ash-melting point by adding SiO_2.

The ash that is produced in the gasifier always has a lower density than the minerals from which they originate, due to loss of water, decomposition of carbonates, and the presence of some carbon. The bulk density of the ash in particular may be low due to the formation of hollow ash particles. This means that special attention has to be given to the transformation and transport of such ashes.[3]

A significant technology challenge is to gasify coals with high AFTs. These coals account for a significant portion of world coal reserves, especially in China where ~50% of the coal reserves are not suitable for gasification feedstocks. The gasification of high AFT coals requires high gasification temperatures to ensure that the slagging system is working continuously. This leads to shortened lives of the nozzle and the refractory bricks, as well as higher demand of pure O_2, resulting in higher operational and maintenance costs.

4.10.2 Increasing the Coal Concentration in the CWS

It is believed that the preparation of the feedstock for coal gasification has a significant effect on the life of the feeding nozzle in the gasifier, and thus affects the process reliability and availability. CWS holds promise as an alternative to oil as fuel. The advantages of CWS as a fuel include that it can be stored without the danger of coal-dust explosions and that it can be pumped and transported in pipelines and combusted like residual fuel oil in an environmentally benign manner. To date, 23 large-scale CWS plants (>30,000 t/year) and almost 100 small- to medium-scale plants (<30,000 t/year) have been built with an annual capacity of about 10 Mt/year (including a 1.5 Mt/year plant in Nanhai, Guangdong Province, with a projected startup date of 2005). In addition, several CWS plants with a total capability of more than 10 Mt/year are being designed.

CWS is a highly loaded suspension of fine coal in water. Typical CWS consists of 60–70 wt % coal, 30–40 wt % water, and about 1 wt % chemical additives. Lowering the water concentration in CWS causes higher thermal efficiency and conversion in gasification, and hence lowers the operating cost. Thus, increasing CWS coal concentration is of great importance in coal gasification. CWS is not economic if the CWS coal concentration is well below 60 wt %.

Every single coal source has its specific slurrying properties, such as moisture, oxygen-containing functional groups, mineral matter content, porosity, hydrophobicity, and organic matter content, which affect its slurriability. However, the inherent relationship between coal properties and its slurriability is not clear yet, so laboratory testing on each candidate coal is a must before industrial application. More work is necessary to get a deeper understanding of the mechanism of CWS formulation.

Among all the coal properties, it is found that inherent moisture and oxygen-containing functional groups have bigger effects on slurriability than others. Most coals contain oxygen-containing functional groups as well as a small fraction of chemically bound water, which could not be removed by air-drying. Low-rank coals, such as lignite, inherently contain more oxygen and bound water than the high-rank coals, such as bituminous and subbituminous coals. Lignites typically contain 15–30 wt % oxygen and 25–40 wt % bound water. In contrast, subbituminous coals generally contain less than 15 wt % oxygen and 20 wt % water, while bituminous coals typically contain less than 10 wt % oxygen and 10 wt % bound water. Higher oxygen content and inherent moisture normally result in lower slurry concentration. Therefore, the most suitable coal types to produce CWS are high-rank coals that contain less oxygen and inherent moisture. At present, most CWS feeding gasifiers use coals whose total moisture (Mt) is below 15 wt % as leverage between their reactivity and ability to form slurries. However, there are large amounts of low-rank coals in the U.S. and in China (12.7% brown coal, ~40% raw soft coal with $Mt > 15\%$). It is quite difficult to produce CWS with coal concentrations of more than 60 wt % using these coals, even in the presence of additives. Consequently, these low-rank coals do not meet the commercial requirements for use in CWS gasifiers. Also, the extra additives to CWS may cause corrosion in downstream pipelines and increase operating costs.

Besides coal properties, a number of other variables also have significant effects on the behavior of CWS, such as coal PSD, the preparation process of the slurry, and the type and volume of chemical additives. Accordingly, CWS coal concentration can be increased by means of seeking effective additives and optimizing PSD and preparation processes. Unfortunately, knowledge in this area is mainly empirical. The inherent relationship between slurry rheology and the preparation process is also unknown.

4.10.3   Improved Performance and Life of Gasifier Nozzles

The improvement or development of new feeding systems for high-pressure gasifiers remains fairly high on the list of priorities. Almost all gasifier users identify the nozzles used in gasifiers as one of the weakest links in the process for achieving high onstream availability factors. The nozzles in gasifiers must feed steam, O_2, and feedstock into a very harsh, high temperature, reducing environment. The life of a typical nozzle is generally between 1 and 6 months, based on industrial experience.

The application of computational fluid dynamics (CFD) modeling around the nozzle may be helpful in elucidating some of the parameters affecting nozzle life. However, there is currently a lack of thorough understanding of all the parameters affecting nozzle life and the fluid dynamics in the vicinity of the nozzle. Additional information may be helpful in designing and extending the life of the injector nozzles.

Materials used in the manufacture of the injector nozzles are cited most often as important parameters. New materials or coatings for existing materials are needed to provide protection from sulfidation and corrosion at high reactor temperatures. Also, better gaskets are needed for equipment that feeds hot O_2 into the nozzle.

It is also believed that nozzle life is affected by feedstock characteristics. Although the dry-feed system may be more difficult to operate at high pressures, nozzle life may be longer in the absence of large amounts of evaporating water. Deeper understanding is needed to identify the effect of feedstock preparation.

In addition, there is also an expressed need for fuel-flexible injector nozzles. In the past, some nozzles have experienced vibration problems because of the load and feedstock changes. Nozzles are desired that can operate with more than one feedstock, with no adverse effect on the operation, and no need to shut down to switch nozzles whenever the conditions require a change in feedstocks.

4.10.4   Gasifier Refractory Brick Life

In order to protect the outer steel shell from high temperature, erosion, and corrosion, the gasification chamber is lined with refractory bricks. Furthermore, the coal

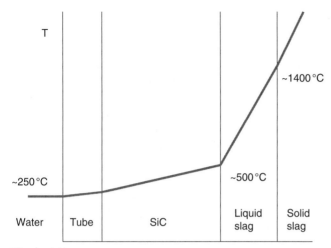

Figure 4.29.   Temperature distribution in a water-cooling wall.

normally contains about 10 wt % ash, and the ash becomes liquid slag at typical operating conditions for gasification. The flowing slag corrodes, penetrates, and reacts with the refractory liner, severely shortening the refractory service and gasifier operation cycle.[48] In practice, the refractory bricks containing high chromium oxide material commonly last from 3 months to 2.5 years.

In summary, the main reasons for the refractory failure in gasifiers include high temperature, large and quick temperature changes, erosion by particulate, molten slag attack, slag variation from the feedstock, hot corrosive gases, and alkali vapor attack.[48]

The replacement of refractory bricks is very expensive; hence, some gasification plants must have a spare gasifier as a backup for the gasifier shutdown schedule, such as brick and nozzle replacement. Refractory brick failure is one of the major problems for gasifiers to achieve high reliability, availability, and maintainability.[47]

Compared with the GE gasifier, Shell and GSP gasifiers employ a water-cooling wall with a refractory lining for the reaction chamber. During the operation, the slag freezes on the surface of the refractory material, restricting slag penetration and corrosion. A scheme of the temperature distribution in the GSP water-cooling wall is shown in Figure 4.29.

Shell claims that the water-cooling wall of its gasifier can last over 20 years and GSP declared 5 years, both of which are much more durable than a CWS gasifier. Another advantage of the long lifetime of the water-cooling wall is that a spare gasifier is unnecessary, which reduces CAPEX of the gasification unit dramatically.

4.10.5 Gasifier Scale-Up

In a commercialized coal-to-liquid (CTL) or IGCC plant, a large amount of coal must be processed by the gasifiers to produce enough syngas. Based on Sasol's experience,[49] minimum indirect plant capacity should be in the range from 40,000 to 80,000 barrels per day to realize the benefits from the economy of scale. This requires the gasifiers to have a capacity of at least 40,000 t of coal per day. Currently, the capacity of the largest GE gasifier is about 2400 t/day, which is operating at Tampa Electric Company's (TECO) Polk County Station in Florida. To reduce the capital cost of a CTL or IGCC plant, it is highly desired to develop an efficient scale-up methodology to further increase the capacity of the gasifier.

Through analyzing the current scale-up method of entrained flow gasifiers, Holt[50] pointed out that it is generally assumed that gasification reactions are first order, and accordingly, gasifiers have usually been sized based on the gas residence time.

However, actual high-temperature gasification reaction rates are not known, and the residence time at a given reactor temperature is usually determined empirically based on small-scale gasifier experiments. It is also generally accepted that reaction rates vary with coal type.

In industrial practice, it has been observed that the scale-up of Texaco (not GE) gasifiers from about 15 t/day (in Montebello) to 150 t/day (in Ruhrkohle) and to 1000–1200 t/day (in Cool Water) was quite satisfactory. However, when conditions were scaled up to about 2400 t/day (Tampa size), lower carbon conversion per pass was observed at the larger diameter and gasifier size, which indicates that the current scale-up method is not sufficient. Therefore, more systematic research work is necessary to establish a more reliable scale-up method.

4.11 SYNGAS CLEANUP

The raw syngas produced by the gasification of various feedstocks is similar. The main gas components are hydrogen (H_2) and carbon monoxide (CO), carbon dioxide (CO_2), water vapor (H_2O), and various impurities. However, the concentrations of these various components depend on the feedstock composition and the specific gasification process employed.

The primary feedstock impurities of concern are the sulfur and ash constituents. In any gasification process, sulfur is converted mainly to H_2S and COS. H_2S is the dominant sulfur species, and approximately 93%–96% of the sulfur is in this form (the rest being COS). A portion of the ash is entrained as particulates; and the mercury is vaporized. The entrained particulate matter also includes unburned carbon. Small amounts of HCN, NH_3, and traces of metal carbonyl compounds are also produced. The required extents of removal of each of the impurities depend on the application, that is, IGCC or chemical synthesis.[3] In order for a gasifier to be integrated with a combustion turbine in an IGCC plant, under current emission regulations, the sulfur content in syngas should be less than 20 ppmv. For chemical synthesis, there is a much more stringent requirement than that for IGCC. To prevent deactivation of methanol and Fischer–Tropsch synthesis catalysts, reduction of the total syngas sulfur content to <0.1 ppmv is required.

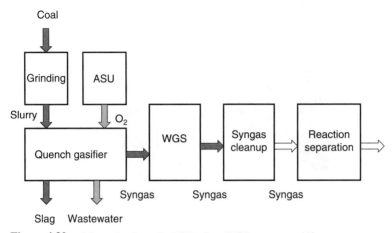

Figure 4.30. Schematic of a typical CTL plant. WGS, water-gas shift.

In quenching system, the process also effectively removes a significant amount of the syngas contaminants, including particulates, chlorides, alkalis, nonvolatile metals, and some ammonia. The only contaminants that remain in the quenched syngas are sulfur gases (H_2S and COS), some ammonia, traces of volatile metals such as mercury and arsenic, and traces of HCl. Figure 4.30 shows a simplified process flow diagram of a typical CTL plant. The temperature of raw syngas from the quench gasifier is in the range of 250~350 °C. However, conventional gas cleanup is typically carried out by scrubbing the syngas using chemical or physical solvents that require cooling the gas below 40 °C. The need to cool the gas to ambient temperature requires the use of a lot of additional equipment (heat exchangers, knockout pots, condensate handling system, etc.). In addition, the cooling of the gas, which must ultimately be reheated before being sent to the reactor, introduces a thermodynamic penalty on the overall system. In a typical CTL plant, the gas cleanup unit takes up a large portion of total CAPEX, as shown in Figure 4.31. Therefore, it is highly desired to develop cost-effective hot/warm desulfurization technologies that can significantly reduce CAPEX and lower the operating and maintenance costs of a CTL plant.

There are numerous commercial processes available to remove the sulfur content from syngas, ranging from throwaway regenerable adsorbent-type to regenerable solvent-type processes. Solvent-type processes are of primary interest for sulfur removal from syngas. The solvent-type processes can be divided into three types:

- chemical solvent,
- physical solvent, and
- mixed chemical/physical.

There are many commercial installations of each type, treating a variety of natural gas and syngas streams. For syngas treatment, the principal chemical-type

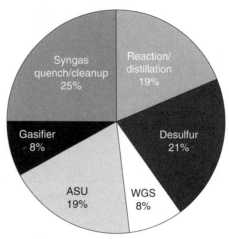

Figure 4.31. CAPEX (%) of a typical 200 kt/year methanol plant. WGS, water-gas shift. See color insert.

solvents of primary interest are aqueous amines, with methyl diethanol amine (MDEA) being the current favorite. The physical solvents coabsorb hydrocarbons to a much greater extent than the amines, causing loss of valuable hydrocarbons. However, since syngas does not contain appreciable quantities of hydrocarbons, physical solvents are also used for syngas cleanup. The most popular physical solvents are methanol and dimethyl ether of polyethylene glycol, as represented by the Rectisol process of Lurgi and the Selexol processes of UOP (Universal Oil Products, now a Honeywell company), respectively.

The principal challenge of the sulfur removal processes is to reduce the sulfur compounds in the syngas to a low enough level to meet the requirements of downstream processes and prevailing emission regulations, and to do so as economically as possible. This would not be a problem if only H_2S were present, since H_2S is easily removed by most processes down to very low levels. However, the solubility of COS in both types of solvents is much lower than that of H_2S, making it more difficult to remove.

Table 4.15 compares common sulfur removal processes. Amine processes are based on the removal of an acid gas by virtue of a weak chemical bond between the acid gas component and the amine. Amine-based sulfur removal processes are generally regarded as a low capital cost option with part CO_2 coabsorption. However, amines do not chemically combine with COS. Only limited amounts of COS are absorbed with a physical solvent. COS can be physically removed only with very high solvent circulation rates. For syngases that contain appreciable quantities of COS, prior removal of the COS is usually required. In addition, for some of amine solvents, degradation and corrosion are also main disadvantages of the process.

At high syngas pressure, physical solvent-based processes become increasingly attractive, for example, the Rectisol and Selexol processes. There are more than 55 Selexol plants worldwide, treating natural and syngas.[51] The Selexol process solvent is a mixture of dimethyl ethers of polyethylene glycol, and has the formulation

Table 4.15.   Common Sulfur Removal Processes

Process	Amine	Rectisol	Selexol	Metal Oxides
Absorbent	Amine solution	Cold methanol	DEPE	Mn, Fe, Cu, Co, Ce, Zn
Pressure (MPa)	<7	5.8	1.6–7.0	700
Temperature (°C)	25–60	−70 to −30	−5 to 25	<0.1
Outlet sulfur content (ppm)	MEA <1, MDEA <0.1	<0.1	<5	High sulfur removal efficiency, high thermal efficiency, high operation temperature
Advantages	Solvent regeneration, low solvent cost, CO_2 coabsorption	High sulfur removal efficiency, complete CO_2 removal, CO_2 coabsorption	Moderate CO_2 slip, moderate cost, CO_2 coadsorption	Mechanical stability, regeneration, diffusion control
Disadvantages	Corrosion, foaming, high solution circulation rate, solution degradation	High cost, toxic, low temperature leads to thermal loss	High circulation rate, high sulfur outlet content	

DEPE, dimethyl ethers of polyethylene glycol; MEA, monoethanolamine.

$CH_3(CH_2CH_2O)_nCH_3$, where n is between 3 and 9. The Selexol solvent is chemically and thermally stable, and has a low vapor pressure that limits its losses to the treated gas. The solvent has a high solubility for CO_2, H_2S, and COS. It also has an appreciable selectivity for H_2S over CO_2. The sulfur content of the purified syngas from the Selexol absorber can be lower than 5 ppmv. For a higher level of sulfur removal for downstream catalytic conversion, a guard bed using ZnO sorbents is usually required.

The Rectisol process, developed by Lurgi, is the most widely used physical solvent gas treating process in the world. More than 100 Rectisol units are in operation or under construction worldwide. Its most prevalent application is for deep sulfur removal from syngas that subsequently undergoes catalytic conversion to such products as ammonia, hydrogen, and Fischer–Tropsch liquids.

The Rectisol process uses cold methanol at a temperature of about −70 to −30 °C. Methanol's selectivity for H_2S over CO_2 at these temperatures is about 6,[52] a little lower than that of Selexol at its usual operating temperature. However, the solubilities of H_2S and COS in methanol, at typical process operating temperatures, are higher than that in Selexol and allow for very deep sulfur removal (<0.1 ppmv).

Rectisol's high selectivity for H_2S over CO_2, combined with the ability to remove COS, is the primary advantage of the process. Rectisol's complex process and the need to refrigerate the solvent are its main disadvantages, resulting in high capital and operating costs.

Various systems for dry removal of sulfur components of syngas over the 250–900 °C temperature range have been tested at the pilot-plant level, and a small number have been installed in commercial-scale IGCC demonstration plants.[53–55] In these processes, several different kinds of sorbents based on zinc (Zn), iron (Fe), copper (Cu), manganese (Mn), and so on have been tested. In the U.S. Department of Energy (DOE) program, "high temperature" typically refers to temperature higher than 600 °C. However, interests in hot gas cleanup (HGCU) have declined, mainly due to process and equipment development challenges, for example, attrition and long-term regeneration of the desulfurization sorbents, and the high cost and unproven status of GT fuel control valves for syngas at high temperatures.

The 260 MW coal-fired IGCC at TECO's Polk County Station in Florida (GE gasification process with a radiant syngas cooler, convective coolers, and water scrubbing) is equipped with both a 100% capacity Selexol system and a 10% capacity HGCU system to be fed by a syngas slipstream. The design placed the absorber column on top of the down-flow regenerator column in one tall column. The regenerated sorbent is transported back to the top of the column. The SO_2 is converted to sulfuric acid for sale.

The HGCU demonstration in Tampa was never demonstrated for the following reasons:[51]

1. The fouling factors in the waste heat boiler (radiant syngas cooler) were not as severe as predicted. Consequently, heat recovery was more efficient and the syngas was cooled to about 700 °F—a much lower temperature than expected. A temperature of at least 900 °F is needed for the HGCU.

2. Cold flow attrition tests on the sorbent showed that sorbent attrition would be very high, leading to extremely high annual sorbent costs.

3. There were also concerns about the potential for chloride stress corrosion cracking with the materials used in the HGCU.

The U.S. DOE's extensive gasification industry interviews in 2002 found that there was not much incentive for gas cleanup operations above 400 °C. Prior engineering analyses have persuaded the industry that the efficiency improvements from operating above 400 °C are not worth the additional capital costs due to materials and the increased equipment sizes resulting from the larger volumetric flows.

Recognizing many of the issues and limitations surrounding hot gas cleaning technologies and their applications, DOE's Gasification Technologies program has transitioned its gas cleaning component away from the development of high-temperature approaches to more moderate temperatures, consistent with downstream applications (e.g., 150–350 °C).

Research Triangle Institute (RTI) is now working with Eastman for installation and long-term testing of the modular warm gas cleanup processes, using a slipstream of raw syngas from Eastman's commercial 1200 t/day coal gasifier in Kingsport, TN.[56] The process uses RTI's attrition-resistant RTI-3 zinc oxide-based sorbent in a transport reactor system consisting of two coupled circulating fluidized-bed reactors for desulfurization and regeneration. Warm syngas enters the absorber at 300–430 °C. Essentially all of the H_2S and COS is removed by the sorbent. This reaction is slightly exothermic. A portion of the sorbent is continuously regenerated in the regeneration reactor. This reaction occurs at about 700 °C. As the reaction is highly exothermic, typically 6–10 vol% O_2 in N_2 is used as the oxidant gas. The oxidant enters the regenerator at about 600 °C, and due to the exothermicity of the reaction, the temperature increases to 700 °C, where regeneration occurs efficiently.

The transport reactor appears ideal for efficiently regenerating the sorbent. The key to the successful operation of this unit is the attrition-resistant, reactive, and regenerable RTI-3 sorbent. Recently, the RTI-3 sorbent has been scaled up to a production level of 10,000 lbs by Süd-Chemie Inc. (SCI).

4.12 INTEGRATION OF COAL GASIFICATION WITH COAL POLYGENERATION SYSTEMS

Due to increased costs of natural gas and petroleum, the generation of chemical products from coal and other low-grade fuels is receiving more consideration. The implementation of an optimal polygeneration system consists of finding the right combination of products and operating conditions. A variety of individual processes are being considered, and the number of candidates for such combinations is very large. For example, coproduction of electricity, hydrogen, MeOH, dimethyl ether (DME), FT fuels, and ammonia could be performed by up to 63 different production schemes, each of which has different designs. Although in practice the number is significantly less, there are still hundreds of possibilities. Indeed, there are already more than 30 designs in recent publications and more can be expected in the future. Therefore, it is very important to find the optimal schemes from the large number of candidates before proceeding to the laboratorial and commercial stages.

Polygen systems are designed to have great operational flexibility. Such flexibility is reflected by two aspects: (a) feed adaptability and (b) product selectivity. Feed adaptability means that the system should be able to effectively process different feedstocks with different coal ranks, ash and water content, AFTs, and so on. Product selectivity means that the system can selectively produce valuable products according to the market needs to achieve maximum profit. One typical polygen example is the electricity–chemical combination, in which the system can switch toward chemical production during off-peak time so that the redundant syngas is not wasted. However, several problems need to be solved to achieve this flexibility, for example, the design of the system, the reliable control during such a dynamic process, and the optimization of product selectivity.

As discussed above, the polygen system is a combination of several individual sub-systems that are already very complex. As a result of the overlapping effects from each subsystem, the complexity of a polygen system is expected to be much more subtle. The following research is important for better understanding the coal polygen system and also important for the integration of coal gasification with a polygen system:

1. Parameter sensitivity on the system performance. In such a complex system, the number of controllable parameters is extremely large. The effects of these parameters on the overall performance are usually interconnected with each other and are difficult to predict without an extremely detailed model.

2. Dynamic behavior and the stability. In a polygen system, in order to achieve operational flexibility, the operation conditions have to be changed frequently. For example, in the electricity–chemical scheme, the fluctuation of the throughput of the chemical subsystem could be as large as 50% in 1 day. Under such a big fluctuation, the dynamic behavior should not be neglected during design and operation. Also, changes in operating conditions can cause system instability, which is a large cause for concern when dealing with high temperature and pressure. Therefore, the demands of a polygen system on the control system are much greater than those in other systems.

REFERENCES

1. National Energy Technology Laboratory and the United States Department of Energy (NETL-DOE). World wide gasification database online. Pittsburgh, PA. **2002**. http://www.netl.doe.gov/publications/brochures/pdfs/SFAgasifi.pdf (accessed September 16, 2009).
2. WILLIAMS, R.H. *IGCC: Next step on the path to gasification-based energy from coal.* Report for the National Commission on Energy Policy. November **2004**.
3. HIGMAN, C. and VAN DER BURGT, M.J. *Gasification.* Burlington, MA: Elsevier, **2003**.
4. WILLIAMS, A., SKORUPSKA, N., and POURKASHANIAN, M. *Combustion and Gasification of Coal.* New York: Taylor & Francis Group, **2000**.
5. SCHILLING, H.D., BONN, B., and KRAUSS, U. *Coal Gasification: Existing Processes and New Developments.* London: Graham & Trotman, **1981**.
6. PERRY, S.T. *A Global free-radical mechanism for nitrogen release during coal devolatilization based on chemical structure.* PhD Dissertation, Chemical Engineering Department, Brigham Young University, Provo, UT, **1999**.
7. ZENG, D. and FLETCHER, T.H. Effects of pressure on coal pyrolysis and char morphology. *Energy & Fuels,* **2005**, 19, 1828.
8. ANTHONY, D.B. and HOWARD, J.B. Coal devolatilization and hydrogasification. *AIChE Journal,* **1976**, 22, 625.
9. SUUBERG, E.M. *Rapid pyrolysis and hydropyrolysis of coal.* ScD Thesis, MIT, **1977**.
10. MANTON, N., COR, J., MUL, G., EXKSTROM, D., and NIKSA, S. Impact of pressure variations on coal devolatilization products. 2. Detailed product distribution from 1.0 MPa. *Energy & Fuels,* **2004**, 18, 508.
11. YU, J., HARRIS, D., LUCAS, J., ROBERTS, D., WU, H., and WALL, T. Effect of pressure on char formation during pyrolysis of pulverized coal. *Energy & Fuels,* **2004**, 18, 1346.
12. CHEN, J.C., CASTAGNOLI, C., and NIKSA, S. Coal devolatilization during rapid transient heating. 2. Secondary pyrolysis. *Energy & Fuels,* **1992**, 6, 264.

13. Brown, A.L. and Fletcher, T.H. Modeling soot derived from pulverized coal. *Energy & Fuels*, **1998**, 12, 757.
14. Fletcher, T.H. Swelling properties of coal chars during rapid pyrolysis and combustion. *Fuel*, **1993**, 72, 1485.
15. Smoot, L.D. and Smith, P.J. *Coal Combustion and Gasification*. New York: Plenum Press, **1985**.
16. Smith, L.K., Smoot, L.D., Fletcher, T.H., and Pugmire, R.J. *The Structure and Reaction Processes of Coal*. New York: Plenum Press, **1994**.
17. Glick, D.C. and Davis, A. Variability in the inorganic element content of U.S. coals, including results of cluster analysis. *Organic Geochemistry*, **1987**, 11, 331.
18. Gale, T.K., Bartholomew, C.H., and Fletcher, T.H. Decreases in the swelling and porosity of bituminous coals during devolatilization at high heating rates. *Combustion and Flame*, **1995**, 100, 94.
19. Benson, S.A., Jones, M.L., and Harb, J.N. Chapter 4. Ash Formation and Deposition. In *Fundamentals of Coal Combustion for Clean and Efficient Use* (ed. L.D. Smoot). New York: Elsevier, pp. 299–373, **1993**.
20. Raask, E. *Mineral Impurities in Coal Combustion, Behavior, Problems, and Remedial Measures*. Washington, DC: Hemisphere Publishing Corp., **1985**.
21. Hoy, H.R., Roberts, A.G., and Wilkins, D.M. Behavior of mineral matter in slagging gasification processes. *Institution of Gas Engineers Journal*, **1965**, 5, 444.
22. Balzhiser, R.E., Samuels, M.R., and Eliassen, J.D. Equilibrium in chemical reacting systems. In *Chemical Engineering Thermodynamics: The Study of Energy, Entropy, and Equilibrium*. Upper Saddle River, NJ: Prentice-Hall, pp. 506–527, **1972**.
23. McBride, B.J. and Gordon, S. Computer program for calculation of complex chemical equilibrium compositions and applications II. User's manual and program description. **2006**. http://www.grc.nasa.gov/www/ceaweb/rp-1311-p2.pdf (accessed September 16, 2009).
24. Niksa, S., Liu, G., and Hurt, R.H. Coal conversion submodels for design applications at elevated pressures. Part I. Devolatilization and char oxidation. *Progress in Energy and Combustion Science*, **2003**, 29, 425.
25. Ergun, S.B. Kinetics of the reaction of carbon dioxide with carbon. *The Journal of Physical Chemistry*, **1956**, 60, 480–485.
26. Gadsby, J., Long, F.J., Sleightholm, P., and Sykes, K.W. The mechanism of the carbon dioxide-carbon reaction. *Proceedings of the Royal Society of London*, **1948**, A193, 357.
27. Roberts, D.G. and Harris, D.J. Char gasification with O_2, CO_2, and H_2O: Effects of pressure on intrinsic reaction kinetics. *Energy & Fuels*, **2000**, 14, 483.
28. Roberts, D.G. and Harris, D.J. A kinetic analysis of coal char gasification reactions at high pressures. *Energy & Fuels*, **2006**, 20, 2314.
29. Wen, C.Y. and Lee, E.S. *Coal Conversion Technology*. Reading, MA: Addison-Wesley, **1979**.
30. Mühlen, H.-J., vanHeek, K.H., and Juntgen, H. Kinetic studies of steam gasification of char in the presence of H_2, CO_2, and CO. *Fuel*, **1984**, 64, 944.
31. Liu, G.-S. and Niksa, S. Coal conversion submodels for design applications at elevated pressures. Part II. Char gasification. *Progress in Energy and Combustion Science*, **2004**, 18, 508.
32. Xu, S., Zhang, D., and Ren, Y. *Coal Gasification Technology*. Beijing: Chemical Industry Press, **2006**.
33. Department of Trade and Industry (DTI). Technological status report: Gasification of solid and liquid fuels for power generation. *Technology Status Report, TSR 008*, **1998**.
34. Zuideveld, P. and Graaf, J. Overview of Shell Global Solutions' worldwide gasification development. *Gasification Technologies*, **2003**, San Francisco, CA.
35. Dongliang, Z. Present status and future development of coal gasification technology in China. *Coal Chemical Industry*, **2004**, 2, 1.
36. De Graff, J.D. and Chen, Q. An update on Shell Licensed Gasification Projects and Performance of Pernis IGCC plant. *Gasification Technologies Conference*. San Francisco, CA, **2000**.
37. He, Y. Chapter 4. In *Modern Coal Industry Handbook* (ed. He, Y.). Beijing: Chemical Industry Publishing House, pp. 506–571, **2004**.
38. Sha, X. and Yang, N. *Coal Gasification and Application*. Shanghai: East China University of Science and Technology Publishing House, pp. 258–261, **1995**.

39. YU, G., NIU, M., WANG, Y., LIANG, Q., and YU, Z. Application status and development tendency of coal entrained-bed gasification. *Modern Chemical Industry (China)*, **2004**, 24, 23.

40. XU, Z., GONG, Y., and JIANG, X. GSP entrained flow gasification process and its application prospects in China. *Clean Coal Technology (China)*, **1998**, 4, 9.

41. SCHINGNITZ, M. and MEHLHOSE, F. The GSP-process entrained-flow gasification of different types of coal. Clean Coal Technology Conference, Sardinia, Italy, **2005**.

42. ECUST. **2005**. Multi-injector coal water slurry gasification. http://icct.ecust.edu.cn/download/cwsg.pdf (accessed September 16, 2009).

43. XU, S. The status and development trends of IGCC in China. *TPRI Report*, **2005**.

44. SMITH, P.V., DAVIS, B.M., VIMALCHAND, P., LIU, G., and LONGANBACH, J. Operation of the PDSF Transport Gasifier. *Gasification Technologies Conference*. San Francisco, CA, **2002**.

45. CONOCO PHILLIPS. Process review. http://www.conocophillips.com/en/tech/downstream/e-gas/pages/process.aspx (accessed September 16, 2009).

46. VAN DYK, J.C., KEYSER, M.J., and COERTZEN, M. Syngas production from South African coal sources using Sasol-Lurgi gasifiers. *International Journal of Coal Geology*, **2006**, 65, 243.

47. CLAYTON, S., STIEGEL, G., and WIMER, J. DOE/FE-0447. Gasification markets and technologies— Present and future. *U.S. Department of Energy Report*, July **2002**.

48. BENNETT, J.P. Refractory liner materials used in slagging gasifiers. *Refractory Applications*, **2004**, 9, 20–25.

49. SICHINGA, J., JORDAAN, N., GOVENDER, M., and VAN DE VENTER, E. Sasol coal-to-liquids development. Presented at the *Gasification Technologies Council Conference*. San Francisco, CA, **2005**.

50. HOLT, N.A.H. Coal gasification research, development and demonstration—Needs and opportunities. Presented at the *Gasification Technologies Council Conference*. San Francisco, CA, **2001**.

51. KORENS, N., SIMBECK, D.R., and WILHELM, D.J. Process screening analysis of alternative gas treating and sulfur removal for gasification. *DOE Report #739656-00100*, **2002**.

52. KOHL, A.L. and RIESENFELD, F.C. *Gas Purification*, 4th ed. Houston, TX: Gulf Publishing Company, **1985**.

53. KARPUK, M.E., COPELAND, R.J., FEINBERG, D., WICKHAM, D., WINDECKER, B., and YU, J. High temperature hydrogen sulfide removal with stannic oxide. *DOE Report DOE/ER/80998-93/c0217*, **1994**.

54. KONTTINEN, J.T. Hot gas desulfurization with zinc titanate sorbents in a fluidized bed. *Industrial & Engineering Chemistry Research*, **1997**, 36, 2332.

55. LIANG, M.S., LI, C.H., and XIE, K.C. Research progress on high temperature desulfurizer. *Coal Conversion*, **2002**, 25, 13.

56. GANGWAL S., TURK, B., COKER, D., HOWE, G., GUPTA, R., KAMARTHI, R., LEININGER, T., and JAIN, S. Warm-gas desulfurization process for Chevron Texaco quench gasifier syngas. Presented at the *2004 Pittsburgh Coal Conference*. Osaka, Japan, **2004**.

Chapter 5

Desulfurization Technologies

Chunshan Song and Xiaoliang Ma

Clean Fuels and Catalysis Program, EMS Energy Institute, and Department of Energy and Mineral Engineering, Pennsylvania State University

5.1 CHALLENGES IN DEEP DESULFURIZATION FOR HYDROCARBON FUEL PROCESSING AND FUEL CELL APPLICATIONS

Deep desulfurization has become an important part of many integrated energy conversion or utilization systems including fuel cells. Fuel cells are attractive for electric power generation because a significantly high efficiency can be achieved by combining hydrogen and oxygen in an electrochemical device that directly converts chemical energy to electrical energy in a more environmentally friendly way. Hydrogen can be obtained from electrolysis of water, and the electric power for electrolysis can be generated using the energy from the sun, wind, or other nonfossil energy. However, while such alternative energy sources are the promise for the future, for now they are not feasible for transportation and power station needs across the globe.[1] For a mass production of hydrogen for fuel cell applications, a transitional solution is required to convert certain hydrocarbon fuels to hydrogen. According to the hydrocarbon fossil storage in the world, for the foreseeable future, the current infrastructure fuels of natural gas, petroleum, and coal will be the major sources of hydrogen. The commercial natural gas, gasoline, and diesel fuel are promising fuels for fuel cell applications because they are readily available, and the existing infrastructure in production, delivery, and storage for these fuels can be used. Some other liquid hydrocarbon fuels, such as JP-8, for fuel cell applications also attract a particular attention, as JP-8 is a logistic fuel for military.[2]

As well known, these hydrocarbon fuels are not suitable for use directly for fuel cells, even for the solid oxide fuel cell (SOFC).[3] A hydrocarbon fuel processor would be required to produce a suitable fuel such as hydrogen or synthesis gas for fuel

Hydrogen and Syngas Production and Purification Technologies, Edited by Ke Liu, Chunshan Song and Velu Subramani
Copyright © 2010 American Institute of Chemical Engineers

219

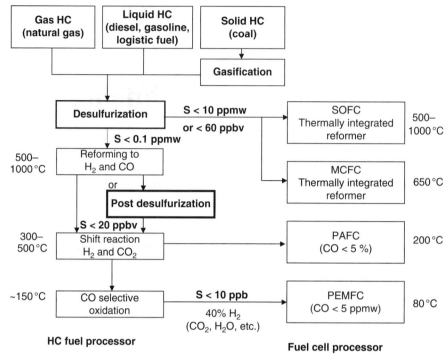

Figure 5.1. Hydrocarbon fuel process for fuel cells. HC, hydrocarbon.

cells. A flowing scheme of the hydrocarbon fuel processing is shown in Figure 5.1. The scheme for the hydrocarbon fuel process is dependent on both the source of hydrocarbon fuels and the properties of the produced fuels required by different types of the fuel cells. If using coal as a source, an additional gasification process is necessary first to convert the solid coal to fuel gas, which will be discussed in detail in Section 5.5. As well known, all these fossil fuels contain sulfur compounds. These sulfur compounds in the fuels and H_2S produced from these sulfur compounds in the hydrocarbon reforming process are poisonous to the reforming catalyst, water-gas shift catalyst, and preferential oxidation (Prox) catalyst in the hydrocarbon fuel processor, the electrode catalysts in fuel cell stacks, even the separation membrane for H_2 purification. Many nickel-containing and noble metal-containing catalysts used in the reforming (such as steam reforming catalyst: Ni/Al_2O_3) and water-gas shift reaction are very sensitive to sulfur, and they are deactivated even at a parts per million sulfur level. Even if a sulfur tolerant catalyst for the reforming and water-gas shift reaction were developed, the remaining sulfur species in the reformate would still poison the anode catalysts in the fuel cells. For SOFC application, the sulfur level in the fuel needs to be reduced, in general, to less than 10 parts per million by weight (ppmw). It was also reported that for Siemens Westinghouse SOFCs, an amount of 100 parts per billion by volume (ppbv) sulfur is the maximum limit permitted.[4] For phosphoric acid fuel cell (PAFC) and proton exchange membrane fuel cell (PEMFC), the gas and liquid hydrocarbons need to be converted to H_2 and CO_2 by reforming followed by water-gas shift reaction. Many developing reforming catalysts and water-gas shift catalysts, such as nickel-containing and noble metal-containing catalysts, are sensitive to sulfur compounds. Consequently, deactivation of the hydrocarbon fuel processing catalysts and the fuel cell anodes via sulfur poisoning is a great concern.

The type of sulfur compounds existing in the commercial hydrocarbons fuels depends on the type and source of the fuels, and the fuel refining processes. The sulfur concentration in the natural gas varies from a couple of ppmv to a hundred ppmv. The major sulfur compounds are hydrogen sulfide (H_2S), carbonyl sulfide, methylmercaptan ($CH_3)HS$, ethylmercaptan ($C_2H_5)HS$, dimethylsulfide ($CH_3)_2S$, diethylsulfide ($C_2H_5)_2S$, proplanylmercaptan, and butylmercaptan.[5,6] It should be noted that some sulfur-containing compounds with 5–20 ppmv, are usually added to commercial fuel gas to act as an odorant to help in the detection of leaks. The typical odorants are blends of butyl- and propylmercaptan isomers, mercaptans, ($CH_3)_2S$, and/or tetrahydrothiophene (THT).[7] The sulfur content of pipeline natural gas in the U.S. is 4–8 ppmw. It usually contains H_2S, COS, methylmercaptan, and trace amounts of dimethylsulfide.[8] The local natural gas distributor usually adds a sulfur odorant package from 5 to 20 ppmw, sometimes containing dimethylsulfide or methylethylsulfide.[7] European natural gas has extremely high concentrations of COS (20 ppmw) because of its higher H_2S concentration and the reaction of H_2S with CO_2 to form COS. Japan imports liquidized natural gas and controls the addition of odorants in a systematic manner, making adsorbent designs simpler than in other parts of the world.[1]

The U.S. Environmental Protection Agency (EPA) Tier II gasoline sulfur regulations are listed in Table 5.1.[9–11] The U.S. EPA sulfur regulation for gasoline has been reduced to 30 ppmw by 2006. The U.S. EPA sulfur regulations for diesel and jet fuels are listed in Table 5.2.[9,12] Highway diesel was 500 ppmw from 1993 to 2006 and has been reduced to 15 ppmw by 2006. The current regulation for jet fuel is 3000 ppmw. The sulfur concentration in commercial jet fuels varies from 500 to 2000 ppmw. Figure 5.2 presents the pulsed flame photometric detector (PFPD) gas chromatograms for the commercial gasoline, jet fuel (JP-8), diesel fuel, and low-sulfur diesel along with the results of sulfur identifications. The major sulfur compounds that exist in the commercial gasoline are thiophene (T), 2-methylthiophene (2-MT), 3-methylthiophene (3-MT), 2,4-dimethylthiophene (2,4-DMT), and benzothiophene (BT). No mercaptans, dialkyl sulfides, and dialkyl disulfides are detected, indicating that these sulfur compounds have higher reactivity and have been removed from gasoline fraction in the current refining processes. The major sulfur compounds existing in jet fuel, JP-8, are 2,3-dimethylbenzothiophene (2,3-DMBT), 2,3,7-trimethylbenzothiophene (2,3,7-TMBT), 2,3,5-trimethylbenzothiophene (2,3,5-TMBT), and/or 2,3,6-trimethylbenzothiophene (2,3,6-TMBT). All these methylbenzothiophenes have two methyl groups at 2- and 3-positions, respectively, implying that the BTs with two alkyl groups at the 2- and 3-positions might be more difficult to be removed than other sulfur compounds in conventional hydrotreating of the jet

Table 5.1. U.S. EPA Tier II Gasoline Sulfur Regulations[9-11]

Category	Year				
	1988	1995	2004	2005	2006
Refinery average (ppmw)	1000 (maximum)	330 (<300 ppm S and <29.2% aromatics required for national certification: <850 ppm S and <41.2% aromatics as national maximum)	–	30	30
Corporate average (ppmw)	–	–	120	90	–
Per gallon cap (ppmw)	–	–	300	300	80

Table 5.2. U.S. EPA Sulfur Regulations for Diesel and Jet Fuels[9,12]

Category	Year			
	1989	1993	2006	2010
Highway diesel (ppmw)	5000 (maximum for no. 1D and 2D, with minimum cetane no. 40)	500 (current upper limit since 1993)	15 (regulated in 2001; exclude some small refineries)	15 (regulated in 2001; apply to all U.S. refineries)
Non-road diesel (ppmw)	20,000	5000 (current upper limit)	500 (proposed in 2003 for 2007)	15 (proposed in 2003 for 2010)
Jet fuel (ppmw)	3000	3000	3000 maximum	<3000 maximum

fuel. No dibenzothiophenes (DBTs) were detected in the JP-8 sample as the boiling point of DBTs (332–333 °C) is beyond the boiling range of jet fuel (160–300 °C). The sulfur compounds in the commercial diesel fuel are alkyl DBTs, but the major sulfur compounds are the DBT derivatives with alkyl groups at the 4- and/or 6-positions, including 4-methyldibenzothiophene (4-MDBT), 4,6-dimethyldiben-zothiophene (4,6-DMDBT), 3,6-dimethyldibenzothiophene (3,6-DMDBT), and 2,4,6-trimethyldibenzothiophene (2,4,6-TMDBT). It implies that the major sulfur compounds remaining in the diesel fuel are the refractory sulfur compounds as they are difficult to be removed by the conventional hydrodesulfurization (HDS) process due to the steric hindrance of the alkyl groups at the 4- and/or 6-positions.[13-16] The major sulfur compounds in the low-sulfur diesel with sulfur concentration of 10 ppmw are the alkyl DBTs with two alkyl groups at the 4- and 6-possitions, respec-

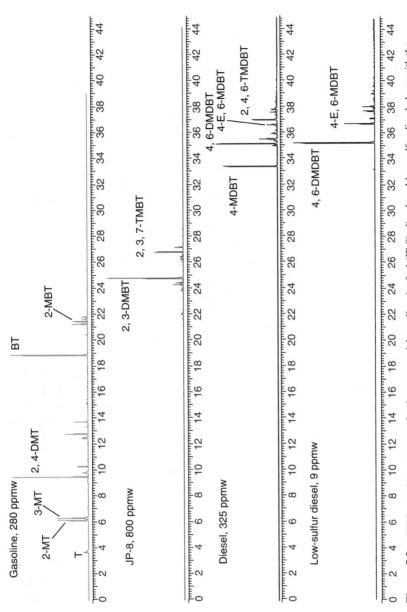

Figure 5.2. PFPD gas chromatograms for the commercial gasoline, jet fuel (JP-8), diesel, and low-sulfur diesel along with the results of sulfur identifications.

223

tively, such as 4,6-DMDBT, 4-ethyl-6-methyldibenzothiophene (4-E-6-MDBT). Thus, it is necessary to pay more attention to these refractory sulfur compounds in deep desulfurization of diesel fuel for fuel cell applications. It should be mentioned that even following the new EPA regulation for low-sulfur fuels, 30 ppmw for gasoline and 15 ppmw for highway diesel, the sulfur contents in the gasoline and the diesel are still too high for the fuel cell applications.

For ultradeep desulfurization of hydrocarbon fuels, the accurate determination of sulfur content in the desulfurized fuels is extremely important in developing and evaluating new adsorbents and catalysts and in monitoring the process. On the other hand, analyzing and quantifying sulfur compounds at very low concentrations (e.g., <10 ppmw) is difficult, and the selection of proper methods requires careful consideration of multiple factors.[17-23] Several American Society for Testing and Materials (ASTM) methods, such as wavelength dispersive X-ray fluorescence spectroscopy (Method D 2622), ultraviolet fluorescence method (Method D5453), and oxidative microcoulometry method (Method D3120), have been specified for the determination of sulfur content in light hydrocarbons and petroleum products.[24] Gas chromatograph (GC) coupled with sulfur-selective detectors such as a flame photometric detector (FPD) or an improved FPD called PFPD and GC coupled with atomic emission detector (AED) or sulfur chemiluminescence detection (SCD) have been used in several research laboratories for the quantitative estimation of sulfur compounds in commercial transportation fuels. However, because of the special characteristics of the FPD and PFPD, care must be taken while treating the GC-FPD/PFPD data. The use of the GC-FPD for the quantitative analysis would require consideration of the nonlinear response and quenching effects of FPD.[20,23,25] Ma et al.[26] studied the analysis of a series of standard fuel samples containing known amounts of sulfur compounds in n-decane and real gasoline samples by using GC coupled with a flame photometric detector (GC-FPD), pulsed flame photometric detector (GC-PFPD), and a total sulfur analyzer (Antek 9000S). The results showed that the GC-FPD and GC-PFPD are not suitable for quantitative estimation of total sulfur concentration in complex hydrocarbon fuels at a low part per million level without considering both the nonlinear response and the quenching effect.

The ultradeep desulfurization of the current infrastructure fuels has become a bottleneck in the production of H_2 for fuel cell applications. It is urgent to develop a more efficient and environmentally friendly process and technology for the ultradeep desulfurization of the hydrocarbon fuels. Consequently, many approaches have been conducted in order to improve the conventional HDS process and to develop new alternative processes. These approaches include catalytic HDS with improved and new catalysts, reactor and/or process, adsorptive desulfurization,[27] oxidative desulfurization (ODS), extractive desulfurization (EDS), and biodesulfurization (BDS) by using special bacteria and others. Some of these works were reviewed recently by Topsoe et al.,[15] Whitehurst et al.,[28] Kabe et al.,[29] Cicero et al.,[30] Babich and Moulijn,[31] Dhar et al.,[32] Song,[33] Song and Ma,[16,34] Bej et al.,[35] Mochida and Choi,[36] Hernandez-Maldonado and Yang,[37,38] Hernandez-Maldonado et al.,[39] Topsoe,[40] Brunet et al.,[41] Gupta et al.,[42] and Ito and van Veen.[43]

The present chapter will review the present state of the art and challenges in the ultradeep desulfurization of natural gas, gasoline, jet fuel, diesel, reformate, and syngas from coal gasifier by HDS, selective adsorption, solvent absorption, and ODS, respectively.

5.2 HDS TECHNOLOGY

5.2.1 Natural Gas

There are many methods for the desulfurization of nature gas, which can be classified into dry desulfurization, wet desulfurization, and catalytic adsorption. In the dry desulfurization, some solid sorbents, such as iron oxide, zinc oxide, activated carbon (AC), zeolites, and molecular sieves, are used. In wet desulfurization method, liquid-phase chemical/physical solvent absorption systems are usually used for scrubbing H_2S; amine-based processes are subject to equipment corrosion, foaming, amine-solution degradation, and evaporation, and require extensive wastewater treatment. As a result, this sulfur removal technology is complex and capital intensive,[44] although the processes are still employed widely in the industry. The desulfurization of coal gasification gas will be reviewed in detail in Section 5.5. In the catalytic-adsorption method, the sulfur compounds are transformed into H_2S by catalytic HDS or into elemental sulfur or SOx by selective catalytic oxidation (SCO), and then, the reformed H_2S and SOx are removed by the subsequent adsorption.

For large-scale hydrogen production from natural gas, currently, HDS is invariably the preferred method of sulfur removal. In this process, the natural gas is mixed with a small quantity of the added hydrogen and passed over a catalyst bed.[6] The typical HDS catalyst is either $Co-Mo/Al_2O_3$ or $Ni-Mo/Al_2O_3$. The organic sulfur is reduced to H_2S and hydrocarbon by reaction of the following kind:

$$(C_2H_5)_2 S + 2H_2 = 2C_2H_6 + H_2S. \qquad (5.1)$$

The H_2S produced can then be removed by adsorption on a bed of ZnO sorbent at a temperature of 300–400 °C.[45]

Because the sulfur compounds in natural gas are nonaromatic and of low molecular weight, HDS can be performed at lower H_2 partial pressures, 1–10 kPa, and temperatures of 200–300 °C depending on the catalyst and the sulfur speciation.[46] Zinc oxide adsorption capacities for H_2S in industrial applications are reported to be high, typically 150–200 mg-S/g sorbent.[47]

The major advantages of the catalytic HDS method are that (a) the method is able to remove all types of the sulfur compounds in natural gas as all these sulfur compounds have higher HDS reactivity in compassion with thiophenic sulfur compounds, which do not exist in natural gas, and (b) ZnO sorption capacity for H_2S is much higher than that of other sorbents, which leads to lowering inventories of sorbent and decreasing sorbent replacement frequency. The major disadvantage of the catalytic HDS method is that the process needs to run at high temperature with adding hydrogen.[48] It might be impractical to use this method in the fuel processors for fuel cell applications mainly because of the energy cost of compressing hydrogen. An additional drawback is the nature of the catalysts themselves. HDS catalysts require activation using an H_2S–H_2 mixture, and they contain priority pollutant metals (e.g., Ni, Co, and Mo) that require special handling and disposal.

5.2.2   Gasoline

In comparison with the diesel desulfurization, it is not very difficult to remove sulfur from gasoline by HDS technology. The challenges to the refinery for gasoline deep desulfurization are to meet the new EPA Tier II regulations on sulfur contents (2006–2010) and aromatic contents, and still produce high-octane gasoline in a profitable manner.

5.2.2.1   Selective HDS of FCC (Fluid Catalytic Cracking) Naphtha

Selective HDS could be achieved by designing catalysts that promote thiophene HDS but do not saturated olefins, or by passivating olefin hydrogenation sites on the catalysts. Some reports indicate that there exist different active sites on hydrotreating catalysts (such as sulfided Co–Mo/Al_2O_3) for thiophene desulfurization and for olefin hydrogenation.[49–55] On the other hand, a recent report suggests that selective HDS of FCC naphtha may be due to competitive adsorption of sulfur compounds that inhibit adsorption and saturation of olefins in naphtha.[56] ExxonMobil's SCANfining[57] and Intitute Francais du Petrole's (IFP) Prime G+ are two representative new processes for gasoline desulfurization based on selective HDS, in which organic sulfur is converted to H_2S but olefinic species are largely preserved for preventing octane loss.

Figure 5.3 shows the scheme of SCANfining (selective cat naphtha hydrofining), first developed by Exxon (now ExxonMobil). It is a catalytic HDS process that

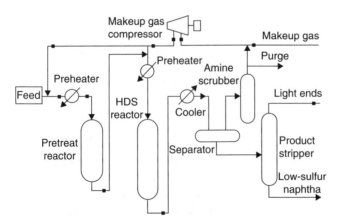

Figure 5.3.   ExxonMobil's SCANfining process for selective naphtha HDS.[57,58]

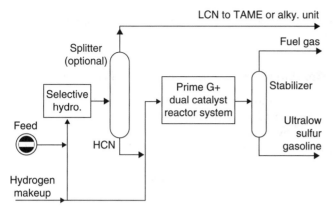

Figure 5.4. Prime G+ process developed by IFP.[59-61] TAME, tertiary amyl methyl ether.

is based on a proprietary catalyst called RT-225.[57,58] The process flow can be described as follows. The feed is mixed with recycle hydrogen, heated with reactor effluent, and passed through the pretreated reactor for diolefin saturation. After further heat exchange with reactor effluent and preheating using a utility, the hydrocarbon/hydrogen mixture enters the HDS reactor containing proprietary RT-225 catalyst. In the reactor, the sulfur is converted to H_2S under conditions that strongly favor HDS, while minimizing olefin saturation. The RT-225 catalyst system was jointly developed by ExxonMobil and AkzoNobel specifically for selective removal of sulfur from FCC naphtha by HDS with minimum hydrogenation of olefins, thus preserving octane.

Figure 5.4 shows the scheme of Prime G+ desulfurization process, which was developed by IFP, and largely preserves olefins as its strategy for diminishing octane loss.[59-61] Prime G+ is based on a combination of a selective hydrogenation unit that removes diolefins and light mercaptans, a splitter, and a selective HDS of mid- and heavy cut naphtha (HCN) cut through a dual catalytic system. It is designed for FCC naphtha ultradeep desulfurization with minimal octane penalty. FCC debutanizer bottoms are fed directly to the first reactor, where under mild conditions diolefins are selectively hydrogenated and mercaptans are converted to heavier sulfur species. The selective hydrogenation reactor effluent is then usually split to produce a light cut naphtha (LCN) and an HCN cut. The LCN stream is mercaptan free with a low-sulfur and diolefin concentration enabling further processing in an etherification or an alkylation unit. The HCN then enters the main Prime G+ section, where it undergoes a deep HDS in a dual catalyst system with very limited olefin saturation and no aromatic losses to produce an ultralow sulfur gasoline. Catalysts used in this HDS section are a combination of HR-806 and HR-841 catalysts developed by IFP and commercialized by Axens, where HR-806 achieves the bulk of desulfurization, and HR-841 is a polishing catalyst that reduces sulfur and mercaptans with no activity for olefin hydrogenation.[62] Prime G+ is less severe and has been commercially demonstrated for over 7 years in two U.S. refineries and in an Asian refinery.[63] There are over 10 Prime G+ units, and the economics are estimated to be as follows: capital investment, US$600–800/bpsd; combined utilities, US$0.32/bbl; H$_2$, US$0.28/bbl; and catalyst, US$0.03/bbl.[64]

5.2.2.2 Deep HDS Combined with Octane Recovery Processing

Another hydrotreating approach is to carry out deep HDS of organic sulfur and saturated olefins, then convert low-octane components such as paraffins to high-octane components for octane gain by isomerization and alkylation. Two representative industrial processes in this category are ExxonMobil's OCTGain and UOP-INTE-VEP's ISAL.

The OCTGain process was first developed and initially commercialized in 1991 by Mobil (now ExxonMobil). Figure 5.5 shows the scheme of the OCTGain process.[58,65] The process, now in its third generation, uses a fixed-bed reactor to desulfurize FCC naphtha while maintaining octane. The process first totally removes sulfur and saturates olefins, and then restores the octane to economically needed levels, thus variation in feed sulfur content does not impact product sulfur and treated products typically contain <5 ppm sulfur and <1% olefins.[66] The basic flow scheme is similar to that of a conventional naphtha hydrotreater. Feed and recycle hydrogen mix is preheated in feed/effluent exchangers and a fired heater then introduced into a fixed-bed reactor. Over the first catalyst bed, the sulfur in the feed is converted to hydrogen sulfide with near complete olefin saturation. In the second bed, over a different catalyst, octane is recovered by cracking and isomerization reactions. The reactor effluent is cooled and the liquid product separated from the recycle gas using high- and low-temperature separators. The vapor from the separators is combined with makeup gas, compressed, and recycled. The liquid from the separators is sent to the product stripper where the light ends are recovered overhead and desulfurized naphtha from the bottoms. The product sulfur level can be as low as 5 ppm. Compared with ExxonMobil's SCANfining, the OCTGain process is run at more severe

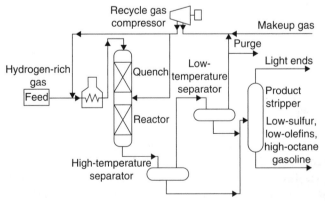

Figure 5.5.   ExxonMobil's OCTGain process for selective naphtha HDS.[58,65]

conditions for it to recover octane, so this process is more appropriate for refiners with higher sulfur levels, which requires severe hydrotreating to reach the sulfur target.[63] While octane loss can be eliminated with the proper operating conditions, some yield loss may result.[63]

OCTGain has been commercially demonstrated at ExxonMobil's refinery in Joliet, IL. More recently, the OCTGain process has been demonstrated for deep desulfurization and octane enhancement of heavy cracked naphtha feeds in its first commercial grassroots application in the Qatar Petroleum OCTGain unit.[67] On the actual refinery feed, an ultralow sulfur (<1 ppm), low-olefin gasoline is produced, while process flexibility allows the adjustment of octane number between −2 and +2 RON (Research *Octane Number*) of the feed, as refinery economics dictate. Naphtha yield loss and hydrogen consumption were much less than design allowances. The desulfurized gasoline is low in mercaptans and has suitable vapor pressure for direct blending into the refinery's gasoline pool.[67]

The ISAL process, jointly developed by INTEVEP, SA, and UOP, is designed as a low-pressure, fixed-bed hydroprocessing technology for desulfurizing gasoline-range feedstocks and selectively reconfigures lower octane components to restore product octane number. Its flow scheme is very similar to that of a conventional hydrotreating process, but a major feature of this process is the catalyst formulation, typically a combination of an HDS catalyst such as Co–Mo–P/Al$_2$O$_3$ and octane-enhancing catalyst such as Ga–Cr/H-ZSM-5 catalysts in two beds.[31,68] The naphtha feed is mixed with H$_2$-rich recycle gas and processed across fixed catalyst beds at moderate temperatures and pressures. Following heat exchange and separation, the reactor effluent is stabilized. The similarity of an ISAL unit to a conventional naphtha hydrotreating unit makes the implementation of ISAL simple and straightforward.[69] There are several revamp and several new units based on the ISAL process.[69]

Researchers from UOP and INTEVEP have discussed the technical aspects of the process and catalyst chemistries leading to these desirable results.[70,71] The ISAL process reduces the naphtha sulfur and nitrogen content, reduces the naphtha olefin content, does not increase the aromatics content, and can maintain or increase the naphtha octane.[72] The ability of the ISAL process to provide both desulfurization and octane flexibility is the result of a development program that was begun by INTEVEP in the early 1990s for possible improvements in fixed-bed isomerization and alkylation technologies. One result of this program was the discovery of a new catalyst system that could increase octane by isomerization of its gasoline-range feedstock. An interesting feature of this system was that an octane boost occurred despite the occurrence of significant olefin saturation.[71] Therefore, INTEVEP and UOP teamed up to develop the ISAL process to enable refiners to hydrotreat highly olefinic feedstocks, such as coker and FCC naphtha, while controlling both the sulfur content and the octane of its product. This flexibility is achieved by the use of a catalyst system that promotes an array of octane-enhancing reactions, including isomerization, conversion, dealkylation, and molecular-weight reduction. The ability of the ISAL process to operate within a wide range of desulfurization and product octane combinations allows the refiner to tailor the unit's operation to the specific processing needs of the refinery. In addition, recent improvements to the ISAL catalyst system and the process configuration allow these tighter gasoline sulfur specifications to be achieved at higher yield and lower capital cost. Its products can meet the most stringent specifications of gasoline sulfur and olefin content. Because the flow scheme and processing conditions of an ISAL unit are similar to a conventional naphtha hydrotreating unit, the process can be implemented as either a new grassroot unit or as a revamp of an existing hydrotreater.[71] The technology is based on typical hydrotreating flow schemes, which imply ease of operation and reliability as a refinery process. The upgraded FCC naphtha from the ISAL process is specifically suited to meet reformulated gasoline specifications in global market.[72]

What are the differences between UOP–INTERVEP's ISAL process and ExxonMobil's OCTGain process? The two processes are similar in terms of process design concept and processing schemes, but the catalysts and processing conditions are different. What are the differences between ExxonMobil's SCANfining and OCTGain processes? SCANfining is used for selective HDS (to <30 ppm sulfur) and has a high content of olefins in the product with little reduction in octane number. It was developed by ExxonMobil and AkzoNobel Catalysts to reduce hydrogen consumption over a low-temperature and a low-pressure, fixed-bed reactor. When the new Exomer process (caustic extraction of sulfur) is added to this, the desulfurization capability is extended further to 10 ppm.[73] The Exomer process has been developed by ExxonMobil and Merichem to extract all the mercaptans from the fuel and to provide catalyst stability. The process does not involve the use of a catalytic naphtha splitter, which reduces the capital and operating costs of the motor gasoline desulfurization unit. OCTGain technology is used for deep HDS and the product has little olefins but a high-octane value due to isomerization of paraffins (alkanes) in the process, where the octane gain is at the expense of some yield loss to liquefied petroleum gas (LPG). The technology can be used to vary product octane on a day-to-day basis, while keeping almost 100% desulfurization.

Isomerization of alkanes in naphtha cuts, as involved in OCTGain and ISAL processes, can be achieved by using catalysts to improve the octane number of gasoline. For example, Jao et al.[74] reported on naphtha isomerization over mordenite-supported Pt catalysts. Their results demonstrated that catalyst performance was determined by Pt dispersion when using pure feed, whereas it was determined by both Pt dispersion and Pt cluster stability for feed containing 500 ppm sulfur. Highest Pt dispersion and, thus, best performance with pure feed were obtained with a catalyst prepared by the ion exchange and pretreated at a low-temperature ramping rate (0.5 °C/min) during calcination and reduction. In addition, this catalyst was pretreated by calcination at 450 °C, followed by reduction at 450 °C. In contrast, the catalyst having the best performance with feed containing 500 ppm sulfur was pretreated by calcination at 450 °C, followed by reduction at 530 °C. The authors suggested that the superior performance may result from a compromise between metal dispersion and metal cluster stability.

As but one example of bimetallic catalysts, Lee and Rhee[75] prepared a series of bifunctional bimetallic M–Pt/H-Beta (M = Cu, Ga, Ni, and Pd) catalysts and examined them for the isomerization of *n*-hexane. The sulfur-containing feed was prepared by the addition of thiophene in pure *n*-hexane to have 500 ppmw of sulfur. Sulfur in the feed brought about a substantial decrease in the catalyst performance, and the sulfur deactivation of bifunctional Pt/H-Beta turned out to be a two-step irreversible process caused by metal poisoning followed by coking. To test their effect on the sulfur tolerance, various second metals (Cu, Ga, Ni, and Pd) were added to monometallic Pt/H-Beta catalysts. Unfortunately, all of these, except for Pd, greatly decreased the sulfur tolerance of the original Pt/H-Beta catalyst. Regardless of the preparation method or the Pd/Pt atomic ratio of the bimetallic Pd–Pt series, all the bimetallic catalysts showed high sulfur tolerance, in comparison to the mono-metallic Pt/H-Beta and Pd/H-Beta.[75] The metal dispersion and the hydrogenation activity decreased in the Pd–Pt series compared with Pt/H-Beta. However, the amounts of sulfur adsorbed and coke deposited on the sulfur-deactivated Pd–Pt/H-Beta were much lower than those on Pt/H-Beta, Pd/H-Beta, and the other M–Pt/H-Beta catalysts. The authors suggested that the Pd–Pt bimetallic interaction in Pd–Pt/H-Beta increased the amount of electron-deficient metal sites, and that Pd–Pt bimetallic interaction inhibits irreversible sulfur adsorption and thereby reduces sulfur-induced coke formation.[75] This is why the Pd–Pt series maintained a high activity under sulfur deactivation conditions.

5.2.2.3 *Catalytic Distillation for Desulfurization*

The catalytic distillation desulfurization process developed by CDTech is significantly different from conventional hydrotreating.[76,77] The most important portion of the CDTech desulfurization process is a set of two distillation columns loaded with desulfurization catalyst in a packed structure. In this process, the LCN, middle cut naphtha (MCN), and HCN are treated separately, under optimal conditions for each. The first column, called CDHydro, treats the lighter compounds of FCC gasoline and separates the heavier portion of the FCC gasoline for treatment in the second column. The second column, called CDHDS (catalytic distillation hydrodesulfurization), removes the sulfur from the heavier compounds of FCC gasoline. Figure 5.6

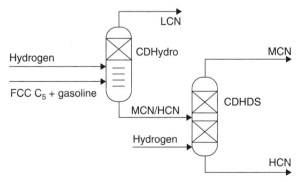

Figure 5.6. Flow scheme of CDTech's CDHydro + CDHDS for naphtha desulfurization.[76,77]

shows the flow scheme of the process.[76,77] The full-range FCC naphtha is fed to the CDHydro column, and the desulfurization begins with fractionation of the light naphtha overhead in the CDHydro column. Mercaptan sulfur reacts quantitatively with excess of diolefins to produce heavier sulfur compounds, and the remaining diolefins are partially saturated to olefins by reaction with hydrogen. Bottoms from the CDHydro column, containing the reacted mercaptans, are fed to the CDHDS column where the MCN and HCN are catalytically desulfurized in two separate zones. HDS conditions are optimized for each fraction to achieve the desired sulfur reduction with minimal olefin saturation. Olefins are concentrated at the top of the column, where conditions are mild, while sulfur is concentrated at the bottom where the conditions result in very high levels of HDS. The temperature and pressure of the CDTech process columns are lower than the fixed-bed hydrotreating processes, particularly in the upper section of the distillation column, which is where most of the olefins are located. These operating conditions minimize yield and octane loss.

While the CDTech process is very different from conventional hydrotreating, the catalyst used for removing the sulfur compounds is the same.[63] CDHydro combines fractionation and hydrogenation and it is designed to selectively hydrogenate diolefins in the top section of a hydrocarbon distillation column. Conventional hydrotreating requires a distillation column after fixed-bed hydrogenation unit, while the CDHydro process eliminates the fixed-bed unit by incorporating catalyst in the column. Proprietary devices containing catalysts are installed in the fractionation column's top section, and hydrogen is introduced beneath the catalyst zone. Fractionation carries the light components into the catalyst zone where reaction with H_2 occurs. Fractionation also sends the heavy materials to the bottom. This prevents foulants and heavy catalyst poisons in the feed from contacting the catalyst. In addition, the clean hydrogenated reflux continuously washes the catalyst zone. These factors combine to give a longer catalyst life. At the bottom of the catalyst zone, mercaptans react with diolefins to form heavy, thermally stable sulfides. These sulfides have higher boiling points than the C5 fraction and are easily fractionated to the bottom product.[77] This can eliminate a separate mercaptan removal step.[76] There are over 14 commercial CDHydro units in operation for C4, C5, C6, and benzene hydrogenation applications.[76]

CDHDS is used in combination with CDHyrdro to selectively desulfurize gasoline with minimum octane loss. Bottoms of CDHydro column, containing the reacted mercaptans, are fed to the CDHDS column where the MCN and HCN are catalytically desulfurized in two separate zones. HDS conditions are optimized for each fraction to achieve the desired sulfur reduction with minimal olefin saturation. Olefins are concentrated at the top of the column, where conditions are mild, while sulfur is concentrated at the bottom, where the conditions results in very high levels of HDS.[76]

Catalytic distillation essentially eliminates catalyst fouling because the fractionation removes heavy coke precursors from the catalyst zone before coke can form and foul the catalyst bed. The estimated ISBL (inside battery limits) capital cost for 35,000 bpd CDHydro/CDHDS unit with 92% desulfurization is US$25 million, and the direct operating cost including utilities, catalyst, hydrogen, and octane replacement is estimated to be US$0.03/gal of full-range FCC naphtha.[76] A recent article discusses the CDHDS technology from CDTech for reliable HDS operation, where the catalyst cycle length for CDTech catalytic distillation technologies can be aligned with the 5-year FCC operating cycles.[9]

It is necessary to point out that the currently developed new deep HDS technologies of gasoline are for meeting the new EPA Tier II regulations on sulfur contents (2006–2010) and aromatic contents and still produce high-octane gasoline in a profitable manner. By these new HDS technologies, it is still difficult to reduce the sulfur in gasoline to the level for the hydrocarbon fuel processing and fuel cell applications. One of the reasons is that the coexisting olefins are easy to react with the produced H_2S in the HDS process to form new mercaptans as shown below:

$$H_2S + H_3CH_2CH_2CHC=CH_2 \leftrightarrow H_3CH_2CH_2CH_2CH_2C-SH. \qquad (5.2)$$

Consequently, the currently developed HDS technologies of gasoline is unsuitable for producing clean gasoline for hydrocarbon fuel processing and fuel cell applications, as the octane number loss is not an issue here. The important thing in modification of the gasoline HDS technologies for producing the low-sulfur gasoline for the fuel cell applications is to reduce the sulfur level in the product, probably by reducing H_2S partial pressure in the HDS process.

5.2.3 Diesel

Approaches to ultradeep HDS include (a) improving catalytic activity by new catalyst formulation for HDS of 4,6-DMDBT, (b) tailoring reaction and process conditions, and (c) designing new reactor configurations. Design approaches for ultradeep HDS focus on how to remove 4,6-DMDBT in gas oil more effectively. One or more approaches may be employed by a refinery to meet the challenges of producing ultraclean fuels at affordable cost.

Figure 5.7 shows the relative volume of catalyst bed required for achieving various levels of diesel sulfur using conventional HDS of gas oil over a commercial hydrotreating catalyst.[16] The modeling is based on HDS using a commercial catalyst reported in 1994,[13,14,78] so it is not a state-of-the-art catalyst. However, it clearly illustrates the key issue. The problem of deep HDS of diesel fuel is caused by the lower reactivity of 4,6-disubstituted DBT, as represented by 4,6-DMDBT, which has much lower reactivity than any other sulfur compounds in diesel blend stocks.[14,78] The methyl groups at 4- and 6-positions create steric hindrance for the interaction between sulfur and active sites on the catalysts. The problem is further exacerbated by the inhibiting effects of polyaromatics and nitrogen compounds in some diesel blend stocks for diesel as well as H_2S that exist in a reaction system on deep HDS. Based on experimental results, polyaromatics compete with sulfur compounds on the surface of hydrotreating catalyst, perhaps more for the flat chemisorption thereby influencing the hydrogenation and subsequent HDS, whereas H_2S compete with sulfur compounds, affecting more of the direct C–S hydrogenolysis route.

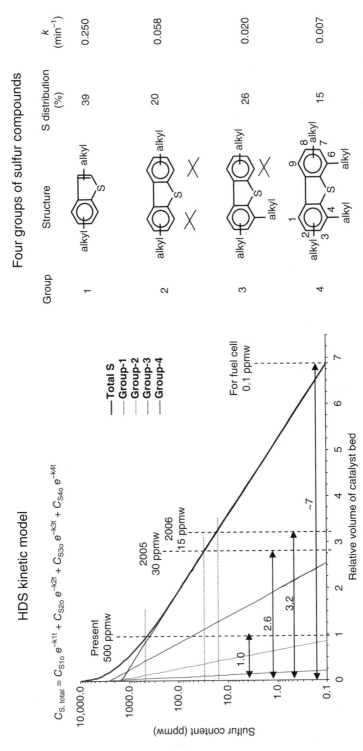

Figure 5.7. Relative volume of catalyst bed required for achieving various levels of diesel sulfur using conventional HDS (single-stage reactor) of gas oil over a commercial hydrotreating catalyst, assuming 1.0 wt % S in feed.[16]

5.2.3.1 Improving Catalytic Activity by New Catalyst Formulation

Design approaches for improving catalytic activity for ultradeep HDS focus on how to remove 4,6-DMDBT more effectively, by modifying catalyst formulations (a) to enhance hydrogenation of aromatic ring in 4,6-DMDBT by increasing the hydrogenating ability of the catalyst, (b) to incorporate acidic feature in catalyst to induce isomerization of methyl groups away from the 4- and 6-positions, and (3) to remove inhibiting substances (such as nitrogen species in the feed, H_2S in gas) and tailor the reaction conditions for specific catalytic functions. The catalytic materials formulations may be improved for better activity and/or selectivity by using different supports (MCM-41, carbon, HY, TiO_2, TiO_2–Al_2O_3, etc.) for preparing supported CoMo, NiMo, and NiW catalysts, by increasing the loading level of the active metal (Mo, W, etc.), by modifying the preparation procedure (using different precursor, using additives, or different steps or sequence of metal loading), by using additives or additional promoters (P, B, F, etc.), by adding one more base metal (e.g., Ni–CoMo or Co–NiMo, Nb, etc.), and by incorporating a noble metal (Pt, Pd, Ru, etc.).

New and improved catalysts and different processing schemes are among the subjects of active research on deep HDS.[28,79–81] For example, some recent studies examined carbon-supported CoMo catalysts for deep HDS.[82–85] Binary oxide supports such as TiO_2–Al_2O_3 have been examined for making improved HDS catalysts.[86–89]

In 1992, novel mesoporous molecular sieve MCM-41 was invented by Mobil researchers.[90,91] The novel mesoporous molecular sieve of MCM-41 type has also been examined as support for Co–Mo/MCM-41 catalyst for HDS. Al-MCM-41 has been synthesized with improved aluminum incorporation into framework[92–94] and applied to prepare Co–Mo/MCM-41 for deep HDS of diesel fuels[95–97] and for HDS of petroleum resid.[94]

The design approach makes use of high surface area of MCM-41 for higher activity per unit weight, uniform mesopore to facilitate diffusion of polycyclic sulfur compounds, and mild acidity of Al-containing MCM-41 to facilitate metal dispersion and possible isomerization.[98] Reddy and Song synthesized MCM-41-type aluminosilicate molecular sieves using different Al sources, and established a proper procedure for making acidic MCM-41.[92–94] Several recent studies have explored the design of new catalysts for HDS of refractory DBT-type sulfur compounds, based on synthesis and application of mesoporous aluminosilicate molecular sieves of MCM-41 type.[95,99] Compared with Co–Mo/Al_2O_3, higher activity for HDS has been observed for Co–Mo/MCM-41 with a higher metal loading. When MCM-41 with proper SiO_2/Al_2O_3 ratio was used to prepare Co–Mo/MCM-41 at suitable metal loading, the catalyst is much more active for the HDS of DBT, 4-methyl-, and 4,6-dimethyldibenzothiopene than a commercial Co–Mo/Al_2O_3 catalyst.[95–97,99]

5.2.3.2 New Commercial HDS Catalysts Developed for Ultralow Sulfur Diesel

For practical applications, Ni–Mo catalysts generally have higher hydrogenation ability for saturating aromatic ring that is connected to thiophenic sulfur, while

Co–Mo catalysts generally have higher selectivity toward C–S bond cleavage without hydrogenation of neighboring aromatic rings. For HDS under higher H_2 pressure, Ni–Mo catalysts tend to be more active than Co–Mo catalysts, and this advantage becomes more apparent with ultradeep desulfurization of diesel fuels. A combination of Ni and Co together with Mo (such as Ni–Co–Mo) is also a practical way to take advantage of both Ni–Mo and Co–Mo catalysts. The catalyst development has been one of the focuses of industrial research and development for deep HDS.[100] For example, new and improved catalysts have been developed and marketed by AkzoNobel, Criterion, Haldor Topsoe, IFP, United Catalyst/Süd-Chemie, Advanced Refining, ExxonMobil, Nippon Ketjen in Japan, and RIPP in China.

AkzoNobel has developed and commercialized various catalysts that can be used for HDS of diesel feed: KF 752, KF 756 and KF 757, and KF 848.[101] KF 752 can be considered to be typical of an AkzoNobel catalyst of the 1992–1993 time frame. KF 756 is a Co–Mo catalyst with high HDS activity; it was jointly developed by AkzoNobel and Exxon Research and Engineering by applying a new alumina-based carrier technology and a special promoter impregnation technique to allow high and uniform dispersion of metals such as Co and Mo on support with moderate density.[52] KF 757 is AkzoNobel's latest Co–Mo with higher HDS activity and optimized pore structure; it was announced in 1998.[101] AkzoNobel estimates that under typical conditions (e.g., 500 ppmw sulfur), KF 756 is 25% more active than KF 752, while KF 757 is 50% more active than KF 752 and 30% more active than KF 756.[102] KF 756 is widely used in Europe (20% of all distillate hydrotreaters operating on January 1, 1998), while KF 757 has been used in at least three hydrotreaters commercially.[64] Under more severe conditions (e.g., <50 ppmw sulfur), KF 757 is 35%–75% more active than KF 756.

KF 757 (Co–Mo) and KF 848 (Ni–Mo) were developed by using what AkzoNobel calls super type II active reaction sites (STARS) technology. Type II refers to a specific kind of catalyst site for hydrogenation, which is more effective for removing sulfur from sterically hindered compounds. KF 848 was announced in 2000.[102] KF 848 Ni–Mo STARS is 15%–50% more active than KF 757 Co–Mo STARS under medium to high pressure. Commercial experience exists for both advanced catalysts at BP refineries. In terms of sulfur removal, AkzoNobel projects that a desulfurization unit, which produces 500 ppmw sulfur with KF 752, would produce 405, 270, and 160 ppm sulfur with KF 756, KF 757, and KF 858, respectively.[64]

As evidenced from the STARS technology, the advances in basic understanding of fundamental reaction pathways of HDS reactions over transition metal sulfides have also resulted in major advances in commercial catalyst developments. AkzoNobel recently reported on the commercial experience of their STARS catalyst for diesel fuel feedstock HDS at two BP refineries (Grangemouth and Coryton) in U.K. The original unit at Grangemouth refinery was designed to produce 35,000 bbl/day of diesel fuel at 500 ppmw, treating mostly straight run material, but some LCO (Light Cycle Oil) was treated as well. AkzoNobel's newest and best catalyst (KF 757 at that time) was dense loaded into the reactor to produce 45,000 bbl/day diesel fuel at 10–20 ppmw (to meet the 50 ppmw cap standard). As the space velocity changed, the sulfur level changed inversely proportional to the change in space velocity. Usually, when the space velocity decreased below 1.0, the sulfur level dropped below 10 ppmw. At that refinery, however, it was not necessary to maintain the sulfur level below 10 ppmw.[64]

More recently, NEBULA catalyst has been developed jointly by ExxonMobil, AkzoNobel, and Nippon Ketjen and commercialized in 2001.[103] NEBULA stands for New Bulk Activity and is bulk base metal catalyst without using a porous support. Figures 5.8–5.10 show the recent results published by AkzoNobel on relative activity of the new NEBULA and STARS catalysts compared with conventional CoMo/Al$_2$O$_3$ developed over the last 50 years.[104] The NEBULA-1 catalyst is even more active than KF 848 STARS catalyst with respect to HDS and hydrodenitrogenation (HDN) and diesel hydrotreating; it has been successfully applied in several

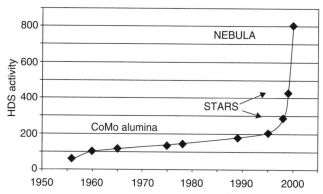

Figure 5.8.   Relative activity of the new NEBULA and STARS catalysts compared with conventional CoMo/Al$_2$O$_3$ developed over the last 50 years.[104] x-axis: year.

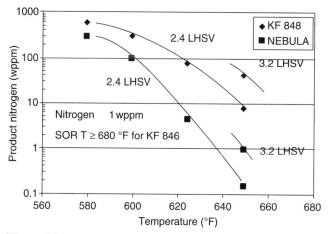

Figure 5.9.   Hydrodenitrogenation (HDN) activity of NEBULA-1 compared with KF 848 NiMo STARS catalysts (R-1 hydrocracking pretreat activity; 2000 psig, 5000 SCF/B).[104]

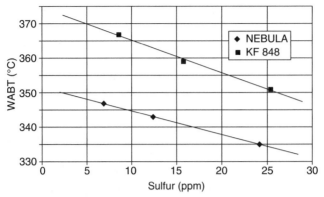

Figure 5.10. HDS activity of NEBULA-1 compared with KF 848 NiMo STARS catalysts.[158] The feed used in this test was an SR gas oil with 1.2 wt% sulfur and a density of 0.86 g/mL. Test conditions are $P = 56$ bar, LHSV = 2.5, and hydrogen to oil = 208 NL/L.[104]

diesel hydrotreaters for months as of early 2002.[103] Hydrocracking pretreatment was the first application where very high activity was found for NEBULA-1. As shown in Figure 5.8, the improvement in the HDN performance over NEBULA-1 compared with KF 848 is indeed very significant. Moreover, KF 848 was already an enormous improvement over the previous generation, KF 846. The advantage in activity is 18 °C (32 °F) compared with KF 848 and about 25 °C (45 °F) compared with KF 846.[104] The performance data shown in Figure 5.8 are for feeds boiling in the LCO boiling range; for typical VGO (vacuum gas oil) applications, the activity advantage is less great.[104] For good hydrocracking pretreatment, one traditionally needs the best possible HDN catalyst, and since HDN usually correlates with good hydrogenation, the typical hydrocracking pretreated catalysts have excellent HDN and HDA (hydrodearomatization) combined. HDS activity used to be of secondary importance for hydrocracking pretreatment, but with the new low-sulfur specifications, it can also limit the performance of the pretreater.[104] A similar improvement in activity is found for HDS and hydrogenation over NEBULA; this will lead to highly improved product qualities like lower sulfur, higher cetane, and lower density. Figure 5.10 compares NEBULA-1 and KF 848 catalysts at equal product sulfur level corresponding to different weighted-average bed temperatures (WABTs), for which long-term tests were run using an SR (straight-run) gas oil with 1.2 wt% sulfur and a density of 0.86 g/mL. Test conditions are $P = 56$ bar, liquid-hourly space velocity (LHSV) = 2.5, and hydrogen to oil = 208 NL/L.[104]

Researchers at Haldor Topsoe and their collaborators in academic institutions have contributed significantly to both the advances in research on fundamental aspects of catalytically active sites of transition metal sulfides and the development of new and more active commercial hydrotreating catalysts and processes.[15,79,80]

Haldor Topsoe has commercialized more active catalysts for HDS. Its TK-554 catalyst is analogous to AkzoNobel's KF 756 catalyst, while its newer, more active catalyst is termed TK-574. For example, in pilot-plant studies, under conditions where TK-554 produces 400 ppmw sulfur in SRGO (straight-run gas oil) TK-574 will produce 280 ppmw. Under more severe conditions, TK-554 will produce 60 ppmw, while TK-574 will produce 30 ppmw, and similar benefits are found with a mixture of straight run and cracked stocks.[64] In addition to catalyst development, Haldor Topsoe has also developed new processes for HDS.

Criterion Catalyst announced two new lines of catalysts. One is called Century, and the other is called Centinel. These two lines of catalysts are reported to be 45%–70% and 80% more active, respectively, at desulfurizing petroleum fuel than conventional catalysts used in the mid-1990s.[64] These improvements have come about through better dispersion of the active metal on the catalyst substrate. The advanced refining technologies LP (ART), a joint venture between Chevron and Grace Davison, has developed the sulfur minimization by ART (SMART) catalyst system for diesel deep HDS.[105] The SMART catalyst takes advantages of different reaction pathways of deep HDS for DBT- and 4,6-DMDBT-type sulfur compounds and is made using a combination of a high-activity Co–Mo catalyst (for C–S bond direct hydrogenolysis of unhindered sulfur such as DBT) and a specialized Ni–Mo catalyst (for aromatic ring hydrogenation followed by C–S bond hydrogenolysis for sterically hindered sulfur such as 4,6-DMDBT). In addition to improving HDS activity over conventional catalysts, the SMART system can be tailored to optimize hydrogen consumption, especially when hydrogen availability is limited.[105]

5.2.3.3 Tailoring Reaction and Processing Conditions

Tailoring process conditions aims at achieving deeper HDS with a given catalyst in an existing reactor without changing the processing scheme, with no or minimum capital investment. The parameters include those that can be tuned without any new capital investment (space velocity, temperature, pressure) and those that may involve some relatively minor change in processing scheme or some capital investment (expansion in catalyst volume or density, H_2S scrubber from recycle gas, improved vapor–liquid distributor).[106,107] First, space velocity can be decreased through increasing the catalyst bed volume or reducing the flow rate of liquid feedstock to increase the reactant–catalyst contact time. More refractory sulfur compounds would require lower space velocity for achieving deeper HDS. Second, temperature can be increased, which increases the rate of HDS. Higher temperature facilitates more of the high activation-energy reactions. Third, H_2 pressure can be increased. Fourth, improvements can be made in vapor–liquid contact to achieve uniform reactant distribution, which effectively increases the use of surface area of the catalyst. Finally, the concentration of hydrogen sulfide in the recycle stream can be removed by amine scrubbing. Since H_2S is an inhibitor to HDS, its buildup in high-pressure reactions through continuous recycling can become significant.

Some of these factors are elaborated further below. It should be noted that conventional approaches for fuel desulfurization in response to the 1993 diesel fuel sulfur regulation (500 ppmw sulfur) in the U.S. were to increase process severity of HDS, increase catalysts-to-fuel ratio, increase residence time, and enhance hydrogenation, or to use additional low-sulfur blending stocks either from separate process streams or purchased. It is becoming more difficult to meet the ultralow sulfur fuel specifications by fuel HDS using the conventional approaches.

LHSV and catalytic bed volume are interrelated parameters that control both the level of sulfur reduction and the process throughput. Increase in catalyst bed volume can enhance desulfurization. UOP projects that doubling reactor volume would reduce sulfur from 120 to 30 ppmw.[64] Haldor Topsoe reports that doubling the catalyst volume results in a 20 °C decrease in average temperature if all other operating conditions are unchanged, and there is a double effect of the increased catalyst volume.[79] The deactivation rate decreases because the start-of-run temperature decreases, and the lower LHSV by itself reduces deactivation rate even at the same temperature.

Increasing the temperature of reaction can enhance the desulfurization of more refractory sulfur compounds. Haldor Topsoe has shown that an increase of 14 °C while processing a mix of SRLGO and LCO with its advanced TK-574 Co–Mo catalyst will reduce sulfur from 120 to 40 ppmw. UOP projects that a 20 °F increase in reactor temperature would decrease sulfur from 140 to 120 ppmw.[64] The downside of increased temperature is the reduction of catalyst life (i.e., the need to change catalyst more frequently).[106] This increases the cost of catalyst, as well as affects highway diesel fuel production while the unit is down for the catalyst change. Still, current catalyst life ranges from 6 to 60 months, so some refiners could increase temperature and still remain well within the range of current industry performance. The relationship between temperature and life of a catalyst is a primary criterion affecting its marketability; thus, catalyst suppliers generally do not publish these figures.[64]

Role of H_2S in deep HDS of gas oils has been discussed in detail by Sie.[108] The decrease in the concentration of hydrogen sulfide in gas phase could reduce the inhibition to the desulfurization and hydrogenation reactions.[14,109,110] H_2S can be removed by chemical scrubbing. Haldor Topsoe indicates that decreasing the concentration of H_2S at the inlet to a cocurrent reactor by 3–6 vol% can decrease the average temperature needed to achieve a specific sulfur reduction by 15–20 °C, or reduce final sulfur levels by more than two-thirds.[64] UOP projects that scrubbing H_2S from recycle hydrogen can reduce sulfur levels from roughly 285 to 180 ppmw in an existing hydrotreater.[64]

The increase in H_2 partial pressure and/or H_2 purity can improve HDS and hydrogenation.[14] Haldor Topsoe indicates that increasing hydrogen purity is preferable to a simple increase in the pressure of the hydrogen feed gas, since the latter will also increase the partial pressure of hydrogen sulfide later in the process, which inhibits both beneficial reactions.[64] Haldor Topsoe projects that an increase in hydrogen purity of 30% would lower the temperature needed to achieve the same sulfur removal rate by 8–9 °C, or temperature could be maintained while increasing the amount of sulfur removed by roughly 40%. Hydrogen purity can be increased through the use of a membrane separation system or a PSA (presssure-swing adsorption) unit. UOP projects that purifying hydrogen can reduce distillate sulfur from 180 to 140 ppmw from an existing hydrotreater.[64]

Increasing the recycle gas/oil ratio (increase in the amount of recycle gas sent to the inlet of the reactor) could increase the degree of desulfurization, but the effect is relatively small so a relatively large increase is needed to achieve the same effect as scribing recycle gas.[9] Haldor Topsoe indicates that a 50% increase in the ratio of total gas/liquid ratio only decreases the necessary reactor temperature by 6–8 °C, or the temperature can be maintained and the final sulfur level reduced by 35%–45%.[64]

The improvement in vapor–liquid contact can enhance the performance of distillate hydrotreaters. As an example, in testing of an improved vapor–liquid distributor in commercial use, Haldor Topsoe and Phillips Petroleum found that the new Topsoe Dense Pattern Flexible Distribution Tray (installed in 1996 to replace a chimney-type distributor installed in 1995 in a refinery) allowed a 30% higher sulfur feed to be processed at 25 °C lower temperatures, while reducing the sulfur content of the product from 500 to 350 ppmw.[79] AkzoNobel estimates that an improved vapor–liquid distributor can reduce the temperature necessary to meet a 50 ppmw sulfur level by 10 °C, which in turn would increase catalyst life and allow an increase in cycle length from 10 to 18 months.[64] Based on the above data from Haldor Topsoe, if temperatures were maintained, the final sulfur level could be reduced by 50%.[64] Maintaining temperature should have allowed an additional reduction in sulfur of more than two-thirds. Thus, ensuring adequate vapor–liquid contact can have a major impact on final sulfur levels.

The above-mentioned individual improvements described cannot be simply combined, either additively or multiplicatively. As mentioned earlier, each existing distillate hydrotreater is unique in its combination of design, catalyst, feedstock, and operating conditions. While the improvements described above are probably indicative of improvements, which can be made in many cases, it is not likely that all of the improvements mentioned are applicable to any one unit; the degree of improvement at one refinery could either be greater than, or less than the benefits that are indicated for another refinery.

5.2.3.4 Designing New Reactor Configurations

Industrial reactor configuration for deep HDS of gas oils in terms of reaction order and effect of H_2S has been discussed by Sie.[108] The reactor design and configuration involve one-stage and two-stage desulfurization. Desulfurization processes in use today in the U.S. generally use only one reactor, due to the need to only desulfurize diesel fuel to 500 ppmw or lower. Hydrogen sulfide strongly suppresses the activity of the catalyst for converting the refractory sulfur compounds, which should occur in the major downstream part of a cocurrent trickle-bed reactor during deep desulfurization. The normally applied cocurrent trickle-bed single reactor is therefore not the optimal technology for deep desulfurization.[108] However, a second reactor can be used, particularly to meet lower sulfur levels. Adding a second reactor to increase the degree of desulfurization is an option, and both desulfurization and hydrogenation in the second reactor can be improved by removing H_2S and NH_3 from the exit gas of the first reactor before entering the second reactor. This last technical change is to install a complete second stage to the existing, one-stage hydrotreater. This second stage would consist of a second reactor with a high pressure, hydrogen sulfide scrubber between the first and second reactor. The compressor would also be upgraded to allow a higher pressure to be used in the new second reactor. Assuming use of the most active catalysts available in both reactors, UOP projects that converting from a one-stage to a two-stage hydrotreater could produce 5 ppm sulfur relative to a current level of 500 ppm today.[64]

A new way of reactor design is to have two or three catalyst beds that are normally placed in separate reactors, within a single reactor shell and have both cocurrent and countercurrent flows.[108] This new design was developed by ABB Lummus and Criterion, as represented by their SynSat process.[111,112] The SynAlliance (consisting of ABB Lummus, Criterion Catalyst, and Shell Oil) has patented a countercurrent reactor design called SynTechnology. With this technology, in a single reactor design, the initial portion of the reactor will follow a cocurrent design, while the last portion of the reactor will be countercurrent.[64] Traditional reactors are cocurrent in nature. The hydrogen is mixed together with the distillate at the entrance to the reactor and flow through the reactor together. Because the reaction is exothermic, heat must be removed periodically. This is sometimes done through the introduction of fresh hydrogen and distillate at one or two points further down the reactor. The advantage of cocurrent design is practical, as it eases the control of gas–liquid mixing and contact with the catalyst. The disadvantage is that the concentration of H_2 is the highest at the front of the reactor and lowest at the outlet. The opposite is true for the concentration of H_2S. This increases the difficulty of achieving extremely low-sulfur levels due to the low H_2 concentration and high H_2S concentration at the end of the reactor. A new solution to this problem is to design a countercurrent reactor, where the fresh H_2 is introduced at one end of the reactor and the liquid distillate at the other end. Here, the hydrogen concentration is highest (and the hydrogen sulfide concentration is lowest) where the reactor is trying to desulfurize the most difficult (sterically hindered) compounds. The difficulty of countercurrent designs in the case of distillate hydrotreating is vapor–liquid contact and the prevention of hot spots within the reactor.

In a two-reactor design, the first reactor will be cocurrent, while the second reactor will be countercurrent. ABB Lummus estimates that the countercurrent design can reduce the catalyst volume needed to achieve 97% desulfurization by 16% relative to a cocurrent design.[64] The impact of the countercurrent design is even more significant when aromatics control (or cetane improvement) is desired in addition to sulfur control. However, operation of countercurrent flow reactor might not be possible in packed beds of the usual catalyst particles because of the occurrence of flooding at industrially relevant fluid velocities. Criterion offers the catalysts for the SynSat and SynShift processes. SynShift involves deep heteroatom removal, ring opening, and aromatic saturation, thus decreasing (shifting) boiling range. Some novel reactor concepts based on special structured packings or monoliths that allow such countercurrent operation have been presented.[108]

Considering deep desulfurization of liquid hydrocarbon fuels for fuel cell applications, the removal of sulfur from a commercial diesel is more difficult than the removal of sulfur from a commercial gasoline, as the sulfur compounds in commercial diesel have much less HDS reactivity. A significant advantage of HDS of diesel is that it has high capacity for removing the majority of sulfur in the diesel. However, even using the currently developed new HDS technologies for preparing ultraclean diesel for fuel cell applications, the process still needs to be conducted at high temperature and pressure with high hydrogen consumption, and it is still difficult or very costly to remove sulfur in diesel fuel to less than 0.1 ppmw for fuel cell applications. A combination of the HDS technology and the adsorption technology might be a promising approach for producing the fuel cell-used liquid hydrocarbon fuel in a large scale from a middle distillate.

5.3 ADSORPTIVE DESULFURIZATION

In terms of technology availability, the current HDS technology is not very efficient for reducing the sulfur content in the commercial hydrocarbon fuels to a level for the fuel cell applications, especially for PEMFC applications. On the other hand, the conventional HDS processes need to work at higher temperature and higher pressure using hydrogen gas, which is not suitable for the on-board or on-site desulfurization for fuel cell applications. Consequently, the development of the selective adsorption processes for ultradeep desulfurization of the commercial hydrocarbon fuels becomes an important research subject in hydrocarbon fuel processing for fuel cell applications.

A common approach to desulfurization of hydrocarbon fuels is to capture the sulfur species in a sorbent bed. In comparison with the HDS process, selective adsorption for removing sulfur (PSU-SARS) in hydrocarbon fuels have some significant advantage: (a) selective adsorption is able to remove sulfur in the hydrocarbon fuels to the level for the fuel cell application; (b) the process is usually conducted at ambient temperature and atmospheric pressure, resulting in more energy efficiency and cost efficiency; and (c) most of these processes do not need to use hydrogen gas, which is the most costly in HDS. These are especially attractive for fuel cell applications. Recently, many selective adsorption (or sorption) processes for removing sulfur from natural gas, gasoline, jet fuel, and diesel fuel have been reported. A key point in developing a successful adsorption process is to develop an adsorbent that has high adsorptive capacity and selectivity, and is facilitative to be regenerated. The adsorbents generally include the reduced metal, metal oxides, metal chlorides, zeolites, ACs, or metal-impregnated high-surface area supports like AC, zeolites, or alumina. The size and composition of a given adsorbent/sorbent bed depend on the composition of the fuel and the servicing period required in the particular application. According to the operating conditions, the reported sorption methods can also be classed into high temperature method (>600 °C), medium temperature method 150–550 °C, and ambient temperature method. The high and medium temperature methods might involve a chemical reaction or absorption, and the sorbents include metal oxides and reduced metals. The ambient temperature method usually involves a physical or chemical adsorption on the surface, and the adsorbents include zeolites, aluminosilicate, and carbon materials.

5.3.1 Natural Gas

Westmoreland and Harrison performed comparative studies on different sorbent systems that consisted almost entirely of different metal oxides.[113,114] These studies can be divided into two groups, with respect to the operating temperatures: high temperature: >600 °C (Ba, Ca, Sr, Cu, Mn, Mo, W) and middle temperature: 300–550 °C (V, Zn, Co, Fe). The high-temperature desulfurization might be only suitable for the desulfurization of fuel gas from coal gasification and desulfurization of SOFC and molten carbonate fuel cell (MCFC) feed considering the energetic efficiency in the process. The middle temperature desulfurization might be used for the desulfurization of SOFC and MCFC feed, also for the desulfurization of reformate (or post desulfurization) before water-gas shift reaction. The post desulfurization will be discussed in detail in Section 5.4.

Research Triangle Institute and National Energy Technology Lab reported their approach in the development of high-temperature desulfurization of natural gas and syngas in a transport reactor over a wide temperature range (260–760 °C) by using regenerable sorbents, primarily based on ZnO.[42] This sorbent technology has also advanced to the point that syngas desulfurization can be carried out in both fixed- and fluidized-bed reactors, depending upon process design needs. As ZnO-based sorbents can work over a wide temperature range, these sorbents are also suitable for reformate desulfurization, which will be discussed further in Section 5.4.

Recently, great attention has been paid to the ambient temperature adsorption for removing sulfur compounds from natural gas, as the system is simple and is quick and easy to be started up. The major adsorbents that were used in the ambient temperature adsorption are AC-based and zeolite-based materials. The Osaka Gas mixed metal and metal oxide catalyst provides a low-temperature method of desulfurization. The catalysts (or adsorbents) is claimed to remove organic sulfur and H_2S at room temperature.[115]

Recently, Israelson of Siemens Westinghouse Power Corporation tested 10 commercially available desulfurization adsorbents, including metal oxides, ACs, and molecular sieves, for pipeline natural gas with 6 ppm volume total sulfur.[4] The primary sulfur species for the tested pipeline natural gas were dimethyl sulfide (DMS), isopropyl mercaptan, tertiary butyl mercaptan, and THT. Most testing was performed at ambient temperature 21 °C to simulate the conditional expected in a commercially viable desulfurizer. Two materials, ZnO and copper oxide/zinc oxide, were tested at elevated temperatures. It was found that DMS always came through a desulfurizer bed first, independent of the adsorption process. Israelson used an index of average ppmv DMS × cubic feet of natural gas/cubic feet of adsorbent to predict the breakthrough performance of the adsorbents. The results are shown in Table 5.3.[4] Three ACs from different suppliers were found to have identical performance in removing DMS. The performance index is around 2500 ppmv. Among the

Table 5.3.   Adsorption Desulfurization Performance of Various Sorbents for Natural Gas[4]

Manufacturer	Active Component	Product Name	Ave. DMS (ppmv)	Ave. Total Sulfur (ppmv)	Performance Index (ppmv) DMT Bed Volumes
United Catalysts Inc. (Süd-Chemie)	ZnO at 350 °C (no preceding CoMo catalyst bed or hydrogen addition)	G-72E	1.5	6.7	668
Norit	Carbon with chromium and copper salts	RGM-3	0.8	5.1	24,000
Calgon Carbon	Carbon	PCB	1.5	6.2	26,550
Süd-Chemie	Carbon with copper oxide	C8-7-01	1.8	6.7	26,550
Süd-Chemie	Nickel, nickel oxide	C28	1.1	4.2	44,992
Grace Davison	Molecular sieve: $13 \times 10\,\text{Å}$ pore) zeolite-X	554HP	0.8	4.4	110,376
Supplier C	Unknown	Proprietary adsorbent S	1.2	5.1	182,573
PNNL	Copper-impregnated zeolites Y substrate	Unnamed	0.8	4.1	2,416,800
Synetix	Copper oxide and zinc oxide, 100 °C top, 170 °C	Puraspec 2084	3.4	8.4	3,661,800
Supplier E	Unknown	Proprietary adsorbent T	2.3	8.2	3,661.800

tested samples, Israelson found three high-performance adsorbents, including a copper-impregnated zeolite Y adsorbent and a copper oxide/zinc oxide adsorbent, which removed about 100–150 times as much DMS as AC did before breakthrough.

Roh et al.[116] investigated the desulfurization of natural gas over AC, Fe/AC, BEA zeolite (BEA), and Fe/BEA for fuel cells. Sulfur capacities were obtained from the breakthrough curves of tert-butyl mercaptan (TBM) and THT. Fe/BEA (Fe: 3%) prepared by the ion exchange method exhibited the highest sulfur capacity of 19.6% g(S-comp)/g(sorb) among the adsorbents tested for a model natural gas containing 50 ppm of TBM and 50 ppm of THT in CH_4. The authors have explained that the high capacity of Fe/BEA is because of the strong chemisorption of sulfur compounds on the Lewis acid sites present in BEA zeolite and the beneficial Fe effect on the Fe/BEA adsorbent through the possibility of complexation. In order to facilitate the comparison with Israelson's data listed in Table 5.3, the performance index for TBM is recalculated on the basis of the breakthrough capacity for TBM. The performance index for TBT is around 6.1×10^6 ppmv as the same order magnitude as the performance index for DMS over the three best adsorbents tested by Israelson.[4]

Recently, Satokawa et al.[117] (Tokyo Gas Co.) reported that the adsorptive desulfurization of DMS and TBM from pipeline natural gas by using silver-exchanged Y zeolites (AgNa–Y) in the presence of water at room temperature and normal pressure. They found that an increase in silver content of AgNa–Y improved the capacity and adsorptivity for DMS and TBM on AgNa–Y. They also found that sulfur adsorption capacity for DMS on Ag(18)Na–Y (1.9 mmol/g) for a model gas was much higher than that for TBM (0.6 mmol/g), although DMS was the first breakthrough sulfur compound in the tests by Israelson.[4] According to their data, the performance index of Ag(18)Na–Y for DMS in pipeline natural gas was about 7.5×10^6 ppmv higher than the best adsorbent tested by Israelson.[4] Satokawa et al.[117] found further that the spent AgNa–Y can be regenerated easily by heat treatment in air at 500 °C. However, only about 54% capacity for Ag(18)Na–Y was recovered, although almost all capacity for Ag(1)Na–Y and Ag(5)Na–Y was recovered.[4]

These studies for desulfurization of natural gas at ambient temperature indicate that metal iron-exchanged zeolites might be promising adsorbents, and more work is necessary for the regeneration of the spent metal ion-exchanged zeolites.

5.3.2 Gasoline

As mentioned in the foregoing, the commercial gasoline not only contains saturated hydrocarbons but also a large amount of unsaturated hydrocarbons, such as aromatics and olefins with concentration even up to 40 and 20 wt %, respectively, especially in FCC gasoline. The molar concentration of these coexisting unsaturated hydrocarbons is even up to 4-order magnitude higher than that of the sulfur compounds in commercial gasoline. Consequently, the challenge in the development of a successfully adsorptive desulfurization process is to develop an adsorbent that has higher adsorptive selectivity and capacity for the sulfur compounds.

Figure 5.11 illustrates the known coordination geometries of thiophene in organometallic complexes, which indicate likely adsorption configurations of thiophenic compounds on the surface of adsorbents.[118] Both thiophenic compounds and nonsulfur aromatic compounds can interact with metal species by π-electrons. However, in Figure 5.11, only two types of interaction of thiophene with metal involve sulfur atom in thiophene: the η^1–S bonding interaction between the sulfur

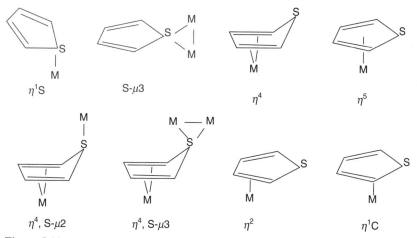

η^1S S-μ3 η^4 η^5

η^4, S-μ2 η^4, S-μ3 η^2 η^1C

Figure 5.11. Known coordination geometries of thiophene in organometallic complexes, indicating likely adsorption configurations of thiophenic compounds on the surface of adsorbents.[118]

atom and one metal atom, and the S–μ3 bonding interaction between the sulfur atom and two metal atoms.

According to the operating temperature, the developing sorbents for gasoline desulfurization can be classified into two groups: mild temperature (150–550 °C) group and low temperature (0–100 °C) group. The former often involves the reduced metals and/or metal oxides as sorbents, and the latter usually involves zeolite-based adsorbents, activated alumina, metal oxides, and/or carbon materials as adsorbents. The adsorptive capacities of different sorbents for different liquid hydrocarbon fuels, which were measured by different researchers in the flowing systems, are listed in Table 5.4.[2,26,38,119–130]

ConocoPhillips Petroleum Company developed a new S-Zorb process for the production of low-sulfur gasoline by reactive sorption of sulfur compounds over a solid sorbent.[131–135] The process uses a fluidized-bed reactor in the temperature range between 377 and 502 °C under H_2 pressures in the range of 7.1–21.1 kg/cm^2. A flow scheme and the principle for the S-Zorb process are shown in Figure 5.12. In this process, FCC gasoline and H_2 are fed to the reactor and heated. Vaporized gasoline is injected into a fluidized-bed reactor, where the proprietary sorbent removes sulfur from the feed. Suspended sorbent is removed from the vapor at a disengaging zone. Sulfur-free vapor exits at the top to be cooled. The spent sorbent is continuously withdrawn, transferred to a regenerator where the sulfur is removed as SO_2 and sent to a sulfur recovery unit. The oxidized sorbent is sent to a sorbent reducer for reduction in the presence of hydrogen gas and then returned to the reactor. A significant advantage for S-Zorb is that the process is able to reduce the sulfur in FCC naphtha to 10 ppmw with almost no change in octane number and no requirement of large volumes of high-purity H_2.

The Research Triangle Institute[136] also reported a TReND process for adsorptive desulfurization of gasoline over a regenerable metal oxide sorbent in a transport

Table 5.4. Adsorptive Capacity of Different Sorbents for Gasoline, Jet Fuel, Diesel, and Model Fuels

Sorbent	Fuel Description	Sulfur in Fuel (ppmw)	Adsorption Conditions	Breakthrough Capacity (mg-S/g)	Saturated Capacity (mg-S/g)	Affiliation	Reference
	For model fuel						
Na–Y	T in n-octane	760	RT	2.24[h]	33.6	UMICH	Hernandez-Maldonado et al.[119]
Na–Y	T in benzene	760	RT	0.64[h]	3.2	UMICH	Hernandez-Maldonado et al.[119]
H–Y	T in n-octane	760	RT	14.7[h]	36.8	UMICH	Hernandez-Maldonado et al.[119]
Ag–Y (LPIE)[a]	T in n-octane	760	RT	17[h]	29	UMICH	Hernandez-Maldonado et al.[119]
Ag–Y (LPIE)	T in benzene	760	RT	1.3[h]	5.5	UMICH	Hernandez-Maldonado et al.[119]
Cu(I)–Y(LPIE)	T in n-octane	760	RT	58.4[h] 42.6[i]	81.8	UMICH	Hernandez-Maldonado et al.[119]
Cu(I)–Y(LPIE)	T in benzene	760		6.1[h]	17.3	UMICH	Hernandez-Maldonado et al.[119]
Cu(I)–Y(LPIE)	T in n-octane	190	RT	21.8[h] 16[i]	41.0	UMICH	Hernandez-Maldonado et al.[119]
Cu(I)–Y(LPIE)	T in 20% benzene, 80% n-octane	190	RT	7.1[h]	14.1	UMICH	Hernandez-Maldonado et al.[119]
Cu(I)–Y(LPIE)	T in isooctane	190	RT, LHSV: 12h^{-1}	17.8[i]	~23[i]	PSU	Ma et al.[26]
Ni(II)–Y(LPIE)	BT, DBT, and 4,6-DMDBT in n-octane	150	RT	40		KIE	Ko et al.[120]
Ni(II)–Y(LPIE)	BT, DBT, and 4,6-DMDBT in n-octane with 5 v/v % benzene	150	RT	0.8[i]		KIE	Ko et al.[120]
Ni(II)–Y(LPIE)	BT, DBT, and 4,6-DMDBT in n-octane with 5000 ppmw H_2O	150	RT	13[i]		KIE	Ko et al.[120]
Ni–Al	T + BT in paraffins with 8% toluene	400	RT, LHSV: 24h^{-1}	12.3[j]	14.2	PSU	Ma et al.[121]

Ni–Al	T + BT in paraffins with 8% toluene + 5.1% olefin	400	RT, LHSV: 24 h^{-1}	2.0[j]	4.0[j]	PSU	Ma et al.[121]
Ni–Al	DBT and 4,6-DMDBT in paraffins with 10% t-butybenzene and 303 ppmw N	687	RT, LHSV: 4.8 h^{-1}	2.2[j]	2.4	PSU	Ma et al.[121]
KYNi30IWI[b] (reduced)	BT + 2-BT + 5-MBT in decane with 19% n-butylbenzene	506	RT	7.1[i,j]	11.0[i]	PSU	Velu et al.[122]
KYNi30IWI (reduced)	Same as the above	506	80 °C	9.7[i,j]	11.5[i]	PSU	Velu et al.[122]
Activated alumina	DBT and 4,6-DMDBT in paraffins with 10% t-butybenzene and 303 ppmw N	687	RT, LHSV: 4.8 h^{-1}	1.6[j]	3.4	PSU	Kim et al.[123]
AC	Same as the above	687	RT, LHSV: 4.8 h^{-1}	7.2[j]	16.3	PSU	Kim et al.[123]
AC (ACNU)	BT, DBT, 4-MDBT, 4,6-DMDBT in paraffins with 10% butybenzene	398	RT, LHSV: 4.8 h^{-1}	4.3[j]	13.1	PSU	Zhou et al.[124]
AC [8]	Same as the above	398	RT, LHSV: 4.8 h^{-1}	6.5[j]	16.7	PSU	Zhou et al.[124]
Real gasoline							
Cu(I)–Y (LPIE-RT)[c]	Gasoline	335	RT	4.57[j]	12.61	UMICH	Hernandez-Maldonado and Yang[126]
Cu(I)–Y (LPIE)	Gasoline	305	RT, LHSV: 4.8 h^{-1}	0.22, 0.49[k]	0.64	PSU	Ma et al.[26]
AC/Cu(I)–Y (LPIE-RT)	Gasoline	335	RT	5.90[j]	15.87	UMICH	Hernandez-Maldonado and Yang[126]
Ni/Si–Al	Gasoline	305	RT, LHSV: 4.8 h^{-1}	0.37[j]	1.5[i]	PSU	Ma et al.[26]

(Continued)

Table 5.4. *Continued*

Sorbent	Fuel Description	Sulfur in Fuel (ppmw)	Adsorption Conditions	Breakthrough Capacity (mg-S/g)	Saturated Capacity (mg-S/g)	Affiliation	Reference
Ni/Si-Al	Gasoline	305	200°C, LHSV: 4.8h⁻¹	0.51ⁱ, 7.3ᵏ	>9.5ⁱ	PSU	Ma et al.[26]
Ni-Al	Gasoline	210	RT	0.3ʲ	>1.7ⁱ	PSU	Ma et al.[121]
Ni-Al	Gasoline	210	200°C, LHSV: 4.8h⁻¹	0.62ʲ	>>4.4ⁱ	PSU	Ma et al.[121]
	Real jet fuel						
Cu(I)–Y (VPIE)ᵈ	Jet fuel	364	RT	12.6ʲ	23.1	UMICH	Hernandez-Maldonado et al.[127]
Zn(II)–Y (LPIE-RT)	Jet fuel	364	RT	1.4ʲ	3.7	UMICH	Hernandez-Maldonado et al.[127]
Zn(II)–X (LPIE-RT)	Jet fuel	364	RT	2.8ʲ	6.3	UMICH	Hernandez-Maldonado et al.[127]
CuCl/AC	JP-5	1172	RT, 2.3h⁻¹	1.0	4.8	UMICH	Wang et al.[128]
PdCl₂/Al₂O₃	JP-5	1172	RT, 2.3h⁻¹	2.1	9.1	UMICH	Wang et al.[128]
PdCl₂/AC	JP-5	1172	RT, 2.3h⁻¹	3.2	19.7	UMICH	Wang et al.[128]
Ni/Si-Al	JP-8	736	220°C, LHSV: 2.4h⁻¹	5.4ⁱʲ, 9.2ˡ		PSU	Velu et al.[2]
Ni/Si-Al	Light JP-8	380	220°C, LHSV: 2.4h⁻¹	7.3ⁱʲ, 13.5ˡ		PSU	Velu et al.[2]
KYNiIE-3 (red)	Light JP-8	380	80°C	4.5ⁱʲ	~6ⁱ	PSU	Velu et al.[122]

KYNi8IWI (red)	Light JP-8 Real diesel	380	80 °C	2.4^{ij}	$\sim 7^{i}$	PSU	Velu et al.[122]
Activated carbon	Diesel		RT	$<0.5^{j}$	9.41	UMICH	Hernandez-Maldonado and Yang[38]
Selexsorb CDX ($-Al_2O_3$)	Diesel		RT	2.63^{j}	12.17	UMICH	Hernandez-Maldonado and Yang[38]
Ce(IV)–Y (LPIE-80)[e]	Diesel		RT	1.04^{j}	3.91	UMICH	Hernandez-Maldonado and Yang[129]
Cu(I)–Y (LPIE-RT)	Diesel		RT	5.33^{j}	11.97	UMICH	Hernandez-Maldonado and Yang[38]
AC/Cu(I)–Y (LPIE-RT)	Diesel		RT	7.57^{i}	13	UMICH	Hernandez-Maldonado and Yang[126]
CDX/Cu(I)–Y (LPIE-RT)	Diesel		RT	7.11^{j}	10.97	UMICH	Hernandez-Maldonado and Yang[38]
CDX/Cu(I)–Y (LPIE-RT)	Diesel		RT	9.33^{j}	13.63	UMICH	Hernandez-Maldonado and Yang[38]
Cu(I)–Y (VPIE)	Diesel		RT	8.89^{j}	13.12	UMICH	Hernandez-Maldonado and Yang[130]
CDX/Cu(I)–Y (VPIE)	Diesel		RT	10.44^{j}	14.02	UMICH	Hernandez-Maldonado and Yang[130]
Ni(II)–Y (LPIE-RT)	Diesel		RT	2.7^{i}	6.53	UMICH	Hernandez-Maldonado and Yang[129]
Ni(II)–Y (LPIE-135)[f]	Diesel		RT	3.82^{i}	6.82	UMICH	Hernandez-Maldonado and Yang[129]

(*Continued*)

251

Table 5.4. *Continued*

Sorbent	Fuel Description	Sulfur in Fuel (ppmw)	Adsorption Conditions	Breakthrough Capacity (mg-S/g)	Saturated Capacity (mg-S/g)	Affiliation	Reference
Ni(II)–X (LPIE-RT)	Diesel	297	RT	4.58[i]	8.03	UMICH	Hernandez-Maldonado and Yang[129]
Ni(II)–Y (SSIE)[g]	Diesel	297	RT	5.06[i]	9.25	UMICH	Hernandez-Maldonado and Yang[129]
CDX/Ni(II)–Y (SSIE)	Diesel	297	RT	6.11[j]	10.59	UMICH	Hernandez-Maldonado and Yang[129]
Ni(II)–Y(LPIE)	Diesel	186	RT	0.46[i]	~0.5[i]	KIE	Ko et al.[120]

[a]Prepared by liquid-phase ion exchange techniques.

[b]30 wt% of Ni impregnated on KY zeolite.

[c]Prepared by liquid-phase ion exchange techniques at room temperature.

[d]Prepared by vapor-phase ion exchange.

[e]Prepared by liquid-phase ion exchange techniques at 80 °C.

[f]Prepared by liquid-phase ion exchange techniques at 130 °C.

[g]Prepared by solid-state ion exchange.

[h]Breakthrough S level: 3.8 or 1.5 ppmw.

[i]Reestimated on the basis of the reported breakthrough curve.

[j]Breakthrough S level: 1 ppmw.

[k]Breakthrough S level: 10 ppmw.

[l]Breakthrough S level: 30 ppmw.

AC, activated carbon; RT, room temperature; LHSV, liquid hourly space velocity; UMICH, University of Michigan; PSU, The Pennsylvania State University; KIE, Korea Institute of Energy.

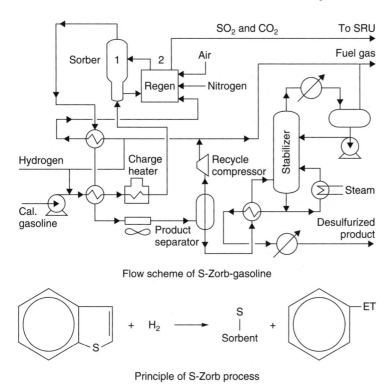

Flow scheme of S-Zorb-gasoline

Principle of S-Zorb process

Figure 5.12. ConocoPhillips's S-Zorb-gasoline process based on solid adsorbent and its continuous regeneration.[131,132] SRU, sulfur recovery unit.

reactor system at 420–540 °C in H_2 atmosphere.[137] For a model fuel containing eight sulfur compounds in isooctane, their results showed that a modified sorbent was able to remove 73.7% of sulfur at 427 °C in the presence of hydrogen gas. The developed sorbent was not efficient for removing thiophene and diethyl sulfide, although it was good for removing BT and DBT. As both S-Zorb process and TReND process involve a process at high temperature and in H_2 atmosphere, these processes might not be suitable for on-site or on-board desulfurization of gasoline for fuel cell applications.

As desulfurization of gasoline for the fuel cell applications prefers a desulfurization process on-board or on-site at low temperatures without using H_2 gas, considerable attention has been paid to developing an adsorptive process under ambient conditions without using H_2 gas for the gasoline-based fuel cell systems.

Activated alumina has been used in chromatographic analysis and separation, and adsorptive separation. The use of activated alumina as an adsorbent in a desulfurization process for low-sulfur gasoline has been reported by Black & Veatch Irritchard, Inc. and Alcoa Industrial Chemicals.[138,139] As the adsorption on activated alumina is dominantly dependent on the molecular polarity, activated alumina might have relatively lower selectivity for the thiophenic sulfur compounds.

Nehlsen et al.[140] reported a developing process for the removal of thiols from a hydrocarbon stream by a heterogeneous reaction with lead oxide at a temperature <50 °C. The principle is heterogeneous reaction of thiols with lead oxide to form insoluble lead thiolates, which is then separated from the hydrocarbon stream. They claimed that the process allows the original thiols to be recovered for other uses, and the lead is also recovered and recycled. Experimental recovery of the lead for recycling exceeds 94%. Thiols are recovered with typical yields of 80%–90%.

Yang and coworkers reported an adsorptive desulfurization of commercial gasoline by selective adsorption via π-complexation with Cu(I)–Y zeolite under ambient conditions.[126,127,141] They reported that the breakthrough adsorption capacity was 0.14 mmol-S/g-A (4.5 mg-S/g-A) for a commercial gasoline with 335 ppmw sulfur. They used GC-FPD (flame photometric detection) for the analysis of the treated commercial gasoline sample and claimed that their FPD detection limit was below 0.28 ppmw. Yang and coworkers emphasized that the selective adsorption is based on a π-complexation mechanism, and the adsorption bond energies via π-complexation with Cu(I)–Y zeolite for thiophene (21.4 kcal/mol) is higher than that for benzene (20.5 kcal/mol) according to their computational results.[142] In a study of the desulfurization of gasoline, diesel, and jet fuels over the π-complexation-based sorbents obtained by ion exchanging faujasite-type zeolites with Cu$^+$, Ni^{2+}, or Zn^{2+} cations using different techniques, including liquid-phase ion exchange (LPIE), vapor-phase ion exchange (VPIE), and solid-state ion exchange (SSIE) techniques, Yang and coworkers found that the π-complexation sorbents desulfurization performance decreases in the order as follows: Cu(I)–Y(VPIE) > Ni(II)–Y(SSIE) > Ni(II)–X(LPIE) > Zn(II)–X(LPIE) > Zn(II)–Y(LPIE).[143]

Richardeau et al. investigated thiophene adsorption from liquid solutions containing hydrocarbons over HFAU zeolites in a stirred batch system at room temperature.[144] They found that the maximum number of thiophene molecules adsorbed per gram of zeolites is equal to their concentration of acidic sites and considered that the acidic sites are the adsorption sites. They further found that the presence of toluene causes a large decrease in the removal of thiophene, and when the concentration of thiophene is high (27.7 wt%), an acid-catalyzed condensation of thiophene occurs to form dimers, trimers, and tetramers, which remain trapped on the zeolite. They concluded that thiophene removal by adsorption on acidic zeolites could only be carried out from diluted solutions containing no olefinic compounds.

Song and coworkers proposed and has been exploring PSU-SARS from liquid hydrocarbon fuels over various adsorbents including metals,[121–123] metal oxides,[145,146] zeolite-based adsorbents,[2,147] carbon materials,[123,124] and other metal compounds[118,148] under ambient conditions without using H$_2$ gas for fuel cell and refinery applications. The PSU-SARS approach aims at removing sulfur compounds in transportation fuels selectively by a direct sulfur-adsorbent interaction. For the desulfurization of commercial gasoline, among several types of adsorbents explored, they found that Ni-based adsorbent exhibited better performance for removing thiophenic sulfur compounds.[26,121] It should be mentioned that nickel was used for the reductive removal of thiophene for purifying benzene and toluene more than half a century ago.[149] The Ni-based sorbent was also widely used for removing sulfur in preparative organic chemistry at room or elevated temperature up to 60–80 °C.[150–152] In 1987, Bailey and Swan[153] filed a patent about using nickel sorbents for sulfur removal from hydrocarbon reformer feeds. Recently, International Fuel Cell[154,155] and UTC Fuel Cell[125,156] filed patents relating to a method for desulfurizing gasoline or diesel fuel over nickel-based adsorbents for use in a fuel cell power plant. According to the breakthrough curves for adsorptive desulfurization of an additive-free gasoline with 21 ppmw of sulfur, as shown in their patent,[156] the capacity is around 0.8 mg-S/g (milligram of sulfur per gram of adsorbent) at a breakthrough sulfur level of 10 ppmw.

Recently, Ma et al.[26,148] comparatively studied the performance of an Ni-based adsorbent and a Cu(I)–Y for the desulfurization of a commercial gasoline with 305 ppmw sulfur in a fixed-bed adsorption system at room temperature and 200 °C. The Cu(I)–Y prepared in this study showed only a breakthrough capacity of 0.22 mg-S/g at room temperature for removing sulfur from a commercial gasoline with 305 ppmw sulfur to less than 1 ppmw. This breakthrough capacity is much less than the breakthrough capacity (4.5 mg-S/g) of the Cu(I)–Y reported by Hernandez-Maldonado and coworkers[127] for another commercial gasoline with 335 ppmw sulfur, although for a model gasoline (190 ppmw of sulfur as thiophene in isooctane), the capacity of the Cu(I)–Y prepared by Ma et al. is almost the same as that reported by Hernandez-Maldonado and coworkers.[37,127] The reason for this significant difference is still unclear. Ma et al.[26,148] also found that at room temperature, the Ni-based adsorbent exhibited a breakthrough capacity of 0.37 mg-S/g, while at 200 °C, the breakthrough capacity of the Ni-based adsorbent was improved significantly. Moreover, the breakthrough capacity of the Ni-based adsorbent corresponding to the outlet sulfur level of 10 ppmw was 7.3 mg-S/g at 200 °C,[26] which was over an order of magnitude higher than that of Cu(I)–Y.

In order to understand the sorption mechanism on the nickel-based sorbents and to evaluate the desulfurization performance of the nickel-based sorbents, Ma et al.[26] also conducted adsorptive desulfurization of a series of model gasoline fuels and a real gasoline over a nickel-based sorbent (Ni–Al) in a flowing adsorption system at a temperature range of 25–200 °C under ambient pressure without using H_2 gas. Adsorptive capacity and selectivity of the Ni–Al adsorbent for various sulfur compounds and the effects of coexisting olefin in gasoline, as well as sorptive conditions on the adsorptive performance, were examined. The observed sorption capacities as a function of sulfur concentration at outlet for different nickel-based sorbents and different feeds at different temperatures are shown in Figure 5.13.[26] It was found that the nickel-based adsorbent shows high capacity and selectivity for the adsorptive desulfurization of gasoline. Olefins in gasoline have a strong inhibiting effect on the desulfurization performance of the nickel-based sorbent at room temperature. Increasing temperature to 200 °C can significantly improve the desulfurization performance of the nickel-based sorbent for real gasoline. They also found that sorption of sulfur compounds on the nickel surface involves C–S bond cleavage, as evidenced by the formation of ethylbenzene from BT in the absence of H_2 gas in the flow adsorption system. The obtained results indicate that the sorption capacity of the nickel-based sorbents can be improved by increasing the Ni surface area and/or by introducing the active hydrogen atoms into the Ni surface.

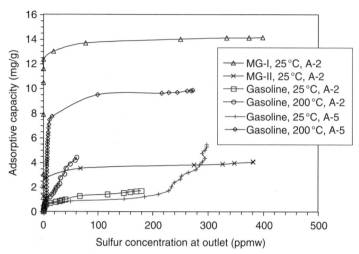

Figure 5.13. Adsorptive capacity of nickel-based adsorbents as a function of sulfur concentration at outlet.[26] See color insert.

Recently, Hu et al.[157] studied the desulfurization mechanism of thiophene on Raney Ni and rapidly quenched skeletal Ni (RQ Ni) in ultrahigh vacuum (UHV) by X-ray photoelectron spectroscopy (XPS). They found that the Raney Ni or RQ Ni could be considered as a hydrogen-preadsorbed polycrystalline Ni–alumina composite. It was found that when thiophene molecularly adsorbs on Raney Ni or RQ Ni at −170 °C and at −100 °C, the adsorbed thiophene contact directly with the metallic Ni in Raney Ni undergoes C–S bond scission, leading to carbonaceous species most probably in the metallocycle-like configuration and atomic sulfur. By 200 °C, the C1s peak has disappeared, leaving nickel sulfide on the surface. The finding by XPS in this study is in agreement with the results obtained by Ma et al.[26]

5.3.3 Jet Fuel

In the study of the desulfurization of a commercial jet fuel by using VPIE Cu(I)–Y adsorbents in a fixed-bed adsorber operated at ambient temperature and pressure, Hernandez-Maldonado and coworkers reported that the best sorbent, Cu(I)–Y(VPIE), which was prepared by the authors, has a breakthrough adsorption capacity of 0.395 mmol-S/g (12.6 mg-S/g) for a commercial jet fuel with 364 ppmw initial sulfur.[127] They further reported that a layered bed with 25 wt% activated alumina followed by Cu(I)–Y was capable of producing 38 cm³ of jet fuel per gram of adsorbent with a weighted average content of 0.07 ppmw-S.[37] The corresponding breakthrough capacity was 0.521 mmol/g (16.7 mg-S/g). They also reported that the adsorption capacity can be fully recovered after regeneration in air at 350 °C for 6–12 h followed by reduction in an inert gas stream containing 5 vol% hydrogen at 231 °C. They emphasized that the high sulfur selectivity and high sulfur capacity of

Cu(I)–Y(VPIE) were due to π-complexation,[127,157] but no selectivity experiment by comparing BT with naphthalene or methyl naphthalene was reported by the same group.

Velu et al.[147] studied PSU-SARS from a model jet fuel (MJF) and a real jet fuel (JP-8) over different zeolite-based adsorbents, which were prepared by the ion exchange of Y zeolites with Cu, Ni, Zn, Pd, and Ce ions. Among the adsorbents tested, Ce-exchanged Y zeolites exhibited a better adsorption capacity of about 10 mg-S/g at 80 °C with an MJF containing 510 ppmw sulfur in a batch test. The same adsorbent exhibited a sulfur adsorption capacity of about 4.5 mg-S/g for the real JP-8 jet fuel containing about 750 ppmw sulfur. They found that Ce-exchanged zeolites exhibited higher selectivity for sulfur compounds as compared with the selectivity of aromatics, for which a comparative study indicated that the direct sulfur-adsorbent interaction plays an important role in the adsorption of sulfur compounds over the Ce-exchanged Y zeolites.

Fukunaga et al.[158] reported using a developed Ni-based sorbent to remove sulfur in a kerosene fuel from 48 ppmw in the fuel to less than 0.05 ppmw. However, no quantitative breakthrough capacity of their adsorbent can be figured out from their reported data.

Recently, Velu et al.[2] studied the optimization of the performance of an Ni/SiO$_2$–Al$_2$O$_3$ adsorbent for the adsorptive desulfurization of jet fuel, including JP-8 jet fuel containing around 736 ppmw sulfur and a light JP-8 jet fuel having around 380 ppmw sulfur, obtained by fractionation. They found that the removal of C3-BTs, such as sterically hindered 2,3,7-TMBT, from the JP-8 jet fuel by fractionation improved the sorbent capacity by 2.5 times. The particle sizes of the sorbent and its bed dimensions were examined and optimized to achieve targeted sulfur sorption capacity of over 10 mg-S/g without encountering pressure drop across the bed. At 220 °C, the sorptive desulfurization of fractionated light JP-8 over the Ni/SiO$_2$–Al$_2$O$_3$ sorbent having particle sizes between 0.15 and 0.25 mm offered a sulfur break-through sorption capacity of about 11.5 mg-S/g at a breakthrough sulfur level of 30 ppmw without developing any significant pressure drop across the beds. High sulfur-sorption capacity could be achieved with Ni/SiO$_2$–Al$_2$O$_3$ when the overall aspect ratio, axial aspect ratio, and radial aspect ratio of the sorbent bed were around 60, 3000, and 100, respectively.

Velu et al.[122] also studied the sorption performance of a series NiY zeolites with different Ni loadings, which were synthesized by incipient wetness impregnation and LPIE methods using NH$_4$Y and KY zeolites. At the sorption temperature of 80 °C, NiY zeolite containing 30 wt % Ni synthesized by incipient wetness impregnation of NH$_4$Y zeolite was able to clean only about 10 mL of an MJF per gram of the adsorbent to produce a desulfurized fuel containing below 1 ppmw sulfur. Under the same experimental conditions, a K-containing NiY zeolite cleaned about 30 mL of the fuel per gram of the adsorbent. A better sulfur sorption performance was observed when the NiY zeolite was synthesized by ion exchange using KY zeolite and reduced before sulfur sorption. The corresponding breakthrough capacity at 1 ppmw sulfur level is about 4.7 mg-S/g, which was recalculated on the basis of the breakthrough curves reported by Velu et al.[122] Temperature program reduction (TPR) studies indicated that the reducibility of Ni-Y was improved when K was present as a co-cation. The *in situ* XPS studies of unreduced and reduced samples revealed that the presence of K as a co-cation in the zeolite matrix helps Ni dispersion at the surface. The promoting effect of K on the sulfur adsorption performance of NiY zeolites was therefore attributed to improved reducibility and surface dispersion of Ni when K was present as a co-cation.

Tawara et al.[159] reported an "adsorptive catalyst" (Ni/ZnO) for removing sulfur in kerosene to less than 0.1 ppmw in the temperature range between 270 and 300 °C in H_2 atmosphere under a pressure of 0.60 MPa. The principle in this process is based on the following reactions:

$$Ni(s) + S \text{ compound} + H_2 \leftrightarrow NiS(s) + \text{hydrocarbon}, \tag{5.3}$$

$$NiS(s) + H_2 \leftrightarrow Ni(s) + H_2S, \tag{5.4}$$

$$ZnO(s) + H_2S \leftrightarrow ZnS(s) + H_2O. \tag{5.5}$$

A patent for desulfurization of petroleum streams containing condensed ring heterocyclic organosulfur compounds on the basis of the similar principle was filed by Research and Engineering Company in 2001.[58]

5.3.4 Diesel

The diesel fuel contains not only saturated hydrocarbons and sulfur compounds, but also a large number of aromatic compounds that have aromatic skeleton structure similar to the coexisting sulfur compounds. As mentioned before, the remaining sulfur compounds in commercial diesel fuel are usually the alkyl DBTs with one or two alkyl groups at the 4- and/or 6-positions, which show strong steric hindrance toward the direct interaction between the sulfur atom in DBTs and the adsorption site. Consequently, a great challenge in the adsorptive desulfurization of diesel fuel is to develop an adsorbent that can selectively adsorb such sulfur compounds. The reported sorbents for removing sulfur from diesel fuel or gas oil include reduced metals, metal oxides, zeolite-based materials, and carbon materials.

5.3.4.1 Metal Sorbents

On the basis of Phillip's existing S-Zorb technology for gasoline, ConocoPhillips has developed an S-Zorb-Diesel technology for lowering the sulfur content of diesel fuel while minimizing the operating costs of the refiners, which will help refiners meet current and future sulfur regulations both in the U.S. and in Europe.[160] The ConocoPhillips S -Zorb process desulfurizes diesel fuel by sorption on a proprietary sorbent. Sulfur-containing molecules are adsorbed, including sterically hindered molecules like 4,6-DMDBT, and the sulfur atom is split from the molecule. The sulfur is retained on the adsorbent while the hydrocarbon desorbs. Sulfur is typically removed to about 5 ppm from feedstock of 500 ppm sulfur (but greater than 2000 ppm sulfur feedstocks are possible). Depending upon the feedstock, the process operates at mild conditions: 275–500 psig about 700–800 °F (371–427 °C). The principle and

Table 5.5.   Diesel Property Changes between Feed and S-Zorb-Diesel Process Product[132]

Property	Feed	Product
Sulfur (ppm)	523	6
API gravity	33.20	33.22
Hydrogen (wt%)	12.72	12.72
Cetane #	43.5	43.5
Cetane index	44.4	44.4
D86 (°F)	Base	Same
Cloud point (°F)	−10.6	−10.6
CFPP (°F)	−13.5	−13.5
Lubricity		
SLBOCLE	3700	3600
HFRR	385	315

HFRR, The high-frequency reciprocating rig; SLBOCLE, scuffing load ball-on-cylinder lubricity evaluator.

process scheme of the S-Zorb-Diesel technology is similar to those for S-Zorb-gasoline (Fig. 5.12). The properties of a typical diesel product from S-Zorb-Diesel process with the properties of feed are listed in Table 5.5.[132] Almost no change in the composition and properties between the product and feed were observed, except the sulfur content.

In the study of adsorptive desulfurization of a model diesel fuel over a nickel-based sorbent in a fixed-bed adsorption system at ambient temperature and atmosphere pressure, Kim et al.[123] found that the direct interaction between the sulfur atom in the sulfur compounds and the surface nickel plays an important role, indicating that the supported nickel adsorbent is good for the selective removal of the sulfur compounds, which have no alkyl steric hindrance, from hydrocarbon streams. However, the sorption affinity of the alkyl DBTs with alkyl groups at 4- and/or 6-positions is reduced due to the steric hindrance. The sorption selectivity of the nickel-based sorbent for various compounds at room temperature increases in the order of Nap $\approx$ 1-MNap < 4,6-DMDBT < DBT < quinoline < indole, as shown in Figure 5.14.[123]

5.3.4.2 *Metal Oxide Sorbents*

Activated alumina has been used widely in chromatographic analysis and separation, and adsorptive separation. In the investigation of desulfurization and denitrogenation of light oils on the basis of adsorption of sulfur- and nitrogen-containing compounds on methyl viologen-modified aluminosilicate (MV^{2+}/AS) adsorbent, Shiraishi et al. found that sulfur and nitrogen compounds were adsorbed on the surface of $MV^{2+}/$ AS via the formation of a charge transfer (CT) complex with MV^{2+} by stirring at room temperature and removed successfully from the oil.[161] However, desulfurization of actual light oils failed. It was ascribed to the presence of aromatic

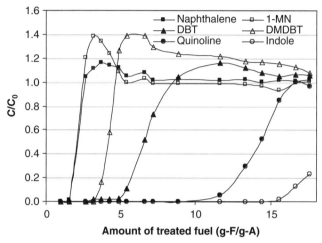

Figure 5.14.   Breakthrough curves of aromatic, sulfur, and nitrogen compounds over Ni/SiO₂–Al₂O₃.[123]

hydrocarbons in the oils, which suppresses the adsorption of sulfur compounds. On the contrary, it was found that adsorption of nitrogen compounds was hardly suppressed, even in the presence of a large quantity of aromatics because of lower ionization potential of the nitrogen compounds, explained by the authors. By employing the process, the nitrogen concentrations of actual light oils were decreased successfully to less than 35% of the feed values.

In the study of adsorptive selectivity of an activated alumina for various compounds, including sulfur compounds, nitrogen compounds, and aromatics over an activated alumina in a fixed-bed adsorption system at ambient temperature and atmosphere pressure, Kim et al.[123] found that the activated alumina is very effective for the selective separation of nitrogen compounds, especially for basic nitrogen compounds, but not very successful for separating the thiophenic compounds from aromatics. They found that the adsorptive selectivity over the activated alumina increased in the order of Nap ≈ 1-MNap < 4,6-DMDBT ≈ DBT << indole < quinoline, as shown in Figure 5.15. The results agree with the findings by Shiraishi et al.[161] In comparison with the molecular properties estimated by the quantum chemical calculations, Kim et al.[123] pointed out that the electrostatic effect and acid–base interaction might play an important role in the adsorption of such compounds on the activated alumina.

5.3.4.3 Zeolite-Based Adsorbents

Yong's group studied the desulfurization of a commercial diesel fuel by different adsorbents in a fixed-bed adsorber operated at ambient temperature and pressure.[38,127,141,142,143] They reported that the adsorbents tested for total sulfur adsorption capacity at breakthrough followed the order AC/Cu(I)–Y > Cu(I)–Y >

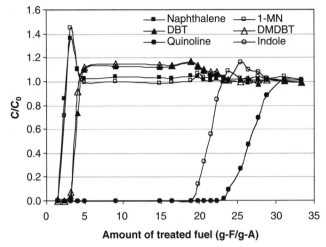

Figure 5.15.   Breakthrough curves of aromatic, sulfur, and nitrogen compounds over activated alumina.[123]

Selexsorb CDX (alumina) > CuCl/r-Al$_2$O$_3$ > AC > Cu(I)–ZSM-5. The breakthrough capacity of the best adsorbent, AC/Cu(I)–Y (layered bed of 15 wt% AC followed by, Cu(I)–Y), was 0.24 mmol/g (7.7 mg-S/g). The breakthrough capacity of Cu(I)–Y was 0.17 mmol/g (5.5 mg-S/g). The added layer of AC significantly improved the adsorption performance. According to the reported breakthrough curves for sulfur compounds over Cu(I)–Y, selectivity of the adsorbent for the sulfur compounds increases in the order of BT < DBT < 4-MDBT < 4,6-DMDBT, indicating that the adsorption is not through the direct interaction between the sulfur atom and active site, as expected by the authors. Unfortunately, no any selectivity data for mono-, di-, and tricycle aromatics in comparison with the sulfur compounds was reported by the authors.

Hernandez-Maldonado and coworkers also compared the desulfurization of diesel fuel by π-complexation sorbents: Cu(I) and Ni(II) zeolites prepared by different ion exchanging methods.[143] Their results showed that the π-complexation sorbents desulfurization performance decreases as follows: Cu(I)–Y(VPIE) > Cu(I)–Y(LPIE-RT) > Ni(II)–Y(SSIE) > Ni(II)–X(LPIE) with the breakthrough capacity of 0.278, 0.167, 0.158, and 0.143 mmol-S/g-A, respectively (or 8.9, 5.3, 5.1, and 4.6 mg-S/g-A, respectively; see Table 5.4).

In a study of selective adsorption of sulfur compounds and aromatic compounds in a hexadecane on commercial zeolites, NaY, USy, HY, and 13X by adsorption at 55 °C and flow calorimetry techniques at 30 °C, Ng et al. found that a linear correlation between the heat of adsorption and the amount of S adsorbed for NaY.[162] Competitive adsorption using a mixture of anthracene, DBT, and quinoline indicates that NaY selectively adsorbs quinoline, while anthracene and DBT have similar affinity to NaY, indicating that NaY is difficult to adsorptively separate sulfur compounds from aromatic hydrocarbons with the same number of the aromatic rings.

Recently, Bhandari et al.[163] studied systematically the effects of coexisting aromatics and moisture on the adsorptive desulfurization of diesel fuel over an Ni(II)–Y adsorbent, which was prepared by an LPIE method. Their results showed that the breakthrough capacity of the Ni(II)–Y for a model diesel fuel containing BT, DBT, and 4,6-DMDBT in n-octane was about 45.6 mg-S/g-A, while the breakthrough capacity of the same adsorbent for a real diesel fuel with 186 ppmw sulfur was only about 0.43 mg-S/g-A (see Table 5.4), more than 100 times less than that for the model diesel fuel. The similar phenomenon was observed by Ma et al. for the adsorptive desulfurization of a model gasoline and a real gasoline over Cu(I)–Y.[26] The breakthrough capacity values of Ni(II)–Y and Cu(I)–Y reported respectively by Ko et al.[164] and Ma et al.[26] are over an order of magnitude less than the values reported by Hernandez-Maldonado et al.[127,143] for the desulfurization of real diesel fuel. In order to explain significant different performance for model fuel and real fuel, Ko et al.[164] further examined the effects of aromatics and moisture. It was found that after adding 5 v/v % benzene into the model fuel, the breakthrough capacity dramatically decreased to about 0.75 mg-S/g, 60 times less than that in the absence of benzene in the feed. According to this finding, Ko et al.[164] attributed the decrease in the breakthrough capacity for the real diesel fuel to the presence of high concentration of aromatics in the commercial diesel fuel. Ko et al.[164] also observed that the presence of 5000 ppmw moisture (H_2O) in the model fuel resulted in the decrease of the breakthrough capacity from 45.6 to 14.7 mg-S/g. These data clearly indicate that the presence of aromatics and moisture greatly reduces the desulfurization performance of Ni(II)–Y significantly. Thus, the question becomes why adsorption performance of Ni(II)–Y (or Cu(I)–Y) prepared from different laboratories is so different. Is it because of the different composition of the tested real fuel (such as different fuel additives), different property of the Ni(II)–Y (or Cu(I)–Y) prepared by different groups or different test methods? In order to answer these questions, further work is necessary, such as measurement and comparison of the adsorption selectivity factors of Ni(II)–Y (or Cu(I)–Y) for the coexisting sulfur and aromatic compounds using a designed model fuel.

5.3.4.4 AC Adsorbents

AC materials as porous materials with very high surface areas and large pore volume have been widely used in deodorization, decolorization, purification of drinking water, treatment of wastewater, and adsorption and separation of various organic and inorganic chemicals. Recently, some carbon materials have been reported for adsorptive desulfurization of liquid hydrocarbon fuels.

Sano and coworkers reported an interesting work on adsorptive desulfurization of real gas oil over AC with surface area from 683 to 2972 m^2/g.[165] They found that using the AC materials can remove sulfur and nitrogen species from gas oil and the AC materials with the larger surface area and higher surface polarity have the better adsorptive performance. They also found that the adsorption pretreatment of gas oil over the AC material significantly improved the HDS performance of the gas oil.

Hernandez-Maldonado and Yang[38] and Hernandez-Maldonado et al.[39] reported that using the AC or activated alumina as an adsorbent in a guard bed can improve the adsorptive performance of Cu(I)–Y zeolites. Haji and Erkey[166] reported using carbon aerogels as adsorbents for desulfurization of a model diesel (DBT in *n*-hexadecane). They found that the saturation adsorptive capacity of a carbon aerogel with pore size of 22 nm was 15 mg-S/g-A and the carbon aerogel selectively adsorbed DBT over naphthalene. In order to develop new adsorbents with high selectivity and high capacity, to modify the commercially available adsorbents, or to design a layered adsorbent bed for a practical application in ultradeep desulfurization, it is critical to fundamentally understand the adsorptive mechanism and selectivity for various species over different adsorbents.

In a study of adsorptive removal of DBT from heptane over ACs, Jiang et al. found that the AC modified by the concentrated H_2SO_4 at 250 °C has much higher adsorption capacities for DBT than the unmodified AC.[167] They ascribed it to the increase of mesopores volume in AC by the H_2SO_4 treatment.

Ania and Bandosz[168] evaluated the performance of various ACs obtained from different carbon precursors as adsorbents for the desulfurization of liquid hydrocarbon fuels. According to their results, they concluded that the volume of micropores governs the amount physisorbed and mesopores control the kinetics of the process. They also found that introduction of surface functional groups enhances the performance of the ACs as a result of specific interactions between the acidic centers of the carbon and the basic structure of DBT molecule.

In order to understand further the adsorptive selectivity and mechanism of sulfur compounds over AC, the adsorptive desulfurization of a model diesel fuel containing the same molar concentration of Nap, 1-MNap, DBT, 4,6-DMDBT, quinoline, and indole over AC at room temperature in flowing adsorption system was studied by Kim et al.[123] They found that the AC shows the highest adsorptive capacity and selectivity for sulfur compounds, especially for the sulfur compounds with methyl groups, such as 4,6-DMDBT. The adsorption selectivity increases in the order of Nap < 1-MNap < DBT < 4,6-DMDBT < quinoline < indole, as shown in Figure 5.16. The adsorptive selectivity trend for the sulfur compounds is quite different from that over the nickel-based adsorbent and the activated alumina. The high adsorption capacity and selectivity of the ACs for the alkyl DBTs with alkyl groups at 4- and/or 6-positions of DBTs, the refractory, and major sulfur compounds in commercial diesel indicate that the ACs are promising adsorbents for deep desulfurization of diesel fuel.

By characterization of the tested ACs and correlation with their adsorption performance, Zhou et al. found that the oxygen-containing functional groups on the surface play an important role in adsorptive desulfurization over the AC.[124] They found further that increasing the oxygen-containing functional groups on the surface appears to increase the sulfur adsorption capacity. Their results show that the adsorption of sulfur compounds on the carbon materials obeys the Langmuir adsorption isotherm and the adsorption equilibrium constant and the adsorption site density on the surface are quite different for different ACs, and the regenerability of the spent ACs is related to the surface affinity and textural properties.

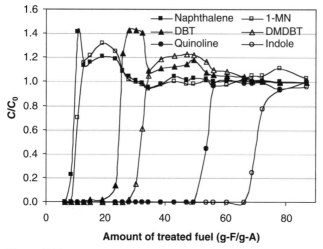

Figure 5.16.   Breakthrough curves of aromatic, sulfur, and nitrogen compounds over activated carbon.[123]

As shown above, since the different adsorbents may be suitable for separating different sulfur compounds from different hydrocarbon streams and some coexisting species, such as polyaromatic hydrocarbons and moisture, in hydrocarbon streams might inhibit the adsorption of sulfur compounds on the sorbent, a combination of two or more sorbents in an adsorptive desulfurization process might be more efficient for a practical ultradeep desulfurization process.

In developing a successful adsorption process for removing sulfur from hydrocarbon fuels, the major challenge is to develop an efficient adsorbent that has higher capacity and selectivity for sulfur compounds and is easy for regeneration. Recently, Mesoscopic Device has developed a new fuel desulfurizer based on adsorption and adsorbent regeneration that promise to continuously treat fuels in an efficient, compact, and portable system.[169] This process allows a dramatic reduction in the amount of required sorbent by making more efficient use of the available sorbent through continuous regeneration. The process makes it possible and practical to use some sorbents with excellent selectivity and stability, but often rejected due to their low capacity, for deep desulfurization of hydrocarbon fuels.

5.4   POST-REFORMER DESULFURIZATION: H₂S SORPTION

If the predesulfurization does not remove all the sulfur compounds in the reforming feed, the remaining sulfur compounds will be converted into H_2S after reforming. Even if some reforming catalysts are able to tolerate the sulfur in the reforming feedstock, the produced H_2S still need to be removed from the reformate, as the H_2S

poisons the downstream catalysts and materials, such as water-gas shift catalysts, Prox catalysts, anode catalysts, and even hydrogen separation membrane. Consequently, in many cases, it is necessary to have a post-reformer desulfurization process to reduce H_2S in reformate to less than 20 ppb, especially for PEMFC applications.

5.4.1 H₂S Sorbents

Candidate sorbents should meet several requirements to be considered for commercial gas desulfurization. The most important feature is that it should be theoretically capable of decreasing gas-phase sulfur concentrations to the low levels required by downstream processes under the relevant conditions. In addition, the absorbent must have acceptable sulfur capacity (in terms of both theoretical value and achievable range), preferably be regenerable, and maintain activity and capacity through a large number of sulfidation/regeneration cycles. The sorbent should also be nonpyrophoric for fuel cell applications.[3,170] Finally, the cost of the sorbent must be reasonable.

According to the working temperatures, the H_2S sorbents can be divided into three groups: high temperature: >600 °C, medium temperature: 300–500 °C, and low temperature: <100 °C. The metal oxides suitable for high temperature desulfurization are usually used for the desulfurization for coal gasification gas. Many of these metal oxides were studied and evaluated by Westmoreland and Harrison[113] and Hepworth et al.[114]

Relatively less attention has been given to the desulfurization of steam-containing gas mixtures with low H_2S at medium temperature suitable for desulfurization of reformate from the reforming, where the temperature of the fuel gas at the outlet of the steam reforming is about 700 °C and at the inlet of high-temperature water-gas shift (HTWGA) is about 400 °C.

Medium-temperature H_2S sorption is efficient on hydrated iron oxide; however, it has been replaced by zinc oxide (ZnO) for environmental and safety reasons.[171,172] ZnO is a promising material because it can provide a desirable level of H_2S removal.[173,174] What is also very important is that ZnO is nonpyrophoric, even after use in a reducing atmosphere at 350–450 °C. Manganese oxide (MnO) is also stable in a reducing atmosphere at 400 °C. The main drawback of manganese-based absorbents is higher outlet H_2S concentrations. The fuel gas that contains 5 vol% H_2O and 1 vol% H_2S (10,000 ppmv) can be desulfurized at 727 °C to 57 ppmv H_2S with MnO and to 6 ppmv H_2S with ZnO.[175] Therefore, ZnO might be a unique oxide that can respond to many requirements of processing and safety.

ZnO has been widely used for more than 30 years as an H_2S removal agent from natural gas.[176] The achieved dynamic capacity was 22–24 wt% (370 °C, 400 h⁻¹), with a maximum possible sulfur loading of 33 wt%, which corresponds to the complete conversion of ZnO to ZnS. If the concentration of sulfur in the feed gas is very small, then on-site regeneration is not necessary for some applications. For such cases, upon breakthrough, the absorbent bed is replaced with a fresh batch of ZnO and then sulfur removal continues. From this point of view, the use of ZnO as a polishing bed is desirable. One of the main challenges for using ZnO as a polishing bed for the post desulfurization is that it must be effective for deep desulfurization in the presence of H_2O, CO, and CO_2 in the reformate. The gas that originates from the steam reformer or autothermal reformer contains a high steam concentration (up to 35 vol%), low H_2S concentration (<10 ppmv in reformate, 1–2 ppmv in the case of natural gas), up to 12 vol% of CO and about 10 vol% of CO_2.

Novochinskii et al. prepared a unique modified ZnO sample with a different morphology and comparatively studied its adsorption performance with a commercially available ZnO sample under various conditions.[177,178] Extremely low H_2S outlet concentration (as low as 20 ppbv) was observed over the modified ZnO sample for extended periods of times. They found that the kinetic sulfur-trap capacity (the amount of H_2S trapped before breakthrough) is also dependent on space velocity, temperature, steam concentration, CO_2 concentration, and particle size. Higher capacity is observed at higher H_2S inlet concentration of 8 ppmv, compared with lower inlet concentrations of 1–4 ppmv. The trap capacity decreases monotonically as the temperature increases. Steam (H_2O) in the reformate inhibits the capture of H_2S by ZnO. The dynamic capacity of the ZnO sorbent at 300 °C and SV = 8775 h^{-1} for a model gas containing 8 ppmv H_2S, 43 vol% H_2, 12 vol% CO_2 (based on dry flow), 20 vol% H_2O with the balance being N_2 was 1.23 mg-S/g-ZnO. For the first time, Novochinskii et al.[177,178] conducted an experimental study on H_2S capture from steam-containing gas mixtures with low H_2S concentrations using washcoated monolith absorbents for fuel cell applications.[177,178] The monolith provides much higher dynamic capacity (the amount of H_2S trapped before breakthrough) under the same conditions and the ZnO-based monolith demonstrated the best performance. An extremely low H_2S outlet concentration (less than 20 ppbv) was observed over ZnO-based monolith samples for extended periods of time, under various conditions relevant for the desulfurization of gas products from the autothermal reforming of hydrocarbon fuels for a PEMFC. The capacity of the H_2S trap is dependent on the monolith characteristics (active component loading per cubic inch and the number of cells per square inch) and operating conditions, including inlet H_2S concentration, space velocity, and temperature. The breakthrough curves for H_2S capture over a washcoated ZnO monolith and 3-mm ZnO-based extrudates are shown in Figure 5.17.[178] For a feed containing 8 ppmv H_2S, 34 vol% H_2, and 13 vol% H_2O with balance of N_2, the dynamic capacity increased from 0.47 mg-S/g-ZnO for the 3-mm extrudates (400 °C; SV = 8775 h^{-1}) to 35 mg-S/g-ZnO for the monolith (400 °C; SV = 11,000 h^{-1}) that was washcoated with the same type of Engelhard ZnO, which was modified by ammonium carbonate pretreatment.[175] Despite the higher space velocity for the test with the washcoated monolith, the dynamic capacity increased by 75 times, indicating that the dynamic factor and support play a very important role in the practical desulfurization over ZnO.

Ikenaga et al. developed a zinc ferrite ($ZnFe_2O_4$) sorbent in the presence of carbon materials such as AC, activated carbon fiber (ACF), and Yallourn coal (YL) for removing H_2S in coal gasification.[179] They found that carbon material-supported $ZnFe_2O_4$ ($ZnFe_2O_4$/AC) exhibited larger desulfurization capacity for H_2S than unsupported ferrites. These sorbents could efficiently remove H_2S from 4000 ppm levels

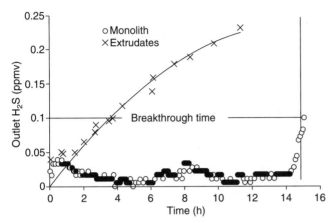

Figure 5.17. Performance of a washcoated ZnO monolith and 3-mm ZnO-based extrudates for H_2S capture. Conditions: 400 °C, 8 ppmv H_2S, 34.4% H_2, and 13% H_2O, with the balance being N_2. The 3-mm extrudates were tested at a space velocity (SV) of 8775 h^{-1}, and the monolith (400 cpsi) with a ZnO loading of 1.07 g/in.^{3} was tested at an SV of 11,000 h^{-1}.[178]

in a simulated coal gasification gas to less than 1 ppm at 500 °C. The absorption capacity of H_2S with $ZnFe_2O_4/AC$, $ZnFe_2O_4/ACF$, and $ZnFe_2O_4/YL$ exhibited nearly 100% of stoichiometric amount of loaded metal species. They also reported that the spent sorbents could be regenerated by an air oxidation in O_2/air (50 vol %) at 450 °C for 30 min. The regenerated ferrite can be used for repeated absorption of H_2S with a slight decrease in the absorption capacity. But, how the structure of the carbon support can be prevented in the oxidation regeneration in the presence of O_2 at 450 °C is unclear.

It has been proven that Zn-containing metal oxides and some other metal oxides have higher sorption capacity for H_2S at high temperature. However, it was found that the sorption desulfurization performance of these metal oxides are quickly degraded within the sorption–regeneration cycles, because the notable structural changes occur during the transformation from the oxidized sorbent to the sulfided sorbent in sorption and from the sulfided sorbent back to the oxidized sorbent in regeneration. The pool regenerability of these metal oxides has limited their wide application in hot gas desulfurization. Recently, Flytzani-Stephanopoulos et al.[180] reported the regenerative adsorption and removal of H_2S from hot fuel gas streams (800 °C) by rare earth oxides. The idea is only use of the surface of the sorbent in sulfidation and regeneration of sorbent upon saturation. By using very high space velocities or short contact times of the gas with the sorbent in regeneration, bulk sulfidation and regeneration of the sorbent with its attendant structure complexities are prevented. Their experimental data indicate that the prepared rare earth oxides were stable in the sorption–regeneration cycle. However, the sorption capacity at breakthrough 1 ppm of H_2S was only around 1.0 mg-S/g, as the sorption only occurred at the surface.

Recently, significant attention has also been paid to the low temperature adsorption of H_2S. Polychronopoulou et al.[181] reported their study in H_2S adsorption from a gas mixture containing 0.06 vol% H_2S, 25 vol% H_2, 7.5 vol% CO_2, and 1 vol% H_2O in the 25–100 °C range over various Zn–Ti-based mixed metal oxides prepared by the sol-gel method. They found that the nominal chemical composition (metal atomic %) of the solid adsorbent have an important effect on the number, chemical composition, and particle morphology of the crystal phases formed. The mixed metal oxides with compositions 20Zn–80Ti–O and 40Zn–60Ti–O presented higher H_2S uptakes than ZnO and TiO_2 solids, which were also prepared by the sol-gel method. In addition, they found that the Zn–Ti–O mixed metal oxides showed higher H_2S uptakes after regeneration with 20% O_2/He in the 500–750 °C range compared with the ZnO and TiO_2 solids. The 10Mn–45Zn–45Ti–O solid results in higher H_2S uptakes than a commercial Ni-based H_2S adsorbent in the 25–50 °C range. They also found that the effectiveness of the regeneration procedure of the 10Mn–45Zn–45Ti–O solid following was in the 45%–170% range depending on the sulfidation temperature and regeneration conditions applied. The solid with composition 10Cu–45Zn–45Ti–O calcined at 200 °C (after synthesis) exhibited three times higher H_2S uptakes at 25 °C than a commercial Ni-based adsorbent, but result was not obtained at higher calcination temperatures.

Very recently, a novel nanoporous H_2S adsorbent that was prepared by loading polyethylenimine (PEI) into the mesoporous molecular sieve MCM-41 was developed by Xu et al.[182] for removing H_2S from a gas mixture at low temperature. They thought that the reasons for loading the PEI into the channels of the mesoporous molecular sieve MCM-41 were twofold. First, because the pore size of the mesoporous molecular sieve MCM-41 is less than 3 nm, PEI nanoparticles can be formed in the channels of MCM-41 that are accessible to gas molecules. The PEI in MCM-41 is expected to show a higher adsorption efficiency for H_2S than the bulk PEI particles. On the other hand, because the mesoporous MCM-41 also acts as a separation medium for the PEI nanoparticles, such nanoparticles are not likely to aggregate in the application process and therefore retain the desired properties that result from nanoparticles. They found that the prepared MCM-41–PEI was able to reduce the H_2S in a hydrogen gas to less than 0.05 ppmv and its adsorption performance was better than that of a commercial ZnO. Another potential advantage of this adsorbent is that the spent adsorbent might be easy for regeneration at low temperature without using hydrogen gas.

5.4.2 H_2S Adsorption Thermodynamics

How low a sorbent is able to reduce the H_2S level in the reformate depends theoretically on the thermodynamic property of the sorbent, composition of the reformate, and the adsorption conditions. As well known, the reformate contains not only H_2 and H_2S, but also H_2O, CO, and CO_2. These coexisting compounds might influence the desulfurization over the metal oxide sorbents via the following reactions:

$$ZnO(s) + H_2S(g) \leftrightarrow ZnS(s) + H_2O(g) \quad \Delta H_r^\circ = -78.9 \text{ kJ/mol}, \quad (5.6)$$

$$CO + H_2S \leftrightarrow COS + H_2 \quad \Delta H_r^{\circ} = -10.9 \text{ kJ/mol}, \tag{5.7}$$

$$CO_2 + H_2S \leftrightarrow COS + H_2O \quad \Delta H_r^{\circ} = 30.3 \text{ kJ/mol}. \tag{5.8}$$

A study in thermodynamic comparison of several sorbent systems for hot coal-derived fuel-gas desulfurization at high temperature (>800 °C) and higher concentration (H₂S > 800 ppm or COS > 800 ppm) was conducted and reported by Hepworth et al.[114] Not much attention has been given to the thermodynamic investigation in the H₂S removal by metal oxide materials at the medium temperature range and low H₂S concentration in the presence of H₂O, CO, and/or CO₂ with their concentrations similar to those in the reformate, although it is critical. Novochinskii et al. did the thermodynamic investigation for the H₂S removal over ZnO at a temperature range from 100 to 700 °C based on Equation 5.6.[177,178] It was found that the presence of H₂O significantly restrained the desulfurization performance of the ZnO sorbent for a model fuel containing 1.2 ppmv H₂S, 34.4 vol% H₂, and 20 vol% H₂O with the balance N₂ because of a shift in the equilibrium of Equation 5.6 toward reactants ZnO and H₂S.

The calculated equilibrium H₂S concentration at a temperature range from 0 to 900 °C in ZnO-ZnS–H₂O-H₂S system in the presence of different concentration of H₂O(g) are shown in Figure 5.18.[183] It is clear that the low temperature is favor for decreasing the equilibrium H₂S concentration, as the reaction is exothermic. The presence of H₂O(g) has a significant effect on the equilibrium H₂S concentration. At 600 °C, when H₂O(g) concentration increases from 0.1 to 20 vol%, the corresponding equilibrium H₂S concentration increases significantly from 3.7×10^{-3} ppbv to

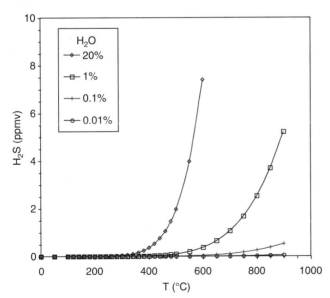

Figure 5.18. The calculated equilibrium H₂S concentration at a temperature range from 0 to 900 °C for ZnO + H₂S ↔ ZnS + H₂O reaction in the presence of different concentrations of H₂O(g).[183]

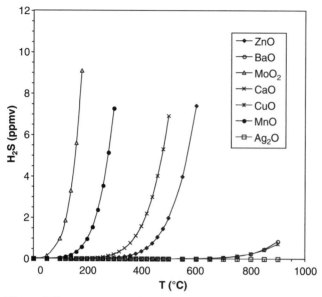

Figure 5.19.   The calculated equilibrium H_2S concentration at a temperature range from 0 to 900 °C for reaction $MeO + H_2S \leftrightarrow MeS + H_2O$ in the presence of 20 vol% $H_2O(g)$ for different metal oxides.[183]

7.4 ppbv. It indicates that H_2O strongly inhibits the degree of the desulfurization, and at 600 °C in the presence of 20 vol% $H_2O(g)$, ZnO is unable thermodynamically to remove H_2S from H_2 gas to less than 7.4 ppbv.

The calculated equilibrium H_2S concentration at a temperature range from 0 to 900 °C in MeO–MeS–H_2O–H_2S system in the presence of 20 vol% $H_2O(g)$ for different metal oxides are shown in Figure 5.19.[183] The results show that the thermodynamic property of different metal oxides is quite different. The equilibrium H_2S concentration in the systems decreases in the order of MoO_2 > MnO > CaO > ZnO > BaO > CuO > Ag_2O. The results show that BaO, CuO, and Ag_2O are able to reduce H_2S to less than 0.1 ppbv at 400 °C in the presence of 20 vol% $H_2O(g)$, indicating that BaO, CuO, and Ag_2O are even thermodynamically better than ZnO for removing H_2S from H_2 gas. These thermodynamic calculations are in agreement with the previous experimental results of the effects of $H_2O(g)$ and temperature observed by Novochinskii et al.[177,178] and the different desulfurization performance of ZnO and MnO reported by Turkdogan.[175] The results imply that the thermodynamic property of the sorbents plays an important role in the desulfurization of the reformate, especially when the target sulfur level is less than 20 ppbv.

The coexisting CO and CO_2 in the reformate might also influence the desulfurization performance of metal oxide-based materials. Novochinskii et al. found that the coexisting CO_2 reduces the kinetic capacity of ZnO.[177,178] They contributed it to competitive adsorption between H_2S and CO_2 in the presence of steam. Another

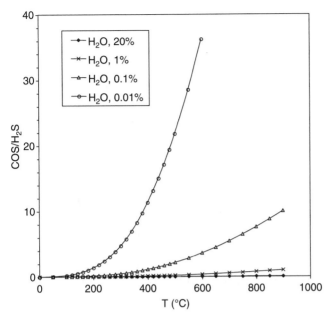

Figure 5.20.   The equilibrium COS/H$_2$S molar ratio as a function of temperature at different H$_2$O concentrations for CO$_2$ + H$_2$S $\rightleftharpoons$ COS + H$_2$O, assuming that CO$_2$ concentration is 20 vol %.[183]

possibility might be that the H$_2$S reacts with the coexisting CO$_2$ to form COS and H$_2$O as shown in Equation 5.8. The equilibrium COS/H$_2$S molar ratio as a function of temperature at different H$_2$O concentration is shown in Figure 5.20, which is calculated on the basis of Equation 5.8.[183] It is clear that the high temperature favors the increase of the equilibrium COS/H$_2$S molar ratio and the presence of H$_2$O represses the formation of COS. The equilibrium COS/H$_2$S molar ratio is about 11 at 400 °C in the presence of 100 ppmv H$_2$O. Consequently, the effect of H$_2$O on H$_2$S removal from the reformate might be complicated.

H$_2$S in the reformate might also react with the coexisting CO to form COS as shown in Equation 5.7. The equilibrium COS/H$_2$S molar ratio as a function of temperature at different CO concentrations is shown in Figure 5.21, which was calculated on the basis of Equation 5.7 and assuming that H$_2$ concentration is 50 vol %.[183] On the contrary, low temperature favors the formation of COS in this reaction.

The thermodynamic calculations on the basis of Equations 5.6–5.8 indicate that ZnO might be unable to reduce the sulfur in the reformate to the level for PEMFC applications when the fuel gas contains a large amount of H$_2$O, CO, and CO$_2$ due to the thermodynamic factor. ZnO is not efficient for removing COS. It needs to catch more close attention to the effects of the coexisting H$_2$O, CO, and CO$_2$ on the H$_2$S removal from the reformate in the design of a post-desulfurization process in a hydrocarbon fuel processor, especially at high temperature.

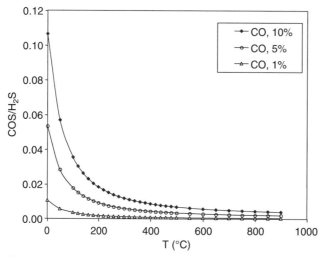

Figure 5.21. The equilibrium COS/H₂S molar ratio as a function of temperature at different CO concentrations for reaction CO + H₂S →← COS + H₂, assuming H₂ concentration is 50 vol %.[183]

5.5 DESULFURIZATION OF COAL GASIFICATION GAS

Among fossil fuels, coal is the most abundant fossil fuel and could meet the human need for more than 200 years from now. In the U.S., coal is also our nation's most abundant domestic fossil fuel resource, about 5800 quadrillion BTUs.[184] Consequently, coal is the most prospective resource for mass hydrogen production in the 21st century. For near- to midterm applications, hydrogen production from coal is expected to be the most economical pathway until other resources, such as renewable sources, become available to produce hydrogen in large scale at lower costs.[184]

The production of synthesis gas and hydrogen from coal is usually through a gasification process. Figure 5.22 shows the simple scheme of the U.S. Department of Energy (DOE) coal gasification power plant.[185] The heart of gasification-based systems is the gasifier. A gasifier converts hydrocarbon feedstock into syngas by applying heat under pressure in the presence of steam. Syngas is primarily hydrogen, carbon monoxide, and other gaseous constituents, the proportions of which can vary depending upon the conditions in the gasifier and the type of feedstock.

Since coal contains heteroatoms other than carbon and hydrogen, the gasification of coal will always produce undesired and even harmful impurities as an intrinsic issue. Table 5.6 shows the typical composition of coal gasification gas. The sulfur content in the raw coal gasification gas can be up to 1 vol % or even more, depending on the source and gasification process. The major sulfur compounds in raw coal gasification gas are hydrogen sulfide (H₂S) and carbonyl sulfide.[5] In order to meet the power plant environmental requirements (SOₓ, NOₓ, and particulate emissions), turbine protection specifications and/or other requirements for different end-use

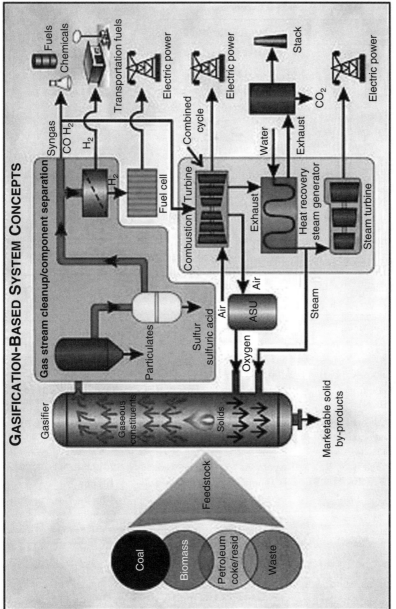

Figure 5.22. Simple scheme of DOE coal gasification power plant.[185] See color insert.

273

Table 5.6.   Typical Gasifier Gas Composition

Gas	Concentration (vol%)
CO	30–60
H_2	25–30
CO_2	5–15
H_2O	2–30
CH_4	0–5
N_2	0.5–4
Ar	0.2–1
H_2S	0.2–1
COS	0–0.1
$HCN + NH_3$	0–0.3
HCl	50–400 ppmv

Table 5.7.   Fuel Gas Requirements[187]

Contaminant	Solid Oxide Fuel Cell	PEMFC	Gas Turbine
Total sulfur (H_2S, COS, etc.)	60 ppbv	10 ppbv	760 ppmv fuel gas
	–	–	20 ppmv for selective catalytic reduction (SCR)
Total halides (Ci, F, Br)	100 ppbv	Not available	5 ppmv fuel gas
Total fuel nitrogen (NH_3, HCN)	Not available	1 ppmv NH_3	Fuel-bound nitrogen
	–	–	200–400 ppmv
Total alkali metals (Na, K, Li vapor, and solid phases)	Not available	Not available	100 ppbv fuel gas
Volatile metals (V, Ni, Fe, Pb, Ca, Ba, Mn, P)	5 ppbv As	Not available	20 ppbw Pb
	0.2 ppmv Se	–	10 ppbw V
	30 ppbv Cd	–	40 ppbw Ca
	–	–	40 ppbw Mg
Particulates	Not available	Not available	0.1–0.5 ppm fuel gas

applications, such as synthesis of hydrocarbon fuels, SOFC, and PEM fuel cell,[184] sulfur content in the raw syngas needs to be removed. For the Integrated Gasification Combined Cycle (IGCC) clean syngas, the total sulfur content has to be removed to 5–25 ppmv for the synthesis of diesel and gas to liquid (GTL) and to 0.1–15 ppmv for methanol.[186] Recently, the DOE has previously defined maximum fuel contaminant specifications for components of the integrated energy plants. For gas turbine application, the total sulfur content in the fuel gas needs to be reduced to less than 750 ppmv, for SOFC application to less than 60 ppbv, and for PEMFC application even to less than 10 ppbv. The fuel gas requirements for the different applications are listed in Table 5.7 in detail.[187] One of the major challenges in the production of hydrogen and synthesis gas from coal is to remove sulfur and other impure gases from the raw coal gasification gas efficiently.

The major techniques and methods for removal acid gases, including H_2S, COS, and CO_2, from the raw coal gasification gas include (a) the absorption by chemical and/or physical solvents, (b) sorption by metals or metal oxides, and (c) membrane separation. The techniques and methods for sulfur removal from the raw coal gasification gas and other streams by solvent absorption at low temperature and by sorption at high or mild temperature were reviewed in detail by Mahin Rameshni,[188] Korens et al.,[189] Newby et al.,[190] and Uhde,[186] respectively.

5.5.1 Absorption by Solvents

Table 5.8 gives a partial list of gasification installation with the year of start-up, feedstock, acid gas removal process employed, and applications.[189] The major demonstration and commercial IGCC and/or cogeneration projects are included as well as many projects producing or coproducing hydrogen, ammonia, Fischer–Tropsch liquids, and chemicals. According to the properties of the used solvents, the absorption processes for the acid gas removal can be further separated into chemical solvent processes, physical solvent processes, and hydride solvent processes. The names and structures of major chemical and physical solvents with some physical properties as well as the process, in which the solvent was used, are shown in Table 5.9.

In general, a solvent applied in absorption desulfurizaton should satisfy the following important characteristics:

- sufficient sulfur solubility or absorption capacity,
- higher selectivity for sulfur compounds,
- ability to be regenerated and recycled,
- no irreversible reactions with precipitated sulfur,
- low reaction heat or solution heat,
- stability under conditions,
- low vapor pressure,
- no or low corrosion,
- ability to separate from water,
- suitable viscosity, and
- simple recovery of the absorbed sulfur.

5.5.1.1 Absorption by Chemical Solvents

The chemical solvents are predominantly the amine-based solvents, which can be classified further into three subgroups: (a) the primary amines, including monoethanolamine (MEA) and diethylene glycolamine (DGA); (b) the secondary amines,

Table 5.8. Partial Listing of Gasification Installations by Acid Gas Removal Process and Year of Start-Up[189]

Plant Owner	Country	Start-up	Feedstock	Process	AGR Process	Application
Sasol Chemical Ind. (Pty.) Ltd./Sasol Ltd.	South Africa	1955	Subbit. coal	Lurgi Dry Ash	Rectisol	FT liquids
Mitsubishi Petrochemicals	Japan	1961	Bunker C oil	Shell	ADIP	C
Lucky Goldstar Chemical Ltd.	South Korea	1969	Bunker C oil	Shell	Sulfinol	Ammonia
Sasol Chemical Ind. (Pty.) Ltd./Sasol Ltd.	South Africa	1977	Subbit. coal	Lurgi Dry Ash	Rectisol	FT liquids
Hydro Agri Brunsbüttel	Germany	1978	Heavy vac. residue	Shell	Rectisol	Ammonia
Sasol Chemical Ind (Pty.) Ltd./Sasol Ltd.	South Africa	1982	Subbit. coal	Lurgi Dry Ash	Rectisol	FT liquids
Gujarat National Fertilizer Co.	India	1982	Refinery residue	Texaco	Rectisol	Ammonia
Eastman Chemical Co.	U.S.	1983	Coal	Texaco	Rectisol	Methanol and other C
Dakota Gasification Co.	U.S.	1984	Lignite and ref. residue	Lurgi Dry Ash	Rectisol	Synthetic natural gas (methane)
SCE Cool Water	U.S.	1984	Bituminous coal	Texaco	Selexol	IGCC
Quimigal Adubos	Portugal	1984	Vacuum residue	Shell	Rectisol	Ammonia
Mitteldeutsche Erdöl-Raffinerie GmbH	Germany	1985	Visbreaker residue	Shell	Rectisol	Methanol
Rheinbraun	Germany	1986	Brown coal	HTW	Rectisol	Methanol
SAR GmbH	Germany	1986	Vacuum residue	Texaco	Sulfinol	H_2 and oxochemicals
LGTI	U.S.	1987	Subbit. coal	E-GAS	MDEA	IGCC/Cogen
China Nat'l Tech. Import Co. (CNTIC)	China	1987	Anthracite	Lurgi Dry Ash	Rectisol	Ammonia
BP Chemicals, Ltd.	U.K.	1989	Natural gas	Texaco	MDEA	Acetyls
NUON (formerly Demkolec BV)	The Netherlands	1994	Coal	Shell	Sulfinol-M	IGCC
Global Energy, Inc.	U.S.	1995	Coal, pet coke	E-GAS	MDEA[a]	IGCC

Company/Owner	Country	Year	Feedstock	Gasifier	Gas cleaning	Application
Dalian Chemical Industrial Corp.	China	1995	Visbreaker residue	Texaco	Rectisol	Ammonia
Frontier Oil & Refining Co. (Texaco, Inc.)	U.S.	1996	Pet coke	Texaco	MDEA	Cogen
Tampa Electric Co.	U.S.	1996	Bit. coal	Texaco	MDEA[a]	IGCC
Schwarze Pump	Germany	1996	Municipal waste	GSP/Noell	Rectisol	IGCC and methanol
Inner Mongolia Fertilizer Co.	China	1996	Vacuum residue	Shell	Rectisol	Ammonia
Juijiang Petrochemical Co.	China	1996	Vacuum residue	Shell	Rectisol	Ammonia
Sokolovska Uhelna, A.S.	Czech Republic	1996	Coal	Lurgi Dry Ash	Rectisol	IGCC/Cogen
Elcogas SA	Spain	1997	Coal and pet coke	PRENFLO	MDEA[a]	IGCC
Shell Nederland Raffinaderij BV	The Netherlands	1997	Visbreaker residue	Shell	Rectisol	IGCC/Cogen, H_2
Unspecified owner	Germany	1997	Visbreaker residue	Texaco	Sulfinol	Methanol
Sierra Pacific Power Co.	U.S.	1998	Coal	KRW	Limestone/ZnO[b]	IGCC
Lanzhou Chemical Industrial Co.	China	1998	Vacuum residue	Shell	Rectisol	Ammonia
ISAB Energy	Italy	2000	Heavy oil	Texaco	MDEA[a]	IGCC, H_2
Motiva Delaware Refinery	U.S.	2000	Pet coke	Texaco	MDEA	IGCC/Cogen
Henan	China	2000	Anthracite	Lurgi Dry Ash	Rectisol	Ammonia
EPZ	The Netherlands	2000	Demolition wood	Lurgi CFB	Scrubber	Fuel gas
Farmland Industries, Inc.	U.S.	2000	Pet coke	Texaco	Selexol	Ammonia

(Continued)

Table 5.8. *Continued*

Plant Owner	Country	Start-up	Feedstock	Process	AGR Process	Application
ExxonMobil Baytown Syngas Project	U.S.	2001	Deasphalter bottom	Texaco	Rectisol	H_2, CO
api Energia S.p.A.	Italy	2001	Visbreaker residue	Texaco	Selexol[a]	IGCC, H_2
SARLUX srl	Italy	2001	Visbreaker residue	Texaco	Selexol[a]	IGCC/Cogen
ExxonMobil	Singapore	2001	Residual oil	Texaco	Flexsorb[c]	IGCC/Cogen
Shin Nihon Sekiyu (Nippon Pet. Ref. Co.)	Japan	2004	Vacuum residue	Texaco	ADIP[a]	IGCC
AgipPetroli/EniPower	Italy	2004[d]	Visbreaker residue	Shell	Amine[a]	IGCC, H_2
PIEMSA	Spain	2006[d]	Visbreaker tar	Texaco	MDEA[a]	IGCC, H_2
Total Fina Elf/Texaco	France	2006[d]	Refinery residues	Texaco	Selexol	H_2

[a]COS hydrolysis precedes the acid gas removal process in this plant.

[b]Commissioning of this demonstration plant was unsuccessful and the project was terminated. Consequently, both the KRW (Kellogg Rust Westinghouse) gasification process and the limestone/ZnO hot gas cleanup process remain unproven.

[c]Version not disclosed—indicated only as "generic Flexsorb."

[d]In planning/engineering/development.

FT liquids, Fischer–Tropsch hydrocarbons; C, chemicals; IGCC, Integrated Gasification Combined Cycle; Cogen, combustion turbine cogeneration; AGR, acid gas removal.

Table 5.9. Chemical and Physical Solvents Employed in the Acid Gas Removal

AGR Process	Solvent Structure	MW	Density (g/mL)	BP (°C)	Typical AGR Process
Chemical solvent processes					
Primary amines					
Monoethanolamine (MEA)	$H_2N–CH_2–CH_2–OH$	61	–	170	–
Diethylene glycolamine (DGA)	$H_2N–CH_2–CH_2–O–CH_2–CH_2–OH$	105	–	218–224	–
Secondary amine		–	–		–
Diethanolamine (DEA)	$HO–CH_2–CH_2–NH–CH_2–CH_2–OH$	105	–	217	–
Diisopropanolamine (DIPA)	$HO–CH(CH_3)–CH_2–NH–CH_2–CH(CH_3)–OH$	133	–	249–250	–
Hindered amines		–	–		Flexsorb
Tertiary amines		–	–		–
Triethanolamine (TEA)	$HO–CH_2–CH_2–N(CH_2–CH_2–OH)–CH_2–CH_2–OH$	149	–	190–193	–
Methyldiethanolamine (MDEA)	$HO–CH_2–CH_2–N(CH_3)–CH_2–CH_2–OH$	119	–	247	MDEA
Potassium carbonate	K_2CO_3	138	2.43		Benfied
Physical solvent process					
Methanol	$H_3C–OH$	32	–	64	Rectisol
Polyalkylene glycol dimethyl ether (PGDE)	$CH_3–O–(CH_2–CH_2–O)n–CH_3$, $n = 3–10$	–	–	>241	Selexol/ Genosorb

(Continued)

Table 5.9. *Continued*

AGR Process	Solvent Structure	MW	Density (g/mL)	BP (°C)	Typical AGR Process
N-methyl pyrrolidinone (NMP)		99.13	–	81–82	Purisol
N-formylmorpholine/*N*-acetylmorpholine (NFM/NAM)		115/129	/1.116	236–237/–	Morphysorb
Propylene carbonate		102	1.189	240	Flour solvent
Hybrid solvent process					
Mixture of aqueous amine (e.g., DIPA)					
Sulfolane (tetrahydrothiophene dioxide)		120	1.26	285	Sulfinol
Mixture of aqueous MDEA and sulfolane		–	–	–	Sulfinol-M

including diethanolamine[144] and diisopropanolamine (DIPA); and (c) the tertiary amines, including triethanolamine (TEA) and methyldiethanolamine (MDEA). The absorption is based on the following chemical reactions:

$$RNH_2 + H_2S \leftrightarrow RNH_2 \cdots H \cdots SH,$$
$$R_2NH + H_2S \leftrightarrow R_2NH \cdots H \cdots SH,$$
$$R_3N + H_2S \leftrightarrow R_3N \cdots H \cdots SH,$$
$$RNH_2 + CO_2 \leftrightarrow RNH_2 \cdots CO_2,$$
$$R_2NH + CO_2 \leftrightarrow R_2NH \cdots CO_2,$$
$$R_3N + CO_2 \leftrightarrow R_3N \cdots CO_2.$$

Since the primary and secondary amines are the stronger base than the tertiary amines, the reaction heat for the primary and secondary amines are higher, leading to higher energy consumption in the regeneration. The primary and secondary amines are also easier to react with some other acid gases to form the heat-stable amine salts (HSAS), resulting in the degradation of the solvent in the cycles. Consequently, the tertiary amine solvents are preferred in the amine scrubbing processes. As a result, MDEA-based acid gas removal has been the predominant process for IGCC application up to the late 1990s and it continues to be selected for new projects.[189] Ten of the projects in Table 5.8, including eight IGCC/cogeneration projects, employ MDEA. Figure 5.23 shows a typical acid gas removal process using amine solvents.[189] Commercial MDEA formulations (i.e., with proprietary additives) have been developed, which are claimed to offer a much enhanced selectivity for H_2S than the case for generic MDEA. UCARSOL is the trade name for a series of formulated amine solvents originally developed by Union Carbide and offered by UOP as a licensed package.[189] More than 500 units of the version called the Amine Guard FS Process have been installed worldwide, mostly treating natural gas, ammonia syngas, and hydrogen streams.[191]

The significant advantages of MDEA solvent are:

- higher absorption selectivity for H_2S and CO_2, as the absorption affinity is based on the acid–base interaction;

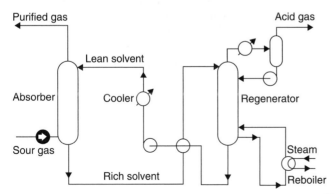

Figure 5.23.   Flow diagram of typical amine acid gas removal process.[189]

- higher capacity at the lower concentration of H_2S;
- less corrosive than the primary and secondary amines, thus allowing a higher MDEA solvent content;
- higher concentration of H_2S from the regenerator, which allows the use of a conventional Claus sulfur recovery process; and
- higher stability of MDEA solvent than other primary and secondary amine solvents.

The major problems in the current liquid amine scrubbing process are the following:

- The liquid amine solution is corrosive to the equipment.
- The reaction with stronger acidic gases (pKa < 6) leads to the HSAS or inorganic salts, which are difficultly regenerated at stripper conditions, resulting in degradation of the amine solvents.
- It is hard to remove sulfur to the level for the fuel cell applications, especially in the presence of appreciable quantities of COS.
- Liquid amines have higher heat of reaction in comparison with the heat of solution of the physical solvent.
- Liquid amines have lower vapor pressure and the solvent loss is serious in the regeneration.

For synthesis gases that contain appreciable quantities of COS, prior removal of the COS is usually required. A catalytic hydrolysis unit is usually employed ahead of the MDEA unit in these cases, as was done both at the Wabash River and Tampa Electric gasification plants.[189] MDEA reacts more strongly with CO_2 than H_2S. If the removal of CO_2 is not required, the presence of CO_2 in the gas reduces greatly the absorption performance of MDEA for H_2S removal. If the removal of both H_2S and CO_2 is the case, the amine scrubbing is preferred. The kinetics of the CO_2 reaction with MDEA can be improved by various additives (activators), which allows for reasonable absorber sizes and solvent circulation rates. Among some of the additives that have been used are MEA, DEA, and DGA (diglycolamine).[192] BASF has used piperazine in its activated MDEA formulations. Piperazine, even at low levels (about 5%) enhances the rate of CO_2 absorption an order of magnitude over nonactivated MDEA.[193]

In order to solve the degradation problem of the amine solvents, ExxonMobil developed the Flexsorb SE process, which is based on a family of proprietary sterically hindered amines (SHAs) in aqueous solutions or other physical solvents. The Flexsorb SHAs are secondary amines that have a large hydrocarbon group attached to the nitrogen group. The large molecular structure hinders the CO_2 approach to the amine. The larger the structure, the more difficult it becomes for the CO_2 to get close to the amine. They also appear to be unstable to the carbamate form of product and revert easily to the carbonate form found in the tertiary amines.[189]

5.5.1.2   Absorption by Physical Solvents

Two of the currently most widely used physical solvent processes for IGCC synthesis gases are Selexol and Rectisol. The Selexol process solvent is polyalkylene glycol dimethyl ether (PGDE), while the Rectisol solvent is methanol. Other physical solvent processes are also offered for license, but are less frequently used commercially. In recent IGCC projects, where physical solvents have been specified for acid gas removal, either the Selexol or the Rectisol processes have been chosen. At least 20 operating commercial gasification plants worldwide use either the Selexol or the Rectisol process for acid gas treating.[189]

5.5.1.2.1   Selexol Solvent and Process.   The Selexol process solvent is a mixture of dimethyl ethers of polyethylene glycol and has the formulation $CH_3(CH_2CH_2O)_nCH3$, where n is between 3 and 10. There are other process suppliers using the same solvent as the Selexol process. For example, Clariant GmbH of Germany offers a family of dialkyl ethers of polyethylene glycol. The Clariant solvents, under the Genosorb name, include dimethyl ether as well as dibutyl ether of polyethylene glycols. The first mentioned Clariant solvent is the same as that used in the Selexol process.[189] The relative solubilities of various gasses in Selexol solvent are listed in Table 5.10,[194] indicating that Selexol solvent has a much higher absorption selectivity for HCN, mercaptans, H_2S, and COS than for CO_2. A typical Selexol processing scheme is shown in Figure 5.24.[189]

The principal benefits of Selexol solvent/process are:

- high selectivity for H_2S and COS over CO_2;
- high loadings at high acid gas partial pressures;
- removal of not only H_2S and COS, but also other trace contaminants, including CS_2, mercaptans, chlorinated hydrocarbons, BTEX (benzene, toluene, ethylbenzene, and xylene), HCN, NH_3, and others;
- no chemical reaction, no degradation, and higher thermal and chemical stability than the chemical solvents;
- easily reclaimed and has low heat requirements in the regeneration because most of the solvent can be regenerated by a simple pressure letdown, leading to high energy efficiency in the process;
- low vapor pressure, resulting in low solvent losses;
- thermally and chemically stable,
- nonfouling; and
- can use air stripping.

As a result, Selexol solvent/process has been used widely in the cleanup of natural gas, landfill/biogas, and synthesis (gasification) gas. The Selexol solvent/process can be configured in various ways, depending on the requirements for the level of H_2S/CO_2 selectivity, the depth of sulfur removal, the need for bulk CO_2 removal, and whether the gas needs to be dehydrated.

Table 5.10.   Relative Solubilities of Various Gases in Selexol Solvent[194]

Component	$R = K'CH_4/K'$ component
H_2	0.2
N_2	0.3
CO	0.43
CH_4	1
C_2H_6	7.2
CO_2	15.2
C_3H_8	15.4
$i\text{-}C_4H_{10}$	28
$n\text{-}C_4H_{10}$	36
COS	35
$i\text{-}C_5H_{12}$	68
C_2H_2	68
NH_3	73
$n\text{-}C_5H_{12}$	83
H_2S	134
C_6H_{14}	167
CH_3SH	340
C_7H_{16}	360
CS_2	360
C_2H_3Cl	400
SO_2	1400
C_6H_6	3800
C_2H_5OH	3900
CH_2Cl_2	5000
CH_2Cl_3	5000
C_4H_4S	8200
H_2O	11,000
HCN	19,000

Where selective H_2S removal is required, together with deep CO_2 removal, two absorption and regeneration columns may be required essentially in a two-stage process, as illustrated in Figure 5.25.[195] H_2S is selectively removed in the first column by a lean solvent that has been deeply stripped with steam, while CO_2 is removed, from the now H_2S-free gas, in the second absorber. The second-stage solvent can be regenerated with air or nitrogen if very deep CO_2 removal is required.[189]

The limitations of Selexol solvent/process are shown below:

- coabsorption of hydrocarbons, which results in hydrocarbon losses in the process;

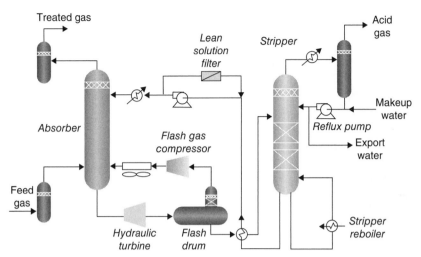

Figure 5.24.   Flow diagram of typical Selexol processing scheme.[189]

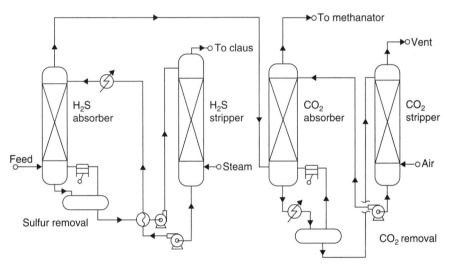

Figure 5.25.   Flow diagram of a two-stage Selexol process for acid gas removal from coal-derived synthesis gas.[195]

- refrigeration often required for lean solution, adding complexity and cost; and
- absorption at high pressure is preferred for the high absorption capacity.

5.5.1.2.2 Rectisol Solvent and Process.   The Rectisol process, developed by Lurgi GmbH, is the most widely used physical solvent gas treating process in the world. More than 100 Rectisol units are in operation or under construction worldwide.[189] As shown in Table 5.8, its most prevalent application is for deep sulfur

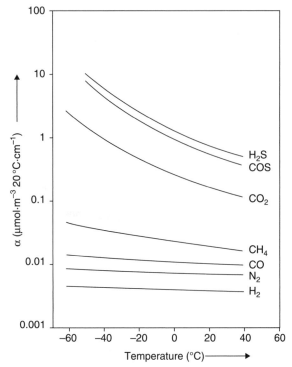

Figure 5.26.   Absorption coefficient α of various gases in methanol (partial pressure: 1 bar).[195]

removal from synthesis gases that subsequently undergo catalytic conversion to such products as ammonia, hydrogen, and Fischer–Tropsch liquids.

The Rectisol process uses chilled methanol at a temperature of about $-40\,°C$ to $-62\,°C$. Figure 5.26 shows the absorption coefficient α of various gases in methanol as a function of temperature. Methanol's selectivity for various gases increases in the order of $H_2 < N_2 < CO < CH_4 \ll CO_2 \ll COS < H_2S$. The selectivity for H_2S over CO_2 at these temperatures is about $6/1$,[195] a little lower than that of Selexol at its usual operating temperature. However, the solubilities of H_2S and COS in methanol, at typical process operating temperatures, are higher than that in Selexol, which allows for very deep sulfur removal (<0.1 ppmv H_2S plus COS). HCN, NH_3, iron and nickel carbonyls, and some other impure gases also have high solubility in methanol.

There are many possible process configurations for Rectisol, depending on process requirements. Different process layouts are used for selective H_2S removal, deep CO_2 removal, and for deep nonselective CO_2 and H_2S removal.[189] Figure 5.27, shows a flow scheme of a basic Rectisol process. In this flow scheme, bulk removal of CO_2 and nearly all of the removal of H_2S and COS take place at the bottom section of the absorber. The methanol solvent contacting the feed gas in the first stage of the absorber is stripped in two stages of flashing via pressure reduction. The regener-

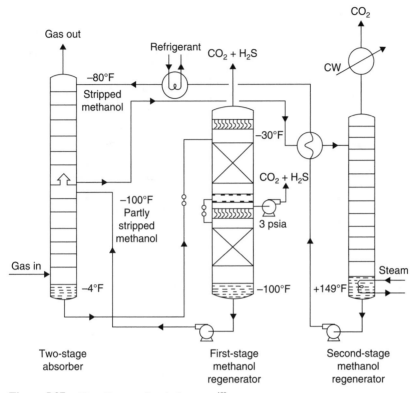

Figure 5.27.   Flow diagram of rectisol process.[189]

ated solvent is virtually free of sulfur compounds but contains some CO_2. The acid gas leaving the first-stage solvent regenerator is suitable for a Claus plant. The second stage of absorption is designed for the removal of the remaining sulfur compounds and CO_2. The solvent from the bottom of the second stage of the absorber is stripped deeply in a steam-heated regenerator and is returned to the top of the absorption column after cooling and refrigeration.[189]

The primary advantages of Rectisol process are the following:

- The process has high selectivity for H_2S and COS over CO_2 with deep sulfur removal (<0.1 ppmv H_2S plus COS).
- The process is also able to remove HCN, NH_3, and iron and nickel carbonyls.
- Cost for absorbent (methanol) is cheaper.
- The Rectisol process is very flexible and can be configured to address the separation of synthesis gas into various components, depending on the final products that are desired from the gas. It is very suitable to complex schemes where a combination of products is needed, for example, hydrogen, carbon monoxide, ammonia and methanol synthesis gases, and fuel gas sidestreams.

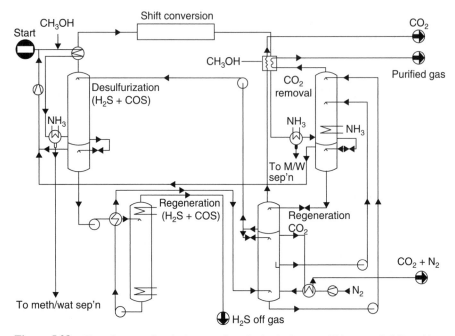

Figure 5.28. Flow diagram of rectisol process for selective hydrogen sulfide removal, followed by carbon dioxide removal.[189]

Figure 5.28 shows a flow scheme of the Rectisol process for selective hydrogen sulfide removal, followed by carbon dioxide removal.[189] H_2S and CO_2 can be removed respectively, but two absorbers and two regenerators are required. The chief drawback of the Rectisol process is its complex scheme and the need to refrigerate the solvent due to its low boiling point, which results in high capital and operating costs.

5.5.1.3 Hydride Solvent Processes

A number of absorption processes use a mixture of amines and physical solvents to take advantage of the best characteristics of both types. A mixed solvent is generally a compromise between H_2S selectivity and the degree of the physical solubilities of the other sulfur compounds. Among the most numerous applications of such processes used commercially are Sulfinol (Shell) and Flexsorb.[58,189]

5.5.1.3.1 Sulfinol and ADIP (Aqueous Di-Isoproponalamine) Processes.
The Sulfinol process, developed by Shell in the early 1960s, is a combination process that uses a mixture of amines and a physical solvent. The Shell Sulfinol process is a regenerable amine process for acid gas removal. As the process uses a mixture of water, sulfolane, and one or more alkanolamine, removal capacity of COS, mercaptans, and organic sulfides from gas streams is excellent by virtue of the improved physical solubility of these compounds in the solvent. A typical flow diagram of sulfinol process is shown in Figure 5.29.[196]

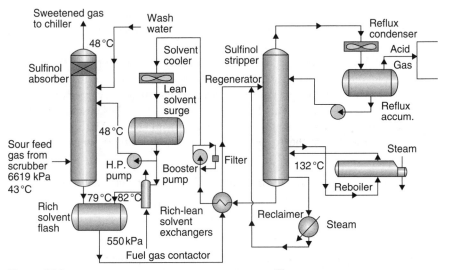

Figure 5.29. Schematic flow diagram of the sulfinol process.[196]

Sulfinol-D uses DIPA, while Sulfinol-M uses MDEA. The mixed solvents allow for better solvent loadings at high acid gas partial pressures and higher solubility of COS and organic sulfur compounds than straight aqueous amines. There are about 200 Sulfinol plants in operation worldwide, most of which use the Sulfinol-D solvent formulations. Sulfinol-D is primarily used in cases where selective removal of H_2S is not of primary concern, but where partial removal of organic sulfur compounds (mercaptans and CS_2) is desired, typically in natural gas and refinery applications. Sulfinol-D is also able to remove some COS via physical solubility in sulfolane and partial hydrolysis to H_2S induced by the secondary amine (DIPA). However, deep removal of COS by Sulfinol-D cannot be guaranteed. Unlike solvents that use other primary and secondary amines (MEA, DEA) Sulfinol-D is claimed not to be degraded by these sulfur compounds.[189]

Sulfinol-M is used when a higher degree of H_2S selectivity is needed. H_2S selectivity in Sulfinol-M is controlled by the kinetics of the reaction of H_2S with MDEA as well as by the physical solubilities of H_2S and CO_2 in the solvent.

The Shell Sulfinol-X process is a regenerable amine process for acid gas removal utilizing a mixture of two or more alkanolamines, in general a base amine such as MDEA or DIPA and an accelerator. As the process uses a mixture of water, sulfolane, and one or more alkanolamine, removal capacity of COS, mercaptans, and organic sulfides from gas streams is excellent by virtue of the improved physical solubility of these compounds in the solvent. It was claimed that the process achieves a higher loading capacity thus enabling a design of smaller absorber columns with reduced number of trays when compared with other Sulfinol processes.

Shell considers that the Sulfinol-X process is the best choice in the applications of removal of H_2S, CO_2, COS, mercaptans, and organic sulfides from gas streams, and Sulfinol-X is highly suitable to revamp deep/bulk CO_2 removal plants. Guaranteed specifications achievable are gas stream H_2S, 100 μbar partial pressure; and gas stream CO_2, 500 μbar partial pressure or 50 ppmv.[197]

The Shell ADIP-X process is a regenerable amine process for acid gas removal utilizing a mixture of two or more alkanolamines, in general a base amine such as MDEA and an accelerator. The process achieves a higher loading capacity compared with single amine solvents. This leads to the design of smaller absorber columns with the reduced number of trays when compared with generic MDEA solvent.[197]

Shell considers that the ADIP-X process is highly suitable for the following applications:[197]

- removal of H_2S and CO_2 from natural gas, refinery gas, and synthesis gas streams;
- bulk removal of CO_2 from gas streams;
- deep removal of CO_2 from gas streams; and
- particularly suitable for natural gas field with CO_2 content in the excess of 5 vol % in the feed gas.

5.5.1.3.2 Others. ExxonMobil also offers two mixed hindered amine/physical solvent versions of the Flexsorb process. The Hybrid Flexsorb SE Process employs a solution of the Flexsorb SE amine, water, and an unspecified physical solvent. Two plants are in operation. The Flexsorb PS solvent consists of a different hindered amine, water, and a physical solvent. Five of these plants are believed to be operating.[198,199]

As reviewed above, it could be summarized that:

- For removal of acid gas from coal gasification gas and natural gas, the scrubbing by using solvents, including the chemical (tertiary amine), physical, and hybrid solvents, is still the commercial mainstay as this technology is relatively mature with lower cost in comparison with others, although the raw gas stream needs to be cooled to room temperature.
- The absorption with the chemical solvents favors the gas streams with lower pressure and lower concentration of the coexisting CO_2, if the capture of CO_2 is not required. The absorption with the physical solvents benefits the gas streams with higher pressure when the higher H_2S selectivity over CO_2 is required at higher concentration of the coexisting CO_2.
- The chemical solvents are able to remove H_2S in the gas to very low level theoretically on the basis of the acid–base interaction. However, it may not be very effective in the removal of COS. Consequently, the COS hydrolysis upstream of the acid gas removal is required for deep total sulfur removal, if there is a higher concentration of COS in the gas stream.
- The absorption with the physical solvents and hybrid solvents improve the absorption selectivity significantly for H_2S and other sulfur compounds, including COS, mercaptans, and CS_2.

- The absorption with the hybrid solvents can combine some advantages of both chemical and physical solvents, which improve the selectivity and absorption rate.
- The most fatal drawback of the cold gas cleanup by solvents is that the raw syngas needs to be cooled to near room temperature, which decreases significantly the heat efficiency of the process.

5.5.2   Hot and Warm Gas Cleanup

The development of hot gas cleanup systems (dry desulfurization) for removing H_2S and COS from the syngas has been pursued since the 1970s. Until about the mid-1990s, this work was primarily focused on syngas from air-blown gasification of coal. Air-blown gasification systems produce over twice the volume of syngas (due to nitrogen dilution) that O_2-blown systems produce, and therefore incur more severe thermal, process efficiency, and capital cost penalties related to syngas cooling to comparable temperature levels. Conventional cold gas cleanup (scrubbing by solvents) needs to reduce the temperature of raw coal gasification gas from 900 to 1600 °C (depending on the gasifier design and the feedstock) to room temperature. If using for SOFC or turbine, the cleaned-up syngas needs to heat up again to 500–1000 °C. Obviously, the process is not efficient energetically. Consequently, significant attention has been paid to research and development of solid sorbent and process working at higher temperature. The motivation for higher temperature desulfurizaton using solid sorbent includes:

- the higher process efficiency achievable without syngas cooling and removal of water from the syngas,
- the elimination of sour water treating (sour water is produced when the syngas is cooled below the dew point of water),
- the elimination of the "black mud" (troublesome ash–char–water mixture) produced in water quenching or wet scrubbing of particulates from the syngas,
- saving electric power for the solvent delivery in the cold gas cleanup, and
- solid sorbents are usually less toxic and corrosive than the liquid amine absorbents used commonly in the cold gas cleanup.

The major sorbents reported for high or mild temperature desulfurization of the raw syngas are the solid metal oxides. Metal oxides, such as ZnO, MnO, Zn–Ti–O, Cu–Zn–Ti–O, Co–Fe–O, CeO_2, and La_2O_3, have been reported to remove H_2S from nature gas, reformate, and other gas mixtures.[4,113,176–178,180] The process is based on the reaction of the metal oxides with H_2S to form metal sulfides. For ZnO sorbent, the reaction is shown below:

$$ZnO(s) + H_2S(g) \leftrightarrow ZnS(s) + H_2O(g) \quad \Delta H_r^\circ = -78.9 \text{ kJ/mol}. \quad (5.9)$$

The regeneration of the spent sorbent is based on the oxidation reaction of the metal sulfide:

$$ZnS(s) + O_2(g) \leftrightarrow ZnO(s) + SO_2(g) \quad \Delta H_r^\circ = -439.11\,\text{kJ/mol}. \quad (5.10)$$

The low temperature is thermodynamically favored. However, since the reaction rate is low at low temperature, the process usually prefers to conduct at higher temperature in the practical application to increase the reaction rate. A review in the gas desulfurizaton sorbents is made in Section 5.4. The major problems in high or mild temperature desulfurization sorbents and processes are summarized below:

- The hot raw syngas at the gasifier outlet contains not only H_2, CO, CH_4, CO_2, sulfur, and nitrogen compounds, but also H_2O, alkali (potassium and sodium), and mercury. In order to prevent alkali corrosion of hot gas cleanup components and to avoid expensive materials and unreliable refractory-lined piping,[200,201] it is necessary to reduce the temperature to less than 540 °C for complete condensation of the alkali vapor on particulates in the hot syngas.

- The prospect of stringent mercury emissions standards for coal-conversion plants seriously dampens the outlook for hot or warm gas cleanup. It is believed that mercury removal becomes more difficult as the syngas temperature increases. If it is necessary to cool the syngas for mercury removal, then the rationale for hot or warm gas desulfurization is gone unless related economic benefits can be demonstrated.[189,201,202]

- As the low temperature is favored for decreasing the equilibrium H_2S concentration due to the exothermic reaction of metal oxides with H_2S and the presence of $H_2O(g)$ has a significant effect on the equilibrium H_2S concentration, it is impossible thermodynamically to remove H_2S from the gas to less than 7.4 ppbv when the temperature is higher than 600 °C over ZnO sorbent in the presence of 20 vol % $H_2O(g)$ in the treated gas. A significant negative effect of H_2O on the sorption performance of ZnO sorbent has been reported by Novochinskii et al.[177,178]

- The removal of ammonia, HCN, COS, and other contaminates with the developed sorbents at the current stage is difficult.

- The high temperature and high content of H_2O (up to 30 vol %) in the raw syngas result in a larger volume of the raw gas than that for the cold gas cleanup. It requires a larger size of the high temperature equipment. As noted by some observers, the physical size of the 10% capacity HGCU (hot *gas cleanup* unit) appears to be much larger than that of the 100% capacity cold gas cleanup section of the plant.[189]

- Regeneration of the spent sorbents is difficult and need to run at high temperature with oxygen-containing gas. The physical and surface structure of the sorbents is usually destroyed in the regeneration, leading to the significant degradation of the sorbent performance in the cycles.

Consequently, none of the proposed sorbent materials have become commercially viable.[203] No any hot or warm gas desulfurization process has been demonstrated.[180] The only two large-scale "hot gas" desulfurization systems installed in the U.S. (both in DOE CCT IGCC demonstration projects), but have never been demonstrated. Both systems were similarly based on the reaction of H_2S with zinc oxide–nickel oxide solid sorbents in an adsorption column, followed by regeneration of the sorbent by contact with air in a separate column. The regenerator off-gas contains SO_2, which must be converted to elemental sulfur or sulfuric acid in a final recovery operation.[189] As the problems shown above, both industry interest and government interest in the hot gas cleanup have declined.

The DOE gasification industry interviews have found that currently there is not much incentive for gas cleanup operations above 370 °C. Currently, DOE Cleaning and Conditioning Program has transitioned its gas cleaning component away from the development of high-temperature approaches to more moderate temperature (150–370 °C) to create more tightly integrated processes for removing multiple contaminants more efficiently.[185,204]

In the outlook of the hot and warm gas cleanup, Korens et al. made the following comments:[189] (a) the development of hot gas cleanup systems for deep cleaning of sulfur and nitrogen components from syngas appear to be long-term prospects. Large-scale demonstrations probably would not be achievable or practical before about 2010; (b) justification for such demonstrations could become difficult if commercial IGCC projects with CGCU continue to proliferate and operate well over the next several years. The prospects of developing hot or warm gas cleanup processes for mercury and CO_2 removal presently are very challenging. Either, or both, the requirement for mercury recovery and/or CO_2 recovery from coal-fired power plants could become the Achilles' heel of dry gas cleanup.

5.6 ODS

The ODS of hydrocarbon fuels consists of the conversion of sulfur compounds in the fuels by oxidation to element sulfur, sulfur oxides, sulfoxides, and/or sulfones followed by adsorption or abstraction separation of the oxidized sulfur compounds from the hydrocarbon fuels. The potential advantages for ODS are (a) the process does not need to use H_2 gas; (b) ODS takes place at relatively mild operating conditions in comparison with HDS or even at ambient conditions; and (c) ODS is able to remove some refractory sulfur compounds, such as 4,6-DMDBT, that are difficult to be removed by HDS or selective adsorption.

5.6.1 Natural Gas

A comparison of fuel-desulfurization technologies (selective adsorption, HDS, ODS) for natural gas are listed in Table 5.11, which is based on the following assumptions: a 2.5-kWe fuel cell (20 L/min natural gas; 6.7 L/min liquid petroleum gas), fuel sulfur levels of 12 ppmv for natural gas, and 120 ppmw for liquid petroleum gas, 1 year onstream.[205] For the selective adsorption, while the simplicity of this approach is attractive, sulfur-adsorption capacities are relatively low (typically less than 10–20 mg-S/g-A). This not only calls for large adsorbent inventories and frequent changeouts, but produces an inventory of spent adsorbents that are hazard-

Table 5.11. Comparison of Fuel Desulfurization Technologies on the Basis of the Following Assumptions: a 2.5-kWe Fuel Cell (20 L/min Nature Gas; 6.7 L/min liquid Petroleum Gas), Fuel Sulfur Levels of 12 ppmv for Natural Fuel, and 120 ppmw for Liquid Petroleum Gas, 1 Year Onstream[205]

	Selective Catalytic Oxidation (SCO)	Passive Adsorption	Hydrodesulfurization (HDS)
Adsorbent volume	3 L (NG) 14 L (LPG)	20 L (NG) 115 L (LPG)	2 L (NG) 75 L (LPG)
Species compatibility	For all sulfur species and fuel compositions	Separate inorganic, organic S adsorbers; sensitive to fuel composition	Ht more difficult to remove
Hazards	None	Spent adsorber is hazardous waste	Catalyst is priority pollutant and requires special activation
Operating requirements	Air addition and elevated temperature (250–280 °C)	Ambient temperature and pressure	Hydrogen recycle and elevated temperature (300–400 °C)

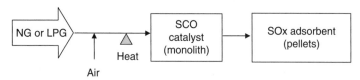

Figure 5.30.   Process flow diagram for selective catalytic oxidation (SCO) developed by Engelhard Corporation.[45]

ous and require special handling considerations.[45] While catalytic HDS produces higher sulfur-adsorption capacities compared with the selective adsorption (typically 50–100 mg-S/g-ZnO) and has lower maintenance needs, it requires greater capital investment to accommodate the required hydrogen recycle and heating of the fuel, and HDS catalysts contain toxic metals (Ni, Co, and Mo).

In order to overcome some of the shortcomings associated with HDS and selective adsorption, Engelhard has developed a new catalytic-adsorption fuel desulfurization technology that does not require hydrogen recycle and whose by-products are nonhazardous.[45] This new method catalytically converts the organic and inorganic sulfur species to sulfur oxides. The sulfur oxides are then adsorbed on a high capacity adsorbent. The process flow diagram for this SCO is shown in Figure 5.30. This technology combines the fuel with a sub-stoichiometric amount of oxygen (from air) and uses a sulfur tolerant monolith catalyst to oxidize selectively the sulfur compounds to sulfur oxides (SO_2 and SO_3, referred to collectively as SOx). The SOx species are then adsorbed downstream by an inexpensive high capacity particulate adsorbent. Sulfur slip from natural gas is below 10 ppbv. For liquid petroleum gas at maximum allowable sulfur levels (120 ppmw or 165 ppmv), the sulfur slip is less than 160 ppbv.[45]

Gardner et al. reported that H_2S catalytic partial oxidation technology with an AC catalyst is a promising method for the removal of H_2S from fuel cell hydrocarbon feedstocks.[206] Three different fuel cell feedstocks were considered for analysis: sour natural gas, sour effluent from a liquid middle distillate fuel processor, and a Texaco O_2-blown coal-derived synthesis gas. Their experimental results indicate that H_2S concentration can be removed down to the part per million level in these plants. Additionally, a power-law rate expression was developed and reaction kinetics compared with prior literature. The activation energy for this reaction was determined to be 34.4 kJ/g mol with the reaction being first order in H_2S and 0.3 order in O_2.

Adsorption-catalytic removal of H_2S from digester gas over the chemically modified/impregnated ACs was studied by Bagreev et al. The results showed differences in the H_2S removal capacities related to the type of carbon and conditions of the experiment.[207] A decrease in H_2S concentration resulted in an increase in a breakthrough capacity, which is linked to slow kinetics of oxidation process. No significant changes were observed when the oxygen content increased from 1% to 2% and the temperature from 38 to 60 °C. On the surface of the studied carbons, hydrogen sulfide was oxidized predominantly to sulfur, which was deposited in micropores, either on the walls or at the pore entrances. The oxidation reaction is shown below:

$$H_2S + \frac{1}{2}O_2 \rightarrow \frac{1}{n}S_n + H_2O. \tag{5.11}$$

Wu et al.[208] studied the removal of low concentrations of H_2S from hydrogen-rich gaseous fuels by SCO, using AC as a catalyst. They found that the capacities of ACs for reducing the H_2S concentration down to the part per billion level are related to their microstructures and impurities. They also found that the complete and exclusive conversion of H_2S to elemental sulfur (S) requires that AC has catalytic activities not only for the oxidation of H_2S, but also for the oxidation of COS and for the reaction between H_2S and SO_2, as the side reactions that form COS and SO_2 are sometimes unavoidable under real fuel processing conditions.

5.6.2   Liquid Hydrocarbon Fuels

As well known, the liquid hydrocarbon fuels contain not only sulfur compounds but also a large number of aromatic compounds that have aromatic skeleton structure similar to the coexisting sulfur compounds. This inherent problem makes a great challenge in development of an effective adsorbent with high adsorptive selectivity for the sulfur compounds. Consequently, many ODS methods for liquid hydrocarbon fuels have been explored. Early work in ODS of thiophenic compounds in the presence of 30% hydrogen peroxide (H_2O_2) was reported by Gilman and Esmay[209] and Heimlich and Wallace.[210] Other reported ODS systems include H_2O_2(oxidant)/poly xometalate(catalyst),[211] H_2O_2/formic acid,[212–214] H_2O_2/phosphotungstic acid,[215] NO_2 oxidation,[216,217] H_2O_2/12-tungstophosphoric acid,[218] H_2O_2/iron complexes (tetra-amido macrocyclic ligand [TAML], activators),[219] H_2O_2/Na_2CO_3,[220] H_2O_2/acetic acid,[221,222] and CF_3COOH/titano silicates,[223] H_2O_2/solid bases,[224] H_2O_2/TiSi,[225] H_2O_2/catalyst,[226–231] H_2O_2/(AC plus formic acid).[232,233] The ODS processes usually consist of two steps. The first step is the oxidation of the thiophenic compounds in the fuels to form their sulfoxides and/or sulfones:

$$(5.12)$$

The second step is the removal of the produced sulfoxides and/or sulfones by adsorption, abstraction, and/or abstraction. The advantages of the ODS are that it avoids the use of hydrogen and allows the process to be conducted at ambient conditions, which is much more efficient energetically. More interesting is that the oxidation reactivity of 4,6-DMDBT in the presence of a mixture of hydrogen peroxide and formic acid is higher than that of DBT, as reported by Otsuki et al.[234]

In 2003, Valero Energy Corp. (San Antonio, TX)[133–135] began producing ULSD at its refinery in Krotz Springs, LA. The 50 bbl/day plant is the first commercial demonstration of a fuel-desulfurization process developed by UniPure Corp. (Houston, TX). Called advanced sulfur removal (ASR), the process converts gasoline and diesel fuel containing 300–3000 ppm sulfur to ultralow sulfur products having less than 5 ppm. The cost was claimed to be "substantially lower" than that for a new hydrotreater. A scheme of ASR process is shown in Figure 5.31. Unlike hydrotreating, the ASR process does not use H_2, so stand-alone plants can be built to treat off-spec products at terminals that do not have a hydrogen infrastructure. Diesel (or other distillate feed) is introduced to an oxidizing reactor at about 100 °C and 1–3 bar of pressure. An aqueous solution of recycled formic acid and some hydrogen peroxide is added, which oxidizes the sulfur compounds to the corresponding sulfones. The formed sulfones are then extracted by the acid and separated from the hydrocarbons in a gravity separator. Diesel from the separator is then washed with water, dried, and passed over a solid alumina bed to extract the remaining sulfones. Two columns operate in tandem, with one performing adsorption while the other is regenerated with methanol. The final diesel product contains less than 5 ppm sulfur.

The National Energy Technology Laboratory (NETL; Pittsburgh, PA), in collaboration with nearby Carnegie Mellon University (CMU), is currently evaluating a selective-oxidation process that uses "green" catalysts.[133–135] The catalyst, developed at CMU, is an iron complex with a hydrogen peroxide-activating ligand called TAML. The process takes place in a biphase system: an oil phase (fuel with sulfur

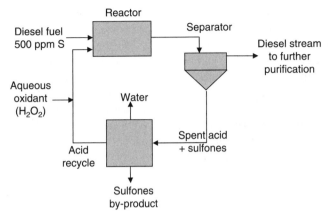

Figure 5.31.   Sulfur removal (ASR) process developed by an H_2-free technology; it reduces the sulfur content in gasoline and diesel fuel from 300 to 3000 ppm to less than 5 ppm.[133]

compounds) and an aqueous phase, containing an extracting agent, catalyst, and hydrogen peroxide. Sulfur compounds are extracted to the aqueous phase and oxidized by H_2O_2 into water-soluble compounds. No H_2 is required and the process is operated at ambient pressure and 60 °C, much milder than those required for conventional hydrotreating.

However, as shown above, most of the reported ODS systems involved use oil-insoluble oxidants, H_2O_2, or peroxides, which results in a biphasic oil–aqueous solution system. This biphasic system limits the mass transfer through the biphasic interface in the oxidation process, which results in a low oxidation rate. The phase separation after the oxidation usually causes a loss of the fuel in the process. The water-soluble acid or base used in biphase system also corrodes the equipments. In order to avoid these problems, some researchers explored the systems using oil-soluble oxidants, such as peracid/Co(II),[235] tert-BuOCl/Mo–Al₂O₃,[234] and tert-BuOOH/Mo–Al₂O₃.[236,237] In a study of oxidation of sulfur compounds existing in a desulfurized light gas oil (LGO: sulfur content: 39 ppm), Ishihara et al.[236] reported using tert-butyl hydroperoxide (t-BuOOH) as the oil-soluble oxidant in the presence of MoO_3/Al_2O_3 catalyst for ODS followed by adsorptive removal of the oxidized sulfur compounds over silica gel.[236,237] They found that the O/S ratio at 15 mol/mol gave the best desulfurization conversion, being 89.5% at 100 °C and weight hourly space velocity (WHSV) of 30 h⁻¹, and the oxidation reactivity of the sulfur compounds decreased in the order of DBT >> 4,6-DMDBT > C-3-DBT. Nevertheless, the reaction safety and the cost for the oil-soluble oxidant are still greatly concerned for developing a commercially feasible ODS process.

Recently, Ma et al.[238] explored a novel ODS method of liquid hydrocarbon fuels, which combines a catalytic oxidation step of the sulfur compounds using molecular oxygen followed by an adsorption step to remove oxidized sulfur compounds in the treated fuel using an AC. The ODS of an MJF and a real JP-8 jet fuel was conducted in a batch system at ambient conditions. They found that the oxidation using molecular oxygen in the presence of catalyst was able to convert the thiophenic compounds in the fuel to the sulfone and/or sulfoxide compounds. The oxidation reactivity of the sulfur compounds decreases in the order of 2-methylbenzothiophene > 5-methylbenzothiophene > BT >> DBT, and the catalytic oxidation of the sulfur compounds to form the corresponding sulfones and/or sulfoxides improved significantly the adsorptive desulfurization performance of the AC, because the AC has much higher adsorptive affinity for the sulfones and sulfoxides due to their higher polarity. The remarkable advantages of this developed ODS method are that the ODS can be run by using O_2 directly as an oxidant at ambient conditions without involving the complicated biphasic oil–aqueous solution system.[211]

5.7 SUMMARY

Deactivation of the catalysts in the hydrocarbon fuel processing catalysis and the fuel cell anodes via sulfur poisoning is a great concern right now in hydrogen production and fuel cell application. The sulfur compounds in the fuels and the H_2S produced from these sulfur compounds in the hydrocarbon reforming process are poisonous to the reforming catalyst, water-gas shift catalyst, and Prox catalyst in the hydrocarbon fuel processor, membranes for hydrogen purification and separation, and electrode catalysts in fuel cell stacks. It is required to reduce the sulfur level to less than 1 ppmw in liquid hydrocarbon fuels and less than 0.20 ppmv in natural gas for reforming. For the final use in fuel cells, the sulfur in the fuel gas needs ideally to be removed to less than 10 ppbv for PEMFC and 20 ppbv for SOFC. Consequently, even using the hydrocarbon liquid fuels that meet the new EPA regulation for low-sulfur fuels, 30 ppmw for gasoline, and 15 ppmw for highway diesel, the sulfur contents in the fuels are still too high for fuel cell applications.

The ultradeep desulfurization of the current commercial fuels has become a bottleneck in hydrogen production for fuel cell applications. It is urgent to develop a more efficient and environmentally friendly process and technology for the ultradeep desulfurization of the hydrocarbon fuels for fuel cell applications. Many approaches have been conducted in the improvement of the conventional HDS process or development of new alterable processes. These approaches include (a) catalytic HDS with improved and new catalysts, reactor, and/ or process; (b) selectively adsorptive desulfurization; and (c) ODS and others.

Current progress in HDS via improvement of conventional catalysts, reactors, and processes or development of new catalysts, reactors, and processes has allowed the refining industry to be able to produce the low-sulfur fuels to meet the new EPA regulation. However, the current HDS technology is still difficult or costly to produce the ultraclean liquid hydrocarbon fuels for fuel cell applications. Adsorptive desulfurization and ODS are two promising alterable technologies for ultradeep desulfurization of hydrocarbons fuels for fuel cell applications.

The selective adsorption/sorption desulfurization of hydrocarbon fuels is becoming a more and more important process, when the requirement for the sulfur concentration in the fuels by the environmental protection and fuel cell application becomes more severe. The major challenge in the development of a successful adsorptive desulfurization technology is to develop an efficient adsorbent/sorbent that has high capacity and selectivity for sulfur compounds and is easy for regeneration without the degradation of the adsorbent/sorbent within cycles. The sulfur compounds in gasoline and jet fuel are relatively easy to be removed selectively, as these sulfur compounds do not have a strong steric hindrance of the alkyl groups. A great challenge is to develop an adsorbent/sorbent that can selectively adsorb/sorb the refractory sulfur compounds from commercial diesel fuel. These refractory sulfur compounds are alkyl DBTs with one or two alkyl groups at the 4- and/or 6-positions, which show a strong steric hindrance toward the direct interaction between the sulfur atom in DBTs and the adsorption site. This steric hindrance makes it difficult to adsorb selectively these sulfur compounds directly through a sulfur-site interaction. As reviewed above, the different adsorbents/sorbents may be suitable for separating different sulfur compounds from different hydrocarbon streams. Some coexisting species, such as polyaromatic hydrocarbons, fuel additives and moisture, in hydrocarbon streams may inhibit the adsorption/sorption of sulfur compounds on the adsorbents/sorbents. Thus, a combination of two or more adsorbents/sorbents in a desulfurization process might be more promising and efficient for a practical ultradeep desulfurization process. A fundamental understanding of the selectively adsorptive mechanism on the adsorbent/sorbent is necessary for designing and developing more efficient and power adsorbents/sorbents.

The selective ODS has shown many potential advantages for deep desulfurization of the fuels for fuel cell applications, because the process usually has higher desulfurization capacity than the adsorption desulfurizaton, and also can run at mild operating conditions without the use of H_2. For ODS of liquid hydrocarbons fuels, direct use of oil-soluble peroxides or O_2 as oxidants in an ODS process is greatly attractive, as the process does not involve a complicated biphasic oil–aqueous solution system. The key in ODS is how to increase the oxidation selectivity for the sulfur compounds.

For the removal of H_2S, COS, and other sulfur compounds from coal gasification gas and natural gas, scrubbing by using solvents, including the chemical (tertiary amine), physical, and hybrid solvents, is still the commercial mainstay, as this technology is relatively mature with lower cost in comparison with others, although the raw gas stream needs to be cooled to room temperature. The absorption by the chemical solvents favors the gas streams with lower pressure and lower concentration of the coexisting CO_2, if the capture of CO_2 is not required. The absorption by the physical solvents benefits the gas streams with higher pressure when higher H_2S selectivity over CO_2 is required at higher concentration of the coexisting CO_2. The chemical solvents are able to remove H_2S in the gas to very low level theoretically on the basis of the acid–base interaction. However, it may not be very effective in the removal of COS. Consequently, the COS hydrolysis upstream of the acid gas removal is required for deep removal of the total sulfur, if there is a higher concentration of COS in the gas stream. The absorption with the physical solvents and hybrid solvents improves the absorption selectivity significantly for H_2S and other sulfur compounds, including COS, mercaptans, and CS_2. The absorption by the hybrid solvents combines some advantages of both chemical and physical solvents, which improves the selectivity and absorption rate. The most fatal drawback of the cold gas cleanup by solvents is that the raw syngas needs to be cooled to near room temperature, which decreases significantly the heat efficiency of the process.

The major motivation for hot and warm gas cleanup by using solid sorbent includes the higher process energy efficiency as the raw syngas is not needed to be cooled and the elimination of sour water treating. However, there are some major problems in hot and warm gas cleanup, including (a) the removal of other contaminants including alkali vapor, mercury, ammonia, HCN, COS, and CO_2 is difficult with this process; (b) the physical and surface structure of the adsorbents/sorbents is usually destroyed in the regeneration, leading to the significant degradation of the sorbent performance in the cycles; and (c) the high temperature and high content of H_2O (up to 30 vol%) in the raw syngas result in a larger size of hot and warm gas cleanup equipment. In the outlook of the hot and warm gas cleanup, the development of the hot gas cleanup systems for deep removal of sulfur components from syngas appears to be long-term prospects. The prospects of developing hot or warm gas cleanup processes for mercury and CO_2 removal presently are very challenging. Either or both the requirement for mercury recovery and/or CO_2 recovery from coal-fired power plants could become the Achilles' heel of hot and warm gas cleanup.

A combination of multidesulfurization technologies together, such as HDS and ADS (adsorptive desulfurization), ODS and ADS, and solvent scrubbing and adsorption, has become a trend in research and development to achieve ultradeep desulfurization technology of the hydrocarbon fuels for environmental protection, hydrogen production, and fuel cell applications.

REFERENCES

1. FARRAUTO, R., HWANG, S., SHORE, L., RUETTINGER, W., LAMPERT, J., GIROUX, T., LIU, Y., and ILINICH, O. New material needs for hydrocarbon fuel processing: Generating hydrogen for the PEM fuel cell. *Annual Review Materials Research*, **2003**, 33, 1.
2. VELU, S., MA, X.L., SONG, C.S., NAMAZIAN, M., SETHURAMAN, S., and VENKATARAMAN, G. Desulfurization of JP-8 jet fuel by selective adsorption over a Ni-based adsorbent for micro solid oxide fuel cells. *Energy & Fuels*, **2005**, 19, 1116.
3. SONG, C. Fuel processing for low-temperature and high-temperature fuel cells—Challenges, and opportunities for sustainable development in the 21st century. *Catalysis Today*, **2002**, 77, 17.
4. ISRAELSON, G.J. Results of testing various natural gas desulfurization adsorbents. *Journal of Materials Engineering and Performance*, **2004**, 13, 282.
5. TUAN, H.P., JANSSEN, H.G., CRAMERS, C.A., SMIT, A.L.C., LOO, E.M. Determination of sulfur components in natural gas: a review. *Journal of High Resolution Chromatography*, **1994**, 17, 373–389.
6. DICKS, A.L. Hydrogen generation from natural gas for the fuel cell systems of tomorrow. *Journal of Power Sources*, **1996**, 61, 113.
7. American Gas Association. *2000 Odorization Manual*. Washington, D.C.: American Gas Association.
8. LISS, W., THRASHER, W., STEINMETZ, G., HOWDIAH, C., and ATTARI, A. 1992 Final Report. Cleveland, OH: American Gas Association Laboratory, 18. *GRI-29/0123*, **1992**.
9. SAE. Diesel Fuels. Warrendale, PA: Society of Automotive Engineers. *SAE J313 June 1989*, **1992**.
10. OWEN, K. and COLEY, T. *Automotive Fuels Reference Book*. Warrendale, PA: Society of Automotive Engineers, **1995**.
11. EPA. Control of Air Pollution from New Motor Vehicles Amendment to the Tier-2/Gasoline Sulfur Regulations. U.E.P. Agency, **2001**.
12. EPA. Reducing Non-Road Diesel Emissions. U.E.P. Agency, **2003**.
13. MA, X.L., SAKANISHI, K., and MOCHIDA, I. 3-Stage deep hydrodesulfurization and decolorization of diesel fuel with CoMo and NiMo catalysts at relatively low-pressure. *Fuel*, **1994**, 73, 1667.
14. MA, X.L., SAKANISHI, K.Y., and MOCHIDA, I. Hydrodesulfurization reactivities of various sulfur-compounds in diesel fuel. *Industrial & Engineering Chemistry Research*, **1994**, 33, 218.
15. TOPSOE, H.C., CLAUSEN, B.S., and MASSOTH, F.E. *Hydrotreating Catalysis, Science and Technology*. Berlin: Springer-Verlag, **1996**.
16. SONG, C. and MA, X.L. New design approaches to ultra-clean diesel fuels by deep desulfurization and deep dearomatization. *Applied Catalysis. B, Environmental*, **2003**, 41, 207.
17. FARWELL, S.O. and BARINAGA, C.J. Sulfur-selective detection with the Fpd—Current enigmas, practical usage, and future directions. *Journal of Chromatographic Science*, **1986**, 24, 483.
18. DRESSLER, M. *Selective Gas Chromatography Detectors*. Amsterdam: Elsevier Science Publishing, **1986**.
19. GAINES, K.K., CHATHAM, W.H., and FARWELL, S.O. Comparison of the Scd and Fpd for Hrgc determination of atmospheric sulfur gases. *HRC. Journal of High Resolution Chromatography*, **1990**, 13, 489.

20. HUTTE, R.S. and RAY, J.D. Sulfur-selective detectors. In *Detector for Capillary Chromatography* (eds. H.H. Hill, D.G. McMinn). New York: John Wiley & Sons, pp. 193–218, **1992**.

21. CHESKIS, S., ATAR, E., and AMIRAV, A. Pulsed-flame photometer—A novel gas-chromatography detector. *Analytical Chemistry*, **1993**, 65, 539.

22. AMIRAV, A. and JING, H.W. Pulsed flame photometer detector for gas-chromatography. *Analytical Chemistry*, **1995**, 67, 3305.

23. Handley, A.J. and Adlard, E.R. (eds.). *Gas Chromatographic Techniques and Applications*. Boca Raton, FL: CRC Press, **2001**.

24. WEISS, M. Analysis of sulfur in motor fuels. **2003**. http://www.controlmagazine.com (accessed January 21, 2003).

25. ZOCCOLILLO, L., CONTI, M., HANEL, R., and MAGRI, A.D. A computerized analysis of the sulfur content in diesel fuel by capillary gas chromatography with flame photometric detector. *Chromatographia*, **1996**, 42, 631.

26. MA, X.L., VELU, S, KIM, J.H., and SONG, C.S. Deep desulfurization of gasoline by selective adsorption over solid adsorbents and impact of analytical methods on ppm-level sulfur quantification for fuel cell applications. *Applied Catalysis. B, Environmental*, **2005**, 56, 137.

27. NISHIOKA, M., BRADSHAW, J.S., LEE, M.L., TOMINAGA, Y., TEDJAMULIA, M., and CASTLE, R.N. Capillary column gas chromatography of sulfur heterocycles in heavy oils and tars using a biphenylpolysiloxane stationary phase. *Analytical Chemistry*, **1985**, 57, 309.

28. WHITEHURST, D.D., ISODA, T., and MOCHIDA, I. Present state of the art and future challenges in the hydrodesulfurization of polyaromatic sulfur compounds. *Advances in Catalysis*, **1998**, 42, 345.

29. KABE, T., ISHIHARA, A., and QIAN, W. *Hydrodesulfurization and Hydrodenitrognation, Chemistry and Engineering*. New York: Kodansha & Wiley-VCH, **1999**.

30. CICERO, D.C., et al. Recent developments in hot-gas purification. In *High-Temperature Gas Cleaning* (eds. G.H.A. Dittler, G. Kasper). Karlsruhe, Germany: Institut fur Mechanische Verfahrenstechnik und Mechanik der Universitat Karlsruhe (TH), p. 525, **1999**.

31. BABICH, I.V. and MOULIJN, J.A. Science and technology of novel processes for deep desulfurization of oil refinery streams: A review. *Fuel*, **2003**, 82, 607.

32. DHAR, G.M., SRINIVAS, B.N., RANA, M.S., KUMAR, M., and MAITY, S.K. Mixed oxide supported hydrodesulfurization catalysts—A review. *Catalysis Today*, **2003**, 86, 45.

33. SONG, C.S. An overview of new approaches to deep desulfurization for ultra-clean gasoline, diesel fuel and jet fuel. *Catalysis Today*, **2003**, 86, 211.

34. SONG, C.S. and MA, X.L., Ultra-deep desulfurization of liquid hydrocarbon fuels: Chemistry and process. *International Journal of Green Energy*, **2004**, 1, 167.

35. BEJ, S.K., MAITY, S.K., and TURAGA, U.T. Search for an efficient 4,6-DMDBT hydrodesulfurization catalyst: A review of recent studies. *Energy & Fuels*, **2004**, 18, 1227.

36. MOCHIDA, I. and CHOI, K.H. An overview of hydrodesulfurization and hydrodenitrogenation. *Journal of the Japan Petroleum Institute*, **2004**, 47, 145.

37. HERNANDEZ-MALDONADO, A.J. and YANG, R.T. Desulfurization of transportation fuels by adsorption. *Catalysis Reviews, Science and Engineering*, **2004**, 46 111.

38. HERNANDEZ-MALDONADO, A.J. and YANG, R.T. New sorbents for desulfurization of diesel fuels via pi-complexation. *AIChE Journal*, **2004**, 50, 791.

39. HERNANDEZ-MALDONADO, A.J., YANG, R.T., and CANNELLA, W. Desulfurization of commercial jet fuels by adsorption via pi-complexation with vapor phase ion exchanged Cu(I)-Y zeolites. *Industrial & Engineering Chemistry Research*, **2004**, 43, 6142.

40. TOPSOE, H. Towards the design of industrial catalysts. *19th North American Catalysis Society Meeting*. Philadelphia, **2005**.

41. BRUNET, S., MEY, D., PEROT, G., BOUCHY, C., and DIEHL, F. On the hydrodesulfurization of FCC gasoline: A review. *Applied Catalysis. A, General*, **2005**, 278, 143.

42. GUPTA, N., ROYCHOUDHURY, P.K., and DEB, J.K. Biotechnology of desulfurization of diesel: Prospects and challenges. *Applied Microbiology and Biotechnology*, **2005**, 66, 356.

43. ITO, E. and van VEEN, J.A.R. On novel processes for removing sulphur from refinery streams. *Catalysis Today*, **2006**, 116, 446.

44. GUPTA, R.P., TURK, B.S., PORTZER, J.W., and CICERO, D.C. Desulfurization of syngas in a transport reactor. *Environmental Progress*, **2001**, 20, 187.
45. LAMPERT, J.J. Selective catalytic oxidation: A new catalytic approach to the desulfurization of natural gas and liquid petroleum gas for fuel cell reformer applications. *Journal of Power Sources*, **2004**, 131, 27.
46. MASUDA, M., OKADA, O., TABATA, T., HIRAI, Y., and FUJITA, H. Method of desulfurization of hydrocarbons. *EP 0 600 406B2*, **2001**.
47. TWIGG, M.V. *Catalyst Handbook*. London: Wolfe Publishing Ltd., **1989**.
48. IKEDA, M. *Recent Developments of Phosphoric Acid Fuel Cell Plant*. Tokyo: Tokyo Gas Co.
49. HATANAKA, S., YAMADA, M., and SADAKANE, O. Hydrodesulfurization of catalytic cracked gasoline. 1. Inhibiting effects of olefins on HDS of alkyl(benzo)thiophenes contained in catalytic cracked gasoline. *Industrial & Engineering Chemistry Research*, **1997**, 36, 1519.
50. HATANAKA, S., YAMADA, M., and SADAKANE, O., Hydrodesulfurization of catalytic cracked gasoline. 2. The difference between HDS active site and olefin hydrogenation active site. *Industrial & Engineering Chemistry Research*, **1997**, 36, 5110.
51. HATANAKA, S., YAMADA, M., and SADAKANE, O. Hydrodesulfurization of catalytic cracked gasoline. 3. Selective catalytic cracked gasoline hydrodesulfurization on the Co-Mo/gamma-Al₂O₃ catalyst modified by coking pretreatment. *Industrial & Engineering Chemistry Research*, **1998**, 37, 1748.
52. KAUFMANN, T.G., KALDOR, A., STUNTZ, G.F., KERBY, M.C., and ANSELL, L.L. Catalysis science and technology for cleaner transportation fuels. *Catalysis Today*, **2000**, 62, 77.
53. HATANAKA, S., SADAKANE, O., and OKAZAKI, H. Hydrodesulfurization of catalytic cracked gasoline—Selective catalytic cracked gasoline hydrodesulfurization using Co–Mo/γ–Al₂O₃ catalyst modified by pyridine. *Journal of the Japan Petroleum Institute*, **2001**, 44, 36.
54. FREDRICK, C. Sulfur reduction: What are the options? *Hydrocarbon Processing*, **2002**, 81, 45.
55. OKAMOTO, Y., OCHIAI, K., KAWANO, M., KOBAYASHI, K., and KUBOTA, T. Effects of support on the activity of Co–Mo sulfide model catalysts. *Applied Catalysis. A, General*, **2002**, 226, 115.
56. MILLER, J.T., REAGAN, W.J., KADUK, J.A., MARSHALL, C.L., and KROPF, A.J. Selective hydrodesulfurization of FCC naphtha with supported MoS2 catalysts: The role of cobalt. *Journal of Catalysis*, **2000**, 193, 123.
57. HALBERT, T.R., STUNTZ, G.F., BRIGNAC, G.B., GREELEY, J.P., ELLIS, E.S., DAVIS, T.J., KAMIENSKI, P., and MAYO, S. A commercially proven technology for low sulfur gasoline. *Akzo Nobel Catalyst Symposium on SCANfining*. Noordwijk aan Zee, The Netherlands, **2001**.
58. ExxonMobil. http://www.exxonmobil.com/refiningtechnologies/fuels/mn_scanfining.html (accessed December 2001).
59. IFP. Gasoline desulfurization, ultra-deep. *Hydrocarbon Processing*, **2000**, 79, 144.
60. NOCCA, J.L., COSYNS, J., DEBUISSCHERT, Q., and DIDILLON, B. The domino interaction of refinery processes for gasoline quality attainment. *The NPRA Annual Meeting*. San Antonio, TX, **2000**.
61. IFP. http://www.ifp.fr/INTF/BI000GF2.html (accessed December 2001).
62. BACO, F., DEBUISSCHERT, Q., MARCHAL, N., NOCCA, J.L., PICARD, F., and UZIO, D. Prime G+ process, desulfurization of FCC gasoline with minimized octane loss. *The Fifth International Conference on Refinery Processing, AIChE 2002 Spring National Meeting*. New Orleans, LA, **2002**.
63. EPA-Gasoline-RIA. Regulatory impact analysis—Control of Air pollution from new motor vehicles: Tier 2 motor vehicle emissions standards and gasoline sulfur control requirements. US Environmental Protection Agency, Air and Radiation. *EPA420-R-99-023*, December **1999**.
64. Diesel-RIA. Regulatory impact analysis: Heavy-duty engine and vehicle standards and highway diesel fuel sulfur control requirements. United States Environmental Protection Agency, Air and Radiation. *EPA420-R-00-026*, December **2000**.
65. SHIH, S.S., OWENS, P.J., PALIT, S., and TRYJANKOWSKI, D.A. Mobil's OCTGain process: FCC gasoline desulfurization reaches a new performance level. *The NPRA Annual Meeting*, **1999**.
66. Refining Technology, ExxonMobil Licensing. Hydrocarbon Asia, p. 25, July/August **2001**.

67. CHITNIS, G.K., RICHTER, J.A., HILBERT, T.L., DABKOWSKI, M.J., AL KUWARI, I., Al KUWARI, N., and SHERIF, M. Commercial OCTGain's unit provides "zero" sulfur gasoline with higher octane from a heavy cracked naphtha feed. *The NPRA Annual Meeting.* San Antonio, TX, **2003**.

68. SALAZAR, J.A., CABRERA, L.M., PALMISANO, E., GARCIA, W.J., and SOLARI, R.B. U.S. Patent 5,770,047, **1998**.

69. UOP. Hydrodesulfurization. *Hydrocarbon Processing*, **2000**, 79, 120.

70. ANTOS, G.J., SOLARI, B., and MONQUE, R. Hydroprocessing to produce reformulated gasolines: The ISAL(TM) process. *Hydrotreatment and Hydrocracking of Oil Fractions*, **1997**, 106, 27.

71. MARTINEZ, N.P., SALAZAR, J.A., TEJADA, J., ANTOS, G.J., ANAND, M., and HOUDE, E.J. Meet gasoline pool sulfur and octane targets with the ISAL process. *The 2000 NPRA Annual Meeting.* San Antonio, TX, **2000**.

72. UOP. ISAL process. http://www.uop.com/framesets/refining.html (accessed December 2001).

73. BRIGNAC, G.B., HALBERT, T.R., SHIH, S.S., ELLIS, E.S., GREELEY, J.P., ROUNDTREE, E.M., GREANEY, M., and COOK, B.R. SCANfining technology meets the challenge of low sulfur gasoline. *Proceedings of the Fifth International Conference on Refinery Processing, AIChE 2002 Spring National Meeting.* New Orleans, LA, **2002**.

74. JAO, R.M., LEU, L.J., and CHANG, J.R. Effects of catalyst preparation and pretreatment on light naphtha isomerization over mordenite-supported Pt catalysts: Optimal reduction temperature for pure feed and for sulfur-containing feed. *Applied Catalysis. A, General*, **1996**, 135, 301.

75. LEE, J.K. and RHEE, H.K. Sulfur tolerance of zeolite beta-supported Pd-Pt catalysts for the isomerization of n-hexane. *Journal of Catalysis*, **1998**, 177, 208.

76. CDTECH. Hydrogenation. *Hydrocarbon Processing*, **2000**, 79, 122.

77. ROCK, K.L. Ultra-low sulfur gasoline via catalytic distillation. *The Fifth International Conference on Refinery Processing, AIChE Spring National Meeting.* New Orleans, LA, **2002**.

78. MA, X.L., SAKANISHI, K., ISODA, T., and MOCHIDA, I. Hydrodesulfurization reactivities of narrow-cut fractions in a gas oil. *Industrial & Engineering Chemistry Research*, **1995**, 34, 748.

79. KNUDSEN, K.G., COOPER, B.H., and TOPSOE, H. Catalyst and process technologies for ultra low sulfur diesel. *Applied Catalysis. A, General*, **1999**, 189, 205.

80. TOPSOE, H., KNUDSEN, K.G., BYSKOV, L.S., NORSKOV, J.K., and CLAUSEN, B.S. Advances in deep desulfurization. *Science Technology in Catalysis 1998*, **1999**, 121, 13.

81. SONG, C.S., HSU, C.S., and MOCHIDA, I. *Chemistry of Diesel Fuels.* New York: Taylor & Francis, **2000**.

82. FARAG, H., SAKANISHI, K. MOCHIDA, I., and WHITEHURST, D.D. Kinetic analyses and inhibition by naphthalene and H_2S in hydrodesulfurization of 4,6-dimethyldibenzothiophene (4,6-DMDBT) over CoMo-based carbon catalyst. *Energy & Fuels*, **1999**, 13, 449.

83. FARAG, H., MOCHIDA, I., and SAKANISHI, K. Fundamental comparison studies on hydrodesulfurization of dibenzothiophenes over CoMo-based carbon and alumina catalysts. *Applied Catalysis. A, General*, **2000**, 194, 147.

84. KALUZA, L. and ZDRAZIL, M. Carbon-supported Mo catalysts prepared by a new impregnation method using a MoO3/water slurry: Saturated loading, hydrodesulfurization activity and promotion by Co. *Carbon*, **2001**, 39, 2023.

85. PAWELEC, B., MARISCAL, R., FIERRO, J.L.G., GREENWOOD, A., and VASUDEVAN, P.T. Carbon-supported tungsten and nickel catalysts for hydrodesulfurization and hydrogenation reactions. *Applied Catalysis. A, General*, **2001**, 206, 295.

86. OLGUIN, E., VRINAT, M., CEDENO, L., RAMIREZ, J., BORQUE, M., and LOPEZ-AGUDO, A. The use of TiO2-Al2O3 binary oxides as supports for Mo-based catalysts in hydrodesulfurization of thiophene and dibenzothiophene. *Applied Catalysis. A, General*, **1997**, 165, 1.

87. LECRENAY, E., SAKANISHI, K., MOCHIDA, I., and SUZUKA, T. Hydrodesulfurization activity of CoMo and NiMo catalysts supported on some acidic binary oxides. *Applied Catalysis. A, General*, **1998**, 175, 237.

88. LECRENAY, E., SAKANISHI, K., NAGAMATSU, T., MOCHIDA, I., and SUZUKA, T. Hydrodesulfurization activity of CoMo and NiMo supported on Al_2O_3-TiO_2 for some model compounds and gas oils. *Applied Catalysis. B, Environmental*, **1998**, 18, 325.

89. SEGAWA, K., TAKAHASHI, K., and SATOH, S. Development of new catalysts for deep hydrodesulfurization of gas oil. *Catalysis Today*, **2000**, 63, 123.

90. KRESGE, C.T., LEONOWICZ, M.E., ROTH, W.J., VARTULI, J.C., and BECK, J.S. Ordered mesoporous molecular-sieves synthesized by a liquid-crystal template mechanism. *Nature*, **1992**, 359, 710.

91. VARTULI, J.C., SHIH, S.S., KRESGE, C.T., and BECK, J.S. Potential applications for M41S type mesoporous molecular sieves. *Mesoporous Molecular Sieves*, **1998**, 117, 13.

92. REDDY, K.M. and SONG, C.S. Synthesis of mesoporous molecular sieves: Influence of aluminum source on Al incorporation in MCM-41. *Catalysis Letters*, **1996**, 36, 103.

93. REDDY, K.M. and SONG, C.S. Synthesis of mesoporous zeolites and their application for catalytic conversion of polycyclic aromatic hydrocarbons. *Catalysis Today*, **1996**, 31, 137.

94. REDDY, K.M., WEI, B.L., and SONG, C.S. Mesoporous molecular sieve MCM-41 supported Co-Mo catalyst for hydrodesulfurization of petroleum resids. *Catalysis Today*, **1998**, 43, 261.

95. SONG, C.S. and REDDY, K.M. Mesoporous molecular sieve MCM-41 supported Co-Mo catalyst for hydrodesulfurization of petroleum resids. *Applied Catalysis. A, General*, **1999**, 176, 1.

96. SONG, C.S. and REDDY, K.M. Chapter 1. In *Recent Trends in Catalysis* (eds. V. Murugesan, B. Arabindoo, M. Palanichamy). New Delhi: Narosa Publishing House, pp. 1–13, **1999**.

97. SONG, C.S., REDDY, K.M., LETA, H., YAMADA, M., and KOIZUMI, N. Mesoporous aluminosilicate molecular sieve MCM-41 as support of Co–Mo catalysts for deep hydrodesulfurization of diesel fuels. In *Chemistry of Diesel Fuels* (eds. C. Song, S. Hsu, I. Mochida). New York: Taylor & Francis, p. 139, **2000**.

98. SONG, C. and REDDY, K.M. Mesoporous zeolite-supported Co–Mo catalyst for hydrodesulfurization of dibenzothiophene in distillate fuels. *American Chemical Society, Division of Petroleum Chemistry*, **1996**, 41, 567.

99. TURAGA, U., WANG, G., MA, X.L., SONG, C.S., and SCHOBERT, H.H. Influence of nitrogen on deep hydrodesulfurization of 4,6-dimethyldibenzothiophene *American Chemical Society, Division of Petroleum Chemistry*, **2002**, 47, 89.

100. SHIFLETT, W.K. and KRENZKE, L.D. Consider improved catalyst technologies to remove sulfur. *Hydrocarbon Processing*, **2002**, 81, 41.

101. AKZO-NOBEL. **2002**. http://www.akzonobel-catalysts.com/html/hydroprocessing/catalysts/hccatze.htm.

102. GERRITSEN, L.A. Production of green diesel in the BP Amoco Refineries. *Akzo Nobel at the WEFA Conference*. Berlin, Germany, **2000**.

103. MEIJBURG, G. Production of ultra-low-sulfur diesel in hydrocracking with the latest and future generation catalysts. *Catalyst Courier*, **2001**, No. 46, Akzo Nobel.

104. PLANTENGA, F.L. Nebula: A hydroprocessing catalyst with breakthrough activity. *Catalysts Courier*, **2002**, No. 47, Akzo Nobel.

105. OLSEN, C., KRENZKE, L.D., and WATKINS, B. Sulfur removal from diesel fuel: optimizing HDS activity through the SMART catalyst system. *Proceedings of the Fifth International Conference on Refinery Processing, AIChE Spring National Meeting*. New Orleans, LA, **2002**.

106. BHARVANI, R.R. and HENDERSON, R.S. Revamp your hydrotreater for deep desulfurization. *Hydrocarbon Processing*, **2002**, 81, 61.

107. LAWLER, D. and ROBINSON, S. Update hydrotreaters to process "green" diesel. *Hydrocarbon Processing*, **2002**, 8, 61.

108. SIE, S.T. Reaction order and role of hydrogen sulfide in deep hydrodesulfurization of gas oils: consequences for industrial reactor configuration. *Fuel Processing Technology*, **1999**, 61, 149.

109. ISODA, T., MA, X.L., NAGAO, S., and MOCHIDA, I. Reactivity of refractory sulfur compounds in diesel fuel. Part 3. Coexisting sulfur compounds and by-produced H_2S gas as inhibitors in HDS of 4,6-dimethyldibenzothiophene. *Journal of the Japan Petroleum Institute*, **1995**, 38, 25.

110. MOCHIDA, I., SAKANISHI, K., MA, X.L., NAGAO, S., and ISODA, T. Deep hydrodesulfurization of diesel fuel: Design of reaction process and catalysts. *Catalysis Today*, **1996**, 29, 185.

111. MAXWELL, I.E. Innovation in applied catalysis. *Cattech*, **1997**, 1, 5.

112. DAUTZENBERG, F. A call for accelerating innovation. *Cattech*, **1999**, 3, 54.

113. WESTMORELAND, P.R. and HARRISON, D.P. Evaluation of candidate solids for high-temperature desulfurization of low-Btu gases. *Environmental Science & Technology*, **1976**, 10, 659.

114. HEPWORTH, M.T., BENSLIMANE, M., and ZHONG, S. Thermodynamic comparison of several sorbent systems for hot coal-derived fuel-gas desulfurization. *Energy & Fuels*, **1993**, 7, 602.

115. OKADA, O.T., TAKAMI, S., HIRAO, K., MASUDA, M., FUJITA, H., IWASA, N., and OKHAMA, T. Development of an advanced reforming system for fuel cells. *International Gas Research Conference*. Orlando, FL, **1992**.

116. ROH, H.S., JUN, K.W., KIM, J.Y., KIM, J.W., PARK, D.R., KIM, J.D., and YANG, S.S. Adsorptive desulfurization of natural gas for fuel cells. *Journal of Industrial and Engineering Chemistry*, **2004**, 10, 511.

117. SATOKAWA, S., KOBAYASHI, Y., and FUJIKI, H. Adsorptive removal of dimethylsulfide and t-butylmercaptan from pipeline natural gas fuel on Ag zeolites under ambient conditions. *Applied Catalysis. B, Environmental*, **2005**, 56, 51.

118. MA, X.L., SUN, L., and SONG, C.S. A new approach to deep desulfurization of gasoline, diesel fuel and jet fuel by selective adsorption for ultra-clean fuels and for fuel cell applications. *Catalysis Today*, **2002**, 77, 107.

119. HERNANDEZ-MALDONADO, A.J. and YANG, R.T. Desulfurization of liquid fuel by adsorption via pi complexation with Cu(I)-Y and Ag-Y zeolites. *Industrial & Engineering Chemistry Research*, **2003**. 42, 123.

120. KO, C.H., BHANDARI, V.M., PARK, J.G., KIM, J.N., HAN, S.S., and CHO, S.H. Adsorption properties change of Ni/Y on sulfur compounds depending on the diesel composition. *Preprints of Papers— American Chemical Society, Division of Fuel Chemistry*, **2005**, 50, 805.

121. MA, X.L., SPRAGUE, M., and SONG, C.S. Deep desulfurization of gasoline by selective adsorption over nickel-based adsorbent for fuel cell applications. *Industrial & Engineering Chemistry Research*, **2005**, 44, 5768.

122. VELU, S., SONG, C.S., ENGELHARD, M.H., and CHIN, Y.H. Adsorptive removal of organic sulfur compounds from jet fuel over K-exchanged NiY zeolites prepared by impregnation and ion exchange. *Industrial & Engineering Chemistry Research*, **2005**, 44, 740.

123. KIM, J.H., MA, X.L., ZHOU, A.N., and SONG, C.S. Ultra-deep desulfurization and denitrogenation of diesel fuel by selective adsorption over three different adsorbents: A study on adsorptive selectivity and mechanism. *Catalysis Today*, **2006**, 111, 74.

124. ZHOU, A.N., MA, X.L., and SONG, C.S. Liquid-phase adsorption of multi-ring thiophenic sulfur compounds on carbon materials with different surface properties. *The Journal of Physical Chemistry. B*, **2006**, 110, 4699.

125. LESIEUR, R.R., TEELING, C., SANGIOVANNI, J.J., BOEDEKER, L.R., DARDAS, Z.A., HUANG, H., SUN, J., TANG, X., and SPADACCINI, L.J. Method for desulfurizing gasoline or diesel fuel for use in a fuel cell power plant. U.S. Patent 6,454,935, **2002**.

126. HERNANDEZ-MALDONADO, A.J. and YANG, R.T. Desulfurization of commercial liquid fuels by selective adsorption via pi-complexation with Cu(I)-Y zeolite. *Industrial & Engineering Chemistry Research*, **2003**, 42, 3103.

127. HERNANDEZ-MALDONADO, A.J., YANG, F.H., QI, G., and YANG, R.T. Desulfurization of transportation fuels by pi-complexation sorbents: Cu(I)-, Ni(II)-, and Zn(II)-zeolites. *Applied Catalysis. B, Environmental*, **2005**, 56, 111.

128. WANG, Y.H., YANG, F.H., YANG R.T., HEINZEL, J.M., and NICKENS, A.D. Desulfurization of high-sulfur jet fuel by pi-complexation with copper and palladium halide sorbents. *Industrial & Engineering Chemistry Research*, **2006**, 45 (22),7649.

129. HERNANDEZ-MALDONADO, A.J. and YANG, R.T. Desulfurization of diesel fuels via pi-complexation with nickel(II)-exchanged X- and Y-zeolites. *Industrial & Engineering Chemistry Research*, **2004**, 43 (4), 1081.

130. HERNANDEZ-MALDONADO, A.J. and YANG, R.T. Desulfurization of diesel fuels by adsorption via pi-complexation with vapor-phase exchanged Cu(I)-Y zeolites. *Journal of American Chemical Society*, **2004**, 126 (4), 992.

131. Phillips-Petroleum. http://www.fuelstechnology.com/szorbdiesel.htm (accessed December 2001).

132. GISLASON, J. Phillips sulfur-removal process nears commercialization. *Oil & Gas Journal*, **2002**, 99, 74.

133. *Chemical Engineering*, July **2003**, p. 17.

134. *Chemical Engineering*, September **2003**, p. 27.

135. *Chemical Engineering*, September **2003**, p. 28.

136. TURK, B.S., GUPTA, R.P., and GANGWAL, S.K. A novel vapor-phase process for deep desulfurization of naphtha. Final report, DOE Cooperative Agreement No. DE-FC26-01BC15282, June **2003**.

137. TURK, B.S. and GUPTA, R.P. RTI's TReND process for deep desulfurization of naphtha. *American Chemical Society, Division of Petroleum Chemistry*, **2001**, 46, 392.

138. IRVINE, R.L. Process for desulfurizing gasoline and hydrocarbon feedstocks. U.S. Patent 5,730,860, **1998**.

139. *Hydrocarbon Processing*, **1999**, 78, 39.

140. NEHLSEN, J.P., BENZIGER, J.B., and KEVREKIDIS, I.G. A process for the removal of thiols from a hydrocarbon stream by a heterogeneous reaction with lead oxide. *Energy & Fuels*, **2004**, 18, 721.

141. YANG, R.T., HERNANDEZ-MALDONADO, A.J., and YANG, F.H. Desulfurization of transportation fuels with zeolites under ambient conditions. *Science*, **2003**, 301, 79.

142. TAKAHASHI, A., YANG, F.H., and YANG, R.T. New sorbents for desulfurization by pi-complexation: Thiophene/benzene adsorption. *Industrial & Engineering Chemistry Research*, **2002**, 41, 2487.

143. HERNANDEZ-MALDONADO, A.J., QI, G.S., and YANG, R.T. Desulfurization of commercial fuels by pi-complexation: Monolayer CuCl/gamma-Al$_2$O$_3$. *Applied Catalysis. B, Environmental*, **2005**, 61, 212.

144. RICHARDEAU, D., JOLY, G., CANAFF, C., MAGNOUX, P., GUISNET, M., THOMAS, M., and NICOLAOS, A. Adsorption and reaction over HFAU zeolites of thiophene in liquid hydrocarbon solutions. *Applied Catalysis. A, General*, **2004**, 263, 49.

145. WATANABE, S., VELU, S., MA, X.L., and SONG, C. S. New ceria-based selective adsorbents for removing sulfur from gasoline for fuel cell applications. *Preprints of Papers—American Chemical Society, Division of Petroleum Chemistry*, **2003**, 48, 695.

146. WATANABE, S., MA, X.L., and SONG, C.S. Selective sulfur removal from liquid hydrocarbons over regenerable CeO$_2$-TiO$_2$ adsorbents for fuel cell applications. *Preprints of Papers—American Chemical Society, Division of Fuel Chemistry*, **2004**, 49, 511.

147. VELU, S., MA, X.L., and SONG, C.S. Selective adsorption for removing sulfur from jet fuel over zeolite-based adsorbents. *Industrial & Engineering Chemistry Research*, **2003**, 42, 5293.

148. MA, X.L., VELU, S., and SONG, C.S. Adsorptive desulfurization of diesel fuel over a metal sulfide-based adsorbent. *Preprints of Papers—American Chemical Society, Division of Fuel Chemistry*, **2003**, 48, 522.

149. BOUGAULT, J., CATTELAIN, E., and CHABRIER, P. New method for obtaining benzene and toluene free of thiophene and methylthiophene. *Bulletin de la Société Chimique de France*, **1940**, 7, 781.

150. Sammes, P.G. (ed.). *Comprehensive Organic Chemistry. Heterocyclic Compounds*. Oxford: Pergamon Press, **1979**.

151. STARTSEV, A. The mechanism of HDS catalysis. *Catalysis Reviews—Science and Engineering*, **1995**, 37, 352.

152. BONNER, W.A. and GRIMM, R.A. Chapter 2. Sulfur compounds, In *The Chemistry of Organic Sulfur Compounds* (eds. N. Kharasch, C.Y. Meyers). New York: Pergamon Press, p. 35, **1996**.

153. BAILEY, G.W. and SWAN, G.A. Nickel adsorbent for sulfur removal from hydrocarbon feeds. U.S. Patent 4,634,515, **1987**.

154. BONVILLR, L.J., DEGEORGE, C.L., FOLEY, P.F., GAROW, J., LESIEUR, R.R., PERSTON, J.L., and SZYDLOWSKI, D.F. System for desulfuri   zing a fuel for use in a fuel cell power plant. U.S. Patent 6,156,084, **2000**.

155. BONVILLR, L.J., DEGEORGE, C.L., FOLEY, P.F., GAROW, J., LESIEUR, R.R., PERSTON, J.L., and SZYDLOWSKI, D.F. Method for desulfurizing a fuel for use in a fuel cell power plant. U.S. Patent 6,159,256, **2000**.

156. LESIEUR, R.R., COCOLICCHIO, B.A., and VINCITORE, A.M. Method for desulfurizing gasoline or diesel fuel for use in a fuel cell power plant. U.S. Patent 6,726,836, **2004**.

157. HU, H.R., QIAO, M.H., XIE, F.Z., FAN, K.N., LEI, H., TAN, D., BAO, X.H., LIN, H.L., ZONG, B.N., and ZHANG, X.X. Comparative X-ray photoelectron spectroscopic study on the desulfurization of

thiophene by Raney Nickel and rapidly quenched skeletal nickel. *The Journal of Physical Chemistry. B*, **2005**, 109, 5186.

158. FUKUNAGA, T., KATSUNO, H., MATSUMOTO, H., TAKAHASHI, O., and AKAI, Y. Development of kerosene fuel processing system for PEFC. *Catalysis Today*, **2003**, 84, 197.

159. TAWARA, K., NISHIMURA, T., IWANAMI, H., NISHIMOTO, T., and HASUIKE, T. New hydrodesulfurization catalyst for petroleum-fed fuel cell vehicles and cogenerations. *Industrial & Engineering Chemistry Research*, **2001**, 40, 2367.

160. PECKHAM, J. *Diesel Fuel News*, August 12, **2002**.

161. SHIRAISHI, Y., YAMADA, A., and HIRAI, T. Desulfurization and denitrogenation of light oils by methyl viologen-modified aluminosilicate adsorbent. *Energy & Fuels*, **2004**, 18 (5), 1400.

162. NG, F.T.T., RAHMAN, A., OHASI, T., JIANG, M. A study of the adsorption of thiophenic sulfur compounds using flow calorimetry. *Applied Catalysis B-Environmental* **2005**, 56, 127.

163. BHANDARI, V.M., KO, C.H., PARK, J.G., HAN, S.S., CHO, S.H., and KIM, J.N. Desulfurization of diesel using ion-exchanged zeolites. *Chemical Engineering Science*, **2006**, 61, 2599.

164. KO, C.H.B., BHANDARI V.M., PARK, J.G., KIM, J.N., HAN, S.S., and CHO, S.H. Adsorption properties change of ni/y on sulfur compounds depending on the diesel composition. *Prepr. Pap.–Am. Chem. Soc. Div. Fuel Chem.* **2005**, 50, 805.

165. SANO, Y., CHOI, K.H., KORAI, Y., and MOCHIDA, I. Effects of nitrogen and refractory sulfur species removal on the deep HDS of gas oil. *Applied Catalysis. B, Environmental*, **2004**, 53, 169; SANO, Y., SUGAHARA, K., CHOI, K.H., KORAI, Y., and MOCHIDA, I. Two-step adsorption process for deep desulfurization of diesel oil. *Fuel*, **2005**, 84, 903.

166. HAJI, S. and ERKEY, C. Removal of dibenzothiophene from model diesel by adsorption on carbon aerogels for fuel cell applications. *Industrial & Engineering Chemistry Research*, **2003**, 42, 6933.

167. JIANG, Z.X., LIU, Y., SUN, X.P., TIAN, F.P., SUN, F.X., LIANG, C.H., YOU, W.S., HAN, C.R., and LI, C. Activated carbons chemically modified by concentrated H_2SO_4 for the adsorption of the pollutants from wastewater and the dibenzothiophene from fuel oils. *Langmuir*, **2003**, 19, 731.

168. ANIA, C.O. and BANDOSZ, T.J. Importance of structural and chemical heterogeneity of activated carbon surfaces for adsorption of dibenzothiophene. *Langmuir*, **2005**, 21, 7752.

169. POSHUSTA, J., SCHNEIDER, E., and MARTIN, J. Continuously regenerating fuel desulfurization system. *Fuel Cell Seminar*. Miami Beach, FL, **2003**.

170. FARRAUTO, R.J. and BARTHOLOMEW, C.H. Fundamentals of industrial catalytic processes. London: Blackie Academic & Professional, **1997**.

171. EDDINGTON, K. and CARNELL, P. Copact catalytic reactor controls vent odors from oil-field. *Oil & Gas Journal*, **1991**, 89, 69.

172. SPICER, G.W. and WOODLAND, C. H_2S control keeps gas from big offshore field on spec. *Oil & Gas Journal*, **1991**, 89, 76.

173. DAVIDSON, J.M., LAWRIE, C.H., and SOHAIL, K. Kinetics of the absorption of hydrogen-sulfide by high-purity and doped high-surface-area zinc-oxide. *Industrial & Engineering Chemistry Research*, **1995**, 34, 2981.

174. ELSEVIERS, W.F. and VERELST, H. Transition metal oxides for hot gas desulphurisation. *Fuel*, **1999**, 78, 601.

175. TURKDOGAN, E.T. *Physical Chemistry of High-Temperature Technology*. New York: Academic Press, **1980**.

176. PHILLIPSON, J.J. *Desulfurization. Catalysis Handbook*. London: Wolfe Scientific Books, **1970**.

177. NOVOCHINSKII, I., SONG, C.S., MA, X.L., LIU, X.S., SHORE, L., LAMPERT, J., and FARRAUTO, R.J. Low-temperature H_2S removal from steam-containing gas mixtures with ZnO for fuel cell application. 1. ZnO particles and extrudates. *Energy & Fuels*, **2004**, 18, 576.

178. NOVOCHINSKII, I., SONG, C.S., MA, X.L., LIU, X.S., SHORE, L., LAMPERT, J., and FARRAUTO, R.J. Low-temperature H_2S removal from steam-containing gas mixtures with ZnO for fuel cell application. 2. Wash-coated monolith. *Energy & Fuels*, **2004**, 18, 584.

179. IKENAGA, N., OHGAITO, Y., MATSUSHIMA, H., and SUZUKI, T. Preparation of zinc ferrite in the presence of carbon material and its application to hot-gas cleaning. *Fuel*, **2004**, 83, 661.

180. FLYTZANI-STEPHANOPOULOS, M., SAKBODIN, M., and WANG, Z. Regenerative adsorption and removal of H2S from hot fuel gas streams by rare earth oxides. *Science*, **2006**, 312, 1508.

181. POLYCHRONOPOULOU, K., FIERRO, J.L.G., and EFSTATHIOU, A.M. Novel Zn-Ti-based mixed metal oxides for low-temperature adsorption of H2S from industrial gas streams. *Applied Catalysis. B, Environmental*, **2005**, 57, 125.

182. XU, X.C., NOVOCHINSKII, I., and SONG, C.S. Low-temperature removal of H_2S by nanoporous composite of polymer-mesoporous molecular sieve MCM-41 as adsorbent for fuel cell applications. *Energy & Fuels*, **2004–005**, 19, 2214.

183. The data was estimated based on the thermal equilibrium.

184. DOE-FE. Hydrogen from Coal Program—Research, Development, and Demonstration Plan, **2004– 2005**. http://www.netl.doe.gov/technologies/hydrogen_clean_fuels/refshelf/pubs/hold/MYRDDP. pdf (accessed August 30, 2009).

185. DOE, NETL. Gasification—Gas cleaning & conditioning. http://www.netl.doe.gov/technologies/ coalpower/gasification/gas-clean/index.html (accessed January 2, 2007).

186. Ein Unternehmen von ThyssenKruppTechnologies (Uhde). Gas conditioning & upgrading: Syngas treating process overview and syngas product routes. *IEA Bioenergy Agreement 2004-2006, Spring 2006 Meeting*. Dresden, **2006**.

187. NETL U.S. DOE. Central hydrogen production pathway. Funding Opportunity Announcement. *DE-PS26-06NT42800*, **2006**.

188. MAHIN RAMESHNI, P.E. *State-of-the Art in Gas Treating*. San Francisco, CA: Worley Parsons, **2000**. http://www.worleyparsons.com/CSG/Hydrocarbons/SpecialtyCapabilities/Documents/State-of-the- Art_in_Gas_Treating.pdf (accessed August 30, 2009).

189. KORENS, N., SIMBECK, D.R., and WILHELM, D.J. Process screening analysis of alternative gas treating and sulfur removal for gasification. SFA Pacific, Inc. Mountain View, CA, **2002**.

190. NEWBY, R.A., YANG, W.C., and BANNISTER, R.L. Fuel gas cleanup parameters in air-blown IGCC. *Journal of Engineering for Gas Turbines and Power*, **2000**, 122, 247.

191. Gas Processes 2002, *Hydrocarbon Processing*, May **2002**.

192. WEILAND, R.H. and DINGMAN, J.C. Effect of solvent blend formulation on selectivity in gas treating. *Laurance Reid Gas Conditioning Conference*. Norman, OK, **1995**.

193. ROCHELLE, G.T. Absorption of carbon dioxide by piperazine activated methyldiethanolamine. *Laurance Reid Gas Conditioning Conference*. Norman, OK, February 25–28, **2001**.

194. ITOH, J. Chemical & physical absorption of CO_2. *RITE International Seminar*. January 14, **2005**.

195. KOHL, A.L. and RIESENTFELD, F.C. *Gas Purification*. Houston, TX: Gulf Publishing Company, **1985**.

196. TECHNOLOGY, S.A.I.O. Sweetening chemicals. http://tlmwebsites.sait.ab.ca/gpo/R2QA9F73/ B48L5JWH/modules/27313820/27313820_content.htm (accessed December 20, 2006).

197. BRUIJN, J.D., HENNEKES, B., KLINKENBIJL, J., KODDE, A., SMIT, K., and van den BORN, I. Treating options for syngas. *Gasification Technologies Conference. San Francisco, CA*, **2003**.

198. FEDICH, R.B., WOERNER, A.C., and CHITNIS, G.K. Selective H_2S removal. *Hydrocarbon Engineering*, **2004**, 9 (5), 89.

199. VOLTZ, B.L.C. and FEDICH, R.B. Solvent changeover benefits. *Hydrocarbon Engineering*, **2005**, 10 (5), 23.

200. TODD, D.M. Clean coal and heavy oil technologies for gas turbines. *38th GE Turbine State-of the Art Technology Seminar*. Schenectady, NY, **1994**.

201. HOLT, N. Coal gasification research, development and demonstration—Needs and opportunities. *2001 Gasification Technologies Conference*. San Francisco, CA, **2001**.

202. STIEGEL, G.J., CLAYTON, S.J., and WIMER, J.G. DOE's gasification industry interviews survey of market trends, issues and R&D needs. *Gasification Technologies Conference*. San Francisco, CA, **2001**.

203. WANG, Z. and FLYTZANI-STEPHANOPOULOS, M. Cerium oxide-based sorbents for regenerative hot reformate gas desulfurization. *Energy & Fuels*, **2005**, 19, 2089.

204. DOE. How coal gasification power plants work. http://www.fe.doe.gov/programs/powersystems/ gasification/howgasificationworks.html (accessed January 10, 2007).

205. FARRAUTO, R.J. Desulfurizing hydrogen for fuel cells. *Fuel Cell Magazine*, December **2003**/ January **2004**.

206. GARDNER, T.H., BERRY, D.A., LYONS, K.D., BEER, S.K., and FREED, A.D. Fuel processor integrated H₂S catalytic partial oxidation technology for sulfur removal in fuel cell power plants. *Fuel*, **2002**, 81, 2157.

207. BAGREEV, A., KATIKANENI, S., PARAB, S., and BANDOSZ, T.J. Desulfurization of digester gas: Prediction of activated carbon bed performance at low concentrations of hydrogen sulfide. *Catalysis Today*, **2005**, 99, 329.

208. WU, X.X., SCHWARTZ, V., OVERBURY, S.H., and ARMSTRONG, T.R. Desulfurization of gaseous fuels using activated carbons as catalysts for the selective oxidation of hydrogen sulfide. *Energy & Fuels*, **2005**, 19, 1774.

209. GILMAN, H. and ESMAY, D.L. The oxidation of dibenzothiophene and phenoxathiin with hydrogen peroxide. *Journal of American Chemical Society*, **1952**, 74, 2021.

210. HEIMLICH, B.N. and WALLACE, T.J. Kinetics and mechanism of oxidation of dibenzothiophene in hydrocarbon solution—Oxidation by Aqueous hydrogen peroxide-acetic acid mixtures. *Tetrahedron*, **1966**, 22, 3571.

211. TE, M., FAIRBRIDGE, C., and RING, Z. Oxidation reactivities of dibenzothiophenes in polyoxometalate/H₂O₂ and formic acid/H₂O₂ systems. *Applied Catalysis. A, General*, **2001**, 219, 267.

212. AIDA, T., YAMAMOTO, D., IWATA, M., and SAKATA, K. Development of oxidative desulfurization process for diesel fuel. *Reviews on Heteroatom Chemistry*, **2000**, 22, 241.

213. OTSUKI, S., NONAKA, T., TAKASHIMA, N., QIAN, W.H., SHIHARA, A., IMAI, T., and KABE, T. Oxidative desulfurization of light gas oil and vacuum gas oil by oxidation and solvent extraction. *Energy & Fuels*, **2000**, 14, 1232.

214. DE FILIPPIS, P. and SCARSELLA, M. Oxidative desulfurization: Oxidation reactivity of sulfur compounds in different organic matrixes. *Energy & Fuels*, **2003**, 17, 1452.

215. COLLINS, F.M., LUCY, A.R., and SHARP, C. Oxidative desulphurisation of oils via hydrogen peroxide and heteropolyanion catalysis. *Journal of Molecular Catalysis. A, Chemical*, **1997**, 117, 397.

216. TAM, P.S., KITTRELL, J.R., and ELDRIDGE, J.W. Desulfurization of fuel-oil by oxidation and extraction. 1. Enhancement of extraction oil yield. *Industrial & Engineering Chemistry Research*, **1990**, 29, 321.

217. TAM, P.S., KITTRELL, J.R., and ELDRIDGE, J.W. Desulfurization of fuel-oil by oxidation and extraction. 2. Kinetic modeling of oxidation reaction. *Industrial & Engineering Chemistry Research*, **1990**, 29, 324.

218. YAZU, K., YAMAMOTO, Y., FURUYA, T., MIKI, K., and UKEGAWA, K. Oxidation of dibenzothiophenes in an organic biphasic system and its application to oxidative desulfurization of light oil. *Energy & Fuels*, **2001**, 15, 1535.

219. HANGUN, Y., ALEXANDROVA, L., KHETAN, S., HORWITZ, C., CUGINT, A., LINK, D.D., HOWARD, B., COLLINS, T.J. Oxidative desulfurization of fuels through TAML activators and hydrogen peroxide. *Preprints of Papers—American Chemical Society, Division of Petroleum Chemistry*, **2002**, 47, 42.

220. DESHPANDE, A., BASSI, A., and PRAKASH, A. Ultrasound-assisted, base-catalyzed oxidation of 4,6-dimethyldibenzothiophene in a biphasic diesel-acetonitrile system. *Energy & Fuels*, **2005**, 19, 28.

221. ZANNIKOS, F., LOIS, E., and STOURNAS, S. Desulfurization of petroleum fractions by oxidation and solvent-extraction. *Fuel Processing Technology*, **1995**, 42, 35.

222. RAPPAS, A.S. Process for removing low amounts of organic sulfur from hydrocarbon fuels. U.S. Patent 6,402,940, **2002**.

223. TREIBER, A., DANSETTE, P.M., ELAMRI, H., GIRAULT, J.P., GINDEROW, D., MORNON, J.P., and MANSUY, D. Chemical and biological oxidation of thiophene: Preparation and complete characterization of thiophene S-oxide dimers and evidence for thiophene S-oxide as an intermediate in thiophene metabolism in vivo and in vitro. *Journal of American Chemical Society*, **1997**, 119, 1565.

224. PALOMEQUE, J., CLACENS, J.M., and FIGUERAS, F. Oxidation of dibenzothiophene by hydrogen peroxide catalyzed by solid bases. *Journal of Catalysis*, **2002**, 211, 103.

225. SHIRAISHI, Y., HIRAI, T., and KOMASAWA, I. Oxidative desulfurization process for light oil using titanium silicate molecular sieve catalysts. *Journal of Chemical Engineering of Japan*, **2002**, 35, 1305.

226. *Chemical Engeneering*, April **2001**, p. 23.
227. NERO, V.P., DECANIO, S.J., RAPPAS, A.S., LEVY, R.E., and LEE, F.M. Oxidative process for ultra-low sulfur diesel. *Preprints of Papers—American Chemical Society, Division of Petroleum Chemistry*, **2002**, 47, 41.
228. SHIRAISHI, Y., NAITO, T., and HIRAI, T. Vanadosilicate molecular sieve as a catalyst for oxidative desulfurization of light oil. *Industrial & Engineering Chemistry Research*, **2003**, 42, 6034.
229. CAMPOS-MARTIN, J.M., CAPEL-SANCHEZ, M.C., and FIERRO, J.L.G. Highly efficient deep desulfurization of fuels by chemical oxidation. *Green Chemistry*, **2004**, 6, 557.
230. LI, C., JIANG, Z.X., GAO, J.B., YANG, Y.X., WANG, S.J., TIAN, F.P., SUN, F.X., SUN, X.P., YING, P.L., and HAN, C.R. Ultra-deep desulfurization of diesel: Oxidation with a recoverable catalyst assembled in emulsion. *Chemistry-A European Journal*, **2004**, 10, 2277.
231. RAMIREZ-VERDUZCO, L.F., TORRES-GARCIA, E., GOMEZ-QUINTANA, R., GONZALEZ-PENA, V., and MURRIETA-GUEVARA, F. Desulfurization of diesel by oxidation/extraction scheme: influence of the extraction solvent. *Catalysis Today*, **2004**, 98, 289.
232. YU, G.X., LU, S.X., CHEN, H., and ZHU, Z.N. Diesel fuel desulfurization with hydrogen peroxide promoted by formic acid and catalyzed by activated carbon. *Carbon*, **2005**, 43, 2285.
233. YU, G.X., LU, S.X., CHEN, H., and ZHU, Z.N. Oxidative desulfurization of diesel fuels with hydrogen peroxide in the presence of activated carbon and formic acid. *Energy & Fuels*, **2005**, 19, 447.
234. OTSUKI, S., NONAKA, T., QIAN, W.H., ISHIHARA, A., and KABE, T. Oxidative desulfurization of middle distillate—Oxidation of dibenzothiophene using t-butyl hypochlorite. *Journal of the Japan Petroleum Institute*, **2001**, 44, 18.
235. MURATA, S., MURATA, K., KIDENA, K., and NOMURA, M. A novel oxidative desulfurization system for diesel fuels with molecular oxygen in the presence of cobalt catalysts and aldehydes. *Energy & Fuels*, **2004**, 18, 116.
236. ISHIHARA, A., WANG, D.H., DUMEIGNIL, F., AMANO, H., QIAN, E.W.H., and KABE, T. Oxidative desulfurization and denitrogenation of a light gas oil using an oxidation/adsorption continuous flow process. *Applied Catalysis. A, General*, **2005**, 279, 279.
237. QIAN, E.W., DUMEIGNIL, F., AMANO, H., and ISHIHARA, A. Selective removal of sulfur compounds in fuel oil by combination of oxidation and adsorption. *Preprints of Papers—American Chemical Society, Division of Petroleum Chemistry*, **2005**, 50, 430.
238. MA, X.L., ZHOU, A.N., and SONG, C.S. A novel method for oxidative desulfurization of liquid hydrocarbon fuels based on catalytic oxidation using molecular oxygen coupled with selective adsorption. *Catalysis Today*, **2007**, 123, 276.

Chapter 6

Water-Gas Shift Technologies

ALEX PLATON AND YONG WANG

Institute for Interfacial Catalysis, Pacific Northwest National Laboratory

6.1 INTRODUCTION

The water-gas shift (WGS) reaction is a reversible, exothermic reaction between carbon monoxide and water (steam) to produce hydrogen and carbon dioxide:

$$CO + H_2O \rightleftarrows CO_2 + H_2. \qquad (6.1)$$

This reaction is typically facilitated by a catalyst (shift catalyst) and is currently being practiced industrially in hydrogen-producing plants based on steam methane reforming. The WGS reaction allows for extra molar hydrogen to be produced from the reaction of carbon monoxide with water. Large-scale, industrial plants that produce hydrogen in this manner are usually integrated with a hydrogen-consuming process such as ammonia production or hydroprocessing of petroleum fractions.

Hydrogen is becoming an increasingly interesting energy carrier for fuel cell applications both at the industrial and the residential and mobile scales. As the currently installed hydrogen production capacity is almost entirely captured by large-scale, high-pressure industrial consumers,[1] an increased focus is placed on new production capacities fitted to smaller-scale applications. In the near future, the increasing hydrogen demands will most likely be met by fossil fuel-based technologies such as steam reforming, autothermal reforming, or catalytic partial oxidation;[1] almost invariably, these processes will produce significant quantities of carbon monoxide, which not only reduces hydrogen productivity but can also poison catalysts in downstream processes. For example, carbon monoxide is a poison for the iron catalyst used in ammonia synthesis and is equally undesirable in fuel cells based on proton exchange membrane, alkaline, and phosphoric acid technologies.[2] However, the CO content can be lowered through the WGS reaction, which produces 1 mol of hydrogen for each mole of CO converted.

Hydrogen and Syngas Production and Purification Technologies, Edited by Ke Liu, Chunshan Song and Velu Subramani
Copyright © 2010 American Institute of Chemical Engineers

Extensive information and several comprehensive reviews exist in the open literature regarding the WGS process, catalysts involved, and current research efforts.[3-7] This information is summarized below and is followed by a discussion of more recent research reports.

6.2 THERMODYNAMIC CONSIDERATIONS

The WGS reaction (Eq. 6.1) is slightly exothermic ($\Delta H = -41.16$ kJ/mol, gas phase) and is a typical example of reaction controlled by equilibrium, especially at higher temperatures. The equilibrium constant is a function of temperature. The reaction proceeds without change in the number of moles and in consequence pressure does not have any significant effect on equilibrium. For pressures between 10 and 50 bar, the following expressions are recommended for the equilibrium constant as a function of temperature:[5]

$$K_p = \begin{cases} \exp(-4.3701 + 4604/T[K]) \text{ at about } 250°C \\ \exp(-4.2939 + 4546/T[K]) \text{ at about } 440°C \\ \exp(-3.670 + 3971/T[K]) \text{ at } 750 - 1050°C. \end{cases} \qquad (6.2)$$

As can be seen in Figure 6.1, the equilibrium constant is approximately 1 at 800°C, about 10 at 415°C, 100 at 240°C, and 330 at 180°C. In order to achieve low CO and elevated hydrogen concentrations in the effluent, without having to use a significant excess of steam, it is desirable that the reaction be run at low temperatures. However, at low temperatures, reaction rates diminish and the reaction becomes kinetically controlled.

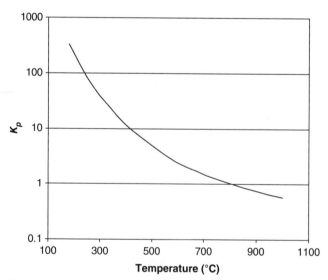

Figure 6.1. Variation of WGS equilibrium constant with temperature.[5]

6.3   INDUSTRIAL PROCESSES AND CATALYSTS

Traditionally, and in most industrial implementations, the WGS reaction is conducted in two steps. In the first step, CO concentration is reduced from 10% to 13% down to 2% to 3% (dry basis) on a chromium-promoted magnetite, high-temperature shift (HTS) catalyst. Most CO_2 that resulted from the reforming reaction is typically removed upstream of the HTS reactor by scrubbing, and a significant excess of steam is then added. This first WGS step operates adiabatically with an inlet temperature in the 350–550 °C range (depending on feed composition). Operating pressures range above 20 bar and sometimes exceed 30 bar (depending on the adjacent plant requirements). Gas hourly space velocities (GHSVs) of 400–1200 h^{-1}, or superficial contact times of 3–9 s (wet basis) are common values.

The second stage, low-temperature shift (LTS), takes advantage of the favorable equilibrium below 250 °C in order to reduce the CO concentration to 0.2%–0.4% in the presence of a Cu/ZnO or CoMo catalyst. With an adiabatic HTS temperature increase of about 50 °C, intermediate cooling is necessary prior to the LTS step. The lower temperature limit for this step, about 200 °C, is dictated by the water dew point under operating conditions, since any condensation would damage the catalyst. A typical dry-basis feed composition is 2%–3% CO, 20% CO_2, and 77%–78% H_2, and a steam-to-dry gas ratio of about 0.4 is practical. The adiabatic temperature rise along the reactor will normally be about 15 °C. Typical pressures ranging between 10 and 30 bar and GHSVs of about 3600 h^{-1} or superficial wet-basis contact times of about 1 s are common.

6.3.1   Ferrochrome Catalyst for HTS Reaction

The HTS process and catalyst were introduced by the German company BASF around 1915. The catalyst used nowadays is very similar to the original formulation and consists of Fe_3O_4 promoted with Cr_2O_3 to prevent sintering. The catalyst is inexpensive, starting from a fresh composition comprising 90% Fe_2O_3 and 10% Cr_2O_3 that is partially reduced *in situ* under a controlled atmosphere. Reduction is performed in a hydrogen and CO atmosphere and is controlled so that a spinel with composition $Fe^{II}Fe^{III}_{2-x}Cr_xO_4$ is formed. Iron carbide and metallic iron formation need to be avoided by using a large excess of steam, since the carbide would catalyze Fischer–Tropsch synthesis, while the metal would promote carbon deposition through the Boudouard reaction. The activated catalyst is pyrophoric and requires careful passivation before being exposed to air. This catalyst is tolerant to sulfur and chlorine compounds; however, sulfur may accumulate as FeS, which is less active and less stable physically. The catalyst is also hydrothermally stable at 400 °C and under 10% steam; however, more severe hydrothermal conditions will induce catalyst fragmentation. Fouling from coke buildup is observed when acetylenes, dienes, and other coke precursors are present in the feed stream from an incomplete reforming operation. Decoking can be performed at 450 °C under a steam atmosphere containing 1%–2% oxygen, with virtually complete recovery of the initial activity.

Under certain conditions, especially when feeds derived from coal gasification are used, irreversible arsenic poisoning of the ferrochrome catalyst is likely to occur. When operated within normal parameters, expected life for this catalyst ranges between 1 and 3 years.

6.3.2   CuZn Catalysts for LTS Reaction

The LTS step was introduced in the industrial practice in the early 1960s. The catalyst was initially derived from CuO, ZnO, and Al_2O_3 precursors (in a typical ICI commercial catalyst formulation, 32–33, 34%–53%, and 15–33 wt%, respectively). In other formulations, alumina is substituted by chromia. A typical composition includes $ZnO : Cr_2O_3 : CuO = 1:0.24:0.24$ with 2%–5% MnO_x, Al_2O_3, and MgO promoters. The catalyst needs to be reduced to form the active species Cu. The reduction reaction is a highly exothermic process and requires a strictly controlled atmosphere of hydrogen diluted in nitrogen, in order to avoid temperature excursions above 230–250 °C that are very likely to cause catalyst sintering. The activated catalyst is pyrophoric and requires passivation with liquid water before being exposed to oxygen. This catalyst is active at temperatures as low as 200 °C and is likely to be operated at mass transfer limited regions above this temperature. The activity was seen to pass through a maximum when the Cu/Zn ratio is varied, with the most favorable ratio ranging between 0.2 and 0.4 for the Cu/ZnO formulation. The CuZn catalyst is very sensitive to sulfur and chlorine poisoning at contaminant concentrations above 0.1 ppm. For this reason, a strict contaminant control is necessary. Because of its affinity to sulfur, ZnO acts as a trap, partially protecting the active sites from poisoning. Sintering is also a major problem, and under typical conditions, Cu particles would double their size over 6 months of operation. Chlorine acts as a sintering promoter and induces the loss of active sites via volatile Cu and Zn chlorides. Irreversible deactivation occurs above a certain temperature (example: 360 °C for a 15–20 wt% CuO, 68–73 wt% ZnO, 9–14 wt% Cr_2O_3 catalyst composition), but Mn- and B-based promoters were found to improve stability (up to 400 °C in the case of coprecipitated Mn). The typical catalyst life is 1–2 years when poisons are strictly controlled.

6.3.3   CoMo Catalyst for LTS Reaction

Sulfur-resistant Co–Mo/Al_2O_3 catalysts have been available commercially since the early 1970s for circumstances where it is necessary to carry out the WGS reaction on coal-derived and heavy petroleum-derived gas streams, with elevated sulfur contamination. An example of commercial as-synthesized formulation is 12.5% MoO_3 + 3.5% CoO on Al_2O_3 with a BET surface area of 270 m^2/g and a pore volume of 0.5 cm^3/g, impregnated with an optimal Cs loading of 0.35 g/cm^3 of pore volume. A suggested pretreatment atmosphere for this catalyst is 6% H_2S in H_2. The active catalyst form is sulfided CoMo, thus being invulnerable to the presence of sulfur compounds. In fact, a minimum ratio of H_2S/H_2O is required in order to preserve the active sulfide form. Sulfur compounds are converted to H_2S, which can subsequently be separated along with CO_2 in a simple, continuous, acid gas scrubbing operation. Alkali promoters are used to enhance activity; the typical promoters are K and Cs, with Cs being cited as the most effective one. The alumina support confers hydrothermal stability, making this class of catalysts suitable for both low- and intermediate-temperature WGS (200–450 °C). One of the few cited examples of operation lists pressures between 5 and 27 bar, temperatures between 250 and 300 °C, and space velocities of 4800–24,000 h^{-1}.[8] The typical catalyst life of the CoMo catalyst is about 2 years.

6.4 REACTION MECHANISM AND KINETICS

Some degree of controversy exists in the literature with respect to the reaction mechanism and the rate-determining steps during WGS catalyzed by the commercial catalysts listed above, especially for the CuZn catalyst. In the case of the HTS ferrochrome catalyst, Rhodes et al.[9] concluded that the most convincing published evidence favors mechanism that involves water dissociation in a redox step on Fe^{2+}/Fe^{3+}. However, for the LTS CuZn catalyst type, evidence suggests that copper active sites are responsible for both dissociative adsorption of water required by a redox mechanism and the formation of a COOH (formate) intermediate that is subsequently decomposed by similar sites.[9]

Significantly less mechanistic information is available for the CoMo WGS catalyst; this is mainly due to the lack of stability of the active sulfided species under experimental conditions employed by the mechanistic study. Hakkarainen and Salmi[10] and Hakkarainen and coworkers[11] observed a consistent loss of sulfur upon establishing the operating conditions for transient response experiments. Nevertheless, the information provided by those studies, as summarized below, is probably capturing the most important mechanistic features relevant to a practical kinetic model.

6.4.1 Ferrochrome Catalyst

Based on analysis of isotope exchange data and stoichiometric number analysis, Oki and Mezaki[12–14] corroborated two most probable mechanisms as shown below, involving the elementary steps in Equations 6.3–6.7 and 6.8–6.13, respectively, with the steps labeled I (CO adsorption) and V (hydrogen desorption) being rate limiting:

$$I: CO \rightarrow CO(a), \tag{6.3}$$

$$II: H_2O \rightarrow 2H(a) + O(a), \tag{6.4}$$

$$III: CO(a) + O(a) \rightarrow CO_2(a), \tag{6.5}$$

$$IV: CO_2(a) \rightarrow CO_2, \tag{6.6}$$

$$V: 2H(a) \rightarrow H_2, \tag{6.7}$$

$$I: CO \rightarrow CO(a), \tag{6.8}$$

$$II: H_2O \rightarrow H(a) + OH(a), \tag{6.9}$$

$$IIIa: CO(a) + OH(a) \rightarrow COOH(a), \tag{6.10}$$

$$IIIb: COOH(a) \rightarrow CO_2(a) + H(a), \tag{6.11}$$

$$IV: CO_2(a) \rightarrow CO_2, \tag{6.12}$$

$$V: 2H(a) \rightarrow H_2. \tag{6.13}$$

From transient response experiments, Hakkarainen et al.[11] found a probable mechanism involving the elementary steps (Eqs. 6.14–6.18) below:

$$I: CO + O^* \rightleftarrows CO_2^*, \tag{6.14}$$

$$II: CO_2^* \rightleftarrows CO_2 + {}^*, \tag{6.15}$$

$$III: H_2O + {}^* \rightleftarrows H_2O^*, \tag{6.16}$$

$$IV: H_2O^* + O^* \rightleftarrows 2OH^*, \tag{6.17}$$

$$V: 2OH^* \rightleftarrows H_2 + 2O^*, \tag{6.18}$$

where * symbolizes an active site. Evidence indicated that CO_2 and H_2 desorption in steps II and V were both rate limiting.

Newsome[3] analyzed the suitability of several rate equations for the HTS reaction on ferrochrome catalysts. The information analyzed pointed to a power rate law-type equation given in Equations 6.19 and 6.20,

$$r = k[CO]^l [H_2O]^m [CO_2]^n [H_2]^q (1 - \beta), \tag{6.19}$$

$$\beta = [CO_2][H_2]/(K[CO][H_2O]), \tag{6.20}$$

with k = rate constant, [] = concentrations, K = equilibrium constant, and exponents $l = 0.90$, $m = 0.25$, $n = -0.60$, and $q = 0$ being partial reaction orders. These values, along with an activation energy of 114.6 kJ/mol, provided acceptable agreement with experimental data at temperatures between 330 and 500 °C and atmospheric pressure. For pressures nearing 20 bar, the above exponents need to be adjusted to $l = 1.00$, $m = 0.25$, $n = -0.60$, and $q = 0$, while ln k increased with ln P linearly with a slope of 0.65. Furthermore, for pressures nearing 40 bar, exponent values of $l = 1.00$, $m = 0$, $n = -0.56 \ldots -0.3$, and $q = 0$ are cited.[3]

More recently, Lloyd et al.[15] pointed out that more theoretically derived Langmuir–Hinshelwood expressions similar to Equation 6.21 provide the best fit of experimental data:

$$r = \frac{kK_{CO}K_{H_2O}(P_{CO}P_{H_2O}P_{CO_2}P_{H_2})^2}{(1 + K_{CO}P_{CO} + K_{H_2O}P_{H_2O} + K_{CO_2}P_{CO_2} + K_{H_2}P_{H_2})^2}. \tag{6.21}$$

In conjunction with this equation form, an intrinsic activation energy value of 122 kJ/mol is listed.[5] Kinetic control is most prevalent in small catalyst pellets, tablets, or rings having approximate volumes up to ~0.3 cm^3 each and internal surface areas of 60–80 m^2/g.[5]

6.4.2 CuZn-Based Catalyst

As noted above, experimental evidence is consistent with both redox-type and for-mate-intermediate mechanisms for the CuZn-catalyzed LTS reaction. It is also noted that slight differences in experimental conditions from one study to another could be the basis of important differences in catalyst surface configuration, leading to different elementary reactions observed.[9]

Kinetic parameters were measured by Koryabkina et al.[16] on a Cu-based LTS catalyst under the conditions potentially relevant for fuel cell applications ($220\,^{\circ}C$, 1 atm, and a feed comprising 7% CO, 8.5% CO_2, 22% H_2O, 37% H_2, and 25% Ar by volume). The data suggested a redox-type mechanism shown in Equations 6.22–6.29, with the rate-determining step involving the formation of adsorbed CO_2 as shown in step VII:

$$\text{I: } H_2O + {}^* \rightleftarrows H_2O^*, \tag{6.22}$$

$$\text{II: } H_2O^* + {}^* \rightleftarrows OH^* + H^*, \tag{6.23}$$

$$\text{III: } 2OH^* \rightleftarrows H_2O + O^*, \tag{6.24}$$

$$\text{IV: } OH^* + {}^* \rightleftarrows O^* + H^*, \tag{6.25}$$

$$\text{V: } 2H^* \rightleftarrows H_2 + 2^*, \tag{6.26}$$

$$\text{VI: } CO + {}^* \rightleftarrows CO^*, \tag{6.27}$$

$$\text{VII: } CO^* + O^* \rightleftarrows CO_2^* + {}^*, \tag{6.28}$$

$$\text{VIII: } CO_2^* \rightleftarrows CO_2 + {}^*. \tag{6.29}$$

The kinetic expression as shown in Equations 6.19 and 6.20 was found applicable, with partial reaction orders $l = 0.8$, $m = 0.8$, $n = -0.7$, and $q = -0.8$ and an activation energy of 77.7 kJ/mol. The values for the exponents n and q suggest a strong inhibition effect form the reaction products H_2 and CO_2.[16]

Ovesen et al.[17] found the evidence of a surface redox mechanism and fitted experimental data with Equations 6.19 and 6.20, finding an activation energy of 78.2 kJ/mol at 20 atm and exponent values $l = 1.0$, $m = 1.2$, $n = -0.7$, and $q = -0.7$. Ayastuy et al.[18] fitted several kinetic models to experimental data obtained on a $Cu/ZnO/Al_2O_3$ catalyst at 3 bar and 180–$217\,^{\circ}C$. Two Langmuir–Hinshelwood-type models provided the best fit, with apparent activation energies of 50.1 and 57.5 kJ/mol. When fitted to a simple power law rate equation given again in Equations 6.19 and 6.20, an activation energy of 79.7 and partial reaction orders $l = 0.47$, $m = 0.72$, $n = -0.38$, and $q = -0.65$ were obtained.

6.4.3 CoMo Catalyst

Hakkarainen and Salmi[10] studied the mechanism and kinetics of WGS on a commercial $CoMo/Al_2O_3$ catalyst, using transient response experiments at $400\,^{\circ}C$. They concluded that the mechanism shown in Equations 6.30–6.32 was the most plausible, with the steps labeled II and III being the rate limiting steps:

$$\text{I: } H_2O + *O* \rightleftarrows 2OH*, \tag{6.30}$$

$$\text{II: } CO + 2OH* \rightleftarrows H_2 + CO(O*)_2, \tag{6.31}$$

$$\text{III: } CO(O*)_2 \rightleftarrows CO_2 + *O*. \tag{6.32}$$

The authors derived a rate expression given in Equations 6.33 and 6.34. Data at 400 °C resulted in a fitted value for k_{H2} = 22.2 mol/(m² atm s):

$$1/r = 1/(k_{H2}P_{CO}) + \alpha', \tag{6.33}$$

$$\alpha' = 1/(k_{H2}K_{H2O}P_{CO}P_{H2O}) + 1/k_{CO2}. \tag{6.34}$$

Park et al.[19] analyzed the effect of K addition to a 3 wt % Co–10 wt % Mo/γ-Al$_2$O$_3$ catalyst on the kinetic parameters obtained by fitting the WGS experimental data at 400 °C. They used a simple, first-order rate expression in CO (Eq. 6.35) and derived activation energy values of 43.1 and 47.8 kJ/mol for the unpromoted and promoted (5 atom % K relative to Mo) catalysts, respectively,

$$-\ln\frac{X_{CO} - X_{CO}^{eq}}{X_{CO}^0 - X_{CO}^{eq}} = k\frac{V}{F}, \tag{6.35}$$

where V = catalyst volume, F = flow rate of reactants, k = first-order rate constant, X_{CO}^0 = initial CO mole fraction, X_{CO} = final CO mole fraction, and X_{CO}^{eq} = CO mole fraction at equilibrium.

6.5 CATALYST IMPROVEMENTS AND NEW CLASSES OF CATALYSTS

While the commercial catalysts and technologies described above are successfully applied in the industry, some major drawbacks exist with these catalysts such as the low activity of the otherwise robust ferrochrome catalyst at low temperatures, and the susceptibility to poisoning and sintering of the CuZn shift catalyst. Additionally, both classes of catalysts are pyrophoric, generating serious safety problems in the case of accidental air exposure. Furthermore, both catalysts require a special, carefully controlled activation treatment in order to achieve the optimal active phase configuration, with the CuZn catalyst being particularly sensitive to accidental shutdowns, accidental water condensation, or temperature or concentration transients.

6.5.1 Improvements to the Cu- and Fe-Based Catalysts

Despite the extensive list of potential drawbacks with existing commercial catalysts, few new WGS catalyst formulations have been commercialized in the past decades. Based on the data compiled by Armor,[20,21] no new WGS catalysts were commercialized during the 1980s, while only two new catalysts were developed during the 1990s by the Süd-Chemie Group: a ferrochrome catalyst promoted with Cu was developed for enhanced activity with suppressed Fischer–Tropsch by-product formation, and a CuZn catalyst promoted with an undisclosed dopant claimed to exhibit improved stability and suppressed methanol by-product formation.

Also notable in the direction of classical catalyst improvement is the research conducted by Rhodes et al.[22] A series of coprecipitated promoters were evaluated on ferrochrome catalyst activity at temperatures between 350 and 440 °C. It was found that activity decreases in the following order: Hg > Ag > Ba > Cu > Pb > unpromoted > B. A noticeable compensation effect observed in the correlation between preexponential factors and apparent activation energies led the authors to conclude that these promoters might only influence the CO adsorption on catalyst rather than the course of surface reactions.

In a recent patent issued to Süd-Chemie,[23] an improvement to the $CuZn-Al_2O_3$ catalyst is described in which substitution of about 20% of the total Al with a (Cu,Zn)-hydrotalcite precursor containing a similar amount of aluminum improved both Cu dispersion and its activity.

6.5.2 New Reaction Technologies

As stated earlier, an increased interest is currently being concentrated on smaller-scale hydrogen production for fuel cell applications ranging from industrial to residential and mobile implementations. An especially important emphasis has been placed on the design of on-board fuel processors capable of generating the hydrogen necessary to power vehicles. Another application of the mobile fuel processor under development is the portable processor/fuel cell system capable of powering electronic devices with liquid hydrocarbon fuels, as replacement for batteries.[24–26]

To meet the requirement of these unique applications, new reaction technologies are being developed in the following areas: identification of WGS process configurations, which provide a high level of integration with reformer and fuel cell units including WGS membrane reactors that can produce pure and fuel cell-grade hydrogen (Fig. 6.2); process intensification in microchannel reactors that operate at optimal temperatures; and oxygen-assisted WGS that combines preferential oxidation (Prox) of CO with the WGS reaction to achieve very low CO levels compatible with fuel cell operation.

Li et al.[27] recently reported relatively active $Cu-Ce(La)O_x$ and $Ni-Ce(La)O_x$ catalysts showing a strong metal–support interaction at a 2–3 wt % loading of metals. Notably, the Cu-based catalyst was active and thermally stable up to 600 °C with no prior activation treatment. Further investigations by the same group on a 10% $Cu-Ce(8\% La)O_x$ catalyst at 300–450 °C, 1 atm, and 80,000 h^{-1} space velocity showed a higher activity than the traditional ferrochrome catalyst.[28] These authors further evaluated the Cu–ceria catalyst under CO_2-rich conditions that are representative of hydrogen-permeable membrane reactor applications. Under these conditions, the ferrochrome catalyst would be deactivated by excess CO_2, whereas the Cu–ceria catalyst under investigation showed remarkable stability and durability, and no carbon formation was detected. For H_2-permeable membrane reactor applications, Ma and Lund[29] demonstrated that for a typical Fe–Cr catalyst in a Pd-based

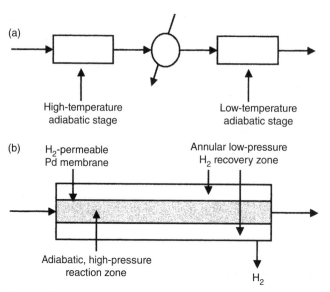

Figure 6.2.   Process configuration for the traditional, two-stage WGS method (a) and for a hydrogen-permeable membrane adiabatic reactor (b). Reproduced from Ma and Lund.[29]

membrane reactor operated at 415 °C, the process would be limited by the global reaction rate rather than the membrane permeation rate, and that the removal of the inhibitory effect of CO_2 on the catalyst would improve the hydrogen yield to a much greater extent than improving the membrane material. However, the partial reaction orders obtained by Qi and Flytzani-Stephanopoulos[28] from power rate law kinetic data fitting ($l = 0.81$, $m = 0.18$, $n = -0.3$, $q = -0.3$) indicated that CO_2 kinetic inhibition is somewhat less important, given the relatively low partial order for the products.

A different approach at scaling down the WGS for mobile and portable fuel cell applications consists of process intensification in microstructured reactors, in which heat and mass transfer limitations are virtually removed, leaving the intrinsic kinetics and thermodynamic equilibrium as the only constrains.[30] Tonkovich et al.[31] used monolith Ru/ZrO_2 catalysts prepared by impregnation of ZrO_2-coated Ni foam with $RuCl_3$ in a preliminary side-by-side comparison with a powdered, commercial Ru/ZrO_2 catalyst. The very fast intrinsic WGS kinetics at 300 °C allowed very short contact times (≥ 25 ms) under which the observed CO conversion (99.8%) was in close proximity to the equilibrium conversion (99.93%), with no noticeable methanation side reactions. In a separate study by Goerke et al.,[32] Ru/ZrO_2 deposited on fecralloy and stainless steel was compared with Au/CeO_2 in a microchannel WGS reactor. It was found that the fecralloy-supported Ru–zirconia was the most active at low temperatures. Between 200 and 275 °C, CO conversion increased from ~18% to ~95% at a residence time of 20 ms with no noticeable methanation. The catalyst durability was studied for at least 100 h with no reported deactivation.

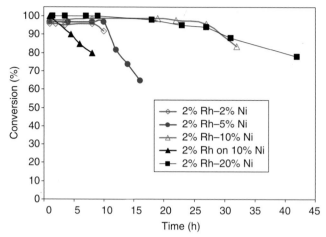

Figure 2.15. Effect of Ni loading on catalytic performance of 2% Rh/CeO$_2$–Al$_2$O$_3$ catalysts in the steam reforming of NORPAR-13 surrogate jet fuel. Adapted from Strohm et al.

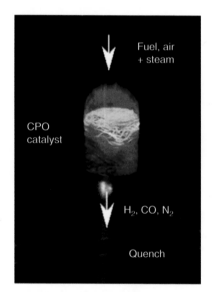

Figure 3.5. Picture of a working CPO catalyst.

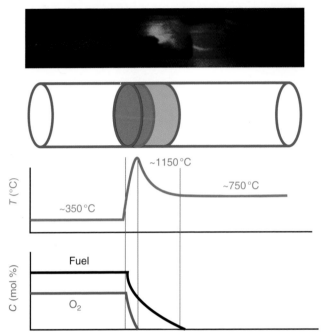

Figure 3.6. Speculated temperature profile of a CPO catalyst bed based on the observation from Figure 3.5.

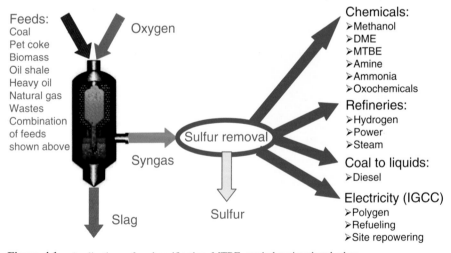

Figure 4.1. Applications of coal gasification. MTBE, methyl tertiary butyl ether.

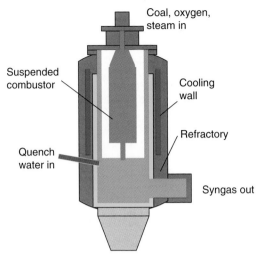

Figure 4.21. Schematics of a typical GSP plus gasifier.

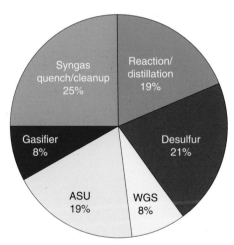

Figure 4.31. CAPEX (%) of a typical 200 kt/year methanol plant. WGS, water-gas shift.

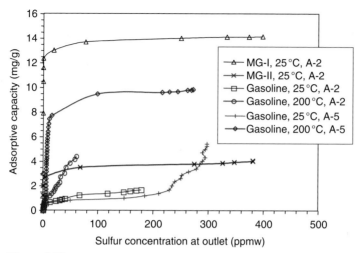

Figure 5.13. Adsorptive capacity of nickel-based adsorbents as a function of sulfur concentration at outlet.

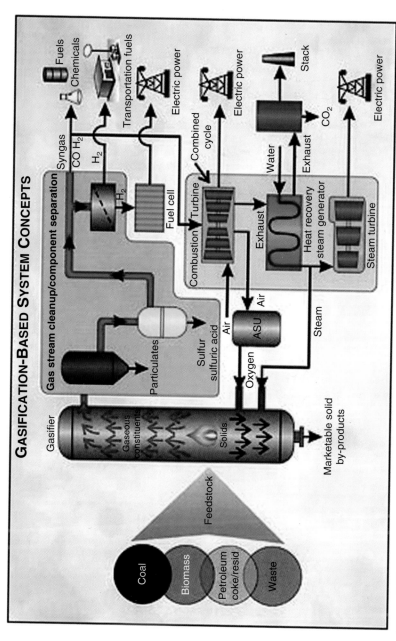

Figure 5.22. Simple scheme of DOE coal gasification power plant.

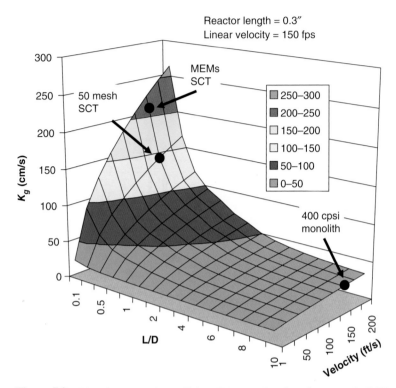

Figure 7.9. Plot of mass transfer coefficient (K_g) versus length-to-diameter ratio (L/D) versus velocity.

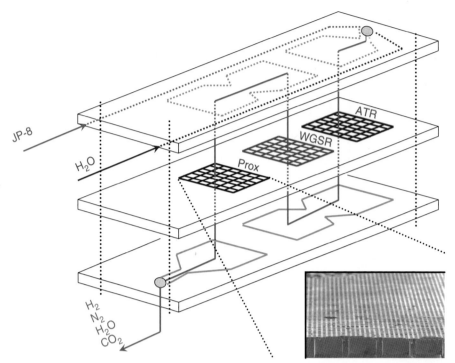

Figure 7.12. MEMs SCT fuel processor. ATR, Autothermal Reformer; WGSR, Water Gas Shift Reactor.

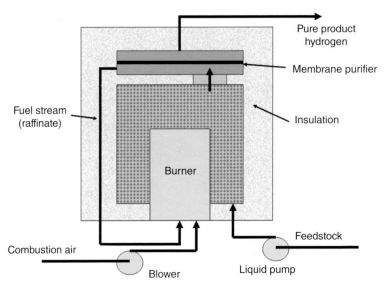

Figure 8.11. Schematic showing integration of a hydrogen-selective membrane with a steam reformer.

Figure 10.2. Aerial photo of Air Products and Chemicals, Inc.; an 80 MMSCFD H_2 plant in Pasadena, TX (SMR H_2 facility and 10-bed H_2 PSA).

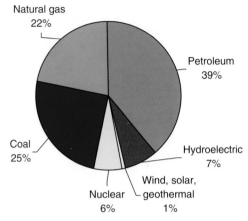

Figure 11.1. Worldwide energy sources distribution.

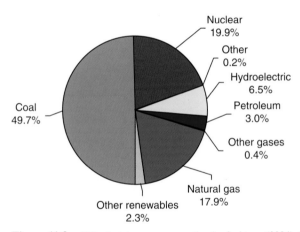

Figure 11.3. U.S. electric power generation by fuel type (2004); http://www.eia.doe.gov/fuelectric.html.

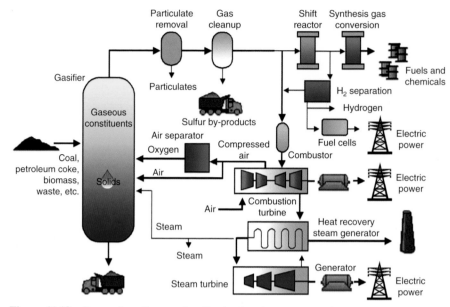

Figure 11.19. Process flow diagram of gasification-based energy conversion system.

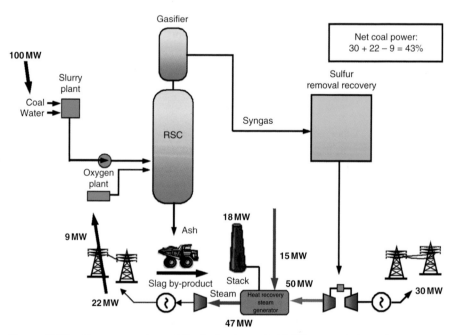

Figure 11.23. Energy conversion and distribution in a typical IGCC process. RSC, radiant syngas cooler.

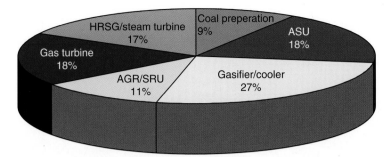

Figure 11.24. IGCC plant capital cost breakdown. Total TPC = $1438/kWe.

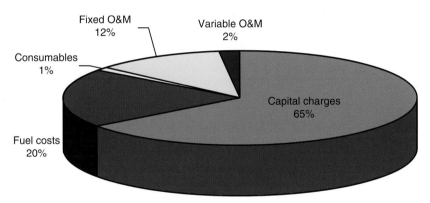

Figure 11.25. Breakdown of cost of electricity of an IGCC plant.

Figure 12.1.   Shenhua CTL plant (China). Reprinted from Chen et al.

Benefiting from the assistance of oxygen to simultaneously remove CO by the Prox reaction, oxygen-assisted WGS on bimetallic Cu–Pd/CeO$_2$ reported by Bickford et al.[33] have achieved an increased CO conversion from 95% (without oxygen) to more than 99.7% (with oxygen). The results were reported for a feed composition of 4 vol% CO, 2% O$_2$, 40% H$_2$O, 10% CO$_2$, and 42% H$_2$ at 210 °C and a GHSV of 17,760 h^{-1}.

6.5.3 New Classes of Catalysts

Scaling down the commercial WGS catalysts to the fuel cell application scale is a nontrivial task given the safety concerns drawn from their pyrophoricity, stability issues drawn from their sensitivity to shutdown/start-up cycles and transients, or the large catalyst volumes necessary due to their low activity at low temperatures where equilibrium is favorable. Small-scale H$_2$ production applications are driving the current research on WGS catalyst innovation. In the identification of new, stable, and nonpyrophoric catalysts, a trade-off between the activity and the cost of the catalyst may exist. Several promising catalyst formulations based on precious and nonprecious metals have been identified. While nonprecious supported metal catalysts have the advantage of a low cost, they usually exhibit acceptable activity only at elevated temperatures where thermodynamic equilibrium is unfavorable. On the other hand, catalysts based on platinum-metal group metals or gold exhibit acceptable activity over a wide range of temperatures, but their cost can become prohibitive for larger-scale applications or mass production.

A number of promising new catalysts have been reviewed recently by Ghenciu,[6] Song,[2] and Farrauto et al.[7] Some of the most recent developments will be reviewed in the discussion below with special emphasis on catalyst performance compared with the classical HTS and LTS catalysts.

6.5.3.1 Nonprecious Metal Catalysts

Among the new nonprecious metal formulations, molybdenum carbide is distancing itself as a material known to have catalytic properties that resemble those of noble metals. In a study performed by Patt et al.[34] on Mo$_2$C catalysts, they found that this material was more active than the traditional Cu–Zn–Al LTS catalyst at temperatures between 220 and 295 °C, atmospheric pressure and space velocities in excess of 10,000 h^{-1}. Additionally, no evidence of methanation activity was observed. No deactivation and no structure change were observed within 48 h of operation. An advantage for compact fuel cell applications, aside from the higher activity, is the higher carbide catalyst density compared with the Cu–Zn catalyst; this is of particular importance in designs where a minimal catalyst bed volume is required. Moon and Ryu[35] studied a similar catalyst in a thermal cycling experiment employing room-temperature exposure to air followed by reduction and reaction at 250 °C. Although the catalyst deactivated by Mo oxidation to MoO$_3$ following repeated cycling, it maintained a better performance and decayed slower than the traditional

Cu–Zn–Al that was subjected to a similar treatment. Most recently, Nagai and Matsuda[36] studied the effect of cobalt promotion on the Mo_2C catalyst at 180 °C. At such a low temperature, the authors reported significantly a lower activity of the unpromoted Mo_2C compared with the traditional Cu–Zn–Al catalyst. Also, deactivation is evident over the first 300 min onstream, with activity dropping by about 75%. Cobalt promotion with a Co/Mo ratio of 1:1 doubled the activity resulting in a catalyst that was slightly more active than CuZn-based catalysts. In addition, catalyst stability was improved as evidenced by only a 33% drop in activity after 300 min onstream. However, the stability is not as good as the CuZn-based catalyst, which maintained fairly stable performance under identical conditions. The authors suggested that amorphous Co–Mo oxycarbide is most likely the active phase in the promoted catalyst.

The promoting effect of cobalt was evidenced earlier by Hutchings and coworkers[37] in a comparative study on chromium oxide and manganese oxide. Major findings from their study showed that Co–Cr with a ratio of 3:1 was the most active and durable, maintaining ca. 98% conversion over 600 h onstream at 400 °C with 58% CO in the feed stream and a CO GHSV of $510 h^{-1}$. The Co–Mn catalyst performed in an almost similar fashion, with evidence pointing to a reaction intermediate of the formate type on these catalysts. Most notably, both catalysts operated with no deactivation in the presence of 240 ppm of sulfur impurities in the gas feed stream, making them promising catalysts for the WGS of coal- and heavy petroleum-derived reformer product streams.

Ceria has been used in numerous catalytic applications requiring strong support–metal interaction and enhanced oxygen mobility. As discussed above, the Flytzani-Stephanopoulos group studied ceria–lanthana-supported Cu and Ni catalysts and concluded that a strong interaction between metal clusters and the support attainable at an optimal metal loading contributed to the surface-enhanced catalytic activity and thermal stability. It was found that a cooperative redox mechanism existed for both catalysts, that is, ceria-supplied oxygen reacts with metal-adsorbed CO, while ceria recovers the oxygen from water that adsorbs on the oxygen vacancy.[27] A 10% $Cu–Ce(8\% La)O_x$ catalyst showed an almost fourfold improvement in CO conversion (85%) at 300 °C, 1 atm, and a space velocity of $80,000 h^{-1}$ when compared with a commercial Fe–Cr catalyst.[28] An initial deactivation was observed on the catalysts with high loadings of Cu and La. Hilaire et al.[38] compared the WGS activity behavior of several ceria-supported precious and base metal catalysts under differential conditions and a wide range of temperatures. It was found that both Ni/CeO_2 and Pd/CeO_2 catalysts were about one order of magnitude less active than a commercial Cu/ZnO/alumina LTS catalyst. Co/CeO_2 and Fe/CeO_2 catalysts were even less active, about one order of magnitude less active than Ni/CeO_2 catalysts. One implication of their results is that, while the strength of the metal–ceria interaction is important for the assumed redox mechanism in which oxygen supplied by ceria will react with metal-adsorbed CO, it is also important that the metal is not readily oxidized under reaction conditions in order to stabilize CO adsorption.[38]

There are some recent literature reports about new commercial catalysts designed to overcome some of the drawbacks associated with the traditional WGS catalysts.

For example, a nonpyrophoric and proprietary catalyst has been developed by Engelhard as a replacement for the CuZn-based LTS catalyst. This catalyst also possesses features such as direct activation under process gas conditions, similar activity as the traditional catalyst, no safety and stability concerns upon air exposure, and good stability against temperature cycling and aging.[39]

6.5.3.2 Precious Metal Catalysts

Platinum-group metals (PGMs) attracted early interest as WGS catalysts due to their nonpyrophoric nature, high activity over a wide range of temperatures, and excellent stability at high temperatures. This became especially true when Pt/CeO_2 was recognized as a good catalyst for WGS.[40] NexTech Materials[41] developed Pt/CeO_2 catalysts by depositing Pt on nanometer-sized ceria particles. These catalysts were found to be more active than the Cu–Zn-type catalyst at temperatures above 270 °C.

Extensive research has also been conducted on the elucidation of reaction and deactivation mechanisms on the PMG–ceria-based catalysts. Bunluesin et al. reported in 1998 that substitution of alumina for ceria in supported Pt, Pd, and Rh catalysts improved their WGS activities by about two orders of magnitude, mainly due to the increased ceria surface area by alumina substitution.[42] The enhancement in activity was also attributed to the oxygen storage capacity of ceria, facilitating a redox type of mechanism. The observed deactivation was initially ascribed to ceria crystallite growth. Further research proved that Pd/ceria and Ni/ceria, although almost one order of magnitude less active than the traditional Cu–Zn catalyst, were much more active than Co–ceria and Fe–ceria catalysts, owing this behavior to the stabilization of CO adsorption on Pd.[38]

More insight into the mechanisms of PMG/ceria catalyst deactivation was provided by data indicating that surface carbonate species are bound to reduced ceria sites in the spent catalysts, possibly limiting the rate of ceria reoxidation. Kinetic data was found to support the assumed redox mechanism, and fitting of experimental data with power rate law equations yielded partial reaction orders of 0 for CO, 0.5 for H_2O, -0.5 for CO_2, and -1 for H_2. These values strongly suggested that the metal surface is saturated with CO under reaction conditions, which can effectively limit the overall reaction rate at elevated CO feed concentrations.[38] In line with the mentioned evidence of carbonate deactivation comes a recent study by Deng and Flytzani-Stephanopoulos[43] who explored the effect of shutdown/start-up cycles on the Pt–ceria and Au–ceria catalysts. The authors found evidence of severe deactivation following such cycling, whereas the addition of a small amount of oxygen (1%) into the gas stream enhanced the stability both under the steady-state WGS operation and the shutdown/start-up transients. Analysis of the deactivated catalysts revealed the presence of cerium (III) hydroxycarbonate, which could be avoided in the presence of 700–900 ppm of oxygen. Liu et al.[44] studied the behavior of Pt–ceria catalysts under simulated shutdown conditions and also found evidence of deactivation from carbonate formation both on the support and on the noble metal. Their data suggested that both CO and CO_2 were responsible for carbonate formation. Regeneration in air at 450 °C recovered the activity to some extent and no metal sintering was observed.

Wang et al.[45] subjected Pd/ceria and Pt/ceria catalysts to accelerated aging under WGS conditions and concluded that the presence of CO caused metal particle growth and subsequent reduction of the number of sites available for CO adsorption during the WGS reaction. The authors concluded that methods generally used for dispersion stabilization should also apply for this class of catalysts. Quite consistent with this conclusion are the results from a study performed by leaching the precious metal prior to activity testing on Pt/ceria and Au/ceria.[46] These results suggested that the metal nanoparticles were not involved in the reaction, but rather precious metal ions associated with the ceria surface were instead responsible for catalyst activity. Based on this finding, methods designed to enhance the populations of metal ions strongly associated with the ceria lattice should lead to enhanced activity.

A major improvement in the activity for this class of catalysts was made by Wang and Gorte[47] when it was discovered that Fe promotion of Pd/ceria resulted in one order of magnitude increase in measured reaction rates, making the Fe–Pd/ceria catalyst as active as the traditional Cu–Zn catalyst at low temperatures (Fig. 6.3). The most significant effect came from one monolayer of Fe_2O_3. Further investigation by the same group into the promotional effect of Fe on Pt, Pd, and Rh supported by ceria[48] prompted to the evidence that the promotion effect of Fe was specific to Pd through the formation of Pd–Fe alloy.

Recently, a new class of WGS catalysts based on gold has attracted increased interests after Andreeva and coworkers showed an enhanced catalytic activity at low temperatures on supported gold catalysts.[49] It was found that the preparation method had a strong effect on activity, and Au/Fe_2O_3, Au/TiO_2, and Au/ZrO_2 had similar activity profiles in the WGS reaction at low temperatures. Au/TiO_2 could achieve 50% CO conversion at temperatures as low as 130 °C. Under standard testing conditions, these catalysts showed better activity than the commercial CuZn-based catalyst. Enhanced activity was attributed to the very small Au particles,

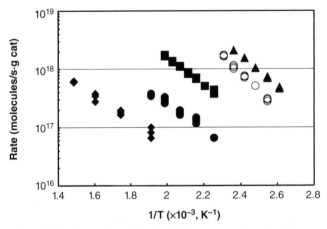

Figure 6.3. Comparison of differential rates of WGS reaction for a commercial Cu/ZnO (▲), Pd/Fe–ceria (○), Pd/ceria (■), Pd/Mo–ceria (●), and ceria (◆). Reproduced from Wang and Gorte.[47]

strong Au/support interaction, and uniform distribution of the metal on support. Optimal Au crystallite size was found to range between 3 and 5 nm. For these catalysts, a mechanism involving H_2O dissociation on Au and spillover on the support was proposed.

While some stability issues have been identified on the supported Au catalysts, a major progress was made by the discovery that Au/CeO_2 showed remarkable performance stability.[50] Among the catalysts with several Au loadings (Fig. 6.4), it was found that 3% Au showed the most stable operation over a 3-week period of tests due to an optimal ratio of surface Au and ceria sites. Evidence of strong metal–support interaction was correlated with the enhanced reducibility of ceria in the presence of Au nanoparticles. Further research into different catalyst preparation methods for Au/CeO_2 showed that Au dispersion and the WGS activity are extremely sensitive to minor variations in the preparation procedure.[51]

As discussed above, the actual sites responsible for WGS activity in Au/ceria are metal ions associated with the ceria lattice.[46] Therefore, catalyst preparation methods could result in highly variable interactions between support and metal, leading to significantly different catalytic activities. Also mentioned before was the

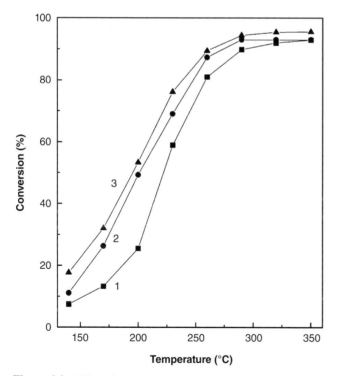

Figure 6.4.   Effect of gold loading on ceria upon Au/CeO_2 catalyst activity. Data were obtained at 12,000 h^{-1} space velocity, atmospheric pressure; Au loading: (1) 1 wt%; (2) 3 wt%; (3) 5 wt%. Reproduced from Andreeva et al.[50]

beneficial effect of low concentrations of oxygen during the WGS reaction on Au/ceria, as well as during shutdown/start-up cycling.[43] Evidence suggesting that the catalyst deactivates through surface carbonate formation can be corroborated with the results from Kim and Thompson[52] reinforcing the deactivation model in which oxygen-deficient sites in ceria promote carbonate and formate intermediates that block the normal reaction WGS cycles. Catalysts initially four times as active as a standard Cu–Al–Zn catalyst decayed to about half the initial activity within 10 h onstream. Following regeneration in air at 400 °C for 4 h, the catalysts fully recovered the activity with no measurable loss in surface area or metal dispersion.[52]

Coupling between the WGS and Prox reactions was suggested in a few instances.[33,43] Since Au/ceria is a recognized CO partial oxidation catalyst, such an approach would enhance at least CO conversion with oxygen. As shown above, this approach has proven itself feasible with bimetallic Cu–Pd deposited on nanosized ceria.

Although the catalysts discussed above are nonpyrophoric and show promising activity at low temperatures where WGS equilibrium is favorable to CO elimination, these precious metal catalysts remain relatively vulnerable to sulfur poisoning. Studies on investigating the catalyst behavior in the presence of sulfur-containing feeds for a series of catalysts showed that Pt/ZrO$_2$ is a potentially promising catalyst. Although some activity of this catalyst was lost upon the addition of 50 ppm sulfur, the initial activity was recovered promptly either with hydrogen or in a sulfur-free feed stream.[53] This catalyst was also more active than the ferrochrome at intermediate temperatures, and the most active under sulfur-containing feed conditions.

REFERENCES

1. ARMOR, J.N. Catalysis and the hydrogen economy. *Catalysis Letters*, **2005**, 101, 131.
2. SONG, C. Fuel processing for low-temperature and high-temperature fuel cells. Challenges and opportunities for sustainable development in the 21st century. *Catalysis Today*, **2002**, 77, 17.
3. NEWSOME, D.S. The water-gas shift reaction. *Catalysis Reviews—Science and Engineering*, **1980**, 21, 275.
4. SATTERFIELD, C.N. *Heterogeneous Catalysis in Industrial Practice*, 2nd ed. Malabar: Krieger Publishing, pp. 442–446, **1996**.
5. FARRAUTO, R.J. and BARTHOLOMEW, C.H. *Fundamentals of Industrial Catalytic Processes*. New York: Blackie Academic & Professional, pp. 357–360, **1997**.
6. GHENCIU, F.A. Review of fuel processing catalysts for hydrogen production in PEM fuel cell systems. *Current Opinion in Solid State and Materials Science*, **2002**, 6, 389.
7. FARRAUTO, R., HWANG, S., SHORE, L., RUETTINGER, W., LAMPERT, J., GIROUX, T., LIU, Y., and ILINICH, O. New material needs for hydrocarbon fuel processing: Generating hydrogen for the PEM fuel cell. *Annual Review of Materials Research*, **2003**, 33, 1.
8. SRIVATSA, N.R. and WELLER, S.W. Water-gas shift kinetics over sulfided catalyst: elevated pressure. In *Proceedings, 9th International Congress on Catalysis* (eds. M.J. Phillips, M. Ternan). Ottawa: The Chemical Institute of Canada, p. 1827, **1988**.
9. RHODES, C., HUTCHINGS, G.J., and WARD, A.M. Water-gas shift reaction: Finding the mechanistic boundary. *Catalysis Today*, **1995**, 23, 43.
10. HAKKARAINEN, R. and SALMI, T. Water-gas shift reaction on a cobalt-molybdenum oxide catalyst. *Applied Catalysis. A, General*, **1993**, 99, 195.

11. HAKKARAINEN, R., SALMI, T., and KEISKI, R.L. Comparison of the dynamics of the high-temperature water-gas shift reaction on oxide catalysts. *Catalysis Today*, **1994**, 20, 395.

12. OKI, S. and MEZAKI, R. Identification of rate-controlling steps for the water-gas shift reaction over an iron oxide catalyst. *The Journal of Physical Chemistry*, **1973**, 77, 447.

13. OKI, S. and MEZAKI, R. Mechanistic structure of the water-gas shift reaction in the vicinity of chemical equilibrium. *The Journal of Physical Chemistry*, **1973**, 77, 1601.

14. MEZAKI, R. and OKI, S. Locus of the change in the rate-determining step. *Journal of Catalysis*, **1973**, 30, 488.

15. LLOYD, L., RIDLER, D.E., and TWIGG, M.V. The water-gas shift reaction. In *Catalysis Handbook* (ed. M.V. Twigg). London: Wolfe, pp. 283–338, **1989**.

16. KORYABKINA, N.A., PHATAK, A.A., RUETTINGER, W.F., FARRAUTO, R.J., and RIBEIRO, F.H. Determination of kinetic parameters for the water-gas shift reaction on copper catalysts under realistic conditions for fuel cell applications. *Journal of Catalysis*, **2003**, 217, 233.

17. OVESEN, C.V., CLAUSEN, B.S., HAMMERSHØI, B.S., STEFFENSEN, G., ASKGAARD, T., CHORKENDORFF, I., NØRSKOV, J.K., RASMUSSEN, P.B., STOLTZE, P., and TAYLOR, P. A microkinetic analysis of the water–gas shift reaction under industrial conditions. *Journal of Catalysis*, **1996**, 158, 170.

18. AYASTUY, J.L., GUTIERREZ-ORTIZ, M.A., GONZALEZ-MARCOS, J.A., ARANZABAL, A., and GONZALEZ-VELASCO, J.R. Kinetics of the low-temperature WGS reaction over a $CuO/ZnO/Al_2O_3$ catalyst. *Industrial & Engineering Chemistry Research*, **2005**, 44, 41.

19. PARK, J.N., KIM, J.H., and LEE, H.I. A study on the sulfur-resistant catalysts for water gas shift reaction IV. Modification of $CoMo/\gamma\text{-}Al_2O_3$ catalyst with K. *Bulletin of the Korean Chemical Society*, **2000**, 21, 1239.

20. ARMOR, J.N. New catalytic technology commercialized in the USA during the 1980's. *Applied Catalysis*, **1991**, 78, 141.

21. ARMOR, J.N. New catalytic technology commercialized in the USA during the 1990s. *Applied Catalysis. A, General*, **2001**, 222, 407.

22. RHODES, C., WILLIAMS, B.P., KING, F., and HUTCHINGS, G.J. Promotion of Fe_3O_4/Cr_2O_3 high temperature water gas shift catalyst. *Catalysis Communications*, **2002**, 3, 381.

23. CAI Y., DAVIES S.L., and WAGNER J.P. Water-gas shift catalyst. U.S. Patent 6,693,057, **2004**.

24. PALO, D.R., HOLLADAY, J.D., ROZMIAREK, R.T., GUZMAN-LEONG, C.E., WANG, Y., HU, J., CHIN, Y.-H., DAGLE, R.A., and BAKER, E.G. Development of a soldier-portable fuel cell power system. Part I: A bread-board methanol fuel processor. *Journal of Power Sources*, **2002**, 108, 28.

25. HU, J., WANG, Y., VANDERWIEL, D., CHIN, C., PALO, D., ROZMIAREK, R., DAGLE, R., CAO, J., HOLLADAY, J., and BAKER, E. Fuel processing for portable power applications. *Chemical Engineering Journal*, **2003**, 93, 55.

26. HOLLADAY, J.D., WANG, Y., and JONES, E. Review of developments in portable hydrogen production using microreactor technology. *Chemical Reviews*, **2004**, 104, 4767.

27. LI, Y., FU, Q., and FLYTZANI-STEPHANOPOULOS, M. Low-temperature water-gas shift reaction over Cu- and Ni-loaded cerium oxide catalysts. *Applied Catalysis. B, Environmental*, **2000**, 27, 179.

28. QI, X. and FLYTZANI-STEPHANOPOULOS, M. Activity and stability of $Cu\text{-}CeO_2$ catalysts in high-temperature water-gas shift for fuel-cell applications. *Industrial & Engineering Chemistry Research*, **2004**, 43, 3055.

29. MA, D. and LUND, C.R.F. Assessing high-temperature water-gas shift membrane reactors. *Industrial & Engineering Chemistry Research*, **2003**, 42, 711.

30. BROOKS, K.P., DAVIS, J.M., FISCHER, C.M., KING, D.L., PEDERSON, L.R., RAWLINGS, G.C., STENKAMP, V.S., TEGROTENHUIS, W., WEGENG, R.S., and WHYATT, G.A. Fuel reformation: Microchannel reactor design. In *ACS Symposium Series Vol. 914: Microreactor Technology and Process Intensification* (eds. Y. Wang, J.D. Holladay). Washington, DC: American Chemical Society, pp.238–257, **2005**.

31. TONKOVICH, A.Y., ZILKA, J.L., LAMONT, M.J., WANG, Y., and WEGENG, R.S. Microchannel reactors for fuel processing applications. I. Water gas shift reactor. *Chemical Engineering Science*, **1999**, 54, 2947.

32. GOERKE, O., PFEIFER, P., and SCHUBERT, K. Water gas shift reaction and selective oxidation of CO in microreactors. *Applied Catalysis. A, General*, **2004**, 263, 11.

33. BICKFORD, E.S., VELU, S., and SONG, C. Nano-structured CeO$_2$ supported Cu-Pd bimetallic catalysts for the oxygen-assisted water-gas-shift reaction. *Catalysis Today,* **2005**, 99, 347.

34. PATT, J., MOON, D.J., PHILLIPS, C., and THOMPSON, L. Molybdenum carbide catalysts for water-gas shift. *Catalysis Letters,* **2000**, 65, 193.

35. MOON, D.J. and RYU, J.W. Molybdenum carbide water-gas shift catalyst for fuel cell-powered vehicles applications. *Catalysis Letters,* **2004**, 92, 17.

36. NAGAI, M. and MATSUDA, K. Low-temperature water-gas shift reaction over cobalt-molybdenum carbide catalyst. *Journal of Catalysis,* **2006**, 238, 489.

37. HUTCHINGS, G.J., COPPERTHWAITE, R.G., GOTTSCHALK, F.M., HUNTER, R., MELLOR, J., ORCHARD, S.W., and SANGIORGIO, T. A comparative evaluation of cobalt chromium oxide, cobalt manganese oxide, and copper manganese oxide as catalysts for the water-gas shift reaction. *Journal of Catalysis,* **1992**, 137, 408.

38. HILAIRE, S., WANG, X., LUO, T., GORTE, R.J., and WAGNER, J. A comparative study of water-gas-shift reaction over ceria supported metallic catalysts. *Applied Catalysis. A, General,* **2001**, 215, 271.

39. RUETTINGER, W., ILINICH, O., and FARRAUTO, R.J. A new generation of water gas shift catalysts for fuel cell applications. *Journal of Power Sources,* **2003**, 118, 61.

40. MENDELOVICI, L. and STEINBERG, M. Methanation and water-gas shift reactions over Pt/CeO$_2$. *Journal of Catalysis,* **1985**, 96, 285.

41. SWARTZ, S.L., SEABAUGH, M.M., HOLT, C.T., and DAWSON, W.J. Fuel processing catalysts based on nanoscale ceria. *Fuel Cells Bulletin,* **2001**, 4, 7.

42. BUNLUESIN, T., GORTE, R.J., and GRAHAM, G.W. Studies of the water-gas-shift reaction on ceria-supported Pt, Pd, and Rh: Implications for oxygen-storage properties. *Applied Catalysis. B, Environmental,* **1998**, 15, 107.

43. DENG, W. and FLYTZANI-STEPHANOPOULOS, M. On the issue of the deactivation of Au-ceria and Pt-ceria water-gas shift catalysts in practical fuel-cell applications. *Angewandte Chemie,* **2005**, 45, 2285.

44. LIU, X., RUETTINGER, W., XU, X., and FARRAUTO, R. Deactivation of Pt/CeO$_2$ water-gas shift catalysts due to shutdown/startup modes for fuel cell applications. *Applied Catalysis. B, Environmental,* **2005**, 56, 69.

45. WANG, X., GORTE, R.J., and WAGNER, J.P. Deactivation mechanism for Pd/ceria during the water-gas-shift reaction. *Journal of Catalysis,* **2002**, 212, 225.

46. FU, Q., SALTSBURG, H., and FLYTZANI-STEPHANOPOULOS, M. Active nonmetallic Au and Pt species on ceria-based water-gas shift catalysts. *Science,* **2003**, 301, 935.

47. WANG, X. and GORTE, R.J. The effect of Fe and other promoters on the activity of Pd/ceria for the water-gas shift reaction. *Applied Catalysis. A, General,* **2003**, 247, 157.

48. ZHAO, S. and GORTE, R.J. The activity of Fe-Pd alloys for the water-gas shift reaction. *Catalysis Letters,* **2004**, 92, 75.

49. ANDREEVA, D. Low temperature water gas shift over gold catalysts. *Gold Bulletin,* **2002**, 35, 82.

50. ANDREEVA, D., IDKAIEV, V., TABAKOVA, T., ILIEVA, L., FALARAS, P., BOURLINOS, A., and TRAVLOS, A. Low-temperature water-gas shift reaction over Au/CeO$_2$ catalysts. *Catalysis Today,* **2002**, 72, 51.

51. TABAKOVA, T., BOCCUZZI, F., MANZOLI, M., SOBCZAK, J.W., IDAKIEV, V., and ANDREEVA, A. Effect of synthesis procedure on the low-temperature WGS activity of Au/ceria catalysts. *Applied Catalysis. B, Environmental,* **2004**, 49, 73.

52. KIM, C.H. and THOMPSON, L.T. Deactivation of Au/CeOx water gas shift catalysts. *Journal of Catalysis,* **2005**, 230, 66.

53. XUE, E., O'KEEFFE, M., and ROSS, J.R.H. Water-gas shift conversion using a feed with a low steam to carbon monoxide ratio and containing sulphur. *Catalysis Today,* **1996**, 30, 107.

Chapter 7

Removal of Trace Contaminants from Fuel Processing Reformate: Preferential Oxidation (Prox)

MARCO J. CASTALDI

Department of Earth and Environmental Engineering, Columbia University

7.1 INTRODUCTION

There are several methods to remove trace contaminants from effluent streams. They fall into the broad categories of adsorption, membranes, scrubbers, and selective reaction. In adsorption, typically, powder activated carbon (PAC) or pressure swing adsorption (PSA) is employed to physically condense or trap the contaminant of interest onto a high specific surface area bed ($\sim 500\,m^2/g$ or higher). Conventionally, the adsorption processes are separated into continuous and batch processes. When using PAC, it is usually operated in batch mode. That is, once the bed is saturated with the contaminant, determined via detection of the contaminant concentration increasing on the downstream side of the bed, the effluent stream is stopped or diverted to another bed and the saturated bed is removed and regenerated.

PSA operates much the same way as PAC beds, but the intent is to have a more continuous process. That is, the effluent stream is never stopped and diverted from one PSA bed to the next. Usually, there is a series of PSA beds in continuous operation at one time to increase the removal efficiency of the contaminant and improve the regeneration capability.

To further improve the adsorption capacity and regeneration ability of the beds, metals and promoters are added in small quantities. For example, a PSA system

Hydrogen and Syngas Production and Purification Technologies, Edited by Ke Liu, Chunshan Song and Velu Subramani

using activated carbon impregnated with $SnCl_2 \cdot 2H_2O$ was used to remove CO from a model H_2/CO mixture representing the steam reformer process gas. The CO adsorptive capacity of impregnated carbon was found to be superior to that of the pure carbon. Typically, 1000 ppm CO can be readily reduced to 10.4 ppm via the impregnated activated carbon PSA system. The species in the impregnated carbon responsible for the improved gas-phase CO adsorption was found to be SnO_2.[1] For an extensive discussion on PSA, please see Chapter 11 of this book.

While adsorption and therefore PAC and PSA are probably the most mature technologies for CO removal from the reformate, they have some drawbacks. The first is the need for high pressure, between 5 and 10 atm, to achieve high removal efficiencies. This leads to design issues in pressurizing the system with large hydrogen concentrations and significant energy consumption. If high-pressure hydrogen is desired, then these beds are well suited, if low pressure is needed, then the energy for pressurization is wasted. Another drawback is the attrition of the adsorption capacity. That is, the amount of material the bed can adsorb deteriorates over time until the entire series of beds needs to be replaced. Finally, the PSA systems necessitate the use of hot valves to divert the reformate flow from one bed to another. Anytime there are moving parts in a hot flow path, there is the potential for component failure.

Another very effective technology is membrane separation. Membranes have been used for a wide variety of chemical purification and have been adapted to reformate purification as well. The most viable membranes currently target only one chemical species, such as hydrogen, thus allowing hydrogen to pass while retaining all the other reformate gases. These are Pd-based materials that can process 0.1 kg/m^3/s of hydrogen. This translates to separator volume of about 0.03 m^3 for a 1 kg/h H_2 production rate, which is a reasonable size.

Membranes are constructed of either inorganic ceramics or precious metal alloys of Pd. Many groups are exploring their use since they can operate at pressures near 10 psi differential. For example, an 11 μ Pd/Ag alloy can operate with a 10 psi pressure differential at 505 °C, maintaining a flux near 0.68 mol H_2 m^2/s. It is also possible to operate at low temperatures, down to 2 °C, albeit at slightly lower fluxes.[2] In addition, with proper backing, that is, metal or ceramic support, the same membrane can sustain a 60 psi differential pressure resulting in higher fluxes.[3] In comparison to PSA, these systems typically operate with 40–50 psi differential pressure at temperatures up to about 200 °C. Currently from a cost perspective, PSA is more affordable than Pd/Ag alloy systems, yet cannot operate at the lower pressure differentials that the membranes can. Therefore, the cost advantage may be offset by the pressurization costs.

One more separation technique that deserves some discussion is the scrubber. Scrubbers are the most mature technology from the chemical separation industry that usually employs liquid–gas interaction to achieve the contaminant removal. While they can be very selective due to the liquid-stripping agent, they are complicated systems that use specialty chemicals and require multiple steps to transfer the contaminant from gas phase to liquid, then from liquid back to gas phase, which is ultimately vented. For very large, stationary installations, scrubbers are usually the technology of choice because of their performance; they can routinely achieve 99.99% removal efficiency, durability, and robustness. However, scale-down is not a trivial issue and as such, scrubbers have not penetrated into the hydrogen purification process. For more on scrubber technology, please see Perry and Green's *Chemical Engineers' Handbook*.[4]

This leads to the last category and the subject of this chapter: selective oxidation or preferential oxidation (Prox) of contaminants. In particular, the focus will be on the removal of trace amounts of CO, from approximately 1% mole fraction or less, from a reformate stream. The reformate typically contains 10% CO_2, 25% H_2O, ~1% CO, 30% H_2, and the balance N_2, which is generated from an autothermal reformer (reforming fuel, air, and water) followed by a water-gas shift (WGS) reactor. It should be noted that if the reformer is a steam reformer, there will be a similar reformate stream but no nitrogen and higher concentrations of hydrogen. The significant aspect of Prox is the ability to react the small amount of CO in the presence of large quantities of H_2. This technology effectively converts the contaminant to an inert and thus can be operated on a truly continuous basis. The required amount of oxygen is fed to the Prox reactor via an air blower. Therefore, the Prox reactor can operate at atmospheric pressure, and the blower, which feeds air, is a low energy consumer.

Since the reaction must be highly selective toward CO and not H_2, it is always done in the presence of a catalyst. Not only does the catalyst reduce the energy necessary to initiate the CO oxidation reaction, but the formulation enables the desired reaction to occur by taking advantage of the strong CO adsorption on the catalyst surface. Unlike, the other technologies described above, Prox is considered a very elegant way to remove the CO since there is no regeneration or bed replacement necessary, air is the only reagent that is supplied by a blower, and the reaction occurs at moderate temperatures, ~200 °C, and low pressures but can operate at higher pressures if desired. Moreover, the reactor size necessary to achieve low CO concentrations, for example, <10 ppmv, is usually one-half to one-fourth the size of the upstream WGS reactor.

There are however concerns with durability or lifetime of the catalyst. There can be contaminants within the reformate, in particular sulfur compounds, that can render the catalyst completely inactive. In addition, if the reactor temperature operates too high or an unexpected over-temperature event occurs, the performance or durability of the catalyst can be seriously compromised. Additionally, the support of the catalyst can sinter over time on stream due to the presence of significant amounts of water. Fouling can be another issue, which impacts durability if very pure water or clean air is not used. While all these factors are present, catalyst formulations have been shown to operate for hundreds of hours with minimal to no degradation.

7.2 REACTIONS OF Prox

The overall targeted reaction in the preferential removal of CO in the reformate stream is the oxidation of CO via a carefully metered amount of air:

$$CO + \frac{1}{2}O_2 \text{ (from air)} \rightarrow CO_2.$$

However, it should be intuitive that this is not the only reaction that occurs, especially in the presence of nearly 30% H_2. The primary undesired reaction to occur is the oxidation of hydrogen:

$$H_2 + \frac{1}{2}O_2 \text{ (from air)} \rightarrow H_2O.$$

Not only is this a consumption of oxygen, thus leaving less for CO oxidation, but it is also a consumption of H_2, which is the desired ultimate product in the reformate. Furthermore, it is an exothermic reaction that raises the probability of an undesired temperature excursion. Aside from the two major reactions above, there are other reactions that impact the ultimate performance and conversion efficiency of the Prox reactor. They are the WGS and methanation reactions. Therefore, the relevant reactions for this system are

$$CO + \frac{1}{2}O_2 \rightarrow CO_2 \quad \Delta H = -283 \text{ kJ/mol}, \tag{7.1}$$

$$H_2 + \frac{1}{2}O_2 \rightarrow H_2O \quad \Delta H = -242 \text{ kJ/mol}, \tag{7.2}$$

$$CO + H_2O \rightleftarrows CO_2 + H_2 \quad \Delta H = -41.2 \text{ kJ/mol}, \tag{7.3}$$

$$CO + 3H_2 \rightleftarrows CH_4 + H_2O \quad \Delta H = -206 \text{ kJ/mol}. \tag{7.4}$$

Other reactions, such as $CO_2 + H_2O$ and $CO_2 + CH_4$, are either not thermodynamically favored or the concentrations of the reactants render them insignificant.

A thermodynamic analysis puts the above reaction sequence into perspective. Representative values of shifted reformate were taken from the *Fuel Cell Handbook*, 5th edition, with O_2 addition to produce an O_2/CO ratio of 1.2 or lambda ($\lambda = 2.4$, where $\lambda = 2*O_2/CO$). Equilibrium calculations and CO conversions done for a range of inlet temperatures for the shifted reformate show that thermodynamics do not favor 100% selectivity of O_2 to convert CO, and reverse water-gas shift (RWGS) reactions need to be avoided. Even at temperatures of 25 °C, the calculations show that all the O_2 is converted with most reacting with H_2 to form water. From that temperature and higher, the RWGS reaction occurs, which works against the reactor designer trying to produce a CO-free effluent. Figure 7.1 shows how quickly the CO conversion drops with rising temperature. At 75 °C, CO conversion is 90%, whereas at 125 °C, conversion is 75% and 175 °C conversion drops to 35%. In fact, CO production, due to the RWGS reaction, begins to occur by 210 °C. However, the levels of H_2, CO_2, and H_2O are only slightly affected by the temperature rise. For example, at 75 °C H_2 concentration is 52.2%, and at 175 °C, the H_2 level is relatively unchanged at 51.9%. This indicates that low operating temperatures are desirable to keep CO conversion high to produce a CO-free effluent. Yet, high temperatures are preferred for higher CO oxidation reaction rates, thus allowing reactors to be reasonably small in size. The equilibrium values can be used as a guide to help define the operating window for allowable CO removal. One way to achieve high CO conversions at temperatures near 200 °C, where the kinetics of the CO oxidation reaction are sufficiently fast, is to operate a reactor at nonequilibrium conditions. The scheme

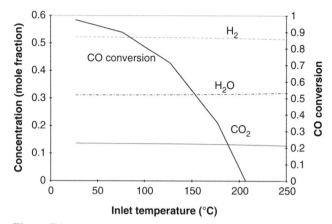

Figure 7.1. Equilibrium calculations for shifted reformate.

leads to the innovative combination of novel catalyst formulations (metal and supports) and reactor design. That is, a catalyst must be employed to drive the oxygen to react with the CO as much as possible. Therefore, the appropriate choice of catalysts and considerations for reactor design are necessary to effectively enable the reaction. These aspects are discussed in Sections 7.4 and 7.3 and 7.5, respectively.

7.3 GENERAL Prox REACTOR PERFORMANCE

Aside from specific variations in catalyst formulations, the operation of Prox reactors has some interesting behavior. To be sure, different catalyst formulations impact the performance of certain parameters to a degree. However, it is likely that they suppress certain operational regimes, such as hydrogen light-off, but do not completely eliminate them. This section discusses some commonly found performance characteristics when operating Prox reactors. In particular, variations in λ, CO concentration, and space velocity are presented. In addition, some other subtle effects such as multiple steady-state operation and reactant synergies such as that between water–oxygen.

The catalyst formulation chosen needs to be selective toward CO oxidation (Eq. 7.1) over water production and methane reactions (Eqs. 7.2 and 7.4), as well as be active at low temperatures where the reverse rate of Equation 7.3 is slow. The formulation will also need to be highly active for Equation 7.1, thus allowing the use of near stoichiometric amounts of O_2. Stoichiometric amounts of O_2 are desirable to keep the selectivity toward CO conversion high. Any O_2 remaining after conversion of CO will react with H_2, thus decreasing the effectiveness of the reactor.

To further illustrate typical Prox performance, results are shown from tests done with an O_2/CO ratio of 1:1 and a reactor space velocity of 440,000 h^{-1} using a representative shifted reformate from an autothermal methane reformer (CO: 500–10,000 ppm; O_2: 1000–20,000 ppm; $H_2O \cong 32\%$; $H_2 \cong 32\%$; $CO_2 \cong 14\%$;

N_2 balance $\cong 20\%$). The catalyst formulation used for this example was a Pt/α-Al$_2$O$_3$. This formulation is very common for the examination of a wide window of temperature operation with no observed methanation.

A CO inhibition effect was observed at higher inlet concentrations of CO leading to a delayed light-off but higher maximum conversions. One cause for the CO inhibition is likely attributable to the strong adsorption of CO on the Pt surfaces. Due to the high concentration of H_2 in the reactants, competing reactions ($H_2 + \frac{1}{2}O_2$) as well as H_2 adsorption on the surface occur. The likely mechanism is that CO strongly adsorbs onto a Pt site $\rightarrow$ H_2 reacts with the adsorbed CO to form H_2O, leaving behind carbon on the surface $\rightarrow$ O_2 may then react with the carbon to reform into adsorbed CO. Clearly, there is a competition between CO adsorption, H_2 reaction with adsorbed CO or O_2, or both and the ability for the O_2 to react with the carbon to form CO_2 and subsequently desorb to free the site for another reaction. Because of the CO inhibition effect, a set of inlet parameters must be optimized to give high conversions of CO with high selectivity, yet have a low light-off temperature to avoid running into RWGS reaction limitations. Figure 7.2 shows a comparison between CO inlet concentrations and amount of O_2 available for reaction, defined as λ; again $\lambda = 2*O_2/CO$. Higher λ indicates more O_2 and $\lambda > 1$ results in excess oxygen for the CO to CO_2 reaction, thus leaving O_2 remaining to react with H_2 once the CO has been exhausted.

Analysis of the data in Figure 7.2 indicates that there is a CO inhibition effect on the platinum catalyst somewhere between 500 and 1000 ppm CO inlet

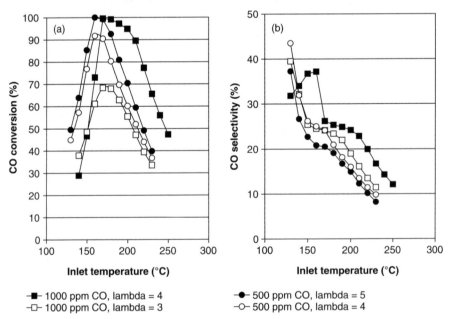

Figure 7.2. Comparison between conversion (a) and selectivity (b) for different CO inlet concentrations and λ using a Pt/α-Al$_2$O$_3$.

concentrations. A comparison between the 1000 ppmv CO inlet (■) and the 500 ppm CO inlet (○) for the same oxygen availability, $\lambda = 4$, shows a higher light-off temperature for the higher CO concentration sample. While the light-off temperatures are higher, the maximum achieved conversion and selectivity are higher for the 1000 ppm case. Further analysis shows that higher λ's lead to more complete conversion but generally lower selectivity, as expected. Another observation is that higher λ's are required with lower CO concentrations to achieve greater than 99% CO conversion. In the temperature range studied, it is evident that higher CO concentrations can unexpectedly affect the reactions. In Figure 7.2b, the 1000 ppm CO (■) sample has poor selectivity at low and high temperatures. At low temperatures, the CO oxidation reaction is probably inhibited, while at high temperatures the catalyst is less discriminating between reacting CO and H_2. Also, notice the rank order of selectivity versus λ for the two CO concentration levels. For the 1000 ppm concentration sample, selectivity gets better as λ is increased, whereas the opposite is true at the 500 ppm level tests. These types of effects are important in reactor design with varying CO inlet concentrations and may lead to cross over into an inhibition regime, where light-off and selectivity are affected. Lastly, it is clear in Figure 7.2a and b that the temperature corresponding to maximum conversion does not correspond to maximum selectivity. The maximum selectivity occurs at lower temperatures, usually before the reactor is fully lit off. This would suggest the catalyst surface might be initially covered by one of the reactants (CO or O_2) at the low temperatures where it is easy for the other reactant to combine with the adsorbed species. As temperature is increased, reaction rates increase, allowing greater conversion, although with decreasing selectivity, indicating possible competition between the initial reactant (CO or O_2) on the surface and H_2. Obtaining and defining a wide operating window therefore presents a significant challenge.

Additional comparisons between CO inlet concentration of 5000 ppm (○) and 10,000 ppm (●) help to determine if much higher CO concentrations had the same effect as that at the lower levels. Figure 7.3 shows the CO conversion comparison for the 5000 and 10,000 ppm inlet concentrations versus temperature. Light-off trends were similar to the low concentration studies, in that the 10,000 ppm sample had a more gradual, higher temperature light-off as compared with the 5000 ppm sample. This suggests that the inhibition mechanism is the same over large changes in CO inlet concentrations. Note that for the data reported in Figure 7.3, $\lambda = 2.0$, which is lower than in the low CO concentration study discussed above in Figure 7.2 and may be why maximum CO conversions were the same. An interesting observation is the ability to achieve approximately 85% conversion here with $\lambda = 2$ whereas at the lower CO concentrations, λ's of 4 or higher were needed to achieve that amount of conversion. This suggests competition between CO and H_2 for O_2, and at the higher CO concentrations, CO is more effective in competing for the available O_2 than at the lower CO concentrations.

The selectivity results, also shown in Figure 7.3, give an indication as to the type of mechanism dominating for the CO oxidation reaction. Notice that the divergence between the selectivities at the low temperatures before light-off is complete. The high concentration, while has a lower conversion rate, has a much

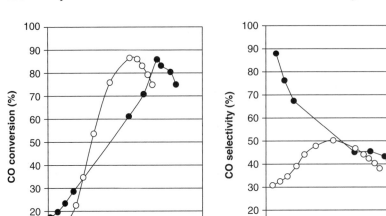

Figure 7.3. A comparison of light-off and maximum conversion versus temperature for high inlet CO concentrations. Study done using $\lambda = 2$ for the same Pt/α-Al$_2$O$_3$ catalyst as in Figure 7.2.

higher selectivity than the 5000 ppm sample. This clearly indicates that the CO initially adsorbs onto the catalyst surface prior to reaction, and the inhibition effect is due to this CO adsorption. Not until a temperature of near 250 °C, when significant activity is taking place on the 10,000 ppm sample, do the selectivities approach each other. The 5000 ppm sample has a selectivity curve that indicates that the amount of CO present at that concentration level is not enough to completely cover the catalyst sites; therefore, there is probably a competition of reactions occurring from the beginning. Again, the temperature of maximum conversion does not correspond to that of maximum selectivity.

Oxygen stoichiometry studies done for the same Pt catalyst help to elucidate the mechanism for CO conversion. For those tests, the CO concentration was kept at 5000 ppm and the oxygen stoichiometry, λ, was varied from 1 to 3 and the Prox activity and selectivity in each case was investigated over a suitable range of temperature. The results are shown in Figure 7.4. As can be seen, with increasing λ, the maximum selectivity decreases but the maximum CO conversion increases. The maximum CO conversion was seen to increase rapidly up to $\lambda = 2$ after which point only incremental increases were obtained by increasing λ. This result suggests the possibility of using a multiple-stage reactor system, which is commonly used.[5] One scheme could be operating with a λ of ~2 in the first stage and using a higher λ in the second stage to get almost complete conversion of the CO.

Since the maximum selectivity occurs for $\lambda = 1$, which is the stoichiometric amount to oxidize CO to CO$_2$, and diminishes from thereon, while simultaneously

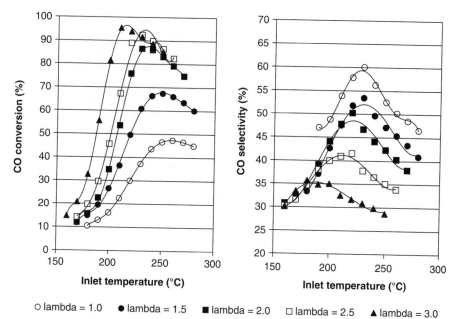

○ lambda = 1.0   ● lambda = 1.5   ■ lambda = 2.0   □ lambda = 2.5   ▲ lambda = 3.0

Figure 7.4. Comparison of oxygen concentration from stoichiometric amounts ($\lambda = 1$) to three times in excess to begin to elucidate the mechanism of CO oxidation.

increasing conversion, this suggests that the rate limiting step is the oxidation of CO. Since the conversion only incrementally increases after $\lambda = 2$ while selectivity decreases more dramatically at that point, increasing O_2 does not efficiently oxidize the CO after stoichiometric amounts have been introduced. Therefore, variables other than stoichiometry, such as reactor design or catalyst formulation must be investigated separately or in combination.

The heat of adsorption of CO on supported platinum is known to be a strong function of platinum particle size with higher strength of adsorption on smaller particles.[6] It has been observed that the operating temperatures for maximum catalyst activity and selectivity were a function of the catalyst loading, with the maximum in activity and selectivity shifted to higher operating temperatures with lower loading as would be expected with a particle size effect. These studies demonstrate the concept of using platinum loading to vary the optimum operating temperature along the length of the reactor. Modifying the catalyst activity by the use of promoters, alloying with other precious group metals (PGM), may also be used to achieve a similar result.

7.3.1 Multiple Steady-State Operation

In addition to the performance variations with reactant concentration and gas hourly space velocity (GHSV), there can be multiple steady states observed. Generally, a reactor in which a single, exothermic reaction is occurring will operate in one of two stable steady states. Additionally, an unstable steady-state solution to the mass and energy balance will be present. However, when multiple reactions are occurring in a reactor, there is the possibility of more than three possible steady-state solutions. This arises from the fact that each reaction has a unique mole balance curve that superimposes upon each other. The energy balance line can then intersect each mole balance curve more than once adding up to four or more stable steady-state conditions where the mole and energy balances are satisfied.[7] This type of phenomena is seen with the Prox reactor being developed, particularly at dry-feed concentrations. Selective oxidation of CO in hydrogen over different catalysts has been extensively examined.

A general understanding of these systems is that at temperatures before the onset of the CO light-off, the surface is covered with adsorbed CO. As the temperature is increased, the fraction of the surface covered with CO decreases, and this opens up sites for oxygen adsorption and subsequent reaction. Above a certain temperature, the fraction of CO occupying the surface decreases even further and hydrogen chemisorbs and reacts on the surface in competition with CO, reducing the selectivity toward CO oxidation. Therefore, a common feature of all these systems is that there exists a window of operation in temperature between the light-off curves for CO and H_2, the object being to operate at a catalyst temperature sufficient for high activity of CO oxidation (for reduced size of catalytic reactor), but below that for significant consumption of the hydrogen. This results in different temperature windows and optimum catalyst temperatures, which need to be identified for different catalyst formulations.

This is shown schematically in Figure 7.5. The figure shows the CO and H_2 light-off curves, which are shifted both in absolute temperature and relative to each other for different catalyst formulations depending on the relative adsorption energies on the different catalyst surfaces. This results in different temperature windows and optimum catalyst temperatures, which need to be identified for different catalyst formulations.

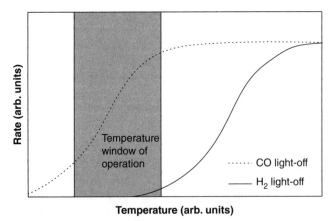

Figure 7.5.   Schematic illustration of CO and H_2 light-off on Prox catalyst.

Table 7.1. Results of the Dual Steady-State Behavior of the Prox Reactor

Inlet Concentrations					Reactor Temperature		Conversion	
O_2	H_2	CO	CO_2	H_2O	Inlet	Surface	O_2	CO
1.69	43.46	1.41	18.08	7.30	91.8	160.1	30	22
1.68	44.40	1.41	18.10	7.30	90.0	253.5	Complete	23

Dual steady-state behavior of the reactor has been observed at dry-feed conditions. However, when the water content is above 15% in the feed, the reactor consistently gives high CO conversion with nearly 45% selectivity. In addition, oxygen has been detected in the product stream at varying levels, depending on the conditions tested. Initial analysis indicates that heat removal at the catalyst surface is a very important parameter in avoiding dual steady-state operation because heat generated from the CO oxidation reaction is quickly transferred into the gas phase, thus moderating the surface temperature.

Table 7.1 displays representative results from experiments that show the dual steady-state nature of the selective CO oxidation reactor. The first row provides the results of the reactor operating in the preferred steady state. The CO conversion is 22% with oxygen conversion at 30%, which results in a selectivity of slightly above 40% with a commensurate temperature rise. The second row shows for the exact same inlet compositions and temperature, the reactor is operating in another, high temperature, and undesired steady state. The data indicate that this state is one in which high hydrogen consumption is occurring. The evidence for that is seen in the complete conversion of oxygen, without a change in CO conversion and the higher surface temperature as compared with the previous state (Fig. 7.6). Since hydrogen is considered the fuel in this case, it is this state that must be avoided to ensure that the efficiency of the reactor system remains high.

Additional experiments have been conducted that varied λ from 0.7 to 3.0, holding all other feed conditions constant.[8] Dual steady-state operation was observed for the entire range tested providing more evidence that the water amount in the feed composition is probably the dominant factor. This type of experiment was repeated for a different CO feed concentration and nearly the same behavior was recorded. This dual steady-state nature persists for a wide range of lambda, which points to the fact that dual steady state is more influenced by feed composition and more specifically by water amount. Since water has high heat capacity, this may temper the heat rise and delay the onset of H_2 oxidation reaction.

7.3.2 Water–Oxygen Synergy

Another result particular to this mix of reactants used in Prox testing with various water concentrations was evidence of a water–oxygen synergy within the reactor[9] (Fig. 7.7). Table 7.2 comprises representative conditions and results from repeat

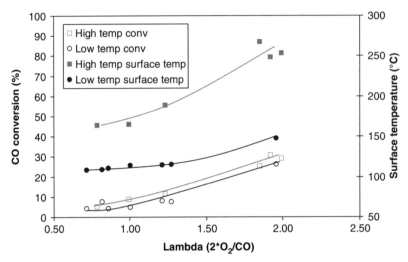

Figure 7.6. CO conversion (conv) for high and low temperature (temp) steady states. Different temperatures give comparable CO conversion.

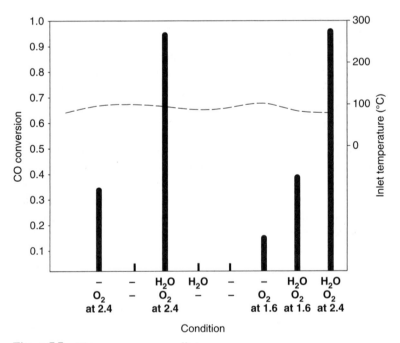

Figure 7.7. Water–oxygen synergy effect.

Table 7.2.   Evidence for a Water–Oxygen Synergy within the Prox Reactor

Inlet Concentrations					Reactor Temperature			Conversion		Lambda
O_2	H_2	CO	CO_2	H_2O	Inlet	Surface	Outlet	O_2	CO	
0.65	31.79	0.55	13.90	32.88	101.2	194.3	192.2	73.33	92.36	2.36
0.65	31.74	0.55	13.91	–	109.6	136.0	135.5	6.15	26.23	2.36
–	31.86	0.55	13.91	32.88	110.4	112.4	109.9	–	–0.50	–

experiments that indicated a water–oxygen synergistic reaction occurring. The objective of these three experiments was to determine the contribution of oxygen and water separately toward oxidation of CO and consumption of hydrogen. The first row of the table provides the conditions and results of a baseline experiment, where the system was operating at its targeted steady-state position. Here, the conversion is, as expected, greater than 90% with 50% selectivity. The temperature rise from the inlet to the outlet is consistent with the amount of reaction. The surface temperature rise also tracks with the gas-phase temperature rise. The second row has the results of the condition where the water is not included in the influent. The third row shows the results of an experiment where the water influent was resumed and the oxygen was turned off. Each time a reactant was turned off, nitrogen was adjusted to compensate for that loss to maintain a constant GHSV. Inspection of the CO conversion shows that the sum of the CO conversions from the individual reactants does not add up to the conversion when they are both present. In fact, the water alone shows nearly zero reaction, which is expected because that condition represents a WGS reaction of which the kinetics are very slow at that temperature. It is clear, however, that it is not only the oxygen that is converting the CO since its conversion is only 26% as compared with a 92% conversion when water is present. It is evident that water somehow initiates or promotes the CO oxidation reaction.

Other water-enhanced reactions have been evidenced in the literature. For example, there was a study that showed a hydrogen–water synergy[10] and a carbon dioxide–water synergy.[11] Both studies indicated that water addition increased the amount of CO oxidation more than expected or predicted. The current experiments, shown in the table, add another piece of evidence to the water-assisted or water-enhanced CO oxidation observations. These results point toward water acting as a catalyst or promoter for the selective CO oxidation on alumina-supported platinum-based catalysts. It is known that water has a high heat capacity, thus its temperature controlling, or heat absorbing, capability is recognized as one way to control reactions that liberate heat. In the case of the selective oxidation of CO this is very important, as discussed above, so as not to activate or light-off the hydrogen oxidation reaction. Yet water plays a significant role as a promoter in this reaction sequence. To this point, the reaction and performance characteristics have been discussed for a Pt/α-Al$_2$O$_3$ formulation. The next section discusses other catalyst formulations found in literature where some of these performance characteristics are observed.

7.4 CATALYSTS FORMULATIONS

This section is focused on the catalyst selection for optimal Prox reactor operation. During the discussion, catalyst formulation will generally cover the active metal (Pt, Rh, Cu, etc.) and support (Al_2O_3, ZnO, etc.) with promoters (La, Ce, etc.); however, there are instances in which studies done are more focused on the active metal regardless of the support used and is indicated where appropriate. It should be kept in mind that catalyst formulation selection must be done in the context of the desired performance outcome, or avoidance of certain characteristics and reformate mix.

In general, different precious metal catalysts have been found to be highly selective for the Prox reaction. Different workers have reported the use of supported Ru, Rh,[12] Pt,[13,14] and Au[15] for this reaction. A general understanding of these systems is that at temperatures before the onset of the CO light-off, the surface is covered with adsorbed CO. As the temperature is increased, the fraction of the surface covered with CO decreases, and this opens up sites for oxygen adsorption and subsequent reaction. Above a certain temperature, the fraction of CO occupying the surface decreases even further and hydrogen chemisorbs and reacts on the surface in competition with CO, reducing the selectivity toward CO oxidation. Therefore, a common feature of all these systems is that there exists a window of operation in temperature between the light-off curves for CO and H_2 (i.e., see Fig. 7.5), the object being able to operate at a catalyst temperature sufficient for high activity of CO oxidation (for reduced size of catalytic reactor), but below that for significant consumption of the hydrogen. The inlet temperature to the Prox reactor should ideally be less than or equal to a low temperature WGS reactor exit temperature generally found just upstream of the Prox reactor. This would allow easy transition, without heat exchangers or control schemes, into the Prox reactor.

Selective oxidation of CO in hydrogen over different catalysts has been extensively examined. Most research to date has occurred with formulations that include a precious metal component supported on an alumina carrier. The catalyst-mediated oxidation of CO is a multistage process, commonly obeying Langmuir–Hinshelwood kinetics for a single-site competitive mechanism between CO and O_2. Initially, CO is chemisorbed on a PGM surface site, while an O_2 molecule undergoes dissociative chemisorption either on an adjacent site or on the support in order for surface reaction between chemisorbed CO and O atoms to produce CO_2.

The precious metals have the high turnover frequencies required for CO oxidation in the presence of hydrogen and water, in particular the chemisorption of both CO and O_2. An optimum range of O_2/CO ratio is required in order to obtain the proper balance of adsorbed CO and adsorbed oxygen on adjacent sites. However, not all pure supported precious metals can achieve selectivity that is required for Prox. One good example of this is palladium because of its strong affinity for hydrogen chemisorption.[16]

Recent literature data suggest that the use of promoters may be beneficial and lead to a significant improvement in CO conversion and selectivity for preferential selective CO oxidation (under an H_2-rich environment).[17] Lanthanum oxide (La_2O_3) is known to be an effective catalytic textural and structural promoter, increasing the thermal stability, the dispersion and the stability of the noble metal (Pt, Pd, or Rh) catalysts.[18–20] The La_2O_3 loading and the type of lanthanum precursor plays an important role on the noble metal or other metal oxide dispersion used as active catalyst component.[21,22] The catalyst configuration (surface and bulk structure) will strongly affect the catalytic activity. Similar promoter effects are expected from Sm_2O_3. Iron (Fe) has also been used as a promoter.[23–27] It is postulated that the precious metal sites chemisorb the CO and the Fe, which, in an oxide state, promotes the dissociated adsorption of oxygen leading to a dual-site mechanism.[26] One study with Pt catalyst promoted with Fe showed that the outlet concentration of CO from Prox is governed by the RWGS reaction as well as temperature and residence time.[28]

CuO–CeO_2 mixed oxide catalysts and Cu/CeO_2 catalysts were studied by Avgouropoulos et al.[11,29,30] The material was shown to have a high selectivity toward CO oxidation, with a minimum operating temperature of 140°C. The advantage of the catalyst is that it does not exhibit RWGS activity. The Cu/CeO_2 catalyst, made using a sol-gel process exhibits both high selectivity and conversion of CO oxidation in the temperature range of 170–190°C range.

Gold has been suggested as a potential Prox catalyst, particularly at low operating temperature, that is, 100°C. Gold supported on manganese oxides, prepared via use of an organogold complex in the liquid phase, albeit with stability concerns of Au/MnOx in H_2, was reviewed by Tanaka et al.[31] Use of Au on iron oxide at 80°C was recommended by Kahlich et al.[32] A recent paper[33] reported that an Au/Fe_2O_3 catalyst is highly selective toward CO oxidation operating at 80°C. A gold on Al_2O_3 catalyst was reported by Ivanova et al.[34]

An Rh/MgO catalyst[35] has been reported to be highly active at 250°C, without exhibiting undesirable side reactions like RWGS and methanation. The highly active catalyst could be used as a low-cost, first-stage Prox catalyst and does not need a heat exchanger directly after the WGS reactor. Kotobuki et al. studied mordenite-supported noble metal catalysts—Pt/γ-Al_2O_3 and Au/α-Fe_2O_3.[36]

It was also reported that the presence of Ru favors CO-selective oxidation under Prox reaction conditions.[12] Ru is clearly favored for first-generation fuel processors because of its low cost relative to Pt or other precious metals. Taylor et al. reported ruthenium/alumina and rhodium/alumina for high CO conversion below 100°C.[37] Echigo et al. reported that selectivity of an Ru Prox catalyst was improved by preconditioning the catalyst in H_2/N_2[38,39] and was able to reduce CO below 10 ppmv. Ru operates between 140–200°C[40] and thus is active for methanation reaction with CO_2. This consumes hydrogen and could potentially generate high temperatures since the methanation reaction is exothermic. Ru will deactivate in the event of air exposure due to many different valence states that the surface can attain. Platinum supported on zeolite and mordenite for selectivity approaching 100% at high H_2 concentrations has been observed by Kotobuki et al.[36] Catalyst surface coverage and reaction kinetics for the Prox reaction over Pt/γ-Al_2O_3, Au/α-Fe_2O_3, and bimetallic PtSn catalysts were described by Gasteiger et al.[41]

The literature cites several other oxidation catalysts that have shown selectivity in the presence of hydrogen. These include Au, Ir, Pt, Pt/Ru, Ru, Rh, and Cu as the active metals, dispersed on various supports. Iridium on cerium oxide has also been explored and found to be reasonably active and selective but somewhat less than Pt.[16] This study was done without H_2O or CO_2 in the reactant stream. Therefore, the performance stability effects from carbonate formation and RWGS reactions are unknown.

A patent was granted to Shore et al.[42] for a process utilizing a Cu/CeO_2 catalyst that also contained Pt. It was found that the precious metal lowered the temperature necessary for the reduction of the base metal from its inactive oxide to the active metal form. The operating temperature is 100°C, compared with the minimum of 140°C required for the reduction of Pt-free copper/ceria in the reformate. One limitation of the copper Prox catalyst is that it is CO inhibited; CO shifts the reduction temperature of the catalyst.

The development of catalyst formulations for Prox is continuing. The discussion just presented should serve as a very good starting point when researching Prox catalyst formulation development, but is not an end. Especially with the rapid development of nanocatalysts, there will likely be future developments for some time.

7.5 REACTOR GEOMETRIES

This section discusses different geometries that many of the formulations discussed above are deposited. It should be noted that the findings discussed in Section 7.2 highlight in general what one could expect. However, different geometries or formulations or a combination will greatly impact some of those performance characteristics.

Prox is the CO removal technology of choice for steam reforming, partial oxidation, and autothermal process routes for fuel cell systems. However, significant issues exist, most notably capital cost and control of the process. Control and performance of the process has been discussed in the previous section (Section 7.2). The cost for Prox units based on 3% wt Pt/Al_2O_3 catalysts is large; typically, the materials are nearly 80% of the entire reactor even when mass produced. The development of cheaper catalyst materials is an important research goal. In addition, the development of precious metal Prox catalysts on reactors that can operate at high space velocities is imperative. While most of the formulations have been studied over pellet bed or monolithic catalysts, there are other reactor substrates or geometries that better lend themselves to the Prox reaction.

Some of these will be discussed here with various attributes that they possess. Aside from the conventional substrates, others that are currently under development are short contact time (SCT) reactors that consist of screens, mesh, or expanded metal that are typically fabricated from high-temperature FeCrAl alloys. Reticulated foams that combine the very low pressure drop of monoliths with the improved transport of SCT reactors are often used. There are flat plate and microchannel reactors and recently microelectromechanical system (MEMS) reactor geometries that have been fabricated. In comparison to monolith or pellet beds, SCT substrates seem to allow Prox reactors to operate at significantly lower water concentrations before the onset of the hydrogen oxidation reaction, that is, the high-temperature steady state.

Since the oxidation of CO and H_2 are highly exothermic (-283 and $-242\,kJ$/ mol, respectively), the control of reactor temperature becomes a critical issue. A calculation of the adiabatic temperature rise for a typical reformate stream exiting a low-temperature shift reactor with 5000 ppm CO and with 2500 ppm added O_2 was ~45°C (assuming 100% selectivity to CO oxidation). Therefore, depending on the operating window of the specific catalyst formulation, the catalyst temperature may have to be controlled prior to the onset of bulk mass transfer limited operation. This can be accomplished by varying the inlet temperature to the Prox reactor to account for the temperature rise along the length of the reactor and/or by using a staged reactor with interstage cooling. As discussed in the next section, the combination of low thermal mass and extremely high convective transport rates provides the ability of maintaining a more uniform gas-catalyst temperature profile. This enables the reactor to run closer to the gas-phase temperature than before the onset of bulk mass transfer controlled reaction.

7.5.1 Monolithic Reactors

Many groups are studying Prox reactions on monolith reactors primarily due to the low pressure drop attributes. Additionally, there is a large experience base with using monolith reactors for mobile, that is, automotive, applications since the tremendous success of the catalytic converter. Moreover, these reactors have the ability to operate at adiabatic conditions and lend themselves to myriad design configurations. Typically, monolith reactors designed for Prox reactions have a high cell density >400 cells in.$^{-2}$ and are fabricated from ceramics such as alumina/silica or cordierite or high-temperature metal, such as FeCrAl alloys. Due to the low pressure drop attributes of monolith reactors typical GHSVs are in the range of 20,000–60,000 h^{-1}.

Technology has now advanced that catalyzed heat exchangers or microchannel reactors are being considered for applications where temperature control is required.[43–45] In one direction of a cross flow heat exchanger, a process gas is heated or cooled by a fluid flowing in the adjacent channel. For catalytic reactions, such as endothermic steam reforming, a washcoat on the walls of the process gas side of the exchanger is heated by catalytic combustion occurring on the opposite side, minimizing heat transfer resistance. The perpendicular channels may contain a cooling or heating fluid, or its walls can be catalyzed with a combustion catalyst to provide endothermic heats of reaction to the process gas. This is a significant step forward in that maintaining isothermal conditions is paramount for the Prox reaction to remain highly selective toward CO oxidation and avoid the consumption of hydrogen.

The U.S. Department of Energy is also investigating the performance of monolithic reactors using the same cell configuration as General Motors. The catalyst used was Selectra® PROX I (BASF Catalysts, LLC), which was used to generate a data set that was incorporated into a correlation that enable them to model different staged reactors. They developed an empirical correlation for selectivity toward CO conversion as a function of the inlet concentration. They further refined the correlation to include a reduced parameter $\lambda X^{1/4}$, which is a combination of oxygen to CO stoichiometry (see Section 7.2) and inlet CO concentration. The correlation was used to determine the optimum operating conditions of a Prox reactor while minimizing parasitic H_2 loss. Their results show to achieve less than 10 ppm CO concentration from the reactor exit, the inlet concentration cannot be more than 1.05% for a single-stage reactor and 3.1% for a two-stage reactor.[46–48]

7.5.2 SCT Reactors

A typical SCT reactor element is shown in Figure 7.8, which shows the expanded metal version. Other elements are made of similar construction using woven wires to form a mesh as well.

The SCT approach to chemical reactor design essentially consists of passing a reactant mixture over a catalyst at very high flow velocities, such that the residence time of the gas mixture inside the catalyst bed is on the order of milliseconds. Such SCT processes have commercially been used for a long time, for example, in ammonia oxidation reaction in nitric acid production, where a mixture of ammonia and air is passed over precious metal gauzes. In addition, SCT reactor technology has been the subject of research since the 1970s.[49] Near 100% conversion of ammonia with very high selectivity to the desired product is achieved. Catalytic systems based upon wire mesh coated with precious metal catalysts have been developed for many applications that need to take advantage of high transport coefficients.[50–56]

Using coated metal screen catalytic systems, it is possible to design reactors operating at very high GHSVs for both mass transfer and kinetically controlled reactions. This provides many advantages over the traditional packed bed or mono-

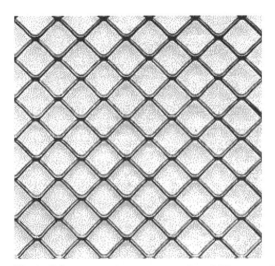

Figure 7.8.   Picture of a short contact time (SCT) reactor fabricated of expanded metal.

lith bed approaches, for example, higher space velocity operation for smaller reactor size and improved selectivity over conventional substrates. For example, in mass transfer limited reactions (e.g., Prox at certain operating conditions), SCT catalyst and substrate designs show higher selectivity to partial oxidation reactions, allowing operation within the material limits of commonly available materials. In the kinetically controlled regime, near-equilibrium operation at high space velocities (i.e., small reactor sizes) with lower selectivity to methanation has been observed. This is due to the higher geometric surface area (GSA) per unit volume of the reactor combined with a high specific surface area of the catalyst support/washcoat.[53,57-59] The ultrashort channel length avoids the boundary layer buildup observed in conventional long-channel monoliths and is justified in the following paragraph.

The breakup of the boundary layer in the SCT substrate (due to the very small L/D ratio) enables higher heat and mass transfer coefficients than that for long channel monoliths or foam substrates, as shown in Figure 7.9. This figure shows the results of a prediction of mass transfer coefficients as a function of channel length to channel diameter (L/D) ratio and flow velocity for monoliths with a channel diameter of 3 mm. The mass transfer coefficients are determined by using Reynold's analogy between heat and mass transfer coefficients and correlation given for heat

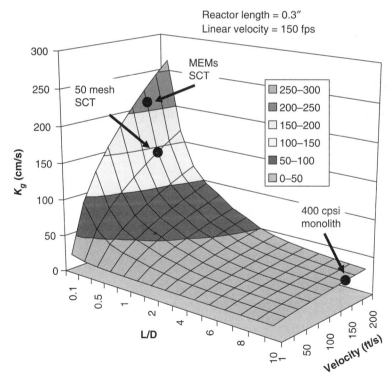

Figure 7.9. Plot of mass transfer coefficient (K_g) versus length-to-diameter ratio (L/D) versus velocity. See color insert.

transfer coefficients in Shah and London.[60] Conventional monoliths have L/D ratio much greater than 10, whereas for an SCT substrate, the L/D ratio is in the range of 0.1–0.5. The three-dimensional plot suggests that while increasing the flow velocity over conventional monolith substrates does provide some increase in mass transfer rate, a similar increase can be obtained over the SCT substrate at much lower flow velocities due to the lower L/D ratio. As an example, consider an SCT substrate with L/D of 0.1 and monolith substrate with L/D of 10. In order to achieve a mass transfer coefficient, K_g, of 100 cm/s, the flow velocity would have to be much greater than 200 ft/s. However, the SCT could operate at a K_g of 100 cm/s at only 75 ft/s. The data also suggest that the higher the velocity, the greater the difference between the mass transfer coefficients for reactor based on SCT and monolith substrate. Higher mass transfer coefficients also imply higher heat transfer coefficients. At a velocity of 150 ft/s, the correlation predicts mass transfer coefficient of 228 K_g (cm/s) for SCT substrate (L/D of 0.1) versus 28 for monolith substrate (L/D of 10). The implication therefore is that SCT coupled with high mass transfer coefficients is available only for low L/D substrates. Convective heat exchange with the gas phase is also strongly dependent on the boundary layer buildup. A lumped-sum capacitance analysis yielded time constants of 0.12 and 3.4 s for an SCT and ceramic monolithic substrate, respectively; which is a 30-fold improvement in thermal response.

To illustrate the effect on Prox performance by changing the substrate from a monolith to an SCT, Figure 7.10 shows the conversion versus selectivity for the exact same Pt/γ-Al$_2$O$_3$ formulation. The SCT reactor was a woven mesh type where a number of interlaced wires are woven together to form a screen. The wire diameters were 0.003 in. and made of Inconel. The monolithic reactor was constructed of 400 cells in.$^{-2}$ with a foil thickness of 0.003 in. and 0.3-in. thick. The material was made from Hastelloy-X, which is a high-temperature metal alloy. The catalyst formulation was identical on both reactors with a nominal loading per piece of 0.5244 g or 178.1 mg/in.$^{-3}$.

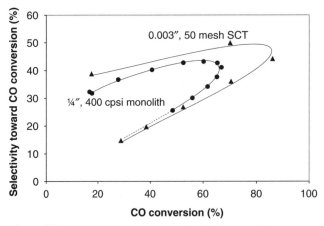

Figure 7.10.   Selectivity versus conversion for Prox reaction.

Figure 7.10 shows a wider operating range of the SCT substrate compared with a conventional monolith. Again, the objective is to have the oxygen from the air react only with the CO and not the hydrogen, yet that is very difficult in light of the disparity between the CO and H_2 concentrations. Clearly, catalyst formulation plays a key role in the selectivity toward certain reactions, yet in this test, both the monolith and SCT had the exact same formulation and loading. Since the SCT has increased transport, both heat and mass, it is likely that the heat of reaction is more quickly dissipated on the SCT, thus delaying or minimizing the light-off of the H_2 oxidation reaction (see Fig. 7.5).

7.5.3 Microchannel Reactors

Microchannel reactors have two basic designs. The first and most common is the fabrication of a unit, similar to a monolith, with very small channels with internal structures on a micrometer scale ($10-500\,\mu$m) and have specific surface areas in the range of $10,000-50,000\,m^2/m^3$. The channels are either chemically or photo etched or consist of a number of stacked plates containing microstructures. The catalysts are then coated onto the walls of the structure. The second type uses very small posts to hold a single catalyst particle in place, yet separated from the other catalyst particles that make up the reactor. In this fashion, the posts produce a microchannel for fluid flow yet enables flow around all sides of the catalyst particle. With the small channel sizes and high surface area-to-volume (SA/V) ratios, these devices are orders of magnitude more efficient than large-scale batch reactors in heat and mass transfer and thus are similar to SCT-type reactor substrates. Fast reactions, along with controlled residence times and temperatures, can generate the desired product without unwanted by-products or side reactions, resulting in higher selectivity and yield. A good overview of microchannel reactors and their potential in the chemical synthesis field is given in an article in the May 30, 2005 volume of *Chemical & Engineering News*.[45]

Combined with the use of precious metal catalyzed washcoats deposited on the walls, microchannel reactors can realize nearly 10 times reduction in reactor size compared with that of a process that utilizes catalyst particles. The washcoat thickness is usually less than $100\,\mu$ and provides greater structural stability. This stability arises from smaller thermal expansion ratios and lower temperature gradients.

Additionally, by microchannel designs with high open frontal areas (80%–90%), the pressure drop is significantly reduced, relative to packed beds. The reduced weight of the structure (relative to the packed bed) and the thin washcoat allow rapid response to transient operation, a requirement for fuel cells with turn down ratios of up to 20 to 1 with varying power demands. Numerous reactions, including many notable and industrially relevant reactions, have been tried out successfully in microreactors. Those reactions range from acrylate polymerization to cumene hydroperoxide rearrangement to electrochemical synthesis, as well as epoxidation reactions, Grignard chemistry, hydrogen/oxygen reduction to hydrogen peroxide and photochemical synthesis.

Microchannel reactors have some significant drawbacks. The most troublesome is clogging of the channels via incoming particulate matter or from fouling during the reaction process. Robustness is another common problem with microreactors. Because the unit is made at such a small internal scale, the resistance to mechanical shock is low. These issues usually render the microchannel reactor unsuitable for reactions that have precipitates as a product. For the Prox reaction, microchannel reactors are suitable provided there is no water condensation and the incoming reformate is particulate free, especially from carbon. Since microchannel reactors are often made from substrates that include stainless steel, Hastelloy, glass, silicon, polymers, and ceramics, another issue that could arise is chemical compatibility.

Currently, there is not much long-term data that exist using microchannel reactors. As such, durability predictions are difficult and extrapolations from large-scale reactors cannot be used. Large-scale reactors are usually tested and examined after millimeters of substrate or catalyst material has been lost to allow for good modeling of the system. If millimeters are lost in a microchannel reactor, the entire unit would be gone; therefore, current research is focused on understanding the long-term performance and degradation mechanisms of these types of reactors.

7.5.4   MEMS-Based Reactors

MEMS reactors are the natural extension to microchannel and SCT reactor geometries. The technology has advanced to the level where myriad flow geometries can be fabricated, from serpentine, single paths to multiple cross-linked paths to flow past cylinders. Besser has demonstrated the serpentine path type while other types are being developed. As an example of MEMS reactor technology, Castaldi et al. (manuscript in preparation) has demonstrated Prox reactor performance and the concept of a complete fuel processor system. The data obtained for the Prox reaction was taken using the MEMS reactor shown in Figure 7.11. The tests were conducted on the exact reactor shown in Figure 7.11 at an inlet temperature ranging between 130 and 150 °C at 100 mL/min total flow rate. The experiments show that the selectivity of oxygen introduced via air toward CO conversion is as high as 69%. This is significantly higher than conventional monolith reactors and is likely due to their ability to immediately remove heat generated from the CO conversion reaction.

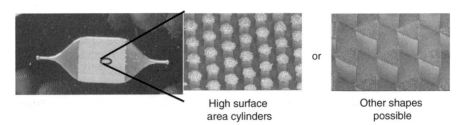

High surface
area cylinders

or

Other shapes
possible

Figure 7.11.   Picture and SEM image of MEMs SCT reactors.

Current pressure drop measurements also show that they are no higher than that of a conventional 400 cpsi monolith.

The specific MEMS reactors to be tested have been fabricated in an SCT-like mode and a bank of cylinders mode. Currently, the cylinders reactor has been tested and has shown operation nearly isothermal for the Prox reaction. Figure 7.11 shows a scanning electron micrograph (SEM) of the tested MEMS reactor where the flow is perpendicular to the cylinders as well as other substrates fabricated. This system differs from microchannel reactors and others in that there are no continuous channels where fully developed flow can exist.[61]

The primary attribute that affords desired behavior is the increased transport due to the SA/V ratios at comparable or lower pressure drop. For example, a 400 cpsi monolith has an SA/V = $2.64 \, m^2/L$ and a 50-mesh SCT has an SA/V = $6.3 \, m^2/L$, whereas the MEMs reactors go as high as $9.62 \, m^2/L$. Figure 7.9 shows this through a calculation of how the mass transfer coefficient increases with channel length to diameter for a given linear velocity. The pressure is dependent on the open area and the flow rate through the units.

The concept design on using MEMS for Prox and thus extending to fuel cell applications is shown in Figure 7.12. Again, since the high SA/V ratio enables more selective operation, this system can be extended to autothermal reforming where perhaps the CO/H_2 ratio could be favored over CO_2/H_2O, leading to more efficient

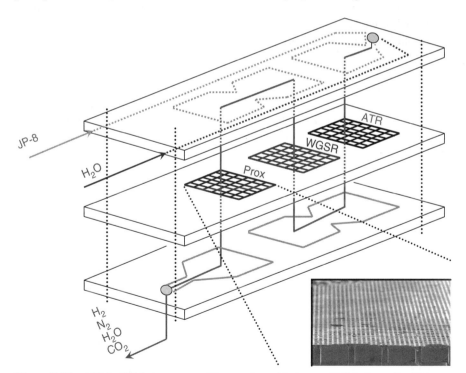

Figure 7.12.   MEMs SCT fuel processor. ATR, Autothermal Reformer; WGSR, Water Gas Shift Reactor. See color insert.

use of the fuel. Figure 7.12 shows the flow path of a microfabricated fuel processor that utilizes the MEMs substrate to afford a very compact lightweight reactor. The SEM image shows the geometry of a mesh-like substrate that can operate similar to a 50-mesh SCT metal substrate.

MEMs reactor technology is still developing. The most common application for these systems is in portable devices that are suitable for military use. Like microchannel reactors, the small internal dimensions can create problems with fouling and plugging of the flow paths.

7.6 COMMERCIAL UNITS

It was stated at the beginning of this chapter that Prox is only one of a number of technologies to purify a hydrogen reformate for fuel cell applications. Since it has the potential to operate at various scales and a range of conditions, especially transient regimes, many companies are actively pursuing commercial development. Most of the focus is on small-scale, portable applications for obvious reasons. The small scale enables rapid evolution and incorporation of lessons learned during demonstration and field operation. The portable power applications target is because Prox is one of only two possible technologies, membrane separation being the other. In the area of stationary applications, power or hydrogen generation, there are many technologies that are comparable or better than Prox. While this section will highlight commercial units being developed since the participants change rapidly, it is likely that it will not be comprehensive.

General Motors Research and Development Center is developing a Pt/Al_2O_3 600 cell/in.$^{-2}$ cordierite monolith reactors to remove the CO concentration in a full cell feed stream. They have developed a reactor model to better understand the interaction between kinetic control and transport limitations during the Prox reactions. Two combined groups of kinetic constants are derived that characterize the rates. The resulting experimentally measured net conversion rates are fit to the model rate expressions.[62,63]

The U.S. Department of Energy is also investigating the performance of monolithic reactors using the same cell configuration as General Motors. The catalyst used was Selectra® PROX I (BASF Catalysts, LLC).

In Los Alamos National Labs (LANL), a four-stage Prox reactor with four oxygen (from air) inlets and five heat exchangers to finely control the temperature in each unit was demonstrated, and the CO concentration was abated to very low with minor parasitic hydrogen consumption. However, to control the temperature by this method is still very limited, and usually, the mixture of oxygen and reformate is not good enough.

Pacific Northwest National Labs (PNNL) designed their microchannel Prox reactor in which reactors are interlayered with the microchannel heat exchanger, so reaction temperature was precisely controlled and high selectivity could be reached.

Moreover, the selectivity and temperature profile could also be controlled by adjusting the amount and type of catalyst in each chamber. In patent US6824904B2, a Prox reactor with a plurality of sections was designed by Brundage et al.[64] All the sections were individually optimized for operating at a preferred reaction temperature, different O_2/CO ratio, different cooling method, and even different catalyst/promotor/support so that it could enable quick light-off of the reactor, enhance the selectivity at the same CO conversion, and limit the RWGS reaction.

Precision Combustion, Inc. (PCI) of North Haven, CT, is developing catalytic reactors for integration into fuel processing and fuel cell power generation systems including using SCT technology for Prox. This platform technology is also extended to WGS and autothermal reformer units.

PCI's Prox technologies are being developed under National Science Foundation (NSF) and internal programs and have demonstrated highly selective conversion of CO in an extremely small package. PCI's Prox technology further reduces the CO concentration to below 10 ppm, typically in two stages while consuming less than 3% of the hydrogen in the reformate. PCI's Prox technology achieves >35 kW/kg and >30 kW/L with a 3-s response to transients from 10% to 100% load variation.

ACKNOWLEDGMENTS

The author would like to acknowledge the insightful discussions with numerous researchers in the field of Prox development, many of whom have been referenced. The collaboration with colleagues from PCI and Engelhard (now BASF Catalysts, LLC) has enabled this chapter to be written. In particular, the author is grateful to Dr. Subir Roychoudhury and Dr. Robert Farrauto. Finally, this chapter could not have been assembled without the excellent research in the many areas that make up Prox technology. It is the hope of the author that the pace of development continues and synergistic connections are made.

REFERENCES

1. IYUKE, S., MOHAMAD, A., DAUD, W., KADHUM, A., FISAL, Z., and SHARIFF, A. Removal of CO from process gas with Sn-activated carbon in pressure swing adsorption. *Journal of Chemical Technology and Biotechnology*, **2000**, 75, 803.
2. LATTNER, J.R. and HAROLD, M.P. Comparison of conventional and membrane reactor fuel processors for hydrocarbon-based PEM fuel cell systems. *Journal of Hydrogen Energy*, **2004**, 29, 393.
3. HAROLD, M.P., NAIR, B., and KOLIOS, G. Hydrogen generation in a Pd membrane fuel processor: Assessment of methanol-based reaction systems. *Chemical Engineering Science*, **2003**, 58, 2551.
4. Perry, R.H. and Green, D. (eds.) *Chemical Engineers' Handbook*, 6th ed. New York: McGraw-Hill, **1984**.
5. KAHLICH, M.J., GASTEIGER, H.A., and BEHM, R.J. Preferential oxidation of CO over $Pt/\gamma\text{-}Al_2O_3$ and $Au/\alpha\text{-}Fe_2O_3$: Reactor design calculations and experimental results. *Journal of New Materials for Electrochemical Systems*, **1998**, 1, 39.
6. ALTAMN, E.I. and GORTE, R.J. A study of small Pt particles on amorphous Al_2O_3 and a-$Al_2O_3\{0001\}$ substrates using TPD of CO and H_2. *Journal of Catalysis*, **1988**, 113 (1), 185–192.
7. FOGLER, H.S. *Elements of Chemical Reactor Engineering*, 2nd ed. Englewood Cliffs, NJ: Prentice Hall, **1990**, p. 447.
8. BARRAI, F. *An Integrated Precious-Metal Catalyzed CO Cleanup Train for PEM Fuel Cells*. New York: Columbia University, **2006**.
9. LYUBOVSKY, M., SMITH, L.L., CASTALDI, M., KARIM, H., NENTWICK, B., ETEMAD, S., LaPIERRE, R., and PFEFFERLE, W.C. Catalytic combustion over platinum group catalysts: Fuel-lean versus fuel-rich operation. *Catalysis Today*, **2003**, 83, 71.

10. SCHUBERT, M.M., GASTEIGER, H.A., and BEHN, R.J. Surface formates as side products in the selective CO oxidation on Pt/?-Al$_2$O$_3$. *Journal of Catalysis*, **1997**, 172, 256–258.
11. AVGOUROPOULOS, G., IOANNIDES, T., PAPADOPOULOU, C., BATISTA, J., HOCEVAR, S., and MATRALIS, H.K. A comparative study of Pt/γ-Al$_2$O$_3$, Au/α-Fe$_2$O$_3$ and CuO-CeO$_2$ catalysts for the selective oxidation of carbon monoxide in excess hydrogen. *Catalysis Today*, **2002**, 75, 157.
12. OH, S.H. and SINKEVITCH, R.M. Carbon monoxide removal from hydrogen rich fuel cell feed streams by selective catalytic oxidation. *Journal of Catalysis*, **1993**, 142, 254.
13. LEMONS, R.A. Fuel cells for transportation. *Journal of Power Sources*, **1990**, 29, 251.
14. WATANABE, M., UCHIDA, H., IGARASHI, H., and SUZUKI, M. Development of Pt/ZSM-5 catalyst with high CO selectivity for preferential oxidation of carbon monoxide in a reformed gas. *Chemistry Letters*, **1995**, 24, 21.
15. TORRES-SANCHEZ, R.M., UEDA, A., and TANAKA, K.J. Selective oxidation of CO in hydrogen over gold supported on manganese oxides. *Journal of Catalysis*, **1997**, 168, 125.
16. MARINO, F., DESCORME, C., and DUPREZ, D. Noble metal catalysts for the preferential oxidation of carbon monoxide in the presence of hydrogen (PROX). *Applied Catalysis. B, Environmental*, **2004**, 54, 59.
17. SON, I.H. and LANE, A.M. Promotion of Pt/γ-Al$_2$O$_3$ by Ce for preferential oxidation of CO in H$_2$. *Catalysis Letters*, **2001**, 76, 151.
18. KIEFFER, R., KIENNEMANN, A., and RODRIGUEZ, M. Promoting effect of Lanthana in the hydrogenation of carbon monoxide over supported rhodium catalysts. *Applied Catalysis. A, General*, **1988**, 42, 77.
19. USMEN, R.K., GRAHAM, G.W., WATKINS, L.H., and McCABE, R.W. Incorporation of La^{3+} into a Pt/CeO$_2$/Al$_2$O$_3$ catalyst. *Catalysis Letters*, **1994**, 30, 53.
20. YING, Y.J. and SWARTZ, W.E. An XPS study of the effect of La$_2$O$_3$ dopant on the dispersion and thermal stability of Pt/Al$_2$O$_3$ catalysts. *Spectroscopic Letters*, **1984**, 17, 331.
21. CRACIUN, R. and DULAMITA, N. Influence of La$_2$O$_3$ promoter on the structure of MnO$_x$/SiO$_2$ catalysts. *Catalysis Letters*, **1997**, 46.
22. LEDFORD, J.S., KIM, Y.M., HOUALLA, M., PROCTOR, A., and HERCULES, D.M. Surface analysis of lanthanum-modified cobalt catalysts. *Analyst*, **1992**, 117, 323.
23. CHIN, P., SUN, X., ROBERTS, G.W., and SPIVEY, J.J. Preferential oxidation of carbon monoxide with iron-promoted platinum catalysts supported on metal foams. *Applied Catalysis. A, General*, **2006**, 302, 22.
24. KOROTKIKH, O. and FARRAUTO, R.J. Selective catalytic oxidation of CO in H$_2$: Fuel cell applications. *Catalysis Today*, **2000**, 62, 249.
25. KOROTKIKH, O., FARRAUTO, R.J., and McFARLAND, A. Method of preparation of catalytic material for selective oxidation and catalyst members thereof. U.S. Patent 6,559,094, May 6, **2003**.
26. LIU, X., KOROTKIKH, O., and FARRAUTO, R.J. Selective catalytic oxidation of CO in H$_2$: Structural study of Fe oxide-promoted Pt/alumina catalyst. *Applied Catalysis. A, General*, **2002**, 226, 293.
27. ROBERTS, G.W., CHIN, P., SUN, X., and SPIVEY, J.J. Preferential oxidation of carbon monoxide with Pt/Fe monolithic catalysts: Interactions between external transport and the reverse water-gas-shift reaction. *Applied Catalysis. B, Environmental*, **2003**, 46, 601.
28. SHORE, L. and FARRAUTO, R.J. Preferential oxidation of CO in H$_2$ streams. In *PROX Catalysts, Handbook of Fuel Cells: Fundamental Technology and Applications* (ed. Vielstich, W., Lamm, A., and Gasteiger, H.A.), Vol. 3, Part 2. New York: John Wiley and Sons, p. 211, **2003**.
29. AVGOUROPOULOS, G. and IOANNIDES, T. Selective CO oxidation over CuO-CeO$_2$ catalysts prepared via the urea-nitrate combustion method. *Applied Catalysis. A, General*, **2003**, 244, 155.
30. AVGOUROPOULOS, G., IOANNIDES, T., MATRALIS, H., BATISTA, J., and HOCEVAR, S. CuO-CeO$_2$ mixed oxide catalysts for the selective oxidation of carbon monoxide in excess hydrogen. *Catalysis Letters*, **2001**, 73, 33.
31. TANAKA, Y., UTAKA, T., KIKUCHI, R., TAKEGUCHI, T., SASAKI, K., and EGUCHI, K. Water gas shift reaction for the reformed fuels over Cu/MnO catalysts prepared via spinel-type oxide. *Journal of Catalysis*, **2003**, 215, 271.
32. KAHLICH, M.J., GASTEIGER, H.A., and BEHM, R.J. Kinetics of the selective low temperature oxidation of CO in H$_2$ rich gas over Au/α-Fe$_2$O$_3$. *Journal of Catalysis*, **1999**, 182, 430.

33. LANDON, P., FERGUSON, J., SOLSONA, B., GARCIA, T., CARLEY, A., HERZING, A., KIELY, C., GOLUNSKI, S., and HUTCHINGS, G. Selective oxidation of CO in the presence of H_2, H_2O and CO_2 via gold for use in fuel cells. *Chemical Communications*, **2005**, 27, 3385.

34. IVANOVA, S., PETIT, C., and PITCHON, V. Application of heterogeneous gold catalysis with increased durabilitiy: Oxidation of CO & hydrocarbons at low temperature. *Gold Bulletin*, **2006**, 39, 3.

35. HAN, Y.F., KAHLICH, M.J., KINNE, M., and BEHM, R.J. CO removal from realistic methanol reformate via preferential oxidation-performance of a Rh/MgO catalyst and comparison to Ru/γ-Al_2O_3 and Pt/γ-Al_2O_3. *Applied Catalysis. B, Environmental*, **2004**, 50, 209.

36. KOTOBUKI, M., WATANABE, A., UCHIDA, H., YAMASHITA, H., and WATANABE, M. High catalytic performance of Pt-Fe alloy nanoparticles supported in mordenite pores for preferential CO oxidation in H2-rich gas. *Applied Catalysis. A, General*, **2006**, 307, 275.

37. TAYLOR, K.C., SINKEVITCH, R.M., and KLIMISCH, R.L. The dual state behavior of supported noble metal catalysts. *Journal of Catalysis*, **1974**, 35, 34.

38. ECHIGO, M., SHINKE, N., TAKAMI, S., HIGASHIGUCHI, S., HIRAI, K., and TABATA, T. Development of residential PEFC cogeneration systems: Ru catalyst for CO preferential oxidation in reformed gas. *Catalysis Today*, **2003**, 84, 209.

39. ECHIGO, M. and TABATA, T. A study of CO removal on an activated Ru catalyst for polymer electrolyte fuel cell applications. *Applied Catalysis. A, General*, **2003**, 251, 157.

40. AOYAMA, S. Apparatus and method for reducing carbon monoxide concentration and catalyst for selectively oxidizing carbon monoxide. European Patent Application #EP1038832 A1, September **2000**.

41. GASTEIGER, H.A., MARKOVIC, N.M., WANG, K., and ROSS, J.P.N. On the reaction pathway for methanol and carbon monoxide electrooxidation on Pt-Sn alloy versus Pt-Ru alloy surfaces. *Electrochimica Acta*, **1996**, 41, 2587.

42. SHORE, L., RUETTINGER, W.F., and FARRAUTO, R.J. Platinum group metal promoted copper oxidation catalysts and methods for carbon monoxide remediation. United States Patent Application #6193739, July 5, **2005**.

43. FARRAUTO, R.J., HWANG, S., SHORE, L., RUETTINGER, W.F., LAMPERT, J., GIROUX, T., LIU, Y., and ILINICH, O. New material needs for hydrocarbon fuel processing: Generating hydrogen for the fuel cell. *Annual Reviews of Material Research*, **2003**, 33, 1.

44. HECK, R., FARRAUTO, R.J., and GULATI, S. The application of monoliths for gas phase reactions. *Chemical Engineering Journal*, **2001**, 82, 149.

45. THAYER, A. Harnessing mircoreactors. *Chemical & Engineering News*, pp. 43–52, May 30, **2005**.

46. AHLUWALIA, R.K., ZHANG, Q., CHMIELEWSKI, D.J., LAUZZE, K.C., and INBODY, M.A. Performance of CO preferential oxidation reactor with noble-metal catalyst coated on ceramic monolith for onboard fuel processing applications. *Catalysis Today*, **2005**, 99, 271.

47. AHLUWALIA, R.K., ZHANG, Q., and INBODY, M.A. Preferential oxidation of CO on a noble-metal catalyst coated ceramic monolith. *Preprints of Symposia—American Chemical Society, Division of Fuel Chemistry*, **2003**, 48, 848.

48. AHLUWALIA, R.K., ZHANG, Q., and INBODY, M.A. Preferential oxidation of CO on a noble-metal catalyst coated ceramic monolith. *Abstracts of Papers, 226th ACS National Meeting*, FUEL-160. New York, September 7–11, **2003**.

49. CATON, J.A. Heterogeneous catalysis of lean ethylene/air mixtures by platinum coated wire screens. American Society of Mechanical Engineers. *Paper No. 76-WA/GT-2*, **1976**.

50. CASTALDI, M., LAPIERRE, R., LYUBOVSKY, M., and ROYCHOUDHURY, S. Effect of water on performance and sizing of fuel-processing reactors. *Abstracts of Papers, 226th ACS National Meeting*, FUEL-094. New York, September 7–11, **2003**.

51. CASTALDI, M.J. Method for reduced methanation. U.S. Patent 6,746,657, **2004**.

52. CASTALDI, M.J., ROYCHOUDHURY, S., BOORSE, R.S., KARIM, H., LAPIERRE, R., and PFEFFERLE, W.C. Fuel Processing Session I. In *Compact, Lightweight Preferential CO Oxidation (PROX) Reactor Development and Design for PEM Automotive Fuel Cell Applications. Proceedings from the 2003 Spring National Meeting and Process Industries Exposition* (ed. AIChE). New Orleans, LA: AIChE, March 30–April 3, **2003**.

53. CASTALDI, M.J., LYUBOVSKY, M., LAPIERRE, R., PFEFFERLE, W.C., and ROYCHOUDHURY, S. Performance of microlith based catalytic reactors for an isooctane reforming system. *SAE Technical Paper 2003-01-1366*, **2003**.

54. PFEFFERLE, W.C., CASTALDI, M., ETEMAD, S., KARIM, H., LYUBOVSKY, M., ROYCHOUDHURY, S., and SMITH, L. Catalysts for improved process efficiency. *Abstracts of Papers, 223rd ACS National Meeting*, CATL-018. Orlando, FL, April 7–11, **2002**.

55. CASTALDI, M.J., BOORSE, R.S., ROYCHOUDHURY, S., MENACHERRY, P., and PFEFFERLE, W.C. Lightweight, Compact, Ultra-fast Short Contact Time Preferential Oxidation Reactor for Automotive PEM Fuel Cell Applications, *NSF National Meeting* (ed. National Science Foundation). San Juan, Puerto Rico: National Science Foundation, January **2002**.

56. DEUTSCHMANN, O., SCHMIDT, L.D., and WARNATZ, J. Detailed modelling of short-contact-time reactors. *Recents Progres en Genie des Procedes*, **1999**, 13, 213.

57. AHMED, S., KOPASZ, J., KUMAR, R., and KRUMPELT, M. Water balance in a polymer electrolyte fuel cell system. *Journal of Power Sources*, **2002**, 112, 519.

58. ROYCHOUDHURY, S., CASTALDI, M., LYUBOVSKY, M., LAPIERRE, R., and AHMED, S. Microlith catalytic reactors for reforming iso-octane-based fuels into hydrogen. *Journal of Power Sources*, **2005**, 152, 75.

59. RUETTINGER, W., ILINICH, O., and FARRAUTO, R.J. A new generation of water gas shift catalysts for fuel cell applications. *Journal of Power Sources*, **2003**, 118, 61.

60. SHAH, R.K. and LONDON, A.L. Laminar flow forced convection in ducts. *Journal of Heat Transfer*, **1974**, 96, 159.

61. SRINIVAS, S., DHINGRA, A., IM, H., and GULARI, E. A scalable silicon microreactor for preferential CO oxidation: Performance comparison with a tubular packed-bed microreactor. *Applied Catalysis. A, General*, **2004**, 274, 285.

62. BISSETT, E.J. and OH, S.H. PrOx reactor model for fuel cell feedstream processing. *Chemical Engineering Science*, **2005**, 60 (17), 4722.

63. BISSETT, E.J., OH, S.H., and SINKEVITCH, R.M. Pt surface kinetics for a PrOx reactor for fuel cell feedstream processing. *Chemical Engineering Science*, **2005**, 60, 4709.

64. BRUNDAGE, M.A., PETTIT, W.H., and BORUP, R.L. Multi-stage, isothermal CO preferential oxidation reactor. U.S. Patent 6,824,904, **2004**.

Chapter 8

Hydrogen Membrane Technologies and Application in Fuel Processing

DAVID EDLUND

Azur Energy

8.1 INTRODUCTION

This chapter will address the use of membrane technology for separating and purifying hydrogen from reformate. Practically speaking, the selection of the reforming process (i.e., the reforming chemical reactions and equilibrium conversion) is of little concern. What is important is that the goal is to produce a relatively pure product stream of hydrogen from a feed stream (reformate) that consists largely of hydrogen, carbon dioxide, carbon monoxide, water, nitrogen (optional), various sulfur species, and various organic species (both depending on feedstock selection and reforming chemistry).

As applied to gas separations, membrane-based processes are nearly always pressure-driven separation processes, similar to pressure swing adsorption (PSA), which has enjoyed significant commercial success. Both membrane-based separation and PSA are universal purification processes as applied to hydrogen separation and purification: meaning that all impurities are rejected in favor of hydrogen preferentially passing into the product stream. Selective methanation and preferential oxidation are examples of two chemical processes for purifying hydrogen from reformate that are selective only for the removal (or reduction in concentration) of carbon monoxide in the reformate feed stream. Hence, selective methanation and preferential oxidation are not universal purification processes as applied to hydrogen separation and purification. However, unlike PSA, membrane processes are typically continuous throughput rather than cyclical in nature.

Hydrogen and Syngas Production and Purification Technologies, Edited by Ke Liu, Chunshan Song and Velu Subramani
Copyright © 2010 American Institute of Chemical Engineers

357

When selecting between membrane and PSA purification processes, it is important to consider scale. Practice has shown that PSA scales up economically and for large stationary applications (e.g., petroleum refining, petrochemical production, coal gasification), this may be the best choice. By the same reasoning, PSA usually does not scale down economically, and here membrane-based processes may be strongly favored. Also, membrane processes are more likely to be selected for applications wherein the platform is moving or otherwise subjected to shock and vibration that would likely have detrimental effects on PSA adsorbent beds.

Yet another potential advantage of membrane purification processes over PSA is that membrane processes are usually more easily controlled. As mentioned above, they are continuous throughput operations as opposed to cyclic, multibed operations (characterized by PSA). Assuming the feed stream composition does not vary too much in composition, control of the membrane process may be as simple as controlling the membrane operating temperature, feed pressure, and permeate pressure.

8.2 FUNDAMENTALS OF MEMBRANE-BASED SEPARATIONS

This section will provide an overview of the principles of hydrogen separation and purification using membranes. More detailed discussions of the theory governing membrane separation processes can be found elsewhere.[1]

There are essentially four different types of membranes, or semipermeable barriers, which have either been commercialized for hydrogen separations or are being proposed for development and commercialization. They are polymeric membranes, porous (ceramic, carbon, metal) membranes, dense metal membranes, and ion-conductive membranes (see Table 8.1). Of these, only the polymeric membranes have seen significant commercialization, although dense metal membranes have been used for commercial applications in selected niche markets. Commercial polymeric membranes may be further classified as either asymmetric (a single polymer composition in which the thin, dense permselective layer covers a porous, but thick, layer) or composite (a thick, porous layer covered by a thin, dense permselective layer composed of a different polymer composition).[2]

Porous membranes, especially ceramic and carbon compositions, are the focus of intense development efforts. Perhaps, the least studied of the group, at least for hydrogen separations, are the ion-conducting membranes (despite the fact that many fuel cells incorporate a proton-conducting membrane as the electrolyte), and this class of membranes will not be discussed further in this chapter.

It is helpful to think of a simple membrane process as shown in Figure 8.1. A hydrogen-selective membrane is sealed within a housing (pressure vessel) to make a membrane module. The feed stream enters the membrane module, and hydrogen selectively permeates the membrane. The hydrogen-depleted raffinate stream exits the membrane module as does the permeate stream (enriched in hydrogen). The hydrogen partial pressure in each stream is denoted by P_{H2} where the subscripts f,

Table 8.1. Comparison of Membrane Types for Hydrogen Separation

Parameters	Membrane Type			
	Polymeric	Nanoporous	Dense Metal	Ion Conducting
Typical composition	Polyimide; cellulose acetate	Silica; alumina; zeolites; carbon	Palladium alloys	Water-swollen, strong-acid, cation exchange membranes; dense ceramics (perovskytes)
Diffusion mechanism	Solution–diffusion	Size exclusion	Solution–diffusion	Solution–diffusion
Driving force	Pressure gradient	Pressure gradient	Pressure gradient	Ionic gradient
Operating temperature	$\leq 110\,°C^3$	$\leq 1000\,°C$	$150–700\,°C$	$\leq 180\,°C$ (polymeric); $700–1000\,°C$ (ceramic)
Relative permeability	Moderate-high	Low-moderate	Moderate	Moderate
Typical selectivity	Moderate	Low-moderate	Very high	Very high
Relative cost	Low	Low	Moderate	Low

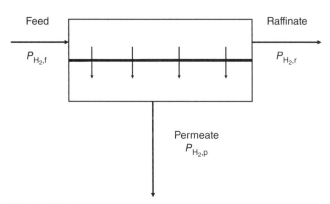

Figure 8.1. Schematic of a hydrogen separation membrane and membrane module.

r, and p indicate respectively the feed, raffinate, and permeate streams. Since hydrogen permeation is a pressure-driven process and by definition hydrogen is passing from the feed stream into the permeate stream, it must be true that $P_{H2,f} > P_{H2,p}$. Moreover, if hydrogen is to permeate across the entire membrane surface into the permeate stream, then $P_{H2,r} > P_{H2,p}$.

Generally, the permeate hydrogen stream will be at a sensible pressure, and it is often preferred that the permeate pressure be greater than ambient pressure (if for no other reason, concerns over safety argue for any hydrogen stream to be at a pressure greater than ambient pressure). The percentage of hydrogen that can be recovered from the feed stream is limited such that $P_{H2,r}$ can only approach, but never equal, $P_{H2,p}$. Indeed, the membrane area increases exponentially as $P_{H2,r}$ approaches $P_{H2,p}$ in value.

The volumetric flow rate at which a gas permeates a membrane is called the flux (J_i) and it is related to membrane thickness (l), permeability (P_i), and differential pressure (ΔP) as shown in Equation 8.1,

$$J_i = P_i \frac{\Delta P}{l}, \tag{8.1}$$

where the subscript i refers to the specific gas that permeates the membrane, for instance hydrogen, and ΔP is the partial pressure of gas i in the feed stream minus the partial pressure of gas i in the permeate stream. Equation 8.1 is true for pure gases permeating a membrane and may be applied to mixed-gas feed streams if the recovery of the permeating component i is very low (refer to Eq. 8.9 when the recovery of component i is moderate to high). The common unit for the permeability constant P_i is the barrer: $10^{-10}\,cm^3$ (STP) cm/(cm^2·s·cm Hg).

For solution–diffusion permeation mechanisms, another useful relationship is

$$P_i = D_i \cdot S_i, \tag{8.2}$$

where D_i is the diffusivity of the permeating gas in the bulk membrane, and S_i is the solubility of the permeating gas in the bulk membrane. This relationship takes on special importance with dense metal membranes for hydrogen separation (see Section 8.5). Diffusivity can be thought of as a kinetic term, and solubility is a thermodynamic term. Indeed, it is the solubility of the permeating gas that establishes the concentration gradient required to have net diffusion from the feed side of the membrane to the permeate side of the membrane.

Selectivity (or separation factor) is another important parameter when evaluating membrane separation processes. The selectivity of a membrane ($\alpha_{i/j}$) is the ratio of the permeability of gas i to the permeability of gas j. Thus, for pure gases with near-zero permeate pressure, the selectivity is simply stated as

$$a_{i/j}^* = \frac{Pi}{Pj}, \tag{8.3}$$

where the * indicates that the selectivity is calculated for the special conditions of near-zero permeate pressure. Realistically, the permeate pressure will be greater than zero and the correct expression for selectivity becomes

$$a_{i/j} = \left(\frac{y_i}{y_j}\right)\left(\frac{x_i}{x_j}\right), \tag{8.4}$$

where y_i and y_j are the concentrations of gas species i and j in the permeate stream, and x_i and x_j are the concentrations of gas species i and j in the feed stream. Ideally, a membrane will exhibit high flux and high selectivity. More often, selectivity decreases as flux increases. However, in the case of defect-free dense metal membranes that only permeate hydrogen and exclude all other gases, it can be seen in Equation 8.4 that in the selectivity for hydrogen over all other gases, j is infinite since the value of y_j would be zero. In this case, selectivity is independent of flux and, as illustrated in Equation 8.4, infinite hydrogen selectivity is independent of the feed stream composition.

Factors that may influence the selectivity of a given hydrogen-purification membrane may be inferred from Equations 8.3 and 8.4. For instance, any factors that influence the permeability of the membrane to hydrogen (i.e., factors that influence the diffusivity and solubility terms) are expected to influence the selectivity provided that those factors do not cause a proportional increase in the permeability of feed-stream components other than hydrogen. Such factors include the composition of the membrane under the operating conditions and the operating temperature. Also, as the hydrogen recovery (see below) approaches the limiting value (i.e., as $P_{H2,r}$ approaches $P_{H2,p}$), the selectivity will decrease unless the membrane has very high (approaching infinite) selectivity for hydrogen. Of course, any membrane defect that allows for hydrodynamic flow of gases through the membrane will have an adverse impact on membrane selectivity.

For all membrane processes, the hydrogen recovery is defined as follows:

$$\text{Recovery} = \left(\frac{M_{H2,p}}{t}\right) \bigg/ \left(\frac{M_{H2,f}}{t}\right), \tag{8.5}$$

where $M_{H2,p}$ and $M_{H2,f}$ are the moles of hydrogen in the permeate stream and in the feed stream, respectively, and t is a constant unit of time (e.g., 1 min or 1 h). Economic considerations almost always argue for the highest possible hydrogen recovery, limited to the requirement to maintain a hydrogen partial-pressure gradient across the membrane. However, as previously mentioned, increasing hydrogen recovery eventually leads to an exponential increase in membrane area that, in turn, causes a substantial increase in capital cost yielding a practical limit to hydrogen recovery that is less than the theoretical limit.

Knudsen diffusion dominates when microporous membranes are used for gas separations. This diffusion mechanism occurs when the membrane pore size is smaller than the mean free path of the gas molecules in the feed stream. Although Knudsen diffusion may allow for relatively high hydrogen flux (assuming a thin membrane that has a high degree of porosity and low degree of tortuosity), the selectivity for hydrogen over other gases is usually low. For instance, in the limiting case of zero permeate pressure, the ideal selectivity for hydrogen over carbon monoxide is only 3.7; in reality, the observed selectivity is likely to be less.

Nanoporous inorganic ceramic membranes show significant promise for hydrogen separation and purification, primarily due to the high selectivity that is afforded by this class of membranes. Development work has focused on zeolites; although other inorganic compositions have been synthesized that yield pore sizes much less than 1 nm.[4] Nanoporous ceramic membranes are typically asymmetric using either microporous ceramic or microporous metal support layers. To date, scale-up and modularization challenges remain to be resolved.

Dense metal membranes offer the interesting characteristic of inherently infinite selectivity for hydrogen. This property is derived from the chemical interaction of hydrogen with the metal membrane—in essence, hydrogen (a metallic element that happens to be a gas) alloys with the bulk metal in the membrane to form a random solution of hydrogen atoms in the bulk metal. Dissolved hydrogen then permeates the metal membrane following a solution–diffusion mechanism. This is shown schematically in Figure 8.2. However, since it is atomic hydrogen (H) and not molecular hydrogen (H$_2$) that dissolves in, and permeates through, the membrane, the expression for hydrogen flux through dense metal membranes is a modified form of Equation 8.1,

$$J_{H2} = P_{H2}\left(\frac{\Delta P}{l}\right) \tag{8.6}$$

and

$$\Delta P = \left[(P_{H2,f})^n - (P_{H2,p})^n \right], \tag{8.7}$$

where the exponent n is a value between 0.5 and 1 (Eq. 8.6 rigorously applies only to a pure hydrogen feed stream or to mixed-gas feed streams if the hydrogen recovery is very low; at moderate to high hydrogen recovery, Eq. 8.9 should be used). Under ideal circumstances (clean membrane surfaces and no interaction of the membrane surfaces with species other than hydrogen), the value of n is 0.5, indicating that the membrane is sufficiently thick that diffusion of atomic hydrogen

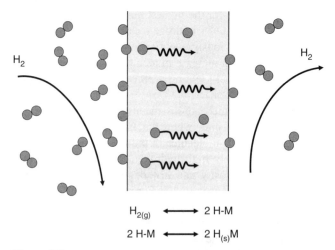

Figure 8.2. Commonly accepted mechanism for the permeation of hydrogen through dense metal membranes.

through the membrane is rate limiting. Under the same ideal circumstances, if the membrane is exceptionally thin such that the surface reactions of molecular hydrogen dissociation to yield solubilized atomic hydrogen ($H_{(S)}M$) are rate limiting, the value of n will be 1. In reality, the value of n is usually found by experiment to be close to 0.6.

In cases where high purity hydrogen is valued, dense metal membranes are an attractive option over polymeric membranes and porous membranes that exhibit much lower selectivities. Two examples where this is true are low-temperature fuel cells (e.g., proton exchange membrane fuel cells [PEMFCs] and alkaline fuel cells [AFCs]) and hydrogen-generating sites where the product hydrogen is to be compressed and stored for future use.

8.3 MEMBRANE PURIFICATION FOR HYDROGEN ENERGY AND FUEL CELL APPLICATIONS

Much has been written about the future of hydrogen energy and fuel cells for a variety of applications. Distributed production of hydrogen for refueling vehicles is being supported on a limited basis through demonstration programs funded by the governments of Canada, the U.S., Japan, and the European Union. These refueling stations are sized to deliver about $50\,Nm^3/h$ up to $500\,Nm^3/h$ of high-purity hydrogen synthesized by reforming natural gas or other feedstocks (including kerosene, methanol, and liquefied petroleum gas [LPG]). The purity of the product hydrogen must be $\geq 99.99\%$ to be compatible with hydrogen fuel cells (PEMFC); this is set by international standards and, to a lesser degree, by the inefficiencies of compressing low-purity hydrogen for storage. Although dense metal membranes can meet this purity requirement in a single pass, polymeric membranes and porous membranes cannot unless multiple membrane stages (with associated recompression) are employed.

Proposed hydrogen fuel cell applications include vehicular propulsion (scooter and motorcycle, automobile, truck, locomotive, boat, and ship); vehicle auxiliary power units (APUs); stationary combined heat and power (CHP) for residential and commercial buildings; backup power and uninterruptible power supplies (UPS); autonomous power supplies for remote instrumentation, navigational aids, and perimeter and border security; and portable power systems for boats, recreational vehicles, and construction and military purposes. Mostly these applications would utilize PEMFC technology, although in some instances, solid oxide fuel cells are under investigation (this class of fuel cells operates at very high temperatures and does not require high-purity hydrogen fuel).

Figure 8.3 is a high-level schematic of the two most common versions of PEMFC systems for the aforementioned applications. Figure 8.3a shows a fuel cell system with an integral fuel reformer and associated hydrogen purifier. In this case, hydrogen is produced and purified in response to demand from the PEMFC stack. The feedstock for the fuel reformer may be any gas or liquid fuel, but practically speaking is either a hydrocarbon or alcohol. Figure 8.3b shows a simpler system

(a)

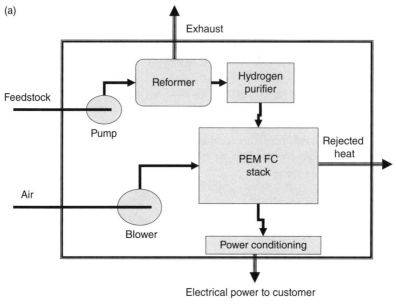

(b)

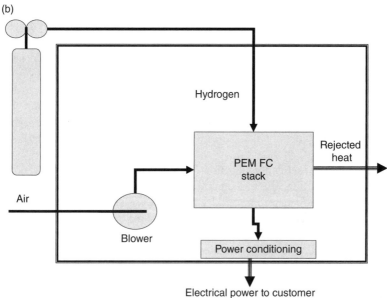

Figure 8.3. Block diagram of a PEMFC system with (a) internal fuel-reforming module and (b) externally supplied merchant-grade hydrogen.

design in which hydrogen is supplied to the PEMFC stack from an external storage device, such as a cylinder of compressed gas.

Current precommercial and commercial PEMFCs use a platinum electrocatalyst on the anode and the cathode.[5] To avoid adverse poisoning of the anode electrocatalyst, the hydrogen fuel must be relatively free of harmful contaminants including carbon monoxide (<1 ppm), sulfur compounds (<10 ppb), and unsaturated organic compounds in general. These contaminants may competitively adsorb onto the platinum electrocatalyst and block the reaction of hydrogen. Other harmful contaminants that are to be avoided include ammonia (<10 ppb) and organic amines; these compounds are bases, and they react with the acid electrolyte in the PEMFC to block the mobility of protons from the anode to the cathode.

When discussing the purity of hydrogen to be used in a fuel cell, it is important to distinguish between the concentration and effects of reactive contaminants (as above) and the concentration and effects of inert contaminants (such as nitrogen, methane, and water). Inert contaminants can be tolerated by PEMFCs at significantly high concentrations, perhaps as high as 50% or so, before adverse impact on the PEMFC performance is observed. This is because the dilutive effect of the inert contaminant is masked by the slow kinetics of oxygen reduction at the cathode. So for the use of membranes to purify hydrogen for fuel cell applications (primarily PEMFC), the focus is on the elimination of reactive contaminants more than on the bulk purity.

Given the many different potential applications of fuel cells, and the range of scale (and economics) represented by distributed generation of hydrogen at one extreme and centralized hydrogen production from fossil fuels at the other extreme, three performance and economic factors dominate the selection of a hydrogen purification process: (a) the required purity of the product hydrogen stream, (b) the process scale, and (c) the required energy efficiency of the hydrogen production and purification process.

8.3.1 Product Hydrogen Purity

Merchant hydrogen is produced in many grades based on the degree of hydrogen purity. The Compressed Gas Association recognizes eight different grades of compressed hydrogen (ranging from 99.8% minimum purity to 99.9991% minimum purity) and four grades of liquid hydrogen (ranging from 99.995% minimum purity to 99.9997% minimum purity).[6] For captive use, hydrogen may be at a lower purity or higher purity that is specific to the application. Captive uses include fuel cell systems wherein the hydrogen is produced by a unit operation within the system (Fig. 8.3a).

If the application requires that the product hydrogen meet or exceed the purity standards for merchant hydrogen, the dense metal membrane technology is suitable with a single pass whereas polymeric or porous membranes will require multiple stages. Between each successive stage, compression and interstage cooling will be necessary. For example, Figure 8.4 illustrates the purification of hydrogen from a

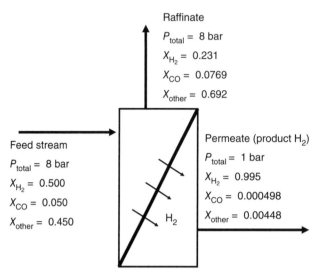

Figure 8.4.   Example purification process using a polymeric membrane and 70% hydrogen recovery.

prototypical reformate feed stream. The membrane is assumed to be a polymeric membrane with a high selectivity of 200. Furthermore, it is assumed that this selectivity is true for all impurities in the feed stream. The hydrogen recovery is 70%.

The product hydrogen bulk purity after a single pass is close to, but does not meet, the lowest grade of merchant hydrogen. Examining the carbon monoxide concentration in the product stream reveals that this impurity is at an unacceptably high concentration (the maximum allowable concentration in merchant hydrogen is 10 ppm).[6] Compressing the product hydrogen stream and passing this into another membrane module with the same selectivity (200) and hydrogen recovery (70%) yields a product hydrogen stream of 99.9975% purity with only 2.5 ppm carbon monoxide. In this two-stage process the overall hydrogen recovery is 49% (0.7×0.7), and there is added capital and operating costs due to the requirement for compression and interstage cooling between the first and second membrane modules.

Instead of a polymeric membrane, if a dense metal membrane is used, the results are quite different due to the inherently high hydrogen selectivity that can be achieved with dense metal membranes. This is illustrated in Figure 8.5 assuming a membrane selectivity of 90,900 (palladium–silver membrane of this selectivity and higher values are available from companies such as Power and Energy, Inc., Ivyland, PA).[7] Even though the change in composition of the raffinate stream is almost imperceptible, there is a significant decrease in trace contaminant concentration in the permeate hydrogen stream. Advantages of the dense metal membrane over the polymeric membrane include substantially higher product hydrogen purity in a single stage and, at the same time, substantially greater hydrogen recovery overall.

When hydrogen is to be used in a captive process, such as the integrated fuel cell system shown in Figure 8.3a, the hydrogen purity requirements may be less

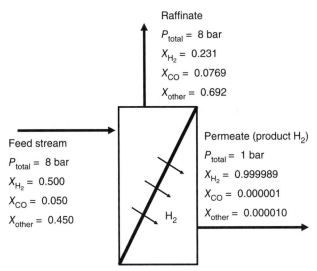

Raffinate

P_{total} = 8 bar

X_{H_2} = 0.231

X_{CO} = 0.0769

X_{other} = 0.692

Feed stream

P_{total} = 8 bar

X_{H_2} = 0.500

X_{CO} = 0.050

X_{other} = 0.450

Permeate (product H_2)

P_{total} = 1 bar

X_{H_2} = 0.999989

X_{CO} = 0.000001

X_{other} = 0.000010

H_2

Figure 8.5.   Example purification process using a dense metal membrane and 70% hydrogen recovery.

rigorously defined. The lifetime of the fuel cell and type of fuel cell (operating temperature, electrolyte composition, and electrocatalyst composition) will dictate the hydrogen purity specifications. For example, if the targeted fuel cell (PEMFC or AFC) lifetime is relatively short (<3000 h) and the electrocatalyst is a platinum–ruthenium composition, then a relatively high carbon monoxide concentration (10–50 ppm) may be tolerated even with low-temperature (PEMFC) fuel cells. But if the electrocatalyst is pure platinum and/or the targeted lifetime is >3000 h, then the carbon monoxide concentration must be <10 ppm and preferably <1 ppm. If the operating temperature of the fuel cell is ≥180 °C, then the carbon monoxide concentration may be ≥0.2%.

8.3.2   Process Scale

A general discussion of large-scale hydrogen purification processes (PSA, polymeric membrane, and cryogenic) has been published by UOP.[8] Although aimed mostly toward refinery and petrochemical operations, the selection guidelines discussed in the paper are also applicable to large-scale hydrogen production by steam-reforming natural gas, coal gasification, and so on. Commercial polymeric hydrogen-selective membrane modules are offered by several global companies including Air Products and Chemicals (PRISM membrane), Air Liquide S.A. (MEDAL membrane), UOP, and Ube Industries, Ltd.

Membranes scale to a hydrogen purification application based on a simple linear relationship: if hydrogen throughput is increased by two times then, the required membrane area will increase by two times (all other factors held constant). This means that the cost of a membrane purification process is nearly linear with the scale of the process. In comparison to volumetric processes (typified by PSA and chemical reactors), membranes generally scale down economically and physically to yield a purification process of smaller size.

Membranes are also simple to operate. This may not be a distinguishing factor for very large-scale applications, but it becomes very important at smaller scales due to limitations of size, cost, and availability of components. Clearly, there is a difference between the relative process controls and complexity that can be accepted in a coal-gasification facility versus a 3-kW fuel cell APU for installation in an automobile. IdaTech, LLC (Bend, OR) has demonstrated compact fuel processors that combine a methanol steam reforming reactor directly with hydrogen separation membranes.[9] (HyPurium™ membrane module utilizing a hydrogen-selective membrane composed of a palladium–copper alloy). The HyPurium™ membrane purifier is smaller than the methanol reformer and requires no active process control since it is thermally integrated with the catalytic methanol steam reformer.

8.3.3   Energy Efficiency

Membrane-based hydrogen purification is a pressure-driven, thermally neutral process. As discussed previously, hydrogen recovery is limited by the need to have a pressure gradient across the membrane. The degree of hydrogen recovery impacts the overall process energy efficiency, since energy efficiency η is defined for a hydrogen-producing process as

$$\eta = (\text{energy content of product } H_2)/(\text{total energy input}). \qquad (8.8)$$

Any convenient and consistent unit may be used to express the energy derived from the product hydrogen as well as that supplied to the process. However, if the product hydrogen is to be used to generate electricity (as in a fuel cell), then care should be taken to only take credit for the electrical-generating potential of the hydrogen and not its heating value (due to inefficiencies in a fuel cell, the electrical-generating potential is usually 50%–70% of the heating value of hydrogen).

The detailed integration of a membrane purification process and a fuel reforming process will also impact the overall energy efficiency. Two example process schematics are discussed in the context of Figure 8.6. Methane steam reforming (MSR) is commonly operated at a temperature of 700–800 °C and it is a net endothermic process. If the membrane is a polymer (e.g., polyimide), it will operate at about 50 °C, requiring substantial heat to be removed from the reformate stream by HEX 1. This thermal energy can be captured by the incoming feed streams (methane and water). The raffinate stream exiting the membrane module has considerable fuel value due to remaining hydrogen, carbon monoxide, and methane. This fuel stream is fed to the burner that supplies heat to the MSR. To maximize the efficiency of combustion, the fuel stream should be preheated prior to being combusted; this is shown by optional HEX 2.

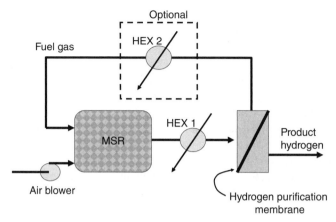

Figure 8.6. Generic membrane purification process integrated with a methane steam reformer (MSR).

Assuming a generic reformate composition and hydrogen recovery of about 70% (consistent with the illustrative examples in Figs. 8.4 and 8.5), the fuel gas stream has a heating value of about 130 kJ/mol and the heat capacity (C_p) is approximately 30 J/(°C·mol). The amount of thermal energy that needs to be added to the fuel gas stream through HEX 2 in order to heat the stream from 40 to 440 °C is about 36 kJ/mol of methane that is reformed. This represents a significant amount of thermal energy—roughly 10% of the energy needed to reform 1 mol of methane.

In contrast, if the membrane is an inorganic composition (e.g., a dense metal membrane or a nanoporous ceramic membrane), the membrane module may be operated at the elevated temperature of 450 °C. In this case, there is no need for optional HEX 2 as the fuel gas stream will exit the membrane module at 450 °C and pass to the burner without further cooling. In addition to a net increase in overall process energy efficiency, the elimination of HEX 2 also represents a reduction in capital cost for the system.

8.4 MEMBRANE MODULES FOR HYDROGEN SEPARATION AND PURIFICATION

A membrane module is the completed device that contains the membranes providing adequate sealing between the feed stream and the permeate stream; a housing that is also a pressure vessel; and appropriate fittings for making the required fluid connections to the feed stream, the raffinate stream, and the permeate stream. Membrane modules are usually cylindrical in shape, although they may alternatively be rectangular. Commercial polymeric membrane modules are either spiral wound (rolled sheet membrane) or shell and tube (hollow-fiber membrane).[10] Ceramic membrane modules and dense metal membrane modules are either plate-and-frame or shell-and-tube designs. The shell-and-tube design differs from that used with polymeric membranes primarily in the size of the membrane "tube" and the sealing means. Ceramic membranes and dense metal membranes are frequently larger in diameter than polymeric hollow-fiber membranes. These inorganic membrane tubes are normally sealed to the module shell using compression fittings, brazing, or welding (polymeric hollow-fiber membranes are glued, or "potted", to the shell).

The highest membrane packing density (total effective membrane area per volume of membrane module) is achieved with hollow-fiber (polymeric) membranes. The lowest packing density results from using large-diameter membrane tubes (approximately 10 mm in diameter or larger). In both cases, the membrane module is a shell-and-tube design. If dimensions of both the membrane and module are assumed as shown in Figure 8.7, then a comparison of membrane packing density as a function of the diameter of the membrane can be conducted. To simplify the analysis, it is assumed that membranes occupy 70% of the internal volume of the membrane module, and that this value does not change throughout the analysis. The results are presented in Table 8.2.

In comparison, a plate-and frame membrane module yields membrane packing densities in the range of several hundred square meters per cubic meter. For example,

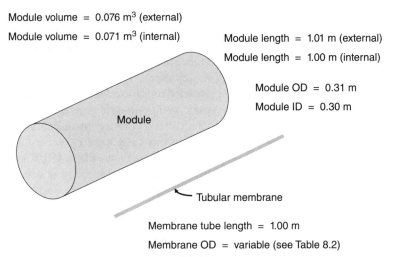

Module volume = 0.076 m³ (external)

Module volume = 0.071 m³ (internal)

Module length = 1.01 m (external)

Module length = 1.00 m (internal)

Module OD = 0.31 m

Module ID = 0.30 m

Module

Tubular membrane

Membrane tube length = 1.00 m

Membrane OD = variable (see Table 8.2)

Figure 8.7. Shell-and-tube membrane module containing either hollow-fiber polymeric membranes or tubular inorganic membranes.

Table 8.2. Membrane Packing Density from Figure 8.7

Membrane Diameter (mm)	Number of Membranes	Total Membrane Area (m²)	Membrane Packing Density (m²/m³)
1	63,600	200	2600
5	2550	40	530
10	637	20	260

using the same membrane module dimensions as in Figure 8.7, but assuming the membranes are planar, stacked 1 mm apart (perpendicular to the radial axis), and have an effective membrane area that is 50% of the module cross-sectional area (to allow for internal manifolding), the membrane packing density is 496 m^2/m^3.

Polymeric membrane modules have been commercially sold for large-scale applications including ammonia synthesis, as well as hydrogen production and recovery from process streams in refineries and petrochemical manufacture. In these applications, polymeric membranes compete well with PSA and cryogenic hydrogen purification. The membrane module is engineered for safe operation at near-ambient temperatures and elevated pressures (perhaps 10–30 bar). Depending on the feed stream composition, high-alloy steels may be required for adequate corrosion resistance.

Inorganic (ceramic, dense metal) membranes will often be operated at temperatures ranging from 200 to 1000 °C. At the extreme temperatures, engineering a membrane module to safely operate at high pressure and in a corrosive environment can be very challenging. Frequently, a compromise must be found between preferred operating conditions and engineering limitations of materials. Conventional 300 series stainless steels (e.g., UNS S31600 and UNS S30400) may be used at temperatures up to about 500 °C, but strength declines rapidly at higher temperatures. Also, these alloys are not acceptable if the feed stream contains sulfur compounds due to excessive corrosion rates.

The importance of using a thin membrane may be inferred from Equation 8.1. Yet, a thin membrane must also remain permselective under the transmembrane operating pressure. Thus, the membrane must not rupture, split, crack, creep, or otherwise develop defects that allow for significant nonselective flow of gases from the feed stream into the permeate stream. Polymeric membranes (asymmetric and composite membranes) are made very thin by incorporating a thick and strong porous polymeric supporting layer. Inorganic membranes are usually constructed in a similar fashion, using a porous and strong support layer to lend adequate strength to a thin permselective membrane layer. For instance, nanoporous ceramic membranes may be made by applying successively finer layers of ceramic sols or suspensions onto either a porous ceramic tube or a porous metal tube.[11–14] Dense metal membranes are frequently made by either depositing a membrane film onto a porous ceramic or porous metal support or by laying thin membrane foils onto a porous support.[15–20]

Thicker, self-supporting, dense metal membranes are known. These are tubular and are usually ≤1 mm in diameter. The first commercially successful palladium–silver hydrogen separation membranes were of this type.[21] Currently, Power and Energy, Inc. also fabricates this type of membrane, although planar membranes are more common due to easier fabrication and a greater variety of fabrication methods.

Because polymeric membranes are operated at near-ambient temperatures, matching the coefficient of thermal expansion (CTE) of materials used to construct the membrane module is not so critical. However, ceramic and metal membranes that will be operated at several hundred degrees may experience unacceptable strain, leading to failure, due to mismatched CTE. A simple example is a stainless steel

(UNS S31600) module housing 1 m in length containing nanoporous alumina membrane tubes sealed to headers at both ends of the module. When heated from 25 to 425 °C, the steel module housing will expand approximately to 6.4 mm (CTE is $16 \times 10^{-6}/°C$),[22] whereas the alumina membranes will expand only to 3.2 mm (CTE is $8.0 \times 10^{-6}/°C$).[23] In this example, it is likely that the alumina membrane tubes will either crack or separate from the seals at each end of the tubes.

Likewise, a thin nanoporous alumina membrane deposited onto a porous stainless steel support will likely fail unless efforts are taken to match the CTE of these diverse layers. This may be done by building up layers on the support such that a gradient in CTE is achieved. Also, thin metal membranes deposited on porous ceramic membranes have historically exhibited poor durability with respect to cracking and general membrane failure due, at least in part, to mismatched CTE.

8.5 DENSE METAL MEMBRANES

Many metals (pure elements and alloys) are permeable to hydrogen to a measurable degree. But the selection of metals that have a sufficiently high permeability to hydrogen is largely limited to palladium and selected alloys of palladium, the metals in groups 3–5 of the periodic table, and alloys of the group 3–5 metals. Pure palladium is not a satisfactory choice for hydrogen separation membranes due to the susceptibility of palladium to hydrogen-induced embrittlement at temperatures below about 300 °C. Most commercialization efforts have focused on alloys of palladium due to the relative chemical inertness of these alloys in the presence of water, hydrocarbons, carbon monoxide, and carbon dioxide. In some cases, alloys of palladium with group 3 metals (e.g., Ce and Y) have yielded exceptionally high hydrogen permeabilities, but stability of these alloys when heated has been problematic due to the preferential oxidation of the group 3 metal (for instance, in the presence of water), leading to depletion of the alloy and enrichment in the concentration of palladium.

There are four general types of metal membrane for hydrogen separation and purification (see Fig. 8.8). They all share one common element: a nonporous, permselective metal layer. If this permselective layer is sufficiently thick to be self-supporting under the design limits of the transmembrane pressure gradient, then the membrane is said to be unsupported. However, economics usually dictate a very thin permselective metal layer. In this case, the membrane is either supported (using a structural layer that is porous and not selective for the permeation of hydrogen) or the membrane is composite in structure. Composite metal membranes may be further classified as either a two-layer composite (coating-metal layer only on the feed surface of the membrane) or a three-layer composite (coating-metal layer on both the feed surface and the permeate surface of the membrane). Composite membranes incorporate a dense base-metal layer that is permselective to hydrogen, but chemically reactive with constituents likely to pass through the membrane module. The purpose of the coating-metal layer is to protect the base-metal layer from corrosion. Clearly, the coating-metal layer must also be permeable to hydrogen.

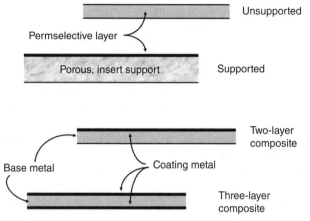

Figure 8.8. Schematic showing key structural features of the four types of dense, hydrogen-permeable metal membranes.

Composite metal membranes are most often the structure of choice when a reactive group 3–5 metal or alloy is the principle constituent of the membrane. The relative chemical reactivity of these metals dictates that an inert coating must be applied to at least the feed surface of the membrane. Palladium, or better yet a palladium alloy, customarily serves as the coating layer. If it can be guaranteed that the permeate side of the membrane will never be exposed to reactive gases (e.g., water, carbon oxides, and hydrocarbons), then a two-layer composite membrane is a satisfactory choice. However, normal operating procedures and the potential for process upsets typically favors the selection of a three-layer composite structure.

The first commercial metal membranes for hydrogen separation and purification were made of palladium alloyed with 23–25 wt% silver. These membrane were of the unsupported type and tubular in shape. Nevertheless, the wall thickness was substantial by current standards—typically at least 100-μm thick. Advances in drawing thin-walled metal tubes has allowed for palladium–silver tubular membranes to be made with much thinner walls, about 20-μm thick. Composite membranes are also usually at least 25-μm thick. REB Research and Consulting (Oak Park, MI) provides tubular composite metal membranes consisting of a palladium coating over a tantalum base metal, although other group 4 or 5 base metals may be used.

If a thinner membrane is required, then one must choose a supported membrane. The permselective metal layer may be palladium or, more commonly, palladium–silver alloy, palladium–copper alloy, or other alloy of palladium. The permselective layer ranges in thickness from about 2–25 μm; thinner than 2 μm is very difficult to achieve without introducing pin holes and other adverse defects into the permselective layer. The support layer is porous and is composed of either metal (such as sintered stainless steel or tightly woven wire cloth) or an inert ceramic; alumina is very common. Since all of the mechanical strength is derived from the support layer, consideration must be given to its shape and thickness.

Supported metal membranes are usually made by one of two general methods. First is the electrochemical plating of the palladium alloy coating layer onto the support layer.[24] This method yields a strong mechanical bond between the two layers. Second is the fabrication (by rolling or etching) of a thin, defect-free palladium-alloy foil that is subsequently placed onto the support layer. In this second technique, there is no initial bond between the coating layer and the support layer, but after some time of operation, a physical lock occurs between the two layers due to plastic deformation of the coating metal layer to conform to the underlying porous support layer. Rolling thin metal foil can be done by a few manufacturers—precision rolling equipment and a clean environment is required. Chemical etching is another method for making very thin membranes. This process is practiced by Hy9 Corporation (Woburn, MA). Figure 8.9 shows examples of both rolled and etched membranes.

Often, the shape of the membrane (flat or tubular) and dimensions are dictated by limits on commercially available supports, or by limits of commercial metal rolling and drawing machines. It is very difficult (and consequently expensive) to roll a flat foil to <20-μm thickness if the width of the rolled foil is greater than 10 cm. Also, drawing thin-walled (<100 μm) tubing is difficult and expensive if the diameter of the tube is 10 times or more than the wall thickness. These general rules apply irrespective of the composition of the metal alloy.

The performance of dense metal membranes may be accurately modeled by using the logarithmic mean driving force to calculate hydrogen flux and infer membrane area and hydrogen recovery. The logarithmic mean driving force is an average driving force for hydrogen permeation that is based on theory for the performance of heat exchangers.[25] For a dense metal membrane, the equation is

$$\text{Logarithmic mean driving force} = \frac{\Delta P_{in} - \Delta P_{out}}{\ln \dfrac{\Delta P_{in}}{\Delta P_{out}}}, \qquad (8.9)$$

Figure 8.9.   Photograph of a 25-μm-thick palladium–copper membrane (left) and an 8-μm-thick palladium–copper membrane (right). Courtesy of Hy9 Corporation.

where $\Delta P_{in} = [(P_{H2,f})^n - (P_{H2,p})^n]$ and $\Delta P_{out} = [(P_{H2,r})^n - (P_{H2,p})^n]$, and the subscripts f, p, and r are, respectively, feed, permeate, and raffinate. Equation 8.9 is valid when gas-phase mass transfer is not limiting the overall performance of the membrane module.

8.5.1 Metal Membrane Durability and Selectivity

As discussed above in Section 8.2, the solution–diffusion mechanism by which hydrogen permeates dense metal membranes theoretically leads to infinite selectivity for hydrogen under all operating conditions. The potential advantage is significant; processes employing metal membranes can deliver ultrahigh purity hydrogen independent of the operating conditions, process upsets, variability in feedstock, and so on. Although thick (about 25–100 μm), unsupported metal membranes will deliver infinite hydrogen selectivity, thinner supported metal membranes often fall short of infinite selectivity for reasons related to membrane durability and integrity of seals in the membrane module.

There are four principle causes for membrane failure under normal operating conditions (in other words, reduced membrane durability). These reasons are (a) hydrogen-induced embrittlement of the membrane, (b) fatigue fracture due to repetitive swelling and contraction of the membrane, (c) mismatch in the CTE of the membrane and underlying support layer, and (d) defects in the underlying support layer that cause a hole or tear to develop in the membrane.

Based on decades of study of hydrogen interactions with metals, the mechanism for hydrogen-induced embrittlement is understood,[26] so there is no reason for this to be a limiting problem for developing and commercializing metal membranes. Essentially, hydrogen-induced embrittlement occurs when hydrogen forms a stoichiometric chemical hydride with the metal constituting the membrane as described in Equation 8.10,

$$M + nH \leftrightarrow MH_n + \Delta H_{rxn}, \tag{8.10}$$

where M is the metal comprising the hydrogen-permeable membrane and ΔH_{rxn} is the enthalpy of reaction (for groups 3–5 metals, the enthalpy of reaction is strongly exothermic, while for palladium and its alloys the enthalpy of reaction is weakly exothermic). The onset of metal-hydride formation is accompanied by a discontinuous and abrupt change in the lattice parameter. The result is a large internal strain in the membrane and severe physical deformation. Rupture of the membrane is inevitable. Like any chemical reaction, the degree to which metal-hydride formation is favorable is dependent on the hydrogen partial pressure and the temperature. The maximum temperature at which the hydride phase is stable under expected hydrogen pressures is called the critical temperature. As mentioned previously, the critical temperature for pure palladium is about 300 °C. Any practical metal selection for use as a membrane should have a critical temperature less than (preferably much less than) room temperature. Palladium–silver and palladium–copper alloys meet this requirement.

A metal membrane should be firmly held in place in the membrane module such that little movement of the membrane can occur. There will always be thermal expansion to contend with, as well as hydrogen-induced metal expansion. Equation 8.2 shows that permeability is the product of diffusivity and solubility. As hydrogen dissolves in a metal, the metal lattice expands to accommodate hydrogen atoms in interstitial tetrahedral or octahedral vacancies. The degree of expansion of the metal lattice increases with increasing solubility of hydrogen. Pure palladium and palladium–silver alloys have a high solubility for hydrogen; hence, these metals swell to a large degree—up to 3%—in the presence of hydrogen (<400 psig). This magnitude of hydrogen-induced swelling of the membrane far exceeds the swelling due to thermal expansion.

The problem with expansion and contraction of a membrane is that it leads to fatigue failure (cracking and pinholes) in membranes. It can also overly stress brazed seals where tubular membranes are joined to a header. Membrane durability with respect to fatigue fractures caused by excessive membrane swelling can be evaluated experimentally. One simply places a sample membrane in a permeation test cell heated to the desired operating temperature. The feed side of the membrane is connected to a supply of pressurized hydrogen and a supply of pressurized nitrogen (or other inert gas) as shown in Figure 8.10 via a three-way valve. The permeate flow rate is measured with a flow meter and monitored for increased flow rate that would indicate a failed membrane. While maintaining constant feed pressure and operating temperature of the membrane, the feed stream is alternately switched from pressurized hydrogen to pressurized nitrogen (e.g., 100 psig), and back to pressurized hydrogen. The membrane should be exposed to hydrogen for a few minutes and then to nitrogen for a few minutes, to provide time for the concentration of dissolved hydrogen in the membrane to equilibrate (the exposure time may need

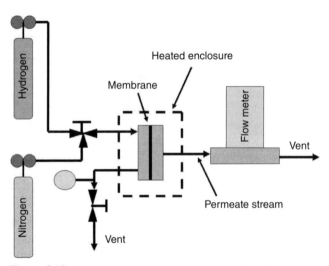

Figure 8.10. Apparatus for testing membrane durability with respect to hydrogen-induced swelling.

to be longer if the membrane is thick). This process is repeated, perhaps as much as 1000 times before examining the membrane for evidence of wrinkles and fatigue fracture.

Experience has shown that a 25-μm-thick membrane (2.5 cm in diameter) composed of Pd-25Ag will fail due to fatigue fracture when tested at 400 °C as described above after only 10–50 cycles. In contrast, a 25-μm-thick membrane (2.5 cm in diameter) composed of Pd-40Cu will survive 1000 cycles without failure and without exhibiting any wrinkles.

Thermal expansion can lead to membrane failure in the case of supported metal membranes if there is a large difference between the CTE of the support layer and the CTE of the permselective metal layer. If the support layer expands more than the permselective metal layer, then the permselective metal layer will be torn or fractured during heat up. On the other hand, if the support layer expands less than the permselective metal layer, then fatigue fracture may occur by the mechanism described above for hydrogen-induced expansion of the permselective metal layer. Mismatch of CTE is often a cause for poor durability in supported membranes comprised of a palladium alloy deposited onto a porous stainless steel support or a porous alumina support.

When considering the influence of thermal expansion or hydrogen-induced expansion on membrane durability, it should be remembered that the absolute magnitude of expansion is dependent on the size (dimensions) of the membrane. Generally, smaller membranes will be more durable than larger membranes.

The last failure mechanism involves defects in the underlying support layer that are transferred to the thin permselective layer. Examples include surface irregularities (such as pits or cracks), and particles on the support surface. Sintered stainless steel supports may be especially prone to residual particles of metal powder that can pierce the thin permselective metal layer during use. Shock and vibration may cause damage (such as cracking) to a ceramic support layer that then results in damage to, and failure of, the permselective layer.

Very little is known about the influence of grain growth, or crystallization if the membrane is composed of an amorphous alloy, on membrane durability. The as-fabricated permselective metal membrane will be polycrystalline or amorphous, depending on the alloy composition and fabrication method. Amorphous, or metallic glass, structures are far less common than are polycrystalline structures. Both amorphous and polycrystalline structures are quasi-stable, meaning that structures are kinetically stabilized and slow to rearrange to the thermodynamically favored structure. In both cases, this would be a single crystal of the metal.

Amorphous metals will crystallize when heated to a sufficiently high temperature. Polycrystalline metals will undergo grain growth as they are heated. These processes require that the metal atoms move, or rearrange, in the solid metal lattice. Hydrogen has been shown to accelerate the mobility of metal atoms[27] and therefore may also accelerate crystallization of an amorphous metal as well as accelerate grain growth in a polycrystalline metal.

Grain boundaries are an interesting and well-studied region of metal structures. Metallic grains are individual crystals, and grain boundaries represent the surface of the crystal. Grain boundaries have free energy characteristics of the surface, and they are locations of high concentrations of defects. If the metal alloy exhibits preferential segregation of one of more alloying components to the surface, this will also occur along grain boundaries.[28] What influence, if any, grain boundaries have on hydrogen permeation and selectivity is largely unexplored. Grain boundaries may provide a path to enhance permeation of hydrogen by either increasing hydrogen solubility or increasing hydrogen diffusion along the grain boundaries. Grain boundaries may provide a pathway for molecules or atoms other than hydrogen to permeate the membrane. For instance, if the permselective metal layer is very thin, on the order of the thickness of individual grains, then grain boundaries will cross from the feed side of the membrane to the permeate side of the membrane. Such grain boundaries may provide a direct pathway for diffusion of impurities across the membrane. Further study of these issues is required.

8.6 INTEGRATION OF REFORMING AND MEMBRANE-BASED PURIFICATION

Fuel cell applications stand to benefit greatly from the close coupling of hydrogen production (from carbon-containing fuels) and hydrogen purification. Hydrogen production from carbon-containing fuels such as hydrocarbons and alcohols may be based on many proven chemical processes such as steam reforming, partial oxidation, and autothermal reforming. These processes are universally conducted at elevated temperatures. Consequently, an inorganic membrane is required to withstand the high temperatures that will be experienced in a closely coupled, or integrated, process. Although any hydrogen-selective membrane that is stable and durable at elevated operating temperatures may be effectively used, only dense metallic membranes offer appreciable selectivity for hydrogen and, thus, the potential for very pure product hydrogen.

The reforming process (as applied to a hydrocarbon or alcohol) yields a product stream that consists predominantly of hydrogen, carbon monoxide, carbon dioxide, water, unconverted feedstock, and trace by-products. This product stream mixture, called reformate, is unsuitable for direct use in low-temperature PEMFC and AFC, and some trace by-products (notably organosulfur compounds) will poison both high-temperature fuel cells and low-temperature fuel cells. A membrane for separating and purifying hydrogen from reformate must also be chemically compatible with the compounds in the reformate stream.

Microporous inorganic oxide membranes are generally compatible with reformate with the qualifier that silica membranes will densify if exposed to hot steam. A better choice would be alumina or titania. Palladium-based hydrogen-selective metal membranes are, for the most part, very sensitive to poisoning by sulfur compounds, but chemically stable to the principle components in the reformate. Alloys of palladium with gold and palladium with copper demonstrate tolerance to sulfur. As expected, the degree of sulfur tolerance typically increases with increasing operating temperature of the membrane.

At low membrane operating temperatures (e.g., <300 °C), metal membranes may experience reversible poisoning by compounds such as carbon monoxide, ammonia, olefins, aromatics, and water (as shown by example in Eq. 8.11). Such reversible poisoning may completely impede hydrogen permeation (as in the case with carbon monoxide). Raising the membrane operating temperature causes the weakly adsorbed molecules to desorb from the membrane surface, thereby allowing the surface reactions (hydrogen chemisorption and hydrogen dissolution) to occur,

$$M + CO \leftrightarrow M(CO) + \Delta H_{rxn}, \tag{8.11}$$

where CO could be any other adsorbing species such as ammonia, olefins, aromatics, and water.

As with any chemical equilibrium, the degree to which the reaction proceeds to the right is dependent not only on temperature but also the chemical nature of the surface metal atoms M and the metal surface structure, the strength of the interaction between the adsorbing species and M, and the partial pressure (or concentration) of the adsorbing species. Fortunately, reforming chemistries commonly employed (including steam reforming, partial oxidation, and autothermal reforming) result in a large dilution of organosulfur impurities that may be present in the feed stream. Consider the following example in which hexadecane (a surrogate for diesel fuel) is steam reformed using a 3:1 steam:carbon ratio. The ideal balanced chemical reaction is

$$C_{16}H_{34} + 48H_2O \rightarrow 49H_2 + 16CO_2 + 16H_2O. \tag{8.12}$$

Addition of steam prior to, or at the onset, of steam reforming results in a 48:1 dilution of the hydrocarbon feed stream. Upon reforming, the feed stream has been effectively diluted by 81:1. Thus, if the feed stream contains refractory organosulfur compounds (e.g., dibenzothiophene) at, say, 300 ppmw, the refractory organosulfur compounds will be diluted to 33 ppmv in the reformate stream. The significance is that the hydrogen-permeable membrane would not be subjected to organosulfur compounds at a concentration of 300 ppmw, rather the membrane would be exposed to organosulfur compounds at a concentration of only 33 ppmv. This is a requirement the membrane is much more likely to meet.

Integrating a membrane with a reformer may be done in one of two ways: the membrane may be close to the reformer, but downstream from the reaction zone, or the membrane may be located within the reaction zone. The latter is referred to as a membrane reactor and this approach offers the advantage of being able to shift chemical equilibrium by removing product hydrogen simultaneously with chemical reaction. The difficulties of membrane reactors include the following:

- The membrane must not be damaged by contact with the catalyst.
- For endothermic steam reforming, an effective way must be engineered to transport heat quickly into the reaction zone.
- Mass transfer from the catalytically active sites to the membrane surface must be exceptionally fast.

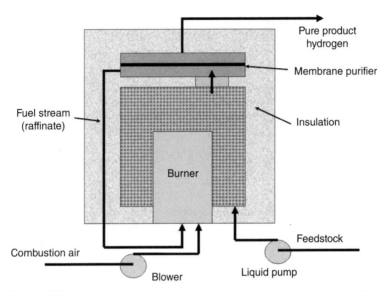

Figure 8.11.   Schematic showing integration of a hydrogen-selective membrane with a steam reformer. See color insert.

- There is a relatively low partial pressure of hydrogen in the vicinity of the membrane, necessitating large membrane areas.

Finally, membrane reactors may not offer cost advantages when applied to reactions that are inhibited by increasing pressure. Obviously, higher pressure favors hydrogen transport across a membrane, thereby yielding a lower capital cost and higher hydrogen recovery. But reforming reaction equilibrium is not favored by increasing pressure, so the hydrogen partial pressure will be suppressed under the conditions (i.e., high pressure) that favor economical membrane performance. Since the water-gas shift (WGS) reaction equilibrium is not pressure dependent, this reaction is well suited to being conducted in a membrane reactor.

Placing the membrane immediately downstream from the reaction zone overcomes all of the above challenges and results in a simpler overall design. An example is shown in Figure 8.11. This configuration is practiced by IdaTech, LLC (Bend, OR) in fuel processors ranging in scale from 6 sLm of product hydrogen to 50 sLm of product hydrogen.[29] In this example, the waste gas stream (raffinate stream) is directed to the burner wherein the fuel gases are combusted to heat the steam reformer.

8.7 COMMERCIALIZATION ACTIVITIES

Polymeric membranes, such as Air Product's PRISM membrane, have not seen use in fuel cell applications, being relegated to large-scale petrochemical, refinery, and hydrogen manufacture applications. The reason for this is probably because fuel cells are very sensitive to certain impurities, often requiring that those impurities be reduced to the parts per million level or lower in the product hydrogen stream. Polymeric membranes simply have insufficient selectivity to achieve adequate purification of hydrogen from the reformate unless multiple stages are employed. The same can be said about porous ceramic or porous metal membranes.

Dense metal membranes, in contrast, offer exceptional selectivity for hydrogen and are being actively used on integrated fuel processors for fuel cell applications. Many companies—Hy9 Corporation; Power and Energy, Inc.; REB Research & Consulting; IdaTech, LLC; Genesis FuelTech, Incorporated (Spokane, WA); Azur Energy (La Verne, CA); and W.C. Heraeus GmbH & Co. KG (Hanau, Germany)— are all engaged in the development and field trials of integrated fuel processors using a palladium-alloy membrane to produce a high-purity product hydrogen stream. Most often, a PEMFC stack receives this product hydrogen directly and without any adverse impact on performance. Figure 8.12 shows two example metal membrane modules, both using palladium-alloy membranes.

Such integrated fuel processors offer the advantages of compact size, ease of operation, and fuel flexibility. Specific applications that are being targeted include stationary primary power for residential and light commercial; vehicle APUs; portable power supplies; backup power supplies for commercial and residential users; and UPS for telecom operators, data centers, and other commercial customers. All of these applications benefit from small fuel cell systems with simplified controls (the latter leads to increased reliability and decreased cost). Another big advantage of an integrated membrane purifier is that many different fuels are preferred, but in all cases, high-purity hydrogen is needed. Only a universal hydrogen purifier can meet this requirement.

Membrane reactors have not yet reached the state of engineering and development where they have demonstrated significant commercial potential. Perhaps, the best known examples of development work have been carried out by Tokyo Gas in Japan and Membrane Reactor Technology in Vancouver, British Columbia, Canada. In both cases, the development work is focused on relatively large membrane reactors to produce hydrogen from natural gas via steam reforming. The target scale is about 50–100 Nm^3/h of product hydrogen for hydrogen refueling stations and on-site manufacture of merchant hydrogen. Tokyo Gas appears to be following a more conventional packed-bed reactor design incorporating palladium-alloy membranes into the catalyst bed. Membrane Reactor Technology uses a fluidized-bed approach, again with palladium-alloy membranes inserted into the bed. Although abrasion of the membrane by the fluidized catalyst particles is a concern, the manufacturer claims this is not a limiting problem.

Hy9 Corporation is also developing membrane reactors based on their thin planar palladium–copper membrane.[30,31] Its work is directed at both MSR and the WGS reaction, and Hy9 Corporation is pursuing several commercial applications today (on-site hydrogen generation from methane and various fuel cell applications). Power and Energy, Inc. is also developing membrane reactor technology using its tubular membrane,[32] but incorporating adequate catalyst (active area) in a

(a)

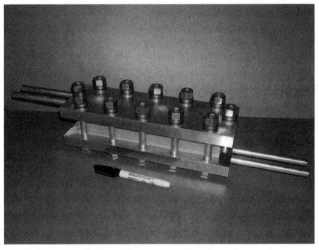

(b)

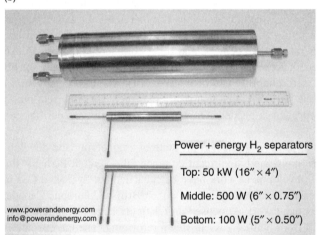

Power + energy H$_2$ separators

Top: 50 kW (16″ × 4″)

Middle: 500 W (6″ × 0.75″)

www.powerandenergy.com
info@powerandenergy.com

Bottom: 100 W (5″ × 0.50″)

Figure 8.12.   (a) Planar membrane module (photo courtesy of Hy9 Corporation) and (b) tubular membrane module (photo courtesy of Power and Energy Incorporated).

small-diameter tubular membrane is questionable unless the catalyst exhibits very high activity.

Despite these efforts, the cost of palladium-alloy membranes is still perceived by many to be prohibitive. However, even without a judicious reclamation and recycling program, the cost of palladium in current state-of-the-art, thin supported membranes is a small fraction of the total cost of the associated reformer and fuel cell system. With further development work from the private sector and government labs[33] (supported by the Department of Energy), the prospect for volume commercial products is encouraging.

REFERENCES

1. Ho, W.S.W. and Sirkar, K.K. (eds.). Chapters 2 and 3. In *Membrane Handbook*. New York: Van Nostrand Reinhold, pp. 19–24 and 25–53, **1992**.
2. Ho, W.S.W. and Sirkar, K.K. (eds.). Chapters 1 and 2. In *Membrane Handbook*. New York: Van Nostrand Reinhold, pp. 3–15 and 19–24, **1992**.
3. Hydrogen recovery and purification. http://www.airproducts.com/Products/Equipment/ PRISMMembranes (accessed October 15, 2005).
4. FREEMANTLE, M. Membranes for gas separation. *Chemical & Engineering News*, pp. 49–57, **October 3, 2005**.
5. LARMINIE, J. and DICKS, A. Chapters 1 and 4. In *Fuel Cell Systems Explained*. New York: John Wiley & Sons, Inc., pp. 1–16 and 61–108, **2000**.
6. Compressed Gas Association. *Handbook of Compressed Gases*, 3rd ed. New York: Van Nostrand Reinhold, **1990**.
7. Power + Energy. Hydrogen separation and purification products based on palladium alloy. http://www.purehydrogen.com (accessed November 3, 2005).
8. MILLER, G.Q. and STÖCKER, J. Selection of a hydrogen separation process. **1999**. http://www.uop.com/objects/SelOfHydroSepProc.pdf (accessed October 15, 2005).
9. Idatech-Advanced Fuel Cell Solutions. http://www.idatech.com (accessed November 3, 2005).
10. Ho, W.S.W. and Sirkar, K.K. (eds.). Chapter 4. In *Membrane Handbook*. New York: Van Nostrand Reinhold, pp. 54–77, **1992**.
11. TSAI, C.-Y., TAM, S.-Y., LU, Y., and BRINKER, C.J. Dual-layer asymmetric microporous silica membranes. *Journal of Membrane Science*, **2000**, 169, 255–265.
12. TROWBRIDGE, L.D. Application of inorganic membrane technology to hydrogen-hydrocarbon separations. *ORNL/TM-2003/139*, June **2003**.
13. AHMAD, A.L., OTHMAN, M.R., and MUKHTAR, H. H_2 separation from binary gas mixture using coated alumina-titania membrane by sol-gel technique at high-temperature region. *International Journal of Hydrogen Energy*, **2004**, 29, 817–828.
14. JUDKINS, R.R. and BISCHOFF, B.L. Hydrogen separation using ORNL's inorganic membranes. *Proceedings of the 29th International Technical Conference on Coal Utilization and Fuel Systems*. Clearwater, FL, April 18–23, **2004**.
15. MARDILOVICH, P.P., SHE, Y., MA, Y.H., and REI, M.-H. Defect-free palladium membranes on porous stainless-steel support. *American Institute of Chemical Engineers*, **2004**, 44, 310–322.
16. KEULER, J.N., LORENZEN, L., and MIACHON, S. Preparing and testing Pd films of thickness 1-2 micrometer with high selectivity and high hydrogen permeance. *Separation Science and Technology*, **2002**, 37, 379–401.
17. GAO, H., LIN, J.Y.S., LI, Y., and ZHANG, B. Electroless plating synthesis, characterization and permeation properties of Pd-Cu membranes supported on ZrO_2 modified porous stainless steel. *Journal of Membrane Science*, **2005**, 265, 142–152.
18. HOANG, H.T., TONG, H.D., GIELENS, F.C., JANSEN, H.V., and ELWENSPOEK, M.C. Fabrication and characterization of dual sputtered Pd-Cu alloy films for hydrogen separation membranes. *Materials Letters*, **2004**, 58, 525–528.
19. ROA, F. and WAY, J.D. Influence of alloy composition and membrane fabrication on the pressure dependence of the hydrogen flux of palladium-copper membranes. *Industrial & Engineering Chemistry Research*, **2003**, 42, 5827–5835.
20. EDLUND, D.J., PLEDGER, W.A., and STUDEBAKER, T. Hydrogen-selective metal membrane modules and method of forming the same. U.S. Patent 6,319,306, **November 30, 2001**.
21. HUNTER, J.B. A new hydrogen purification process. *Platinum Metals Review*, **1960**, 4, 130–131.
22. Boyer, H.E. and Gall, T.L. (eds.). *Metals Handbook Desk Edition*. Metals Park, OH: American Society for Metals, **1985**.
23. Goodfellow Alumina Material Information. http://goodfellow.com/A/Alumina.html (accessed September 14, 2009).

24. ROA, F., WAY, J.D., McCORMICK, R.L., and PAGLIERI, S.N. Preparation and characterization of Pd-Cu composite membranes for hydrogen separation. *The Chemical Engineering Journal*, **2003**, 93, 11–22.
25. SHAH, R.K. and SEKULIĆ, D.P. Chapter 3, Basic thermal design theory for recuperators. In *Fundamentals of Heat Exchanger Design*. Hoboken, NJ: John Wiley & Sons, **2003**, pp. 186–190.
26. PHILPOTT, J. and COUPLAND, D.R. Chapter 26, Metal membranes for hydrogen diffusion and catalysis. In *Hydrogen Effects in Catalysis: Fundamentals and Practical Applications* (ed. Paál, Z. and Menon, P.G.). New York: Marcel Dekker, **1988**, pp. 679–694.
27. EDLUND, D.J. and McCARTHY, J.M. The relationship between intermetallic diffusion and flux decline in composite-metal membranes: Implications for achieving long membrane lifetime. *Journal of Membrane Science*, **1995**, 107, 147–153.
28. FOILES, S.M. Chapter 3, Calculation of the surface segregation of alloys using the embedded atom model. In *Surface Segregation Phenomena* (ed. Dowben, P.A. and Miller, A.). Boca Raton, FL: CRC Press, **1990**, pp. 79–105.
29. EDLUND, D.J. and PLEDGER, W.A. Hydrogen producing fuel processing system. U.S. Patent 6,221,117, **April 24, 2001**.
30. KRUEGER, C., BOMBARD, T., FIORA, S., and JUDA, W. Membrane reactors in hydrogen production applications. Poster presented at the *National Hydrogen Association Meeting*, **2005**.
31. KRUEGER, C.W. Steam-reforming catalytic structure and pure hydrogen generator comprising the same and method of operation of same. U.S. Application Publication US2003/0172589 A1, September 18, **2003**.
32. METTES, J. A novel approach to energy independence. Poster presented at the *2005 Fuel Cell Seminar*, **2005**.
33. KAMAKOTI, P., MORREALE, B.D., CIOCCO, M.V., HOWARD, B.H., KILLMEYER, R.P., CUGINI, A.V., and SHOLL, D.S. Prediction of hydrogen flux through sulfur-tolerant binary alloy membranes. *Science*, **2005**, 307 (5709), 569–573.

Chapter 9

CO$_2$-Selective Membranes for Hydrogen Fuel Processing

JIN HUANG, JIAN ZOU, AND W.S. WINSTON HO

*William G. Lowrie Department of Chemical and Biomolecular Engineering,
Department of Materials Science and Engineering, Ohio State University*

9.1 INTRODUCTION

As a very promising energy conversion device, fuel cells have attracted worldwide interest in power generation for transportation in the recent years.[1,2] In the conversion of chemical energy directly into electricity, fuel cells are about twice as efficient as the internal combustion engine in terms of gas mileage.[3] On the other hand, fuel cells are also environmental friendly devices; water is the only emission when using hydrogen as the fuel.

Although pure hydrogen is a superior fuel for fuel cells, currently there are issues on its storage and distribution.[4] One potentially practical way of producing hydrogen to be used in an automotive fuel cell is by reforming of a readily available fuel such as methanol, natural gas, gasoline, jet fuel, or diesel. However, the purity of hydrogen is a critical issue for fuel cell applications. Even traces of CO in the hydrogen, for example, >10 ppm, will deteriorate the fuel cell performance by poisoning platinum, which works as the electrocatalyst in polymer electrolyte membrane fuel cells.[5]

CO$_2$-selective membranes have the potential to remove CO to a very low concentration by enhancing the water-gas shift (WGS) reaction with *in situ* CO$_2$ separation from the synthesis gas, thus producing high purity H$_2$ on the high-pressure feed side.[6] Compared with the conventional CO$_2$ separation processes, such as amine scrubbing, membrane separation represents a prospective approach of capturing and concentrating CO$_2$ with reduced energy consumption, enhanced weight and space efficiency, and process simplicity.

Hydrogen and Syngas Production and Purification Technologies, Edited by Ke Liu, Chunshan Song and Velu Subramani

There are primarily two parameters to characterize the separation performance of a membrane. One is the selectivity (or the separation factor), which is defined as

$$\alpha_{ij} = \frac{y_i/y_j}{x_i/x_j}. \tag{9.1}$$

Another parameter is the permeability P_i, which is defined as

$$P_i = \frac{J_i}{\Delta p_i/l}. \tag{9.2}$$

The common unit of P_i is Barrer (1 Barrer $= 10^{-10}$ cm^3 (STP)·cm/ cm^2·s·cm Hg $= 0.76·10^{-17}$ m^3 (STP)·m/m^2·s·Pa). The ratio P_i/l is referred to as the permeance, and its common unit is the gas permeation unit (GPU), which is 10^{-6} cm^3 (STP)/(cm^2·s·cm Hg). If the downstream pressure is negligible compared with the upstream pressure, the selectivity can be expressed as

$$\alpha_{ij} = \frac{P_i}{P_j} = \left(\frac{D_i}{D_j}\right)\left(\frac{S_i}{S_j}\right), \tag{9.3}$$

where D_i/D_j is the mobility selectivity, which is the ratio of the diffusion coefficients of components i and j. The solubility selectivity, S_i/S_j, is the ratio of the solubility of components i and j.[7]

Current CO$_2$-selective membranes are based on either solution–diffusion mechanism or facilitated transport mechanism. For glassy polymeric membranes based on the solution–diffusion mechanism, it is very challenging to achieve a high CO$_2$/ H$_2$ selectivity since H$_2$, as a much smaller molecule, exhibits an unfavorably higher diffusion coefficient than CO$_2$. Even though CO$_2$ usually has higher solubility than H$_2$ in most conventional polymeric membranes, diffusivity selectivity dominates the overall selectivity, thereby resulting in a CO$_2$/H$_2$ selectivity of less than 1.[8] Recently, Lin et al.[9] synthesized highly branched cross-linked poly(ethylene glycol) membranes and obtained significantly enhanced solubility selectivity due to the strong affinity of the polar ether linkage to CO$_2$. A CO$_2$/H$_2$ selectivity of 31 and a CO$_2$ permeability of 410 Barrer were achieved with a CO$_2$ partial pressure of 17 atm but at a very low temperature of -20°C, which is not common for the practical H$_2$ fuel processing.

Facilitated transport membrane offers an attractive alternative to achieve high selectivity and high flux simultaneously.[10,11] This type of membrane is based on the reversible reaction of the targeted gas with the reactive carrier contained in the membrane. There are two main types of reactive carriers: the mobile carrier, which can move freely across the membrane, and the fixed carrier, which is covalently bonded on the polymer backbone and only has limited mobility. In mobile-carrier membranes, the carrier reacts with a targeted component on the feed side of a membrane, and the reaction product moves across the membrane and releases this component on the sweep side. As a result, the component being facilitated permeates through the membrane preferably, and the other components, which are not affected by facilitated transport, are retained on the retentate side. In fixed-carrier membranes, the targeted component reacts at one carrier site and then hops to the next carrier site along the direction of the concentration driving force via the "hopping" mechanism.[12]

Facilitated transport membranes for CO_2 separation have been studied since the 1960s.[10,11] The membranes reported in the literature include supported liquid membranes (SLMs), ion-exchange membranes, and membranes with reactive carriers bonded in the matrix of membranes. For SLMs, Ward and Robb immobilized an aqueous bicarbonate–carbonate solution into a porous support and obtained a CO_2/O_2 separation factor of 1500.[13] Meldon et al.[14] investigated the facilitated transport of CO_2 through immobilized alkaline liquid film. Their experimental results confirmed that weak acid buffers significantly increased the CO_2 transport. However, such SLMs have two major problems: loss of solvent and loss or degradation of carriers. The loss of solvent is caused by its evaporation especially at a high temperature and/or by its permeation through the support under a high transmembrane pressure. The loss of the carrier occurs when the carrier solution is forced to permeate through the support ("washout"), and the degradation of carriers is led by the irreversible reaction of the carrier with impurities or the feed gas stream.[15] To address the instability issue of SLMs, Quinn et al.[16] developed membranes consisting of molten salt hydrates, which were nonvolatile and immobilized in microporous polypropylene supports. Teramoto et al.[17–19] proposed a "bulk flow liquid membrane," in which a carrier solution was forced to permeate through a microporous membrane and then was recycled continuously.

In ion-exchange membranes, ionic carriers were retained inside the membranes by electrostatic forces; therefore, the washout of carriers was minimized. LeBlanc et al.[20] first reported ion-exchange facilitated transport membranes in the early 1980s. Since then, a number of papers have been published on ion-exchange facilitated transport membranes for CO_2 separation. Way et al.[15] and Yamaguchi et al.[21] used perfluorosulfonic acid ionomer cation-exchange membranes containing amines as the carriers. The ion-exchange membrane used by Langevin et al.[22] was sulfonated styrene-divinylbenzene in a fluorinated matrix, and the transport model based on the Nernst–Planck equation was used to correlate the experimental results. Matsuyama et al.[23,24] grafted acrylic acid and methacrylic acid on different substrates and used various diamines, diethylenetriamine, and triethylenetetramine as the carriers. They also blended poly(acrylic acid) with poly(vinyl alcohol) (PVA) to prepare membranes and introduced monoprotonated ethylenediamine into the membranes by ion-exchange and used it as the carrier.[25]

Membranes, with reactive carriers covalently bonded on their polymer backbone, were synthesized by several researchers and were believed to have better stability than SLMs. Yamaguchi et al.[26] developed a membrane with poly(allylamine) and compared it with ion-exchange membranes, which had amines as counter ions. Matsuyama et al.[27] heat-treated the PVA-polyethylenimine membrane to improve its stability and to increase the amount of polyethylenimine retained inside the membrane, thus increasing the water content, which in turn increased the diffusivity of the carrier complex. Quinn and Laciak[28] developed polyelectrolyte membranes based on poly(vinylbenzyltrimethylammonium fluoride) (PVBTAF) and achieved a

CO_2/H_2 selectivity of 87 at 23 °C. They also blended fluoride-containing organic and inorganic salts, like cesium fluoride (CsF) into the PVBAT membranes and obtained a CO_2 permeance four times more than that of PVBAT, while CO_2/H_2 and CO_2/CH_4 selectivities were comparable.[29] Ho and coworkers[30–32] synthesized cross-linked PVA membranes containing glycine salts as mobile carriers and polyethylenimine as the fixed carrier, and found that both CO_2 permeability and CO_2/H_2 selectivity increased as temperature increased in the temperature range of 50–100 °C. Zhang et al.[33] prepared membranes with poly(N-vinyl-γ-sodium aminobutyrate), which was obtained from the hydrolysis of poly(vinyl pyrrolidone). Kim et al.[34] prepared composite membranes by casting polyvinylamine on various supports and achieved a CO_2/CH_4 selectivity of 1100.

Typically, WGS reaction shown below is used to convert CO in synthesis gas and generate more H_2:

$$CO(g) + H_2O(g) \rightleftharpoons CO_2(g) + H_2(g) \quad \Delta H_r = -41.1 \text{ kJ/mol.} \quad (9.4)$$

But this reaction is reversible, exothermic, and thermodynamically unfavorable at elevated temperatures, and it results in a high CO concentration (~1%). Consequently, the reaction requires a bulky and heavy reactor. However, a membrane reactor can be used to improve the reaction performance with the *in situ* separation of products. It is therefore possible to overcome thermodynamic constraint and increase the CO conversion significantly.

By using palladium or other inorganic H_2-selective WGS membrane reactors, many researchers have achieved high CO conversion values beyond the equilibrium ones or close to 100%.[35–40] However, the difficulty to prepare thin, flawless, and durable membranes is still the remaining challenge for the commercial application of this type of membrane reactor.[41]

It has been shown that using a CO_2-selective membrane, it is possible to shift the WGS reaction equilibrium to right (Eq. 9.4) to clean up CO and generate more H_2.[6,30–32] With the continuous removal of CO_2, the CO_2-selective WGS membrane reactor is a promising approach to enhance CO conversion and increase the purity of H_2 under relatively low temperatures (~150 °C). In comparison with the H_2-selective membrane reactor, the CO_2-selective WGS membrane reactor is more advantageous because (a) an H_2-rich product is recovered at high pressure (feed gas pressure) and (b) air can be used to sweep the permeate, CO_2, on the low-pressure side of the membrane to obtain a high driving force for the separation.

In the present study, a new solid facilitated transport membrane has been prepared by incorporating both fixed and mobile carriers in cross-linked PVA. Based on the membrane transport properties, we have also developed a mathematical model to study the performance of the CO_2-selective WGS membrane reactor.

9.2 SYNTHESIS OF NOVEL CO_2-SELECTIVE MEMBRANES

Polymeric CO_2-selective membranes with the thin-film-composite structure were prepared by casting an aqueous solution onto microporous BHA Teflon® (a trademark of DuPont, Wilmington, DE) supports (thickness: 60 μm, average pore size:

0.2 μm, BHA Technologies, Kansas City, MS) or GE E500A microporous polysulfone supports (thickness: about 60 μm excluding nonwoven fabric support, average pore size: 0.05 μm, GE Infrastructure, Vista, CA). The aqueous solution was prepared by mixing PVA, formaldehyde (cross-linking agent), 2-aminoisobutyric acid (AIBA), potassium salt, potassium hydroxide, and poly(allylamine) with deionized (DI) water as the solvent.[42] The active layer was dense and about 20 to 80 μm thick. The thickness of a membrane to be mentioned hereafter all refers to the thickness of the active layer.

The transport properties of the membrane were measured by using a membrane permeation unit as described in our previous papers.[32,42] Two gas mixtures were used as the feed gas for the gas permeation tests: one consisting of 20% CO_2, 40% H_2, and 40% N_2, and the other consisting of 1% CO, 17% CO_2, 45% H_2, and 37% N_2 (both on dry basis). The second composition was used to simulate the composition of the synthesis gas from autothermal reforming (ATR) of gasoline with air. Argon was used as the sweep gas for the ease of gas chromatography analysis. Unless otherwise stated, the feed pressure was maintained at about 2.0 atm, while the permeate pressure was set at approximately 1.0 atm.

9.3 MODEL DESCRIPTION

As one of the two common types of membrane modules, the hollow-fiber membrane module has shown excellent mass transfer performance due to its large surface area per unit volume (about 1000–3000 ft^2/ft^3 for gas separation). In the modeling work, the WGS membrane reactor was configured to be a hollow-fiber membrane module with catalyst particles packed inside the fibers.

The catalyst packed was assumed to be the commercial Cu/ZnO catalyst for lower-temperature WGS reaction. A number of studies on the reaction kinetics of the commercial WGS catalyst, $CuO/ZnO/Al_2O_3$, have been published.[43–48] Based on the experimental data of the commercial catalyst (ICI 52-1), Keiski et al.[47] suggested two reaction rates for the low-temperature WGS reaction in the temperature range 160–250 °C. The first was dependent only on CO concentration and gave an activation energy of 46.2 kJ/mol. The second reaction rate was dependent on CO and steam concentrations with a lower activation energy of 42.6 kJ/mol. Because of the proximity of our operation conditions to theirs and the fact that steam is in excess in most of the membrane reactors, Keiski and coworkers' first reaction rate expression was chosen for this work. The reaction rate is given in Equation 9.5,

$$r_i = 1.0 \times 10^{-3} \frac{\rho_b p_f}{n_t R T_f} \exp\left(13.39 - \frac{5557}{T_f}\right) n_{CO}\left(1 - \frac{n_{f,H_2} n_{f,CO_2}}{K_T n_{f,CO} n_{f,H_2O}}\right), \quad (9.5)$$

where the expression for K_T [43,47] is as follows:

$$K_T = \exp\left(-4.33 + \frac{4577.8}{T_f}\right). \quad (9.6)$$

Based on the schematic diagram of the WGS hollow-fiber membrane reactor illustrated in Figure 9.1, the molar and energy balances were performed on both feed

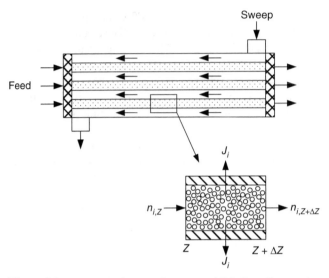

Figure 9.1. Schematic diagram of water-gas shift hollow-fiber membrane reactor. (Reprinted with permission from Huang et al.,[6] Copyright 2005 Elsevier.)

(lumen) and sweep (shell) sides, respectively. The details of this model are provided in a previous paper.[6]

1. Molar balance
   - Feed or lumen side

$$\frac{dn_{\text{fi}}}{dz} = \frac{1}{4}\pi d_{\text{in}}^2 r_i - \pi d_{\text{in}} J_i,\qquad(9.7)$$

where the permeation flux, J_i, is calculated by

$$J_i = P_i \frac{\Delta p_i}{\ell}.\qquad(9.8)$$

   - Sweep or shell side

$$\frac{dn_{\text{si}}}{dz} = -\pi d_{\text{in}} J_i\qquad(9.9)$$

2. Energy balance
   - Feed or lumen side

$$\frac{d\sum(n_{\text{fi}}c_{\text{pfi}}T_f)}{dz} = \frac{1}{4}\pi d_{\text{in}}^2 r_i \Delta H_r T_f - \pi d_{\text{in}}\left(U_i + c_{p\text{CO}_2} J_{\text{CO}_2} + c_{p\text{H}_2} J_{\text{H}_2}\right)\Delta T\qquad(9.10)$$

   - Sweep or shell side

$$\frac{d\sum(n_{\text{si}}c_{\text{psi}}T_s)}{dz} = -\pi d_{\text{in}}\left(U_i + c_{p\text{CO}_2} J_{\text{CO}_2} + c_{p\text{H}_2} J_{\text{H}_2}\right)\Delta T\qquad(9.11)$$

Table 9.1. Composition of Autothermal Reforming Syngas

	CO	H_2O	H_2	CO_2	N_2
Molar fraction	1%	9.5%	41%	15%	33.5%

Steam reforming (SR), partial oxidation (POX), and ATR are three major commercial reforming processes. In this work, ATR syngas was chosen as the feed gas, and the feed inlet molar flow rate, n_{t0}, was 1 mol/s. With the composition given in Table 9.1, this flow rate was chosen because a sufficient H_2 molar flow rate would hence be provided to generate a power of 50 kW via the fuel cell for a five-passenger car.[4] Heated air was used as the sweep gas.

The *bvp4c* solver in Matlab® was used to solve the above differential equations of the boundary value problem with given boundary conditions. During the calculation, the hollow-fiber number was adjusted to satisfy the constraint of feed exit CO concentration, that is, <10 ppm.

9.4 RESULTS AND DISCUSSION

9.4.1 Transport Properties of CO_2-Selective Membrane

Figure 9.2 shows the scanning electron microscopic (SEM) image of a cross-section of the membrane on the GE E500A microporous polysulfone support. It can be seen that the membrane consisted of two portions. The top portion was a dense active layer, which provided separation, and the bottom portion was a microporous polysulfone support, which provided mechanical strength. This composite structure minimizes the mass transfer resistance while maintaining sufficient mechanical strength.

The membranes used in the present study contained 50.0 wt% PVA (60 mol% cross-linked by formaldehyde), 20.7 wt% AIBA-K, 18.3 wt% KOH, and 11.0 wt% poly(allylamine), unless otherwise stated. Figure 9.3 presents a schematic of the CO_2 transport mechanism in the membranes. The membranes synthesized contained both AIBA-K and $KHCO_3$-K_2CO_3 (converted from KOH) as the mobile carriers, and poly(allylamine) as the fixed carrier for CO_2 transport. AIBA is a sterically hindered amine, and its reaction with CO_2 is depicted in Equation 9.12.[49] Poly(allylamine) is a nonhindered amine, and its reaction is shown in Equation 9.13. The reaction mechanism of the CO_2 with $KHCO_3$-K_2CO_3 is presumably similar to that of hindered amine-promoted potassium carbonate described in Equation 9.14:[50]

$$R\text{-}NH_2 + CO_2 + H_2O \rightleftharpoons R\text{-}NH_3^+ + HCO_3^-, \tag{9.12}$$

$$2R\text{-}NH_2 + CO_2 \rightleftharpoons R\text{-}NH\text{-}COO^- + R\text{-}NH_3^+, \tag{9.13}$$

$$CO_3^{2-} + CO_2 + H_2O \rightleftharpoons 2HCO_3^-. \tag{9.14}$$

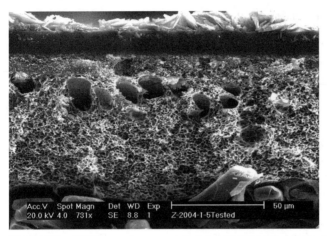

Figure 9.2. SEM image of a cross-section of the membrane synthesized (on GE E500A microporous polysulfone support).

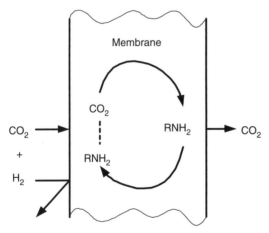

Figure 9.3. Schematic of CO$_2$ transport mechanism in the membranes synthesized.

The transfer of CO$_2$ across the membrane is enhanced by the facilitated transport with reactions shown above, and the flux equation for the CO$_2$ transport can be expressed as:[51]

$$J_A = D_A(C_{A|p1} - C_{A|p2})/1 + D_{AB}(C_{AB|p1m} - C_{AB|p2m})/l. \qquad (9.15)$$

In this equation, the first term on the right-hand side is the flux contributed by the solution–diffusion mechanism, while the second term is due to the facilitated transport mechanism. The nonreacting gases, like H$_2$, N$_2$, and CO, do not have chemical association with carriers and therefore can only be transported by diffusion, which is limited by their low solubility in the highly polar sites in the membranes.[16]

For these nonreacting gases, the flux equation for the diffusion step in the membrane is the first term on the right-hand side of Equation 9.15 only.

9.4.1.1 Effects of Feed Pressure on Separation Performance

The effects of feed pressure on CO_2 flux and permeability, H_2 flux, and CO_2/H_2 selectivity were investigated using a membrane with a thickness of ~60 μm on the BHA microporous Teflon support. The feed pressures ranged from 1.5 to 2.8 atm. Temperature was maintained at 110 °C, and water rates were kept at 0.03 cc/min for both the feed side and sweep side. The feed gas consisted of 20% CO_2, 40% H_2, and 40% N_2 (on dry basis) with a feed gas rate of about 60 cc/min in the gas permeation experiments.

Figure 9.4 illustrates the effects of feed pressure on CO_2 flux and permeability. As illustrated in this figure, CO_2 flux increased rapidly with the feed pressure first and then approached a constant value. This can be explained with the carrier saturation phenomenon, which is a characteristic of facilitated transport membranes. As described by Ho and Dalrymple,[51] when the partial pressure of CO_2 is equal to or higher than a critical CO_2 partial pressure, p_{1c}, the carrier saturation occurs, in which the concentration of CO_2-carrier reaction product attains its maximum value, $C_{AB, max}$, and becomes a constant. In other words, further increase in the partial pressure of CO_2 will not increase the concentration of CO_2-carrier reaction product. This can be expressed as follows:

$$C_{AB|p1} = H_{AB|p1} p_1 = C_{AB,max} = \text{const. when } p_1 \geq p_{1c}. \qquad (9.16)$$

Compared with its facilitated transport, the transport of CO_2 via the solution–diffusion mechanism is negligible. Therefore, total CO_2 flux becomes constant eventually as the feed pressure increases.

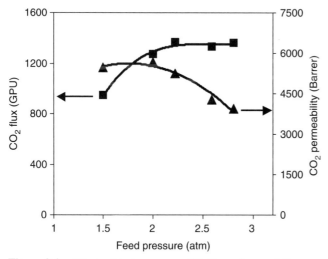

Figure 9.4. Effects of feed pressure on CO_2 flux and permeability.

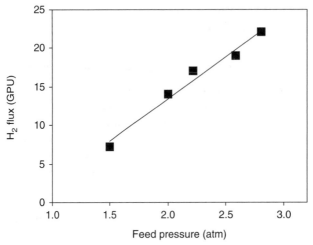

Figure 9.5.   Effect of feed pressure on H$_2$ flux.

As also demonstrated in this figure, CO$_2$ permeability decreased when the feed pressure increased. This can be explained by Equation 9.2. Since CO$_2$ flux did not increase proportionally with the CO$_2$ partial pressure, p_1, CO$_2$ permeability, or the pressure- and thickness-normalized flux dropped with the increase of the feed pressure.

Figure 9.5 depicts the effect of feed pressure on H$_2$ flux. Unlike CO$_2$ flux, H$_2$ flux increased linearly with the feed pressure. This is because H$_2$ has no chemical association with carriers, and the permeability is independent of the feed pressure.[7] Therefore, the flux increased linearly with the feed pressure.

Figure 9.6 shows the effect of feed pressure on CO$_2$/H$_2$ selectivity. As shown in this figure, CO$_2$/H$_2$ selectivity reduced as the pressure increased. Again, this can be explained using the carrier saturation phenomenon as described earlier. Since H$_2$ flux increased much faster than CO$_2$ flux as the pressure increased, the selectivity decreased.

9.4.1.2 Effects of Gas Water Content on Separation Performance

In the experiments, both the feed and the sweep gases were mixed with controlled amounts of water before they entered the permeation cell. The effects of water content in gases on the membrane separation performance at 120 and 150 °C were investigated using a membrane with a thickness of ~70 μm on the BHA microporous Teflon support. The feed gas consisted of 20% CO$_2$, 40% H$_2$, and 40% N$_2$ (on dry basis). Figure 9.7 depicts the CO$_2$ permeability as a function of the water content on the sweep side at 120 and 150 °C. As the water content on the sweep side increased, CO$_2$ permeability increased almost linearly. When the water mole concentration increased from 58% to 93%, the CO$_2$ permeability increased from 3700 Barrer to as high as 8200 Barrer at 120 °C and from 920–2700 Barrer at 150 °C.

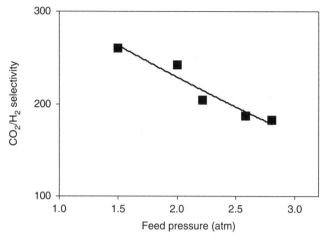

Figure 9.6. Effect of feed pressure on CO_2/H_2 selectivity.

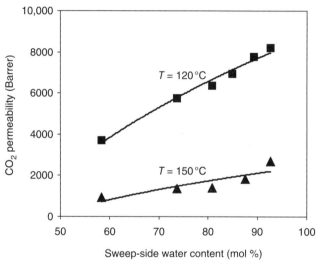

Figure 9.7. CO_2 permeability versus water content on the sweep side (feed water content = 41 mol% and feed pressure = 2.0 atm).

These were presumably due to two reasons: (a) higher water content on the sweep side raised the water retention in the membrane, thus increased the mobility of both mobile and fixed carriers and enhanced the reaction rates of CO_2 with the carriers (Eqs. 9.12–9.14) and (b) higher water content also increased the driving force for CO_2 transport by diluting CO_2 in the permeate. The increase of CO_2 permeability with increasing gas water content was also reported in the literature.[28,29]

Figure 9.8 shows the CO_2/H_2 selectivity as a function of the water content on the sweep side at 120 and 150 °C. The CO_2/H_2 selectivities at both temperatures

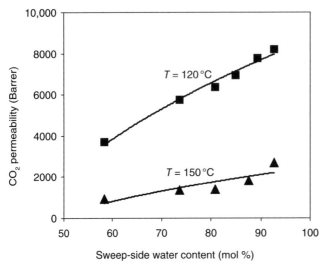

Figure 9.8. CO$_2$/H$_2$ selectivity versus water content on the sweep side (feed pressure = 2.0 atm).

Table 9.2. Separation Performance versus Feed Gas Water Content

Feed Water Rate (cc/min)	Feed Water Content (mol %)	P_{CO2} (Barrer)	α (CO$_2$/H$_2$)
0.03	41	6950	400
0.06	58	9710	520

$T = 120\,°C$; feed pressure = 2.0 atm; sweep water rate = 0.12 cc/min (85 mol %).

increased as the water content on the sweep side increased. When the sweep water content was 93%, the CO$_2$/H$_2$ selectivities at 120 and 150 °C reached 450 and 270, respectively. This increase could be explained by the rise of CO$_2$ flux while the transport of H$_2$ was not affected much by the increase of water content.

The water content on the feed side also had significant effects on CO$_2$ permeability and CO$_2$/H$_2$ selectivity. As shown in Table 9.2 for 120 °C, both CO$_2$ permeability and CO$_2$/H$_2$ selectivity increased as the feed water content increased. This might be caused by the higher water retention in the membrane with the higher water content on the feed side. Consequently, the CO$_2$ transport was enhanced by the increased mobility of both mobile and fixed carriers and the CO$_2$-carrier reaction rates, while the transport of H$_2$ was not affected significantly.

In the present study, argon was used as the sweep gas for the ease of gas chromatography analysis. In real applications, air or steam can be used to sweep CO$_2$ on the low-pressure side of the membrane to obtain a high driving force for the separation. If steam is used as the sweep gas, the permeate CO$_2$ can be easily separated from steam condensation and then be ready for sequestration.

9.4.1.3 Effects of Temperature on Separation Performance

The effects of temperature on CO_2 permeability, CO_2/H_2 selectivity, and CO_2/CO selectivity were studied in a temperature range of 100–180 °C. As described earlier, increasing the feed and the sweep water contents can effectively improve the separation performance; therefore, in order to maintain the membrane performance at elevated temperatures, the feed and sweep water rates were increased gradually with temperature. A membrane consisting of 40 wt % PVA, 20 wt % AIBA-K, 20 wt % KOH, and 20 wt % poly(allylamine) was used to investigate the temperature effects. This membrane had a thickness of ~26 μm on the BHA microporous Teflon support. Table 9.3 lists the water rates used at different temperatures. As shown in Figure 9.9, both CO_2 permeability and CO_2/H_2 selectivity decreased as temperature increased. This was presumably due to the reduction of water retention in the membrane as temperature increased. As described earlier, with the loss of water in the membrane, the mobility of mobile and fixed carriers and the reaction rates of CO_2 with the carriers were reduced, while the transport of H_2 was not affected significantly.

However, the membrane still showed good separation performance in the temperature range of 110–160 °C. At 150 °C, the CO_2 permeability and CO_2/H_2 selectivity were 2500 Barrer and 80, respectively. The CO_2/H_2 selectivity reduced slightly as the temperature increased to 170 °C, and it decreased significantly to about 10 at 180 °C due to the significant swelling of the membrane, thus resulting in a sharp increase of H_2 flux at this temperature.

Table 9.3. Feed and Sweep Water Contents Used for Figure 9.9

Temperature (°C)	100	110	120	130	140	150	160	170	180
Feed water content (mol %)	41	41	41	41	59	68	74	74	74
Sweep water content (mol %)	58	58	58	58	74	81	85	85	85

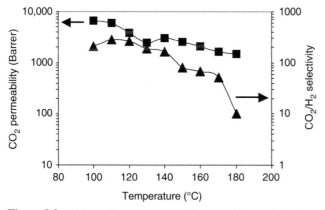

Figure 9.9. Effects of temperature on CO_2 permeability and CO_2/H_2 selectivity (feed pressure = 2.0 atm with increasing water rates at elevated temperatures).

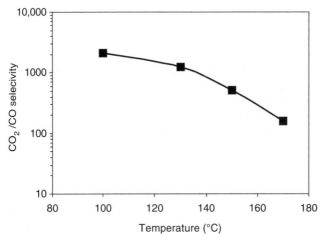

Figure 9.10. CO_2/CO selectivity versus temperature (feed pressure = 2.0 atm with increasing water rates at elevated temperatures).

Table 9.4.   Sweep Water Rates and Contents Used for Figure 9.10

Temperature (°C)	110	130	150	170
Sweep water rate (cc/min)	0.03	0.03	0.09	0.15
Sweep water content (mol %)	58	58	81	88

Feed water rate = 0.03 cc/min (water content = 41 mol %).

The CO_2/CO selectivity was investigated with a feed gas of 1% CO, 17% CO_2, 45% H_2, and 37% N_2 from 100 to 170 °C by employing the same membrane used for Figures 9.7 and 9.8. As shown in Figure 9.10, the selectivity reduced as temperature increased, which can be explained by the decrease of CO_2 permeability at elevated temperatures. Yet at 170 °C, the CO_2/CO selectivity still reached 160. The water rates and contents used for Figure 9.10 are listed in Table 9.4.

9.4.1.4   Comparison with the Literature Data

The results shown in Figures 9.7–9.9 were compared with the results reported in the literature [9,28,29] and our previous results [32] as shown in Table 9.5. As we can see from the comparison, the results from this work are gratifying. The improvement in operating temperature could be mostly attributed to better cross-linking of PVA as the polymer matrix. The high permeability and selectivity were presumably attributable to better carriers, AIBA-K and $KHCO_3$-K_2CO_3, and poly(allylamine) contained in the membranes.

Gas permeation results presented in this work, especially above 150 °C, showed that the polymeric membranes that we prepared were capable of applications at high temperatures, such as WGS membrane reactors. The membrane reactor incorporates both CO_2 removal and WGS reaction to produce high-purity H_2.[6]

Table 9.5. Comparison of Membrane Performance Reported in the Literature

Membrane	T (°C)	CO_2 Pressure (atm)	Feed Gas	P_{CO_2} (Barrer)	α (CO_2/H_2)	Reference
Poly(vinylbenzyltrimethylammonium fluoride)	23	0.42	31% CO_2, 35% H_2, 34% CH_4	~120	87	Quinn and Laciak[28]
Poly(vinylbenzyltrimethylammonium fluoride)-CsF	23	0.40	33% CO_2, 34% H_2, 33% CH_4	~510	127	Ho and Dalrymple[51]
Highly branched cross-linked poly(ethylene glycol)	−20	17	80% CO_2, 20% H_2	410	31	Lin et al.[9]
Cross-linked PVA/ dimethylglycine-Li	90	0.24	20% CO_2, 40% H_2, 40% N_2	1700	50	Tee et al.[32]
Cross-linked PVA/AIBA-K/ poly(allylamine)	120	0.24	20% CO_2, 40% H_2, 40% N_2	8200	450	This study (Figs. 9.7 and 9.8)
	110	0.24		6500	210	This study (Fig. 9.9)
	150	0.24		2500	80	

9.4.2 Modeling Predictions

9.4.2.1 Reference Case

A reference case for the CO$_2$-selective WGS membrane reactor was chosen with the CO$_2$/H$_2$ selectivity of 40, the CO$_2$ permeability of 4000 Barrer, the inlet sweep-to-feed molar flow rate ratio of 1, the membrane thickness of 5 μm, 52,500 hollow fibers (a length of 61 cm, an inner diameter of 0.1 cm, and a porous support with a porosity of 50% and a thickness of 30 μm), both inlet feed and sweep temperatures of 140 °C, and the feed and sweep pressures of 3 and 1 atm, respectively. With respect to this case, the effects of CO$_2$/H$_2$ selectivity, CO$_2$ permeability, sweep-to-feed ratio, inlet feed temperature, inlet sweep temperature, and catalyst activity on the reactor behavior were then investigated.

Figure 9.11 depicts the profiles of the feed-side mole fractions of CO and CO$_2$ along the length of the countercurrent membrane reactor. The modeling results showed that this membrane reactor could convert CO via the WGS reaction and then decrease CO concentration from 1% to 9.82 ppm along with the removal of almost all the CO$_2$ from the hydrogen product. In the membrane reactor, the removal of CO$_2$ enhanced the WGS reaction.

Figure 9.12 depicts the profiles of feed-side H$_2$ concentrations on the dry and wet bases. As depicted in this figure, the membrane reactor could enhance H$_2$ concentration from 45.30% to 54.95% (on dry basis), that is, from 41% to 49.32% (on wet basis). In this case, the H$_2$ recovery calculated from the model was 97.38%. With the advancement of the high-temperature proton exchange membrane fuel cell (120–160 °C), it is expected that the constraint of CO concentration can be relaxed to about 50 ppm in the near future. Then, the required hollow-fiber number could be reduced significantly to 39,000 based on the modeling results.

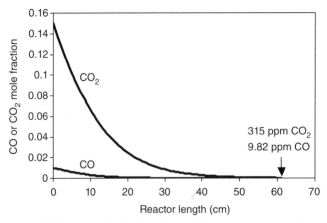

Figure 9.11.   Feed-side CO and CO$_2$ mole fraction profiles along the length of membrane reactor for autothermal reforming syngas. (Reprinted with permission from Huang et al.,[6] Copyright 2005 Elsevier.)

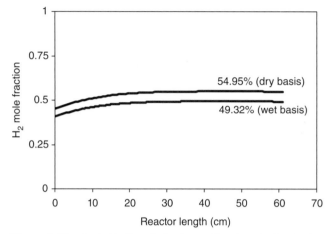

Figure 9.12.   Feed-side H_2 mole fraction profiles along the length of membrane reactor for autothermal reforming syngas. (Reprinted with permission from Huang et al.,[6] Copyright 2005 Elsevier.)

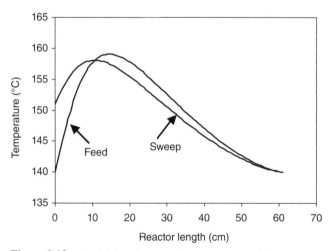

Figure 9.13.   Feed-side and sweep-side temperature profiles along the length of the membrane reactor for autothermal reforming syngas. (Reprinted with permission from Huang et al.,[6] Copyright 2005 Elsevier.)

The temperature profiles for both feed and sweep sides are shown in Figure 9.13 with a maximum for each profile. Since the overall module was adiabatic, the feed gas was heated by the exothermic WGS reaction. The highest feed-side temperature was 158 °C at about $z = 15$ cm. Beyond that, the feed-side temperature reduced, and it became very close to the sweep-side temperature at the end of membrane reactor. This was due to the efficient heat transfer provided by the hollow-fiber configuration.

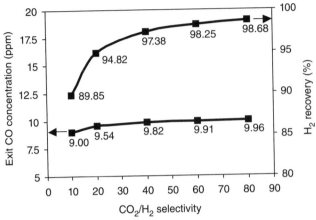

Figure 9.14.   Effects of CO$_2$/H$_2$ selectivity on feed-side exit CO concentration and H$_2$ recovery for autothermal reforming syngas. (Reprinted with permission from Huang et al.,[6] Copyright 2005 Elsevier.)

Higher temperatures enhance WGS reaction rates but are unfavorable for CO conversion. Thus, it is important to use air with appropriate temperature, that is, 140 °C as the sweep gas to keep the feed gas within 150 ± 10 °C.

9.4.2.2 Effects of CO$_2$/H$_2$ Selectivity

In order to study the impact of CO$_2$/H$_2$ selectivity on the membrane reactor performance, $\alpha = 10, 20, 40, 60,$ and 80 were applied in the model while the other parameters for the reference case were kept constant. As shown in Figure 9.14, the feed-side exit CO concentration increased slightly as CO$_2$/H$_2$ selectivity increased. This was due to the fact that higher selectivity caused lower H$_2$ permeability and thus a lower H$_2$ permeation rate or higher H$_2$ concentration on the feed side, which was unfavorable for the WGS reaction rate. Also shown in this figure, the H$_2$ recovery increased from 89.85% to 98.68% as the CO$_2$/H$_2$ selectivity increased from 10 to 80. This indicated that the higher selectivity decreased the H$_2$ loss because of the reduction in H$_2$ permeation through the membrane. In addition, the modeling results showed that a CO$_2$/H$_2$ selectivity of 10 was the minimum value required for an H$_2$ recovery of about 90%.

9.4.2.3 Effects of CO$_2$ Permeability

The membrane areas required for the exit feed CO concentration of <10 ppm in the H$_2$ product were calculated with five different CO$_2$ permeabilities ranging from 1000 to 8000 Barrer, while the other parameters for the reference case were kept constant. As demonstrated in Figure 9.15, the required membrane area or hollow-fiber number dropped rapidly as permeability increased from 1000 to 4000 Barrer. Beyond that,

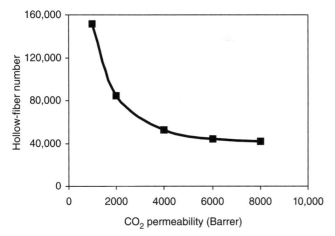

Figure 9.15.   Effect of CO_2 permeability on required membrane area for autothermal reforming syngas. (Reprinted with permission from Huang et al.,[6] Copyright 2005 Elsevier.)

it approached a constant value gradually. Increasing CO_2 permeability increased the CO_2 permeation rate and enhanced the CO_2 removal, which shifted the WGS reaction toward the product side. However, after exceeding the permeability over 6000 Barrer, the overall system became more reaction controlled. Hence, the influence of the permeability became less significant.

9.4.2.4   Effects of Sweep-to-Feed Ratio

The inlet sweep-to-feed molar flow rate ratios of 0.5, 1, 1.5, 2, and 2.5 were used in the calculation while the other parameters for the reference case were kept constant. Figure 9.16 illustrates the effect of sweep-to-feed ratio on the feed-side exit CO concentration. As illustrated in this figure, increasing the sweep-to-feed ratio decreased the exit CO concentration first and then increased it slightly. A higher sweep-to-feed ratio resulted in a lower CO_2 concentration on the sweep side and then a higher CO_2 permeation driving force. However, it also enhanced heat transfer and then decreased the feed-side temperature, which was unfavorable to the WGS reaction rate. Therefore, an optimal sweep-to-feed ratio of about 1 existed as a result of the trade-off between the effects on the CO_2 permeation rate and the WGS reaction rate. Also illustrated in this figure is the effect of sweep-to-feed ratio on H_2 recovery. The sweep-to-feed ratio did not have a significant effect on the H_2 recovery. This was due to the fact that the resulting CO concentrations were very low (<30 ppm) and did not affect the H_2 recovery.

9.4.2.5   Effects of Inlet Feed Temperature

In order to study the impact of inlet feed temperature on the membrane reactor performance, $T_{f0} = 80, 100, 120, 140, 160, 180,$ and $200\,°C$ were applied in the model

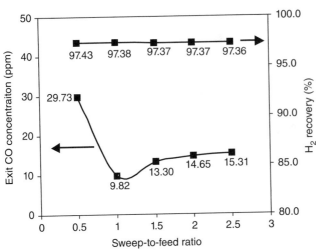

Figure 9.16.   Effects of sweep-to-feed ratio on feed-side exit CO concentration and H$_2$ recovery for autothermal reforming syngas. (Reprinted with permission from Huang et al.,[6] Copyright 2005 Elsevier.)

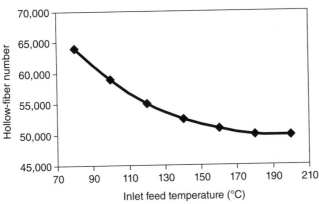

Figure 9.17.   Effect of inlet feed temperature on required membrane area for autothermal reforming syngas.

while the other parameters for the reference case were kept constant. As shown in Figure 9.17, the required membrane area or hollow-fiber number decreased as the inlet feed temperature increased. It approached an asymptotic value gradually. The feed-side temperature profiles for different feed inlet temperatures are presented in Figure 9.18. The feed-side temperature increased as the inlet feed temperature increased, especially at the entrance section. The higher feed-side temperature gave a higher WGS reaction rate, and thus a less reactor or catalyst volume, that is, a lower membrane area, was required. The unfavorable WGS equilibrium at high temperatures was compensated by the simultaneous CO$_2$ removal.

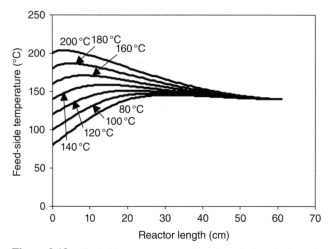

Figure 9.18. Feed-side temperature profiles along the length of membrane reactor for autothermal reforming syngas with different inlet feed temperatures.

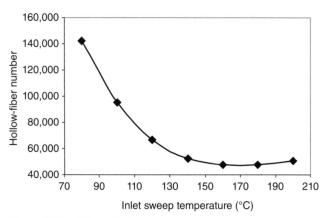

Figure 9.19. Effect of inlet sweep temperature on required membrane area for autothermal reforming syngas.

9.4.2.6 Effects of Inlet Sweep Temperature

The membrane areas required for the exit feed CO concentration of <10 ppm in the H_2 product were calculated with seven different inlet sweep temperatures ranging from 80 to 200 °C, while the other parameters for the reference case were kept constant. As demonstrated in Figure 9.19, the required membrane area or hollow-fiber number dropped rapidly as the inlet sweep temperature increased from 80 to 160 °C. Beyond 160 °C, it increased slightly. Figure 9.20 depicts the feed-side temperature profiles along the membrane reactor with different inlet sweep temperatures. Increasing the inlet sweep temperature increased the feed-side temperature

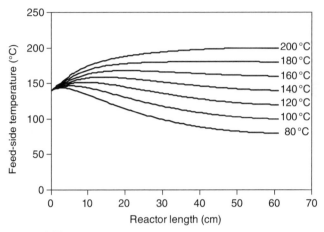

Figure 9.20.   Feed-side temperature profiles along the length of membrane reactor for autothermal reforming syngas with different inlet sweep temperatures.

significantly over a longer reactor length in comparison with increasing the inlet feed temperature as shown in Figure 9.18. A higher feed-side temperature resulted in a higher WGS reaction rate and thus a lower membrane area as described earlier. When the inlet sweep temperature exceeded about 160 °C, the WGS reaction equilibrium became less favorable, and the overall system became more mass transfer controlled. Hence, more membrane area was needed to remove the generated CO$_2$ to achieve <10 ppm CO in the H$_2$ product.

9.4.2.7   Effects of Catalyst Activity

The effect of catalyst activity on the required membrane area was studied by assuming several WGS reaction kinetics based on the Cu/ZnO kinetics equation proposed by Keiski et al.[47] In Figure 9.21, the number on the horizontal x-axis indicates the reaction kinetic rate in terms of the times of the Cu/ZnO kinetics, for example, 1 represents the Cu/ZnO kinetics and 2 represents a kinetics of 2 times the Cu/ZnO kinetics. As illustrated in this figure, increasing catalyst activity reduced the required membrane area significantly. The higher catalyst activity resulted in a higher reaction rate, which also increased the CO$_2$ permeation rate because of a higher CO$_2$ partial pressure on the feed side and thus a higher driving force across the membrane. Hence, with the advancement of a more active WGS catalyst, the membrane reactor would become more compact.

9.4.2.8   Experimental Study of CO$_2$-Selective WGS Membrane Reactor

In the experiments, a rectangular flat-sheet membrane reactor using the commercial Cu/ZnO catalyst (C18-AMT-2, Süd-Chemie Inc., Louisville, KY) was set up. Figure

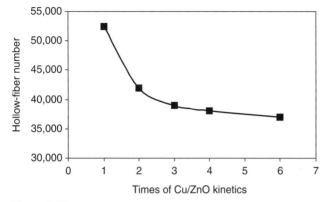

Figure 9.21.   Effect of catalyst activity on required membrane area for autothermal reforming syngas.

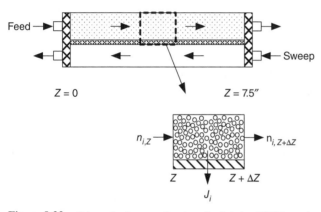

Figure 9.22.   Schematic diagram of rectangular flat-sheet WGS membrane reactor.

9.22 shows the schematic of this membrane reactor. This reactor was designed to provide well-defined flows on both the feed and sweep sides, which made possible an accurate comparison between experimental data and modeling prediction. The gas mixture with the composition of 1% CO, 17% CO_2, 45% H_2, and 37% N_2 (on dry basis) was used as the feed gas, and argon was used as the sweep gas for the ease of gas chromatography (GC) analysis.

By varying the feed flow rate, the performance of the reactor was investigated. Figure 9.23 shows the results obtained from this rectangular WGS membrane reactor. As shown in this figure, the CO concentration in the exit stream, that is, the H_2 product, was <10 ppm (on dry basis) for the various feed flow rates of the syngas ranging from 20 to 70 cc/min (at ambient conditions). The data agreed reasonably with the prediction by the nonisothermal mathematical model that we have developed based on the material and energy balances, membrane permeation, and the

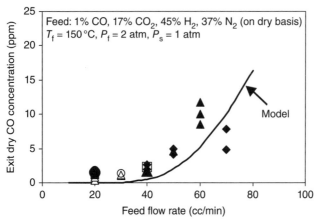

Figure 9.23.   The results of CO in the H_2 product for the inlet 1% CO feed gas at various flow rates from the rectangular WGS membrane reactor.

low-temperature WGS reaction kinetics for the commercial catalyst ($Cu/ZnO/Al_2O_3$) reported by Moe[43] and Keiski et al.[47] as described earlier. If the feed pressure of the synthesis gas was higher than 2 atm, a higher feed gas rate could be processed to obtain <10 ppm CO in the H_2 product for the given membrane area of the rectangular reactor, as a result of a higher driving force for the CO_2 transport.

Gas hourly space velocity (GHSV) is defined in the following equation:

$$\text{GHSV} = \frac{\text{Gas flow rate (L/h)}}{\text{Reactor volume (L)}}. \tag{9.17}$$

Thus, the unit of GHSV is h^{-1}. The GHSV values for the data shown in Figure 9.23 were calculated according to this equation by taking into account the membrane thickness and packing density. The calculated GHSV results corresponding to the experimental data given in Figure 9.23 are shown in Figure 9.24. As shown in the figure, a high GHSV of about $4600 h^{-1}$ is achievable. As mentioned above, if the feed pressure of the synthesis gas was higher than 2 atm, a higher feed gas rate could be processed, that is, a higher GHSV would be achievable.

9.5   CONCLUSIONS

Polymeric CO_2-selective membranes consisting of both mobile and fixed carriers in cross-linked PVA were synthesized. The membranes showed good CO_2/H_2 and CO_2/CO selectivities, and high CO_2 permeability up to 170 °C. The effects of feed pressure, water content, and temperature on transport properties were investigated. The CO_2 permeability and CO_2/H_2 selectivity decreased with increasing feed pressure,

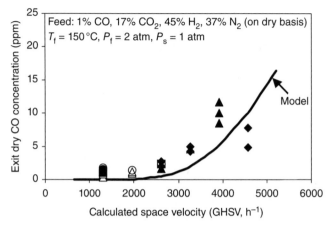

Figure 9.24.   The calculated GHSV results for the data shown in Figure 9.23 at various flow rates from the rectangular WGS membrane reactor.

which could be explained with the carrier saturation phenomenon, a characteristic of facilitated transport membranes. The CO_2 permeability and CO_2/H_2 selectivity increased significantly with increasing water contents in both feed and sweep, which suggested that water played an important role in the facilitated transport. The overall gratifying results obtained were mostly attributed to the improvement in crosslinking and better carriers.

A one-dimensional nonisothermal model was developed to predict the performance of the novel WGS membrane reactor. The modeling results have shown that H_2 enhancement (>54% H_2 for the ATR of gasoline with air on a dry basis) via CO_2 removal and CO reduction to 10 ppm or lower are achievable for the autothermal synthesis gas. With this model, we have elucidated the effects of system parameters on the membrane reactor performance. As the CO_2/H_2 selectivity increased, the recovery of H_2 increased, without affecting the membrane area requirement and the low CO attainment significantly. Higher membrane permeability resulted in the reduction of the required membrane area. Increasing sweep-to-feed ratio enhanced the permeation driving force but decreased the feed-side temperature and thus the reaction rate, resulting in a net effect balanced between them and an optimal ratio of about 1. As either of the inlet feed and sweep temperatures increased, the membrane area requirement decreased. However, the temperatures greater than about 170 °C would be unfavorable to the exothermic, reversible WGS reaction. Increasing catalyst activity enhanced WGS reaction and CO_2 permeation. The modeling study showed that both WGS reaction and CO_2 permeation played an important role on the overall reactor performance. In the experimental study, we confirmed the <10 ppm CO result using a rectangular WGS membrane reactor with well-defined flows. The data from the rectangular WGS membrane reactor agreed well with the model prediction.

GLOSSARY

C_A CO_2 concentration (cm^3 (STP)/cm^3 polymer)

C_{AB} CO_2-carrier reaction product concentration (cm^3 (STP)/cm^3 polymer)

c_p heat capacity (J/mol/K)

d hollow-fiber diameter (cm)

D_A diffusivity coefficient for CO_2 (cm^2/s)

D_{AB} diffusivity coefficient for CO_2-carrier reaction product (cm^2/s)

H_{AB} Henry's law constant for CO_2-carrier reaction product (cm^3 (STP)/(cm^3 polymer atm)

ΔH_r heat of reaction (J/mol)

J permeation flux (mol/cm^2/s)

K_T reaction equilibrium constant (atm^{-2})

l membrane thickness (cm)

n molar flow rate (mol/s)

p pressure (atm)

p_1 CO_2 partial pressure on the high-pressure side of the membrane (atm)

p_2 CO_2 partial pressure on the low-pressure side of the membrane (atm)

p_{1c} critical CO_2 partial pressure at which carrier saturation occurs (atm)

p_{1m} CO_2 partial pressure in the membrane on the high-pressure side of the membrane (atm)

p_{2m} CO_2 partial pressure in the membrane on the low-pressure side of the membrane (atm)

P permeability (Barrer)

r volumetric reaction rate (mol/cm^3/s)

R ideal gas constant (atm·cm^3/mol/K)

T temperature (°C)

U_i overall heat transfer coefficient (W/cm^2/K)

x feed-side molar fraction

y sweep-side molar fraction

z axial position along the length of reactor (cm)

Greek Letters

α CO_2/H_2 selectivity

ρ_b catalyst bulk density (g/cm^3)

Subscripts

0 initial

f feed side

i species

in inside of the hollow fiber

j species

s sweep side

t total

ACKNOWLEDGMENTS

This research was supported by the U.S. Department of Energy (Grant/Contract No. DE-FC36-03AL68510), the Ohio Department of Development (Wright Center of Innovation

Grant No. 342-0561), and the Ohio State University. We would like to thank Chris Plotz and BHA Technologies, and Debbie de la Cruz and GE Infrastructure for giving us the BHA microporous Teflon support and GE E500A microporous polysulfone support, respectively, used in this work.

REFERENCES

1. ACRES, G.J.K. Recent advances in fuel cell technology and its applications. *Journal of Power Sources*, **2001**, 100, 60.
2. CROPPER, M.A.J., GEIGER, S., and JOLLIE, D.M. Fuel cells: A survey of current developments. *Journal of Power Sources*, **2004**, 131, 57.
3. BARNETT, B.M. and TEAGAN, W.P. The role of fuel cells in our energy future. *Journal of Power Sources*, **1992**, 37, 15.
4. BROWN, L.F. A comparative study of fuels for on-board hydrogen production for fuel-cell-powered automobiles. *International Journal of Hydrogen Energy*, **2001**, 26, 381.
5. AHMED, S. and KRUMPELT, M. Hydrogen from hydrocarbon fuels for fuel cells. *International Journal of Hydrogen Energy*, **2001**, 26, 291.
6. HUANG, J., EL-AZZAMI, L., and HO, W.S.W. Modeling of CO_2-selective water gas shift membrane reactor for fuel cell. *Journal of Membrane Science*, **2005**, 261, 67.
7. Zolandz, R.R. and Fleming, G.K. Gas permeation. In *Membrane Handbook* (eds. W.S.W. Ho, K.K. Sirkar). New York: Chapman & Hall, p. 26, **1992**.
8. LIN, H. and FREEMAN, B.D. Materials selection guidelines for membranes that remove CO_2 from gas mixture. *Journal of Molecular Structure*, **2005**, 739, 57.
9. LIN, H., WAGNER, E.V., FREEMAN, B.D., TOY, L.G., and GUPTA, R.P. Plasticization-enhanced hydrogen purification using polymeric membranes. *Science*, **2006**, 311, 639.
10. Way, J.D. and Noble, R.D. Facilitated transport. In *Membrane Handbook* (eds. W.S.W. Ho, K.K. Sirkar). New York: Chapman & Hall, p. 834, **1992**.
11. Ho, W.S.W. and Sirkar, K.K. (eds.). *Membrane Handbook*. New York: Chapman & Hall, **1992**.
12. CUSSLER, E.L., ARIS, R., and BHOWN, A. On the limit of facilitated diffusion. *Journal of Membrane Science*, **1989**, 43, 149.
13. WARD, W.J. and ROBB, W.L. Carbon dioxide-oxygen separation: Facilitated transport of carbon dioxide across a liquid film. *Science*, **1967**, 156, 1481.
14. MELDON, J.H., SMITH K.A., and COLTON C.K. The effect of weak acids upon the transport of carbon dioxide in alkaline solutions. *Chemical Engineering Science*, **1977**, 32, 939.
15. WAY, J.D., NOBLE, R.D., REED, D.L., GINLEY, G.M., and JARR, L.A. Facilitated transport of CO_2 in ion exchange membranes. *AIChE Journal*, **1987**, 33, 480.
16. QUINN, R., APPLEBY, J.B., and PEZ, G.P. New facilitated transport membranes for the separation of carbon dioxide from hydrogen and methane. *Journal of Membrane Science*, **1995**, 104, 139.
17. TERAMOTO, M., TAKEUCHI, N., MAKI, T., and MATSUYAMA, H. Gas separation by liquid membrane accompanied by permeation of membrane liquid through membrane physical transport. *Separation and Purification Technology*, **2001**, 24, 101.
18. TERAMOTO, M., TAKEUCHI, N., MAKI, T., and MATSUYAMA, H. Facilitated transport of CO_2 through liquid membrane accompanied by permeation of carrier solution. *Separation and Purification Technology*, **2002**, 27, 25.
19. TERAMOTO, M., KITADA, S., OHNISHI, N., MATSUYAMA, H., and MATSUMIYA, N. Separation and concentration of CO_2 by capillary-type facilitated transport membrane module with permeation of carrier solution. *Journal of Membrane Science*, **2004**, 234, 83.
20. LeBLANC, O.H., WARD, W.J., MATSON, S.L., and KIMURA, S.G. Facilitated transport in ion-exchange membranes. *Journal of Membrane Science*, **1980**, 6, 339.
21. YAMAGUCHI, T., KOVAL, C.A., NOBLE, R.D., and BOWMAN, C.N. Transport mechanism of carbon dioxide through perfluorosulfonate ionomer membranes containing an amine carrier. *Chemical Engineering Science*, **1996**, 51, 4781.

22. LANGEVIN, D., PINOCHE, M., SELEGNY, E., METAYER, M., and ROUX, R. CO$_2$ facilitated transport through functionalized cation-exchange membranes. *Journal of Membrane Science*, **1993**, 82, 51.
23. MATSUYAMA, H., TERAMOTO, M., and IWAI, K. Development of a new functional cation-exchange membrane and its application to facilitated transport of CO$_2$. *Journal of Membrane Science*, **1994**, 93, 237.
24. MATSUYAMA, H., TERAMOTO, M., SAKAKURA, H., and IWAI, K. Facilitated transport of CO$_2$ through various ion exchange membranes prepared by plasma graft polymerization. *Journal of Membrane Science*, **1996**, 117, 251.
25. MATSUYAMA, H., TERAMOTO, M., MATSUI, K., and KITAMURA, Y. Preparation of poly(acrylic acid)/poly(vinyl alcohol) membrane for the facilitated transport of CO$_2$. *Journal of Applied Polymer Science*, **2001**, 81, 936.
26. YAMAGUCHI, T., BOETJE, L.M., KOVAL, C.A., NOBLE, R.D., and BROWN, C.N. Transport properties of carbon dioxide through amine functionalized carrier membranes. *Industrial & Engineering Chemistry Research*, **1995**, 34, 4071.
27. MATSUYAMA, H., TERADA, A., NAKAGAWARA, T., KITAMURA, Y., and TERAMOTO, M. Facilitated transport of CO$_2$ through polyethylenimine/poly(vinyl alcohol) blend membrane. *Journal of Membrane Science*, **1999**, 163, 221.
28. QUINN, R. and LACIAK, D.V. Polyelectrolyte membranes for acid gas separations. *Journal of Membrane Science*, **1997**, 131, 49.
29. QUINN, R., LACIAK, D.V., and PEZ, G.P. Polyelectrolyte-salt blend membranes for acid gas separations. *Journal of Membrane Science*, **1997**, 131, 61.
30. HO, W.S.W. Membranes comprising salts of amino acids in hydrophilic polymers. U.S. Patent 5,611,843, **1997**.
31. HO, W.S.W. Membranes comprising amino acid salts in polyamine polymers and blends. U.S. Patent 6,099,621, **2000**.
32. TEE, Y.H., ZOU, J., and HO, W.S.W. CO$_2$-selective membranes containing dimethylglycine mobile carriers and polyethylenimine fixed carrier. *Journal of the Chinese Institute of Chemical Engineers*, **2006**, 37, 1.
33. ZHANG, Y., WANG, Z., and WANG, S.C. Selective permeation of CO$_2$ through new facilitated transport membranes. *Desalination*, **2002**, 145, 385.
34. KIM, T., LI, B., and HAGG, M. Novel fixed-site-carrier polyvinylamine membrane for carbon dioxide capture. *Journal of Polymer Science. Part B, Polymer Physics*, **2004**, 42, 4326.
35. UEMIYA, S., SATO, N., ANDO, H., and KIKUCHI, E. The water gas shift reaction assisted by a palladium membrane reactor. *Industrial & Engineering Chemistry*, **1991**, 30, 585.
36. BASILE, A., CRISCUOLI, A., SANTELLA, F., and DRIOLI, E. Membrane reactor for water gas shift reaction. *Gas Separation & Purification*, **1996**, 10, 243.
37. XUE, E., O'KEEFFE, M., and ROSS, J.R.H. Water-gas shift conversion using a feed with a low steam to carbon monoxide ratio and containing sulphur. *Catalysis Today*, **1996**, 30, 107.
38. CRISCUOLI, A., BASILE, A., and DRIOLI, E. An analysis of the performance of membrane reactors for the water-gas shift reaction using gas feed mixtures. *Catalysis Today*, **2000**, 56, 53.
39. GIESSLER, S., JORDAN, L., DINIZ DA COSTA, J.C., and LU, G.Q. Performance of hydrophobic and hydrophilic silica membrane reactors for the water gas shift reaction. *Separation and Purification Technology*, **2003**, 32, 255.
40. TOSTI, S., BASILE, A., CHIAPPETTA, G., RIZZELLO, C., and VIOLANTE, V. Pd-Ag membrane reactors for water gas shift reaction. *Chemical Engineering Journal*, **2003**, 93, 23.
41. ARMOR, J.N. Applications of catalytic inorganic membrane reactors to refinery products. *Journal of Membrane Science*, **1998**, 147, 217.
42. ZOU, J. and HO, W.S.W. CO$_2$-selective polymeric membranes containing amines in crosslinked poly(vinyl alcohol). *Journal of Membrane Science*, **2006**, 286, 310.
43. MOE, J.M. Design of water-gas-shift reactors. *Chemical Engineering Progress*, **1962**, 58, 33.
44. CAMPBELL, J.S. Influences of catalyst formulation and poisoning on the activity and die-off of low temperature shift catalysts. *Industrial and Engineering Chemistry Process Design and Development*, **1977**, 9, 588.
45. FIOLITAKIS, E., HOFFMANN, U., and HOFFMANN, H. Application of wavefront analysis for kinetic investigation of water-gas shift reaction. *Chemical Engineering Science*, **1980**, 35, 1021.

46. SALMI, T. and HAKKARAINEN, R. Kinetic study of the low-temperature water-gas shift reactor over a Cu-ZnO catalyst. *Applied Catalysis*, **1989**, 49, 285.

47. KEISKI, R.L., DESPONDS, O., CHANG, Y.F., and SOMORJAI, G.A. Kinetics of the water-gas-shift reaction over several alkane activation and water-gas-shift catalysts. *Applied Catalysis. A, General*, **1993**, 101, 317.

48. AMADEO, N.E. and LABORDE, M.A. Hydrogen production from the low temperature water-gas shift reaction: kinetics and simulation of the industrial reactor. *International Journal of Hydrogen Energy*, **1995**, 20, 949.

49. SARTORI, G., HO, W.S.W., SAVAGE, D.W., CHLUDZINSKI, G.R., and WIECHERT, S. Sterically-hindered amines for acid-gas absorption. *Separation and Purification Methods*, **1987**, 16, 171.

50. SARTORI, G. and SAVAGE, D.W. Sterically hindered amines for CO_2 removal from gases. *Industrial and Engineering Chemistry Fundamentals*, **1983**, 22, 239.

51. HO, W.S.W. and DALRYMPLE, D.C. Facilitated transport of olefins in Ag^+-containing polymer membranes. *Journal of Membrane Science*, **1994**, 91, 13.

Chapter 10

Pressure Swing Adsorption Technology for Hydrogen Production

SHIVAJI SIRCAR[1] AND TIMOTHY C. GOLDEN[2]

[1]Department of Chemical Engineering, Lehigh University
[2]Air Products and Chemicals, Inc.

10.1 INTRODUCTION

The current global production capacity of gaseous hydrogen is about 17 trillion standard cubic feet per year.[1] Approximately 95% of that hydrogen is captively produced and used by petroleum refiners and by ammonia and methanol manufacturers. The balance is available as "merchant hydrogen," which is produced by various industrial gas companies and sold to different customers. The primary applications of merchant hydrogen are in food, electronics, chemicals, metal refining, and petrochemical industries. The use of hydrogen as a clean fuel for transportation is also an emerging market.

The production of high-purity hydrogen from a gas mixture containing 60–90 mol % hydrogen by using pressure swing adsorption (PSA) processes has become the state-of-the-art technology in the chemical and petrochemical industries. Over 85% of current global hydrogen production units use PSA technology for hydrogen purification, and several hundred H_2 PSA process units have been installed around the world.

The basic concept of a H_2 PSA process is relatively simple. The impurities from the H_2-containing feed gas mixture are selectively adsorbed on a micro- and mesoporous solid adsorbent (zeolites, activated carbons, silica and alumina gels) at a relatively high pressure by contacting the feed gas with the solid in a packed column of

Hydrogen and Syngas Production and Purification Technologies, Edited by Ke Liu, Chunshan Song and Velu Subramani

the adsorbent in order to produce a high-purity H_2 product gas at feed pressure. The adsorbed components are then desorbed from the solid by lowering their superincumbent gas-phase partial pressures inside the column so that the adsorbent can be reused. Lowering the column pressure (depressurization) and flowing a part of the impurity-free H_2 product gas over the adsorbent at a lower pressure (purging) are two common methods of carrying out the desorption process. The desorbed gases are enriched in the more strongly adsorbed components of the feed gas. Although simple in concept, the actual design of a practical H_2 PSA process can be fairly complex because it generally involves the use of a multicolumn adsorption system operating under a cyclic steady state using a series of sequential, nonisothermal, nonisobaric, and non-steady-state process steps. These include the steps of adsorption, desorption, and a multitude of complementary steps that are designed to increase the H_2 product gas purity and recovery, to decrease the adsorbent inventory, and to optimize the overall separation efficiency.[2,3]

The two most common feed gas streams used for this application are (i) the steam methane reformer off-gas (SMROG) after it has been further treated in a water-gas shift (WGS) reactor, and (ii) the refinery off-gas (ROG) from various sources. The typical feed gas compositions to the PSA system for these cases are (i) 70%–80% H_2, 15%–25% CO_2, 3%–6% CH_4, 1%–3% CO, and trace N_2, and (ii) 65%–90% H_2, 3%–20% CH_4, 4%–8% C_2H_6, 1%–3% C_3H_8, and less than 0.5% C_4+ hydrocarbons. Both feed gases are generally available at a pressure of 4–30 atm (1 atm = 101.3 kPa) and at a temperature of 21–38 °C (70–100 °F), and they are generally saturated with water. The PSA processes are designed to produce a dry hydrogen-rich product stream at the feed gas pressure containing 98–99.999 mol % H_2 with a H_2 recovery of 70%–95%. A waste gas stream containing the unrecovered H_2 and all of the impurities of the feed gas is also produced at a pressure of 1.1–1.7 atm. This waste stream is typically used in combustion applications for its fuel value. Several specially designed PSA processes simultaneously produce a secondary product stream containing 99+ mol % CO_2 at a near-ambient pressure when the SMROG is used as the feed gas. Other PSA processes are designed to directly produce an ammonia synthesis gas containing a N_2–H_2 mixture in the mole ratio of 1:3 from the SMROG feed gas with or without a by-product stream of CO_2.

Figure 10.1 is a schematic box diagram for the most commonly used steam methane reforming (SMR)–WGS–PSA route of H_2 production.[4] Steam and natural gas in the ratio of 2.5:6.0 are reacted ($CH_4 + H_2O \leftrightarrow CO + 3H_2$) at a high temperature (750–900 °C) and pressure (4–30 atm) over a nickel catalyst. The endothermic SMR reactor effluent gas containing 70%–72% H_2, 6%–8% CH_4, 8%–10% CO, and 10%–14% CO_2 (dry basis) is cooled to 300–400 °C (steam produced) and subjected to the exothermic WGS reaction ($CO + H_2O \leftrightarrow CO_2 + H_2$), cooled again, and purified in a multicolumn PSA unit. The H_2 product contains 99.999+ H_2 (<15 ppm CO_x), and the H_2 recovery from the PSA unit is typically 70%–95%. The PSA waste gas is recycled as fuel to the reformer, fresh natural gas providing the balance of the fuel.

Figure 10.2 shows an aerial photograph of a H_2 production facility by the above-described route. It is located in Pasadena, TX, and it belongs to Air Products

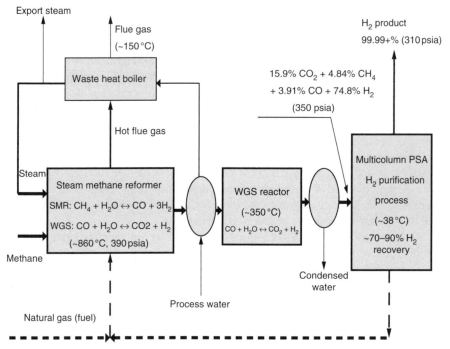

Figure 10.1.   Conventional steam methane reforming (SMR) route for hydrogen production.

Figure 10.2.   Aerial photo of Air Products and Chemicals, Inc.; an 80 MMSCFD H_2 plant in Pasadena, TX (SMR H_2 facility and 10-bed H_2 PSA). See color insert.

and Chemicals, Inc. The plant produces 80 million standard cubic feet of H_2 per day. The multicolumn H_2 PSA unit can be seen at the upper right-hand quadrant of the picture.

Recovery of H_2 from ROGs by a PSA process has also found a major niche for this technology. The driving force for this application is that the refineries are becoming more and more deficient in H_2 due to higher demands to meet sulfur removal and reformed gasoline regulations, and they are looking for alternative low cost source of H_2.[5-7]

The cost of hydrogen from an integrated SMR/PSA system or an integrated ROG/PSA system is impacted by both the capital and operating costs of the PSA process. The capital cost is primarily determined by the number and size of the PSA beds used, the amount of adsorbent in each bed, and the number of switch valves required. The bed size and the adsorbent inventory decrease as (i) the hydrogen productivity (moles of H_2 produced/total volume of the adsorbent in the system/ cycle) increases and (ii) the hydrogen recovery (moles of H_2 produced/mole of feed H_2/cycle) increases. Hydrogen productivity can be increased by using (i) faster PSA cycle times and/or (ii) adsorbents with higher PSA working capacities for the impurities being removed from H_2. Hydrogen recovery by the PSA process can be increased by (i) creative process design to preserve the void H_2 in the adsorption column before the depressurization steps of the cycle, and (ii) judicious selection of the adsorbents used to decrease the coadsorption of H_2 and to facilitate desorption of the adsorbed impurities during the product H_2 purge step of the cycle. Higher H_2 recovery also results in (i) significant capital and operating cost savings for production of H_2 by SMR due to downsizing of the reformer and reduction in natural gas requirement, and (ii) substantial cost incentive in H_2 recovery from the ROG.

The research and development activities in this field have been extensive during the last 30 years. Figure 10.3 shows the year-by-year breakdown of the number of issued U.S. patents on H_2 PSA processes between 1978 and 2005.[8] A total of 275 basic patents were granted to 73 corporations around the world. The figure also shows a recent upsurge in the patent activity in this area. The primary reason for this is the design of new PSA process configurations to increase the primary and secondary product recoveries while maintaining their high purities and to reduce the adsorbent inventory, as well as the development of new adsorbents for this application. A common practice is to use more than one type of adsorbent in the PSA columns (as different adsorbent layers in the same vessel) in order to (i) obtain optimum working adsorption capacities and selectivities for the feed gas impurities while reducing the coadsorption of H_2, and (ii) achieve efficient desorption of the impurities using minimum amount of product H_2 as purge gas.

This chapter provides a brief review of (i) several commercial H_2 PSA processes and their separation performances, (ii) characteristics of adsorbents used in these processes, and (iii) recent research and developments on the subject of H_2 production and purification by PSA.

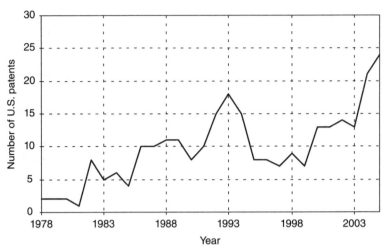

Figure 10.3.   Survey of U.S. patents on H_2 PSA (number of U.S. patents vs. year).

10.2 PSA PROCESSES FOR HYDROGEN PURIFICATION

10.2.1 PSA Processes for Production of Hydrogen Only

The most frequently used PSA process concept for sole production of hydrogen is patented by Union Carbide Corporation.[9,10] One version of this process consists of nine cyclic steps:[9]

(a) Adsorption: The feed gas is passed at pressure P^F through an adsorber and an essentially pure H_2 stream is produced through the product end at the feed gas pressure. A part of this gas is withdrawn as the primary H_2 product stream. The step is stopped when the impurity mass transfer zones are somewhere in the middle of the adsorber and the rest of the adsorber is essentially clean.

(b) Cocurrent depressurization I: The adsorber is then cocurrently depressurized to a pressure level of P_1. An essentially pure H_2 stream is again produced through the product end which is used to pressurize a companion adsorber undergoing step (h).

(c) Cocurrent depressurization II: The adsorber is further depressurized cocurrently to a pressure level of P_2. The effluent gas through the product end is again a high-purity H_2 stream, which is used to pressurize another companion adsorber undergoing step (g).

(d) Cocurrent depressurization III: The adsorber is again depressurized cocurrently to a pressure level of P_3. The effluent gas through the product end is still a high-purity H_2 stream, which is used to countercurrently purge another companion adsorber undergoing step (f).

(e)   Countercurrent depressurization: The adsorber is then countercurrently depressurized to the lowest pressure level of the cycle (P^D). The effluent gas from this step, which contains a part of the desorbed gases and most of the column void gases, is wasted.

(f)   Countercurrent purge: The adsorber is then countercurrently purged with a gas stream of essentially pure H_2 at pressure P^D obtained from a companion column undergoing step (d). The effluent gas from this step contains the remaining part of the desorbed impurities and it is wasted.

(g)   Countercurrent pressurization I: The adsorber is then pressurized from pressure P^D to P_2 by countercurrently introducing the gas produced by a companion column undergoing step (c).

(h)   Countercurrent pressurization II: The adsorber is further pressurized from P_2 to P_1 by countercurrently introducing the effluent gas from step (b).

(i)   Countercurrent pressurization III: Finally, the adsorber is countercurrently pressurized from P_1 to P^F with a part of the H_2 product gas produced by a companion column undergoing step (a). The adsorber is now ready to start a new cycle.

Other modifications to the above-described PSA cycle with only two cocurrent depressurization steps (which produce the high-purity H_2 effluent gases for only one countercurrent pressurization step and the countercurrent purge step) are also practiced.[10] Typically, multicolumn PSA units containing 4–12 parallel columns are used to accommodate these steps. Several adsorbers can be simultaneously receiving the feed gas mixture (step a) when the H_2 production capacity is very large. Figure 10.4 is a schematic flowsheet for a 10-column PSA process using the 11-step process. This process is generally known as the Poly-bed process. Union Carbide's H_2 PSA process technology was sold to UOP LLC in the 1980s. More detailed information on the performance of Poly-bed processes, their control systems, and market applications can be found elsewhere.[11]

The most distinguishing features of the Poly-bed process consist of (i) stopping the adsorption step (step a) while a substantial adsorption capacity for the feed gas impurities remain unused near the product end of the adsorber and then (ii) carrying out a series of cocurrent depressurization steps (steps b–d). The feed-like void gas from the section of the column which holds the feed impurities at the end of step (a) expands towards the product end during steps (b)–(d) and the impurities are adsorbed in the clean section of the column. Consequently, high-purity H_2 streams at different pressure levels are produced from the column during these steps, which are used to countercurrently purge (step f) and pressurize (steps g and h) other companion columns. These sequences of steps extract valuable H_2 from the column void gas at the end of step (a) as pure H_2 and use it efficiently for impurity desorption and partial pressurization steps. The net result is higher H_2 recovery from the feed gas. This is, however, done at the cost of increased adsorbent inventory per

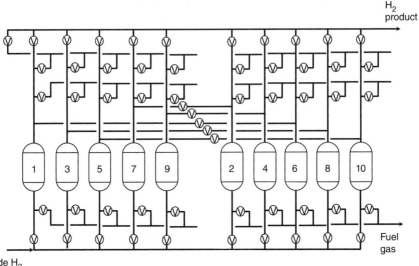

Figure 10.4.   Schematic flow diagram for the Poly-bed PSA process.

unit amount of feed processed because the entire adsorber is not used to process the feed gas in step (a).

A Poly-bed system consisting of 10 parallel columns, each containing a layer of activated carbon in the feed end and a layer of 5A zeolite in the product end, could process an SMROG (77.1% H_2, 22.5% CO_2, 0.35% CO, and 0.013% CH_4) at 20.7 atm and 21 °C to produce a 99.999% pure H_2 product at near the feed gas pressure with a H_2 recovery of 86.0%.[9] The lowest desorption pressure (P^D) for the process was 1 atm and the composition of the waste gas was 32.0% H_2, 66.8% CO_2, 1.0% CO, and 0.04% CH_4. The volume fractions of the carbon and the zeolite layers in the adsorbers were 0.762 and 0.238, respectively. The total cycle time (11 steps) used was 13.33 min, and the feed processing capacity for the process was 34.9 ft³ of feed gas (1 atm, 15 °C/ft³) of total adsorbent in the system/cycle.

A very interesting variation of the above-described process is patented by Toyo Engineering Corporation of Japan.[12,13] The process is used for the recovery of H_2 from a ROG. It consists of nine sequential steps:

(a) Adsorption: This step is identical to step (a) of the Poly-bed process. It produces a high-purity H_2 stream at feed gas pressure (P^F), a part of which is withdrawn as the primary H_2 product.

(b) Cocurrent depressurization I: This step is also identical to step (b) of the Poly-bed process. The column pressure at the end of this step is P_1 and the effluent gas is high-purity H_2, which is used to pressurize a companion column undergoing step (h).

(c) Cocurrent depressurization II: The effluent gas from this step is initially high-purity H_2 but the step is continued until some of the less strongly adsorbed impurities from the feed gas break through the product end. The column pressure at the end of this step is P_2.

   The entire effluent gas from step (c) is stored in a separate tank packed with nonporous inert solids. As this effluent gas enters the storage tank, it displaces the previously stored gas from the tank (received from another column that underwent step (c). The previously stored gas leaves the tank through the same end that was used to introduce the gas into the tank. Thus, the effluent gas from the tank during this step initially contains a H_2 stream containing some impurities of the feed gas followed by an essentially pure H_2 stream. This gas is used to countercurrently purge a column at pressure P^D for the removal of the adsorbed impurities. Thus, the gas is introduced into the tank through one end during step (c) for one cycle and then through the other end for the next cycle and so on.

(d) Cocurrent depressurization III: The column is then further depressurized cocurrently to pressure P_3 and the effluent gas is used to pressurize a companion column undergoing step (g).

(e) Countercurrent depressurization: The column is countercurrently depressurized to a pressure of P^D. The effluent gas forms a part of the waste gas from this process.

(f) Countercurrent purge: The column is countercurrently purged with the effluent gas from the storage tank at pressure P^D. The column effluent is wasted.

(g) Countercurrent pressurization I: The column is countercurrently pressurized to pressure P_3 using the effluent gas from a companion column undergoing step (d).

(h) Countercurrent pressurization II: The column is countercurrently pressurized to a pressure level of P_1 by introducing the effluent from a companion column undergoing step (b).

(i) Countercurrent pressurization III: The column is finally pressurized to P^F by using a part of the H_2 product being produced by a companion column undergoing step (a).

The distinguishing feature of this process is described in step (c). The countercurrent purge gas is produced by allowing some impurities to break through the column (substantially increases purge gas quantity), but the purge efficiency is maintained by reversing the order of flow of this gas into the column being purged. This process is thus called the last out, first in (LOFIN) process. The use of a larger quantity of purge gas, partly with slightly impure H_2, increases the overall H_2 recovery by the process and reduces the adsorbent inventory.[14]

Figure 10.5 is a schematic process diagram for the LOFIN process using four parallel adsorbers and a gas storage tank. It could produce a 99.96% H_2 product at

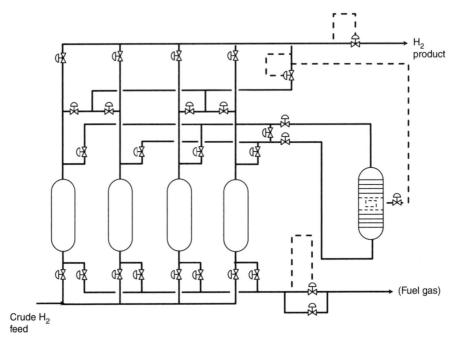

Figure 10.5. Schematic flow diagram for the LOFIN PSA process.

feed gas pressure from an ROG containing 78.8% H_2, 15.3% CH_4, 4.0% C_2H_6, 1.4% C_3H_8, 0.3% C_4H_{10}, and 0.1% C_5H_{12} (dry basis) at a pressure of 28 atm and 15 °C with a H_2 recovery of 86.3%.[12] The adsorbers were packed with a layer of silica gel (25%) near the feed end and a layer of activated carbon (75%) near the product end. The total cycle time (nine steps) for the process was 30.0 min and the feed processing capacity for the process was 153.0 ft^3 of feed gas/ft^3 of total adsorbent in the system/ cycle. The waste gas containing 33.8% H_2, 47.9% CH_4, 12.5% C_2H_6, 4.4% C_3H_8, and 1.25 C_4+ hydrocarbons was produced at a pressure of 1.3 atm.

Several other PSA process configurations have been proposed for purification and recovery of H_2 from various ROG streams.[15–19]

10.2.2 Process for Coproduction of Hydrogen and Carbon Dioxide

As the concern about greenhouse gas emissions increases, methods to sequester CO_2 from H_2 generation systems are of commercial interest. The total CO_2 emission from a 100 million standard cubic foot per day (MMSCFD) H_2 plant is about 2500 t of CO_2/day. A PSA process that coproduces both high-purity CO_2 and high-purity H_2 is the Gemini-9 process developed by Air Products and Chemicals, Inc.[20,21] Figure 10.6 shows a schematic flow diagram for the process. It consists of six parallel adsorbers (called A beds) connected in series with three other parallel adsorbers

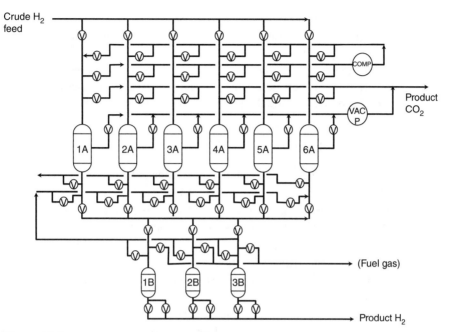

Figure 10.6.   Schematic flow diagram for the Gemini PSA process.

(called B beds). The A and B beds undergo the following six- and seven-step cycles, respectively:

Cycle steps for A beds:

(a)   Adsorption: The SMROG is passed through a train of A and B beds in series and a stream of pure H_2 at feed pressure (P^F) is produced through the product end of the B bed. A part of this gas is withdrawn as the primary H_2 product.

(b)   Cocurrent CO_2 rinse: The A and B beds are then disconnected and a stream of essentially pure CO_2 at feed gas pressure is cocurrently passed through the A bed. The effluent gas from the A bed is feed like, which is recycled as feed gas to another A bed. The A bed is saturated with CO_2 at the end of this step.

(c)   Countercurrent depressurization: The A bed is then depressurized to a near-ambient pressure level. The effluent gas is pure CO_2. A part of this gas is recompressed to P^F and used as CO_2-rinse gas to a companion A bed. The balance is withdrawn as the secondary CO_2 product.

(d)   Countercurrent evacuation: The A bed is then evacuated to the lowest subatmospheric pressure level of the cycle (P^D). The effluent is again pure CO_2, which is withdrawn as the secondary product.

(e)   Countercurrent pressurization I: The column is then pressure-equalized to a pressure level of P_1 with a B bed that has just finished its adsorption step (a).

(f)   Countercurrent pressurization II: The A bed is finally repressurized with a part of the H_2-rich gas produced by an A–B tandem in series undergoing step (a) by introducing the gas to the A bed through a companion B bed in series.

Cycle steps for B beds:

(a)   Adsorption: The B bed is connected with an A bed in series undergoing step (a).

(b)   Countercurrent depressurization I: The B bed is connected with an A bed that has just finished step (d) in order to pressure equalize the two beds.

(c)   Countercurrent depressurization II: The B bed is connected with another B bed that has completed step (f) in order to pressure equalize those beds.

(d)   Countercurrent depressurization III: The B bed is depressurized to near-ambient pressure level and the effluent gas is wasted.

(e)   Countercurrent purge: The B bed is purged with essentially pure H_2 obtained from another B bed undergoing step (a). The effluent is wasted.

(f)   Cocurrent pressurization: The B bed is then connected with another B bed undergoing step (c) in order to pressure equalize the two beds.

(g)   Countercurrent pressurization: The B bed is finally pressurized to P^F by introducing a part of H_2-rich gas produced by another B bed undergoing step (a). The B bed is connected with an A bed undergoing step (f) during this step.

The A beds are packed with activated carbons that selectively remove CO_2 and H_2O from the SMROG. The B beds are packed with zeolites for the selective removal of the remaining CO_2, CH_4, CO, and N_2 from H_2. The process is designed in such a fashion that very little CO_2 breaks through the A beds during step (a).

The most distinguishing features of this process are (i) cocurrent CO_2 rinse at feed pressure, (ii) separation of A and B beds during column regeneration steps, (iii) use of different regeneration methods for A (depressurization and evacuation) and B (depressurization and purge) beds, and (iv) pressure equalization between A and B, and B and B beds to conserve the void gases. These features permit the production of two pure products (CO_2 and H_2) from SMROG with high recoveries of both components. The process, however, requires rotating machinery (vacuum pumps and CO_2 recycle compressor) for its operation.

The process is called Gemini because of its ability to produce two products from a multicomponent feed gas. It could simultaneously produce a primary H_2 product at a purity of 99.999% with an H_2 recovery of 87.1% and a secondary CO_2 product at a purity of 99.4% with a CO_2 recovery of 94.0% from an SMROG feed gas containing 75.4% H_2, 19.9% CO_2, 0.96% CO, and 3.73% CH_4 at a pressure of 18 atm at 21 °C.[21] The H_2 product was produced at the feed gas pressure and the CO_2 product was produced at ambient pressure. The final evacuation level in step (d) of A beds was 0.13–0.20 atm. The waste gas was produced at near-ambient pressure, and it consisted of 8.1% CO_2, 5.6% CO, 20.8% CH_4, and 65.4% H_2. The absence of large amounts of CO_2 in the waste gas of Gemini process compared with those for the Poly-bed and the LOFIN processes makes it a fuel gas with higher calorific value. More detailed information about the operation and performance of this process can be found elsewhere.[21]

There are other options for recovering CO_2 from SMROG. It can be removed by using an amine scrubber prior to its introduction (CO_2 free) into a PSA system for the production of H_2 in order to reduce the size of the PSA unit and to improve the H_2 recovery.[22] It is also possible to recover the CO_2 from the vent gas of an existing H_2 PSA process by using an additional PSA system, which uses near-ambient pressure adsorption of CO_2 followed by its desorption by evacuation.[23] NaX or NaY zeolite can be conveniently used to adsorb about 60% CO_2 from a gas stream at near-ambient pressure and produce a high-purity CO_2 stream as the desorbed gas by evacuation.

Capture of the CO_2 produced by combustion of natural gas as the reformer fuel is not addressed in the above-described scenarios. The possible approaches to sequester CO_2 from the reformer furnace would be (i) oxycombustion or (ii) flue gas cleanup. For oxycombustion, natural gas would be burned in pure O_2, producing a waste gas containing only CO_2 and H_2O. The water can be removed by compression and cooling to produce a pure CO_2 stream at pressure for sequestration. Removal of CO_2 from the flue gas would involve either scrubbing of the waste gas with aqueous or nonaqueous solvents or the use of another adsorption-based CO_2 removal process.

10.2.3 Processes for the Production of Ammonia Synthesis Gas

A very important modification of the Poly-bed and Gemini process cycles described above can be used to directly produce an ammonia synthesis gas (a N_2 and H_2 mixture in the molar ratio of ~1:3) from SMROG as feed to the PSA systems.

For the case of the Poly-bed cycle, it is achieved by using N_2 from an external source to (i) countercurrently purge (step f) and pressurize (step i) the adsorbers. These steps introduce weakly adsorbing N_2 into the adsorbers before the adsorption step (a) begins. The N_2 is expelled out from the adsorber in conjunction with H_2 from the feed gas as the effluent stream during step (a) of the Poly-bed system.[24]

For the Gemini process, the ammonia synthesis gas can be produced as the effluent gas from step (a) by (i) carrying out the final countercurrent repressurization (step f) of the A beds by introducing N_2 from an external source, (ii) eliminating steps (c) and (f) for the B beds, and (iii) countercurrently purging (step e) and repressurizing (step g) the B beds with an external source of N_2.[25,26] The A and B beds remain disconnected during step (g) in this case. Thus, the modified Gemini process simultaneously produces an ammonia synthesis gas stream at feed gas pressure and a by-product CO_2 stream. Only four A beds and two B beds are needed to operate the modified cycle.[25,26] The A beds are again filled with an activated carbon and the B beds are filled with a zeolite. A H_2 recovery of 95% in the product ammonia synthesis gas, and a by-product CO_2 recovery of 94% at a CO_2 purity of 99.4% can be achieved by this process using SMROG feed gas (same composition as that for the Gemini process) at a pressure of 18 atm and a temperature of 21 °C. The ammonia synthesis gas is free of carbon oxides. About 75% of N_2 used in pressurizing and purging the A and B beds is recovered in the ammonia synthesis gas.[25,26] The waste gas has a composition of 6.9% CO_2, 5.5% CO, 21.3% CH_4, 45.7% N_2, and 20.6% H_2. The process is especially attractive for urea production by reacting NH_3 and CO_2:

$$[2NH_3 + CO_2 \leftrightarrow NH_2CONH_2 + H_2O]. \tag{10.1}$$

The above examples demonstrate the versatility and flexibility of PSA processes designed for the purification of H_2 from a bulk feed gas containing 70%–95% H_2. All of these processes are designed to meet certain product specifications and to increase the product recoveries at high purities.

10.3 ADSORBENTS FOR HYDROGEN PSA PROCESSES

The selection of adsorbents is critical for determining the overall separation performance of the above-described PSA processes for hydrogen purification. The separation of the impurities from hydrogen by the adsorbents used in these processes is generally based on their thermodynamic selectivities of adsorption over H_2. Thus, the multicomponent adsorption equilibrium capacities and selectivities, the multicomponent isosteric heats of adsorption, and the multicomponent equilibrium-controlled desorption characteristics of the feed gas impurities under the conditions of operation of the ad(de)sorption steps of the PSA processes are the key properties for the selection of the adsorbents. The adsorbents are generally chosen to have fast kinetics of adsorption. Nonetheless, the impact of improved mass transfer coefficients for adsorption cannot be ignored, especially for rapid PSA (RPSA) cycles.

The use of zeolites, activated carbons, silica gels, and activated aluminas for the production of pure H_2 from various feed gases by using a PSA process has been recognized from the early days of the development of this technology.[27] The use of different adsorbents in layers of different volume fractions in a single column for optimizing the separation process has also been recognized very early.[10,27] The following sections briefly describe some of the logic behind adsorbent selections as well as new developments in this field.

10.3.1 Adsorbents for Bulk CO_2 Removal

Activated carbons are universally chosen as the preferred adsorbent for the removal of bulk CO_2 from SMROG. A layer of an activated carbon is used at the inlet end of the adsorption column for this purpose. The commercial activated carbons provide good PSA working capacities for CO_2 adsorption from SMROG and acceptable selectivities for CO_2 over CO, CH_4, N_2, and H_2. At the same time, the strength of adsorption of CO_2 on the activated carbon is moderate, which permits relatively easy desorption of CO_2 under the conditions of a H_2 PSA process. Figure 10.7 shows the isotherms for the adsorption of CO_2 and other components of SMROG at 30 °C on BPL activated carbon produced by Calgon Corp.[28] Figure 10.8 shows the adsorption isotherms of the same gases on a 5A zeolite at 30 °C.[28] These isotherm data were measured in the laboratories of Air Products and Chemicals using a volumetric adsorption apparatus.[29]

Table 10.1 compares the binary gas selectivities of the components of SMROG on the BPL carbon and 5A zeolite in Henry's law region (limit of zero pressure) at 30 °C.[28] The first named component is the more selective species. The polar zeolite offers higher selectivities of adsorption of the polar gases (CO_2, CO, N_2) over H_2 than the nonpolar carbon, as expected. However, some of the interesting features include (i) the selectivity of adsorption of nonpolar CH_4 over H_2 is practically the same on both adsorbents, (ii) nonpolar CH_4 is more selectively adsorbed over weakly polar N_2 on the carbon than the zeolite, and (iii) CH_4 is selectively adsorbed over CO on the carbon while CO is preferentially adsorbed over CH_4 on the zeolite.

Table 10.2 compares the isothermal CO_2 working capacities (between CO_2 partial pressures of 5.0 and 0.5 atm at 25 °C) and the isosteric heats of adsorption of CO_2 on the BPL carbon and the 5A zeolite. It can be seen that the isothermal working

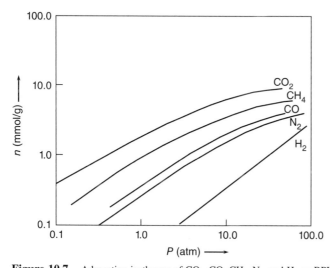

Figure 10.7. Adsorption isotherms of CO_2, CO, CH_4, N_2, and H_2 on BPL carbon at 30 °C.

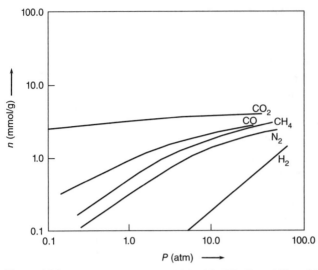

Figure 10.8. Adsorption isotherms of CO_2, CO, CH_4, N_2, and H_2 on 5A zeolite at 30 °C.

Table 10.1. Henry's Law Selectivities on BPL Carbon and 5A Zeolite at 30 °C

Gas Mixture	Binary Adsorption Selectivity	
	BPL	5A
$CO_2–CH_4$	2.5	195.6
$CO_2–CO$	7.5	59.1
$CO_2–N_2$	11.1	330.7
$CO_2–H_2$	90.8	7400.0
$CO–CH_4$	0.33	3.3
$CO–N_2$	1.48	5.6
$CO–H_2$	12.11	125.0
$CH_4–N_2$	4.5	1.7
$CH_4–H_2$	36.6	37.8
$N_2–H_2$	8.2	22.3

Table 10.2. PSA Working Capacities and Heats of Adsorption of CO_2

Property	Activated Carbon	5A Zeolite
CO_2 working capacity (wt %) (25 °C, P_{CO2} = 5–0.5 atm)	14.0	3.0
Heat of CO_2 adsorption (kcal/mol)	5.5	9.5

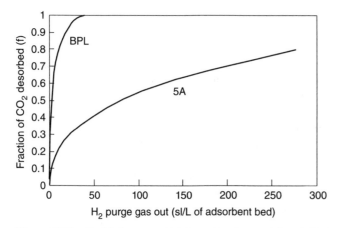

Figure 10.9. Equilibrium-controlled desorption characteristics of CO_2 by H_2 purge.

capacity of CO_2 is much larger on the carbon. It is also adsorbed much less strongly on the carbon (lower heat of adsorption).

Figure 10.9 shows the isothermal desorption characteristics of pure CO_2 from a column (packed with BPL carbon or 5A zeolite) by purge with pure H_2 at 30 °C and 1.0 atm.[30] The figure plots the fraction of CO_2 desorbed from a column that was initially equilibrated with pure CO_2 at 1.0 atm and 30 °C as a function of the specific amount of H_2 leaving the column.

These data show that even though the 5A zeolite could exhibit a higher absolute adsorption capacity for CO_2 (Fig. 10.8) than the BPL carbon (Fig. 10.7), particularly in the low-pressure region, and offers much larger selectivities of adsorption for CO_2 over the other components of the SMROG (Table 10.1), the net PSA working capacity for CO_2 on the carbon is much larger than that on the zeolite (Table 10.2) and a significantly lower amount of H_2 purge gas is wasted to desorb the CO_2 from the carbon than the zeolite (Fig. 10.9). That is because CO_2 is much less strongly adsorbed on the carbon than the zeolite as indicated by the lower isosteric heat of adsorption of CO_2 on the carbon (Table 10.2). The high heat of adsorption of CO_2 on the zeolite is created by the strong pole–pole interactions between the charged sites of the polar zeolite structure and the permanent quadrupole of CO_2. The interaction is much weaker between the CO_2 and the relatively nonpolar activated carbon. Since higher PSA working capacity and the easier desorption by purge are the two most desirable properties for improved performance of a PSA system, the nonpolar activated carbon is the preferred adsorbent for bulk CO_2 removal by PSA.

10.3.2 Adsorbents for Dilute CO and N_2 Removal

The commercial performance of a H_2 PSA process (online time, productivity, product purity) is generally dictated by the adsorption characteristics of the most weakly adsorbing component (other than H_2) in the feed gas. The adsorption step of the PSA process is continued until the composition of that impurity in the effluent gas reaches a preset limit (usually a part per million level when a high-purity H_2 product is desired). That impurity is typically N_2 for production of H_2 from SMROG. The selectivities of adsorption of the constituents of SMROG increase in the order $CO_2 > CH_4 > CO > N_2 > H_2$ on the carbon, and in the order $CO_2 > CO > CH_4 > N_2 > H_2$ on the zeolite (Table 10.1). Consequently, a layer of 5A zeolite, which exhibits a much higher N_2 adsorption capacity (Figs. 10.7 and 10.8) and selectivity of adsorption over H_2 (Table 10.1) at low partial pressures of N_2 due to the stronger interactions between the charged sites of the zeolite and the permanent quadrupole of N_2, is typically placed at the product end of the adsorber. The 5A zeolite layer is also used to remove the dilute CO from the SMROG because of its high adsorption capacity (Fig. 10.8) and selectivity over H_2 (Table 10.1) caused by the strong interactions between the permanent dipole and quadrupole of CO and the polar zeolite sites.

The isothermal desorption characteristics of pure (a) N_2 and (b) CO from 5A zeolite and BPL carbon by H_2 purge at 1.0 atm and 30 °C are shown in Figure 10.10.[31] The column was initially equilibrated with pure CO or N_2 at 1.0 atm and 30 °C, and then purged with pure H_2 at the same conditions. The figure plots the fraction of CO or N_2 desorbed as a function of the specific amount of H_2 leaving the column. It may be seen that CO and N_2 can be very easily desorbed from the carbon by H_2 purge because they are very weakly adsorbed on the nonpolar carbon (low adsorption capacity and selectivity over H_2). However, these gases can also be desorbed reasonably well (~80% of impurities removed) from the 5A zeolite without consuming a large volume of H_2 as purge gas. In fact, a comparison between Figures 10.9 and 10.10 shows that desorption of CO_2 from the BPL carbon, and not desorption of CO and N_2 from the zeolite by H_2 purge, will set the limit on the amount of H_2 purge quantity required in the PSA process. Consequently, 5A zeolite is the preferred adsorbent for the removal of CO and N_2 from SMROG.

There has been a significant amount of R & D to identify improved adsorbents for this application. Zeolite CaX is found to be a better adsorbent than 5A because of its higher N_2 capacity when very low levels of N_2 is required in the product H_2.[32] Another study suggested the use of LiX and Li–CaX zeolites as better materials for the removal of CO and N_2 from H_2 streams because the CO and N_2 adsorption isotherms on LiX zeolite are more linear than those on CaX zeolite in the low-pressure region, which offer higher CO and N_2 working capacities in a PSA process.[33] The use of BaX zeolite has also been suggested for this application because it exhibits extended linear isotherm in the low-pressure region and it has a high bulk density.[34]

An extensive study of optimum adsorbents for CO and N_2 removal from SMROG for H_2 production revealed that the use of highly Ca-exchanged, binderless A zeolite in the product end of the PSA column could increase the H_2 productivity by ~10% and the overall H_2 recovery by ~1.0 percentage point compared with those for a standard 5A zeolite (~70% Ca-exchanged NaA zeolite) when N_2 in the product

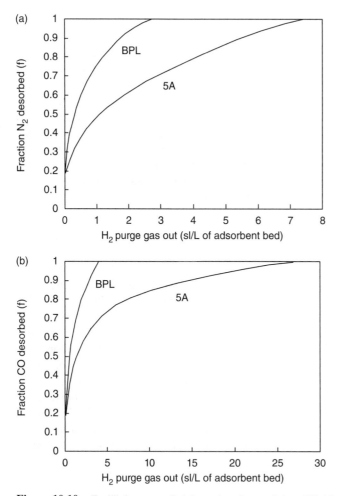

Figure 10.10. Equilibrium-controlled desorption characteristics of N_2 (a) and CO (b) by H_2 purge.

gas is the limiting impurity.[35] The improvement was caused by higher N_2 capacity and faster adsorption of N_2 on the modified type-A zeolite. On the other hand, the study showed that the use of NaX zeolite instead of 5A zeolite as the adsorbent layer in the product end of the adsorber is preferable when N_2-free H_2 is not required, and CO is the limiting impurity in the product gas. The desorption of CO from the NaX zeolite by H_2 purge is easier than that from the 5A zeolite because CO is held less strongly by the NaX zeolite, which results in a better CO working capacity on that material. The hydrogen productivity is increased by ~10% and its recovery is improved by ~0.5 percentage point by the use of NaX in the product end instead of conventional 5A zeolite.[35]

10.3.3   Adsorbents for Dilute CH₄ Removal

The capacity of adsorption of nonpolar CH_4 and its selectivity of adsorption over H_2 are about the same on both the 5A zeolite and the BPL activated carbon as shown in Figures 10.7 and 10.8 and Table 10.1. However, CH_4 is selectively adsorbed over CO on the carbon and CO is selectively adsorbed over CH_4 on the zeolite (Table 10.1). In general, the majority of CH_4 removal from the SMROG is accomplished in the activated carbon layer because of its higher selectivity over CO. Any CH_4 that breaks through the carbon layer will be less selectively adsorbed than CO in the zeolite layer. Consequently, CO and CH_4 often break through the product end of a H_2 PSA process simultaneously. Nevertheless, both the zeolite and the carbon layers are used to remove the CH_4 from the SMROG.

10.3.4   Adsorbents for C₁–C₄ Hydrocarbon Removal

The key impurities present in a typical ROG for the recovery of H_2 by a PSA process are bulk C_1 and C_2 and dilute C_3 and C_4 hydrocarbons. Figures 10.11 and 10.12 describe the pure gas adsorption isotherms of the components of ROG at 30 °C on the BPL activated carbon and a silica gel sample (Sorbead H produced by Engelhard Corp.), respectively.[31] These data were also measured in Air Products and Chemicals, Inc. laboratories. It may be seen that the carbon adsorbs C_3+ hydrocarbons very strongly. Consequently, desorption of these hydrocarbons from the carbon by H_2 purge becomes rather impractical requiring a large volume of purge gas.

Figure 10.13 compares the isothermal desorption characteristics of pure (a) C_2H_6 and (b) C_3H_8 by H_2 purge from the BPL carbon and the silica gel samples at 1.0 atm and 30 °C.[31] These data were also generated in the same fashion as those for the components of the SMROG. They show that desorption of C_3H_8 from the carbon by

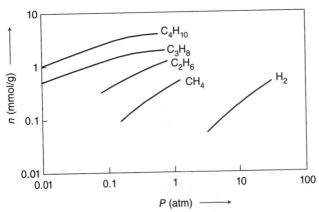

Figure 10.11.   Adsorption isotherms of C_4H_{10}, C_3H_8, C_2H_6, CH_4, and H_2 on BPL carbon at 30 °C.

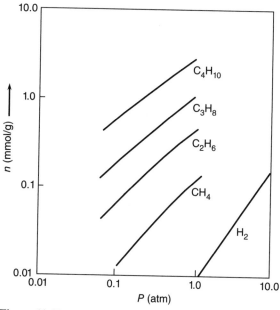

Figure 10.12. Adsorption isotherms of C_4H_{10}, C_3H_8, C_2H_6, CH_4, and H_2 on Sorbead H silica gel at 30 °C.

H_2 purge requires a large volume of H_2, while desorption of C_2H_6 from the carbon can be achieved by using a moderate amount of H_2 purge. Consequently, the silica gel, which offers relatively lower absolute adsorption capacities and selectivities over H_2 for C_3+ hydrocarbons than the BPL carbon is the preferred material for the removal of the higher hydrocarbons from the ROG because it can be more easily regenerated by H_2 purge (Fig. 10.13). The carbon, on the other hand, is favored for the removal of CH_4 and C_2H_6 from H_2 because it exhibits relatively higher capacities and selectivities of adsorption over H_2 for these gases and yet these gases are not very strongly adsorbed on the carbon. The advantage of using silica gel as the preferred adsorbent for the C_3+ hydrocarbons in purification of H_2 from ROG by PSA has been experimentally evaluated and numerically simulated using a process model.[36] The study demonstrated the difficulty in desorbing butane from a PSA column containing activated carbon alone. A column packed with a layer of silica gel at the feed end followed by a layer of activated carbon at the product end resolved the problem although the productivity of the PSA process decreased with the use of the dual-sorbent system. A column containing three layers of different adsorbents, such as a low surface area activated alumina at the feed end (2%–20% of bed volume), an intermediate surface area silica gel in the middle (25%–40% of bed volume), and a high surface area activated carbon in the product end (40%–78% of bed volume), has been found to be an optimum arrangement for the purification of H_2 from an ROG.[37] It should be understood that various forms of activated carbons and silica gels can be employed for the production of H_2 from ROG.

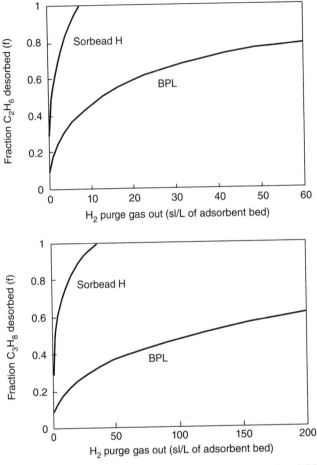

Figure 10.13.   Equilibrium-controlled desorption characteristics of C_3H_8 by H_2 purge.

10.3.5   Other Adsorbent and Related Improvements in the H_2 PSA

The majority of the published literature on improved adsorbents for H_2 purification by PSA deals with equilibrium adsorption properties (adsorption capacities of the impurities and their selectivities over H_2) of the materials. The adsorbents are generally chosen in such a way that the kinetics of adsorption of the impurities into the adsorbents are relatively fast, primarily being controlled by macro- and mesopore diffusion within the adsorbent particles. The kinetics of adsorption may, however, become an issue for the removal of the trace amounts (ppm) of a relatively weakly adsorbed impurity (N_2 or CH_4) at the product end of an H_2 PSA due to the existence of a very low driving force for the adsorption process. It was suggested that a layer

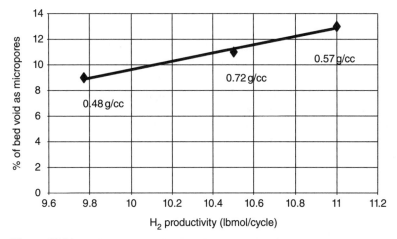

Figure 10.14.   Effect of volume fraction of microporosity of activated carbon on PSA performance.

of smaller adsorbent particles be used in the product end of the H_2 PSA adsorber in order to improve the impurity mass transfer coefficients (inversely proportional to the square of particle radius for macropore diffusion control).[38] For example, ~10% increase in the H_2 productivity and 2.0 percentage point increase in H_2 recovery in a PSA process for treating SMROG were claimed by lowering the diameter of the zeolite particles from 0.2 to 0.1 cm (used in a shallow layer at the product end of the adsorber).[38] The increase in the column pressure drop due to the use of smaller particle size was minimal due to the shallowness of the layer.

The physical properties of an activated carbon are also important for improving the performance of a PSA process.[39] Generally, carbons with higher bulk densities are preferable because they increase the working capacity of an adsorbate per unit volume of the adsorbent, and thus lower the adsorber volume. The denser carbon also improves H_2 recovery of a PSA process by reducing the void volume of the column. The micropore volume of a carbon is also a critical variable. Figure 10.14 shows that higher H_2 productivity (pound moles of H_2 produced/cycle for a given bed volume) in a PSA process can be achieved by increasing the ratio of the carbon micropore volume to the total void volume in the column.[39] The bulk densities of the carbons are also given in the plot. It may be interesting to note that the ratio of micropore volume to total void volume in the column reached a maximum at an intermediate bulk density.

10.4 FUTURE TRENDS FOR HYDROGEN PSA

The H_2 production capacities of the current commercial single-train hydrogen PSA process units range from ~1 to 130 MMSCFD. It is expected that single-train PSA units with a capacity of ~200 MMSCFD H_2 will be built in the near future. One approach to achieve this goal will be the reduction in the total cycle time of the existing process design and employing improved adsorbents, which will be compatible with faster cycles. The other significant development in the hydrogen PSA technology will be to scale down the production unit to support the forthcoming "hydrogen economy," which calls for efficient smaller size (0.05–0.5 MMSCFD) systems. They will service numerous hydrogen-based applications including hydrogen-powered fuel cells or internal combustion vehicles, stationary power applications, backup power units, power grid management, power for remote locations, and portable power applications in consumer electronics, business machinery, and recreational equipments.

10.4.1 RPSA Cycles for Hydrogen Purification

The conventional H_2 PSA process designs typically use total cycle times of 10–30 min.[2,3,10–26] This is generally needed to accommodate the multistep cyclic process needed for improving product recovery at high product purity. Longer cycle times, and particularly longer adsorption step time, increases the adsorber size and cost, and reduces the bed size factor (BSF) defined by the volumetric H_2 production rate per unit volume of the adsorbent (ft^3 of H_2/h/ft^3 of adsorbent). For example, a Polybed H_2 PSA process operating with a total cycle time of ~15 min yielded a BSF of ~100 SCFH/ft^3.[9]

Faster PSA cycle times can significantly reduce the capital cost of the process by decreasing vessel sizes and increasing the BSF. However, significantly shorter total cycle time and consequently shorter adsorption step time severely decrease the gas contact time in the adsorber for a given feed gas or product H_2 flow rate. This requires adsorbents with very large impurity mass transfer coefficients in order to retain the separation efficiency of a longer cycle time conventional PSA process. Decreasing adsorbent particle size may provide the required enhancement in impurity adsorption rate but at the cost of increased column pressure drop and the possibility of local fluidization of the adsorbent bed, both of which adversely affect the product purity and productivity. High column pressure drop during the desorption of impurities by product H_2 purge step of the process increases purge H_2 quantity, which reduces H_2 recovery by the process.[30] Faster cycles may not also allow the time required for employing the multitude of complementary steps used for enhancing the separation efficiency in a conventional H_2 PSA cycle. Despite these prospects of diminishing returns, rapid or ultrarapid H_2 PSA cycles have been designed by various corporations. Questair Industries, Inc. developed a H_2 PSA process that utilizes a conventional seven-step PSA cycle[10] but operates using a total cycle time of about 60 s.[40] The hardware consists of six parallel adsorbers (packed with a layer of an activated carbon and a layer of a zeolite particles) and two rotating valves (instead of multiple switch valves) for the introduction and removal of gases to and from the adsorbers. The process could produce a product stream of 99.95% H_2 (<2 ppm CO_x) from a feed gas at 8.5 atm containing 30% H_2, 50% N_2, 17% CO_2, and 3% CO on a dry basis. The H_2 product rate from the unit was 0.012–0.11 MMCFD depending on the feed flow rate (S. Sircar and Questair Industries, pers. comm.). Another Questair rapid H_2 PSA unit employing six to nine adsorber beds and two specially designed rotary valves can process an SMROG to produce a high-purity H_2 gas (<1 ppm CO) with a H_2 recovery of about 80% at a much higher H_2 productivity (4–10 times) than a conventional PSA unit. The unit can be designed to produce 5–5000 Nm^3/h of H_2 using a total cycle time of 0.5–1.5 min (S. Sircar and Questair Industries, pers. comm.).

Very compact and low cost H_2 PSA units are being developed with the use of rotary valve technology[41–45] and rapid cycles (less than 30 s adsorption time). A photograph of a small PSA unit developed by Air Products and Chemicals, Inc. is shown in Figure 10.15. The compact unit has a volume of only 40 ft^3 and it can produce 0.1 MMSCFD H_2 with a typical SMROG feed gas at 120 psig feed pressure using a simple four-bed PSA process.[10] This commercial system has a BSF of about 1000 SCFH/ft^3, which is about 10 times more than that of a conventional H_2 PSA process. The H_2 product recovery at a high H_2 purity (<1 ppm CO) is about 80%. The increased BSF is primarily due to the reduction in cycle time and use of adsorbents with fast sorption kinetics.

Figure 10.15. Air Products and Chemicals 100 Nm^3/h rapid H_2 PSA unit.

Recent patents by Questair Industries claim to use ultrarapid PSA cycle times using adsorbent laminate modules. Cyclic frequency of 100 cycles per minute (~cycle time = 0.6 s) to possibly 600 cycles per minute (~cycle time = 0.1 s) are claimed.[46,47] The ultrarapid cycles can potentially increase the BSF further by two orders of magnitude and significantly decrease the adsorber size. It is not known to the authors whether these ultrarapid PSA units have been used for H_2 purification applications.

It should be noted that (i) the productivity (pound moles of product/pound of adsorbent/time) cannot be increased indefinitely by lowering the cycle time, there being a finite limiting value of productivity for a finite value of the adsorbate mass transfer coefficient,[48] and (ii) instantaneous thermal equilibrium between the gas and the solid adsorbent inside an adsorber cannot be achieved when the cycle times are very short and adversely affect the working capacity of the adsorbent.[49] These findings were demonstrated by simplified analysis of idealized PSA processes on a single adsorbent particle. Nevertheless, the development of RPSA processes opens up further research and development opportunities on (i) novel adsorbent configurations such as structured adsorbents and (ii) innovative mechanical devices for operating the rapid cycles.

10.4.2 Structured Adsorbents

The adsorbent particles are normally used as beads, extrudates, or granules (~0.1 –0.3 cm equivalent diameters) in conventional H_2 PSA processes. The particle diameters can be further reduced to increase the feed gas impurity mass transfer rates into the adsorbent at the cost of increased column pressure drop, which adversely affects the separation performance. The particle diameters, however, cannot be reduced indefinitely and adsorption kinetics can become limiting for very fast cycles.[48] New adsorbent configurations that offer (i) substantially less resistance to gas flow inside an adsorber and, thus, less pressure drop; (ii) exhibit very fast impurity mass transfer coefficients; and (iii) minimize channeling are the preferred materials for RPSA systems. At the same time, the working capacity of the material must be high and the void volume must be small in order to minimize the adsorber size and maximize the product recovery. Various materials satisfy many of the requirements listed above, but not all of them simultaneously.

The two promising candidates are adsorbent monoliths and adsorbent sheets. The fabrication of activated carbon and zeolite monoliths are reported in the literature. Zeolite monoliths have also been tested for air separation application by PSA.[50,51] However, the use of monoliths for use in H_2 PSA is not known to the authors. Monoliths having very high cell density (several hundred to thousand cells per square inch) will be necessary in order to have fast adsorption kinetics as well as reasonable bulk density for a PSA application. Manufacture of such monoliths is complex, and they are not yet commercially available. Gas channeling through the monoliths can also be a problem.[52] Adsorbent sheets have been used for air separation by RPSA.[53,54] The thickness of the adsorbent sheets and the space between the adjacent sheets must be very small, requiring reinforcement matrix and spacing systems. Methods of creating such spacing systems involve creating embossing or ridges on the laminates, corrugating the laminates and alternating corrugated and noncorrugated layers, and using an external spacing device between the adjacent layers.

PSA units using adsorbent sheets have been operated by Questair Industries, Inc.[46,47,53–55] The sheets were coated with a fine powder of zeolite particles ($1-10\,\mu m$) and there were spacers between the sheets to establish flow channels in a flow direction tangential to the sheets and between adjacent pairs of sheets. The sheets could be made in various configurations (rectangular, annular stack, spiral-wound, etc.) and included a support for the adsorbent in the form of an aluminum foil, a metallic mesh, or a matrix that could be woven, nonwoven, ceramic, or wool. These structured materials did not fluidize at high gas velocities and exhibited equilibrium and mass transfer properties of the powdered adsorbent.

The use of activated carbon fabric as the adsorbent in an H_2 PSA process has also been reported.[56] The advantage of the fabric is that it does not require additional supports like metal mesh, paper, and aluminum foil, and the bulk density of the adsorbent material in the PSA adsorber is increased.

10.4.3 Sorption-Enhanced Reaction Process (SERP) for H_2 Production

The concept that the removal of an undesired reaction product by selective adsorption from the reaction zone of an equilibrium-controlled reaction increases the conversion and the rate of formation of the desired component (based on Le Chatelier's principle) was used to develop a novel PSA process concept called SERP for direct production of fuel cell-grade hydrogen by steam reforming of methane ($CH_4 + 2H_2O \leftrightarrow CO_2 + 4H_2$).[57–61] The concept uses a physical admixture of a reforming (noble metal on alumina) catalyst and a chemisorbent (K_2CO_3 promoted hydrotalcite), which selectively and reversibly chemisorbs CO_2 from a gas at a temperature of ~450 °C in the presence of steam. The cyclic SERP steps consisted of the following:

(a) Sorption reaction: A mixture of steam and methane at superatmospheric pressure P_A and temperature T is passed through a reactor–sorber column packed with the catalyst–chemisorbent admixture and a H_2 product stream at pressure P_A is withdrawn.

(b) Countercurrent depressurization: The reactor–sorber pressure is countercurrently lowered to a near-ambient pressure. The effluent gas is vented.

(c) Countercurrent evacuation with steam purge: The column is then evacuated countercurrently to a subatmospheric pressure level of P_D while purging it with steam at temperature T.

(d) Countercurrent pressurization: The column is finally pressurized with steam at temperature T from P_D to P_A.

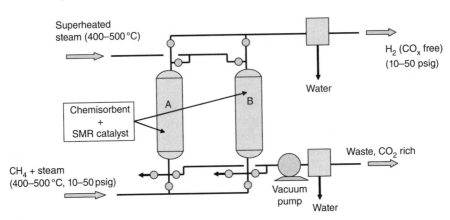

Figure 10.16.   Schematic flow diagram of the SERP concept.

Table 10.3.   Example of Performance of SERP Concept

Process	Product Purity (Dry Basis) (mol %)				CH_4 to H_2 Conversion (%)
	H_2	CH_4	CO_2	CO	
SERP concept	94.4	5.6	40 ppm	30 ppm	73.0
Conventional SMR reactor	67.2	15.7	15.9	1.2	52.6

The reactor–sorber is now ready to start a new cycle. Figure 10.16 is a two-column embodiment of the SERP concept.

It is a compact, small footprint system compared with the conventional route for the production of H_2 by SMR shown in Figure 10.1. This is because the reaction and the subsequent separation steps of the conventional route are simultaneously carried out in a single-unit operation. The concept allows the equilibrium-controlled, endothermic reforming reaction to be carried out at a significantly lower temperature of 400–500 °C instead of the conventional high temperature (800–900 °C) without sacrificing the conversion of CH_4 to H_2. This potentially saves energy and capital cost by not requiring high-temperature metallurgy.

Table 10.3 provides an example of the cyclic steady-state performance of the SERP concept using a 6:1 H_2O + CH_4 feed gas at a pressure of 11.4 psig and a temperature of 490 °C.[61] The process can directly produce an essentially CO_x-free H_2 product (~94.4% H_2 + 5.6% CH_4), which is suitable for H_2 fuel-cell use. The conversion of CH_4 to H_2 was ~73.0%. The table also shows the equilibrium compositions of the H_2 product from a conventional plug-flow reforming reactor operating under identical conditions. Both the H_2 conversion and product purity were rather poor in the latter case, which demonstrates the advantage of the SERP concept. Theoretical models of the above-described SERP concept and its variations for H_2 production by SMR have been developed, and theoretical parametric studies of the process have been conducted by various authors.[62,63]

10.5 PSA PROCESS RELIABILITY

Prior to the acceptance of H_2 PSA processes as the industry standard for H_2 purification, the most common method for the production of high purity H_2 from SMROG consisted of a chain of separation units including (i) CO_2 removal by absorption using a physical or chemical solvent, (ii) CO removal by catalytic methanation that consumed some of the product H_2, and (iii) CH_4 removal by cryogenic condensation. The H_2 product still contained some N_2 as impurity. The advent of H_2 PSA nearly 35 years ago significantly simplified the H_2 purification process by the use of a single-unit operation while delivering a high-purity (99.999+%) product H_2 gas at feed gas pressure. The H_2 PSA process also permitted the use of only one WGS reactor (high temperature) instead of two such reactors (high and low temperatures) needed by the earlier systems because the PSA process can remove higher concentrations of CO from the feed gas.

Despite these major advantages, a key challenge with the earlier PSA systems was their operational reliability, particularly in view of the use of numerous switch valves, control systems for unsteady-state cyclic process, and so on. The "onstream" time of the earlier systems was not very good (middle to high 90%). However, the reliability of H_2 PSA systems has dramatically increased in recent years. This high reliability is due to a good preventive maintenance program, skilled operators, improved PSA components (e.g., butterfly switch valves), and "reduced bed mode of PSA cycle operation" in the event of a valve failure. Highly reliable operation has also been achieved through efficient and automated controls of production capacity, product purity, turn down, and so on.[11] As the reliability of H_2 supply increases, industry acceptance of PSA as a general tool for gas purification continues to increase. The new generation of RPSA hydrogen systems described earlier must meet the current reliability standards set by today's larger H_2 PSA units.

10.6 IMPROVED HYDROGEN RECOVERY BY PSA PROCESSES

The waste gases from the H_2 PSA processes contain low to medium purity H_2 (25%–60%), and they are produced at a pressure of 1.1–1.7 atm. About 8%–30% of the feed gas H_2 is typically wasted in these streams.[64] It is not generally attractive to recover H_2 from these waste gases because H_2 is often a minor component in these gases and its partial pressure is low. Consequently, these waste gases are combusted to recover their fuel values.[64]

10.6.1 Integration with Additional PSA System

An earlier proposal to partially recover H_2 from these waste gases was to recompress the gas to a pressure of ~7–8 atm and to employ a two-column, four-step Skarstrom PSA cycle[65] consisting of adsorption, countercurrent depressurization, countercurrent purge, and pressurization with a part of the pure H_2 product steps.[66] About

60%–70% of H_2 from the main PSA waste gases can be recovered as pure H_2 by this route, and the recovered H_2 can be used to purge the main PSA adsorbers.[26,66] The net result is increased overall H_2 recovery by the integrated process. For example, the H_2 recovery of the Gemini process described earlier can be increased from 87% to 95% by this option.[26] This increase in H_2 recovery becomes more important as the costs of natural gas and hydrogen continue to rise.

10.6.2 Hybrid PSA-Adsorbent Membrane System

More recently, it has been shown that the selective surface flow (SSF) membrane, which was originally developed by Air Products and Chemicals, Inc., can be integrated with a H_2 PSA process to increase the overall H_2 recovery by the hybrid system.[64,67] The SSF membrane consists of a thin nanoporous carbon layer supported on a macroporous (pore diameters 1–2 μm) alumina tube. The pore diameters of the carbon membrane are in the range of 5–7 Å.[68] When the PSA waste gases (SMROG or ROG feed) are passed through the high-pressure side of the SSF membrane, the larger and more polar molecules (CO_2, CO, C_1–C_5 hydrocarbons) are selectively adsorbed over H_2 on the pore walls of the carbon membrane. A guard carbon bed may be needed to filter out heavy hydrocarbons like aromatics and double-ring compounds from the feed gas to the SSF membrane in order to prevent the nanopores from being plugged. The adsorbed molecules then selectively diffuse toward the low-pressure side of the membrane where they desorb into the permeate stream. Thus, the SSF membrane produces a H_2-enriched gas stream as the high-pressure effluent gas (retentate). This gas can be compressed to the feed gas pressure level of the main H_2 PSA process and recycled by mixing it with the fresh feed to the PSA system in order to increase the overall H_2 recovery by the hybrid PSA–SSF membrane process.

The adsorption-surface diffusion–desorption mechanism of transport through the SSF membrane can simultaneously provide high separation selectivity between H_2 and the impurities of the PSA waste gas and high flux for the impurities even when the gas pressure in the high-pressure side of the membrane is low to moderate (3–5 atm).

Figure 10.17 shows an example of the separation efficiency of the SSF membrane for SMROG–PSA waste gas.[64] It plots the rejection (β_i) of the more selectively adsorbed components of the gas mixture ($i = CO_2$, CH_4/CO) as a function of H_2 recovery (a_{H_2}). The rejection of component i is defined by the ratio of the molar flow rate of that component in the low-pressure permeate stream to that in the feed stream. The recovery of H_2 is defined by the ratio of the molar flow rate of H_2 in the high-pressure effluent stream to that in the feed stream. The plot also shows the ratio of the membrane area (A) needed to process a given flow rate (F) of the feed gas. These data are sufficient to design the membrane for a given feed gas composition and flow rate.[69]

The data in Figure 10.17 were measured using a feed gas composition of 52% CO_2, 37% H_2, and 11% CH_4 at a pressure of 3 atm. This gas composition represents

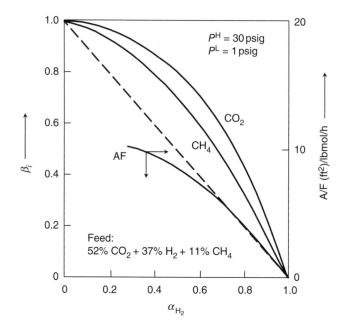

Figure 10.17. Separation performance of SSF membrane for PSA waste gas (SMROG feed).

a typical H_2 PSA waste gas from the SMROG feed. It may be seen from Figure 10.17 that about 90% CO_2 and 80% CH_4 + CO can be rejected by the SSF membrane from the above-described feed gas at a very moderate feed gas pressure when the H_2 recovery is 40%. There is practically no difference between the separation characteristics of CH_4 and CO by this membrane. The corresponding (A/F) value is about $10\,ft^2/lbmol/h$ ($1\,ft^2/lbmol/h = 0.20\,m^2/kgmol/h$). The high-pressure effluent gas composition from the membrane under these conditions will be 25.7% CO_2, 1.1% CH_4 + CO, and 73.2% H_2, which is comparable to that of the fresh SMROG feed to the PSA system.

It can be shown that an increase in the overall H_2 recovery of 7–10 percentage points can be achieved by using the hybrid PSA–SSF membrane system.[64] The compression duty and the membrane area for the hybrid process can be significantly reduced by (i) fractionating the PSA depressurization waste gas; (ii) directly passing the initial part of the PSA depressurization waste gas, which is richer in H_2, through the SSF membrane without additional compression; and (iii) compressing the PSA purge waste gas to the pressure level needed to process it through the same membrane.[64]

Figure 10.18 shows a schematic flow diagram for the PSA–SSF hybrid concept for increased H_2 recovery using the process scheme described above. The fresh feed to the PSA process is SMROG containing 72.8% H_2, 22.6% CO_2, and 4.6% CH_4 + CO at a pressure of 19.4 atm. The PSA process cycle is similar to that for the Poly-bed process except that only two cocurrent depressurization steps are used (eliminate

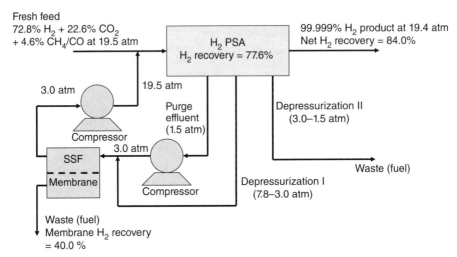

Figure 10.18. Schematic flow diagram for hybrid H_2 PSA–SSF membrane separation system.

steps e and g). The H_2 recovery by the PSA process is 77.6%. The countercurrent depressurization (step e) effluent gas is fractionated. The initial part of this gas, which is richer in H_2, is directly fed to an SSF membrane at a pressure of 3 atm. The countercurrent purge (step f) effluent gas is compressed to 3 atm and sent to the same membrane. The H_2-enriched high pressure effluent gas from the membrane is recompressed to 19.4 atm and recycled as feed gas to the PSA process. This increased the overall H_2 recovery of the hybrid process to 84.0%. More detailed description of this concept can be found elsewhere.[64] The SSF membrane can also be used to enrich and recycle H_2 from the waste gas of a PSA process, which uses ROG as feed.[70,71]

10.7 ENGINEERING PROCESS DESIGN

The key performance variables for an industrial H_2 PSA process are the H_2 product purity, H_2 recovery from the feed gas, and H_2 productivity (amount of H_2 produced per unit amount of adsorbent per cycle). Productivity determines the adsorber size. The required accuracy of a commercial process design can be very stringent. Table 10.4 lists the product H_2 purities required by various applications. A H_2 purity of 99.995+% is demanded in most cases.

The H_2 recovery from a PSA process is another critical variable. A difference of ~±2 percentage points in the estimation of H_2 recovery (typically between 80% and 93%) for a medium-sized PSA plant (30 MMSCFD H_2) can create an error of ~±$9.5 MM in the revenue ($H_2$ price at $3.00/1000 SCFT) over the plant life (350 days/year, 15 years), which is ~±150% of the cost of the PSA plant itself (neglecting the time value of money). Such dramatic impact can make or break the economics of a process design.[72]

Table 10.4.   Required H_2 Purities for Various Applications

Application	Required H_2 Purity
Ammonia synthesis gas	<10 ppm CO_x
Pipeline hydrogen	<10 ppm CO_x, <100 ppm CH_4, <200 ppm N_2
Proton exchange membrane fuel cells	<30 ppm CO
Electronic gas	<10 ppb N_2, O_2 CH_4, CO, hydrocarbons
Food-grade H_2	99.9–99.9995% H_2

The commercial design and optimization of a H_2 PSA process still largely remains an empirical effort. A priori design of a practical H_2 PSA process without the use of supporting performance data from a pilot-scale process unit may not yet be feasible for two reasons:[2,72]

1. Most practical H_2 PSA processes described earlier are fairly complex, involving a number of sequential but interacting unsteady-state cyclic steps. Theoretical design of these processes requires numerical integration of rigorous, nonisothermal, and nonisobaric PSA process models (coupled partial differential equations describing the mass, the heat, and the momentum balances within the adsorber) using appropriate initial and boundary conditions for the process steps. It is often difficult and time-consuming to solve the equations of these models with the accuracy and reliability needed for industrial design.

2. The key input data for solving the PSA process model equations are multicomponent gas adsorption equilibria, kinetics, and heats for the system of interest.[2,72] Basic understanding of these multicomponent adsorptive properties that govern the performance of the PSA process is often limited, particularly for heterogeneous adsorbents used in practice.[72] An accurate knowledge of these interactions under all conditions of pressure, temperature, and gas compositions prevailing inside the adsorber during the process cycle is needed for a reliable solution of the process models. These conditions can vary over a wide range in a practical adsorption process.

A common practice is to develop a specific model for the adsorptive process of interest and use simplistic descriptions (models or empirical) of pure and multicomponent gas adsorption equilibria and kinetics in order to describe the effects of various operating variables to obtain an optimum design. The effort is always closely tied to experimental verification and empirical fine-tuning using actual process data from pilot plants. A comprehensive set of data on pure and multicomponent adsorption equilibria of the components of SMROG on an activated carbon and a 5A zeolite is available in published literature.[73]

The mathematical process simulation models can, however, be extremely valuable for screening new ideas and adsorbents, parametric study of the processes for optimization, and establishing process limitations, process scale-up, and design of control schemes. Many corporations designing and selling H_2 PSA systems develop their own proprietary PSA process models and database.

There are only a few publications describing multicomponent, nonisothermal H_2 PSA process models for producing a high-purity H_2 product gas using feed gases consisting of the components of the SMROG or ROG, and where the simulated process performance is compared with the corresponding experimental PSA process data.[36,74–77] Figure 10.19 shows two examples of the comparison between the calculated H_2 purity and recovery obtained by a PSA process model and those measured experimentally.[36,77] Figure 10.19a corresponds to the case of an ROG feed (~17.0% CH_4, 6.5% C_2H_6, 3.0% C_3H_8, 1.0% C_4H_{10}, and 72.5% H_2 at a pressure of 18.0 atm), where the adsorbers were packed with a layer of silica gel in the feed end (24.5% of bed length) and a layer of an activated carbon in the product end.[36] The PSA process steps were identical to those of the Poly-bed process described earlier. The solid and the dashed lines, respectively, represent the model calculations under the nonisothermal and the isothermal conditions of operation. The points represent the performance data from an industrial plant. Figure 10.19b corresponds to the case of an SMROG feed (~21.2% CO_2, 2.5% CO, 3.9% CH_4, and 72.4% H_2 at a pressure of 23.6 atm), where the adsorber was packed with a layer of an activated carbon in the feed end (83.3% of bed length) and a layer of 5A zeolite at the product end.[77]

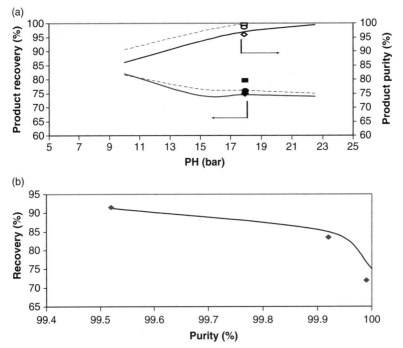

Figure 10.19.   Comparison of model calculations of H_2 purity and recovery with experimental data for two H_2 PSA processes: (a) ROG feed and (b) SMROG feed.

The PSA process steps again followed those of the Poly-bed process except that the order of steps (c) and (d) were interchanged. The solid line represents the results of the nonisothermal model simulations and the points are experimental data from a bench-scale unit. Both figures show that the model simulations are in fair to good agreement with experimental data but the accuracy demanded by industrial designs may still be lacking.

10.8 SUMMARY

Production of high-purity hydrogen from a feed gas containing 70%–90% H_2 using a PSA process has become a common industrial practice. About 85% of the 17 trillion standard cubic feet of hydrogen produced yearly is purified by PSA. Numerous PSA processes are developed for this purpose. The commonly used feed gases are the SMROG and the ROG. The basic research and development objectives in this area are (i) increasing the hydrogen recovery at high purity (99.999% H_2) and (ii) decreasing the adsorbent inventory and the hardware costs.

Several commercially developed PSA processes are described, and their key distinguishing features and separation performances are reviewed. Some of these processes are designed to produce a by-product stream (CO_2 from SMROG) along with the primary product stream (H_2). A few of these processes can also be modified to directly produce an ammonia synthesis gas as the primary product (from SMROG feed) with or without the production of CO_2.

The pure gas equilibrium ad(de)sorption characteristics of the components of the SMROG and ROG feed streams on BPL activated carbon, 5A zeolite, and Sorbead H (silica gel) are reported and the criteria for adsorbent selection in these PSA processes are described.

Recent developments in rapid pressure swing H_2 PSA processes are described. The rapid cycles are used to reduce adsorbent inventories, to reduce vessel size, and to lower the overall cost of the PSA unit. Research in the area of rotary valves and structured adsorbents, which enable the RPSA cycles, is outlined.

Other recent ideas on (i) simultaneous sorption–reaction process concepts using the principles of a novel PSA technology for the production of fuel cell-grade hydrogen and (ii) improving the H_2 recovery from existing H_2 PSA processes by integrating it with additional PSA units or nanoporous SSF adsorbent membrane systems are reviewed.

Mathematical PSA process simulation models are used in conjunction with process performance data obtained from pilot-scale PSA units for the design and optimization of H_2 PSA processes. These models are also useful for screening adsorbents and conceptual process designs.

REFERENCES

1. SURESH, B., SCHLAG, S., and INOGUCHI, Y. Hydrogen. *CEH Marketing Research Report*, p. 6, August **2004**.
2. SIRCAR, S. Pressure swing adsorption: Commentaries. *Industrial & Engineering Chemistry Research*, **2002**, 41, 1389.
3. SIRCAR, S. Pressure swing adsorption technology. In *Adsorption Science and Technology* (eds. A.E. Rodrigues, D.M. Levan, D. Tondeur). Dordrecht, The Netherlands: Kluwer Academic Publishers, p. 285, **1989**.
4. LEIBY, S.M. Options for refinery hydrogen. Menlo Park, CA: Process Economics Program, SRI International. *PEP Report No. 212*, February **1994**.

5. TOWLER, G.P., MANN, R., SERRIERE, J.L., and GABAUDE, M.D. Refinery hydrogen management: Cost analysis of chemically integrated facilities. *Industrial & Engineering Chemistry Research*, **1996**, 35, 2378.
6. DEY, R.D. and MALIK, Z.I. Technology advances improve liquid recovery from refinery off gases. Presented at *NPRA Annual Meeting*. San Antonio, TX, **2000**.
7. FARAJI, S., SOTUDEH-GGAREBAGH, R., and MOSTOUFI, N. Hydrogen recovery from refinery off-gases. *Journal of Applied Sciences*, **2005**, 5, 459.
8. *World Patent Index*. London: Derwant Publication.
9. FUDERER, A. and RUDELSTORFER, E. Selective adsorption process. U.S. Patent 3,986,849, **1976**.
10. WAGNER, J.L. Selective adsorption of gases. U.S. Patent 3,430,418, **1969**.
11. STÖCKER, J., WHYSALL, M., and MILLER, G.Q. 30 years of PSA technology for hydrogen purification. UOP LLC Web site, Des Plaines, IL. **1998**. http://www.uop.com/objects/30YrsPSATechHydPurif. pdf (accessed September 20, 2005).
12. YAMAGUCHI, T. and KOBAYASHI, Y. Gas separation process. U.S. Patent 5,250,088, **1993**.
13. YAMAGUCHI, T., OHKAMO, U., and KOBAYASHI, Y. Hydrogen recovery and purification by LOFIN PSA. In *Proceedings of 10th World Hydrogen Energy Conference* (eds. D.L. Block, V.T. Nejat). Florida Solar Energy Center, 2, p. 853, **1994**.
14. OKAMA, U.V. Increased hydrogen recovery with advanced PSA technology. *PTQ Summer*, **1996**, 1, 5.
15. BUDNER, Z., RETERSKA, Z., and MORAWIEC, B. Technology of hydrogen recovery from refinery off-gases by PSA method. *Polish Przemysl Chemiczny*, **1990**, 69, 533.
16. MALEK, A. and FAROOQ, S. Study of a six bed pressure swing adsorption process. *AIChE Journal*, **1997**, 43, 2509.
17. DE SOUZA, G. Production and purification of hydrogen-rich gases from petroleum refinery streams. French Patent 2,836,062, **2003**.
18. DRNEVICH, R.F. and HERZOG, J.O. Integrated olefin recovery and hydrogen production from refinery off-gases. U.S. Patent Application 2004073076, **2004**.
19. ZHOU, L., LUE, C.Z., BIAN, S.J., and ZHOU, Y.P. Pure hydrogen from the dry gas of refineries via a novel pressure swing adsorption process. *Industrial & Engineering Chemistry Research*, **2002**, 41, 5290.
20. SIRCAR, S. Separation of multicomponent gas mixtures. U.S. Patent 4,171,206, **1979**.
21. SIRCAR, S. and KRATZ, W.C. Simultaneous production of hydrogen and carbon dioxide from steam reformer off-gas, PSA. *Separation Science and Technology*, **1988**, 23, 2397.
22. SATISH, R. Hydrogen and carbon dioxide co-production. U.S. Patent 6,500,241, **2002**.
23. RARIG, D.L., GOLDEN, T.C., and WEIST, E.L. Purification of CO_2 from H_2 PSA vent gas. *National AIChE Meeting*. Indianapolis, IN, November 8, **2002**.
24. FUDERER, A. Selective adsorption process for production of ammonia synthesis gas mixtures. U.S. Patent 4,375,363, **1983**.
25. SIRCAR, S. Production of nitrogen, hydrogen and carbon dioxide from hydrocarbon reformate. U.S. Patent 4, 813,980, **1989**.
26. SIRCAR, S. Production of hydrogen and ammonia synthesis gas by PSA. *Separation Science and Technology*, **1990**, 25, 1087.
27. KIYONAGA, K. Adsorption separation process. U.S. Patent 3,176,444, **1962**.
28. SIRCAR, S., GOLDEN, T.C., and RAO, M.B. Activated carbon for gas separation and storage. *Carbon*, **1996**, 32, 1.
29. GOLDEN, T.C. and SIRCAR, S. Gas adsorption on silicalite. *Journal of Colloid and Interface Science*, **1994**, 162, 182.
30. SIRCAR, S. and GOLDEN, T.C. Isothermal and isobaric desorption of carbon dioxide by purge. *Industrial & Engineering Chemistry Research*, **1995**, 34, 2881.
31. SIRCAR, S. and GOLDEN, T.C. Purification of hydrogen by PSA. *Separation Science and Technology*, **2000**, 35, 667.
32. REISS, G. Molecular sieve zeolite for producing hydrogen by pressure variation adsorption technique. U.S. Patent 4,477,267, **1984**.
33. PLEE, D. PSA process for production of hydrogen. European Patent 0 855 209, **1998**.

34. GOLDEN, T.C., KUMAR, R., and KRATZ, W.C. Hydrogen purification. U.S. Patent 4,957,514, **1990**.
35. JOHNSON, L.M., FARRIS, T.S., GOLDEN, T.C., WEIST, E.L., and OCCHIALINI, J.M. Optimal adsorbents for hydrogen recovery by pressure and vacuum swing adsorption. U.S. Patent 6,302,943, **2001**.
36. MALEK, A. and FAROOQ, S. Hydrogen purification from refinery fuel gas by pressure swing adsorption. *AIChE Journal*, **1998**, 44, 1985.
37. GOLDEN, T.C. and WEIST, E.L. Layered adsorbent zone for hydrogen pressure swing adsorption. U.S. Patent 6,814,787, **2004**.
38. MILLER, G.Q. Multiple zone adsorption process. U.S. Patent 4,964,888, **1990**.
39. GOLDEN, T.C., FARRIS, T.S., KRATZ, W.C., WALDRON, W.E., and JOHNSON, C.H. Adjusted density carbon for hydrogen PSA. U.S. Patent 6,027,549, **2000**.
40. KEEFER, B.G. and DOMAN, D.G. WIPO International Publication No. WO97/39821. World International Property Organization published under the Patent Corporation Treaty (PCT), **1997**.
41. LEMCOFF, N.O., FRONZONI, M.A., GARRETT, M.E., GREEN, B.C., ATKINSON, T.D., and LA CAVA, A.I. Process and apparatus for gas separation. U.S. Patent 5,820,656, **1998**.
42. CONNOR, D.J., DOMAN, D.G., JEZIOROWSKI, L., KEEFER, B.G., LARISH, B., MCLEAN, C., and SHAW, I. Rotary pressure swing adsorption apparatus. U.S. Patent 6,406,523, **2002**.
43. WAGNER, G.P. Rotary sequencing valve with flexible port plate. U.S. Patent 6,889,710, **2005**.
44. HILL, T.B., HILL, C.C., and HANSEN, A.C. Rotary valve assembly for pressure swing adsorption system. U.S. Patent 6,311,719, **2001**.
45. MALLAVARAPU, K., RUHL, J.B., and GITTLEMAN, C.S. Control of a hydrogen purifying pressure swing adsorption unit in fuel processor module for hydrogen generation. U.S. Patent Application 20050098033, May 12, **2005**.
46. KEEFER, B.G. High frequency pressure swing adsorption. U.S. Patent 6,176,897, **2001**.
47. KEEFER, B.G. and MCLEAN, C.R. High frequency rotary pressure swing adsorption apparatus. U.S. Patent 6,056,804, **2000**.
48. SIRCAR, S. and HANLEY, B.F. Production of oxygen enriched air by rapid pressure swing adsorption. *Adsorption*, **1995**, 1, 313.
49. SIRCAR, S. Influence of gas-solid heat transfer on rapid PSA. *Adsorption*, **2005**, 11, 509.
50. JAIN, R., LACAVA, A.I., MAHESWARY, A., AMBRIANO, J.R., ACHARYA, D., and FITCH, F.R. Air separation using monolith adsorbent bed. European Patent Application EP 1070531, **2001**.
51. KOPAYGORODSKY, E.M., GULIANTS, V.V., and KRANTZ, W.B. Predictive dynamic model of single-stage, ultra-rapid pressure swing adsorption. *AIChE Journal*, **2004**, 50, 953.
52. LI, Y.Y., PERERA, S.P., and CRITTENDEN, B.D. Zeolite monoliths for air separation. Part 2 oxygen enrichment, pressure drop and pressurization. *Chemical Engineering Research & Design*, **1998**, 76, 931.
53. KEEFER, B.G. Apparatus and process for pressure swing adsorption separation. U.S. Patent 4,702,903, **1989**.
54. KEEFER, B.G. Extraction and concentration of a gas component. U.S. Patent 5,082,473, **1992**.
55. KEEFER, B.G. and DOMAN, D.G. Modular pressure swing adsorption with energy recovery. U.S. Patent 6,051,050, **2000**.
56. GOLDEN, T.C., GOLDEN, C.M.A., and ZWILLING, D.P. Self-supported adsorbent for gas separation. U.S. Patent 6,565,627, **2003**.
57. SIRCAR, S., NATARAJ, S., and HUFTON, J.R. Sorption enhanced reaction process for hydrogen production. U.S. Patent 6,103,143, **2000**.
58. ANAND, M., CARVILL, B.T., and SIRCAR, S. Process for operating equilibrium controlled reaction. U.S. Patent 6,303,092, **2001**.
59. CARVILL, B.T., HUFTON, J.R., ANAND, M., and SIRCAR, S. Sorption enhanced reaction process. *AIChE Journal*, **1996**, 42, 2765.
60. HUFTON, J.R., MAYORGA, S., and SIRCAR, S. Sorption enhanced reaction process for hydrogen production. *AIChE Journal*, **1999**, 45, 248.
61. WALDRON, W.E., HUFTON, J.R., and SIRCAR, S. Production of hydrogen by cyclic sorption enhanced reaction process. *AIChE Journal*, **2001**, 47, 1477.

62. YING, D. and ALPAY, E. Adsorption enhanced steam methane reforming. *Chemical Engineering Science*, **2000**, 55, 3929.
63. XIU, G.H., PING, L., and RODRIGUES, A.E. New generalized strategy for improving sorption enhanced reaction process. *Chemical Engineering Science*, **2003**, 58, 3425.
64. SIRCAR, S., WALDRON, W.E., RAO, M.B., and ANAND, M. Hydrogen production by hybrid SMR-PSA-SSF membrane system. *Gas Separation and Purification*, **1999**, 17, 11.
65. SKARSTROM, C.W. Method and apparatus for fractionating gaseous mixtures by adsorption. U.S. Patent 2,944,627, **1960**.
66. SIRCAR, S. Fractionation of multi-component gas mixture by PSA. U.S. Patent 4,790,858, **1998**.
67. SIRCAR, S., WALDRON, W.E., ANAND M., and RAO, M.B. Hydrogen recovery by PSA integrated with adsorbent membranes. U.S. Patent 5,753,010, **1998**.
68. RAO, M.B. and SIRCAR, S. Performance and pore size characterization of nano-porous carbon membranes for gas separation. *Journal of Membrane Science*, **1996**, 110, 109.
69. PARANJAPE, M., CLARKE, P.F., PRUDEN, B.B., PARRILLO, D.J., THAERON, C., and SIRCAR, S. Separation of bulk carbon dioxide—Hydrogen mixtures by SSF membrane. *Adsorption*, **1998**, 4, 355.
70. RAO, M.B. and SIRCAR, S. Nano-porous carbon membranes for separation of gas mixtures by selective surface flow. *Journal of Membrane Science*, **1993**, 85, 253.
71. NAHEIRI, T., LUDWIG, K.A., ANAND, M., RAO, M.B., and SIRCAR, S. Scale-up of SSF Membrane for gas separation. *Separation Science and Technology*, **1997**, 32, 1589.
72. SIRCAR, S. Basic research needs for design of adsorptive gas separation processes. *Industrial & Engineering Chemistry Research*, **2006**, 45 (16), 5435.
73. SIEVERS, W. and MERSMANN, A. Single and multi-component adsorption equilibria of CO_2, N_2, CO, and CH_4 in hydrogen purification process. *Chemical Engineering & Technology*, **1997**, 17, 325.
74. CEN, P. and YANG, R.T. Separation of a five-component gas mixture by PSA. *Separation Science & Technology*, **1985**, 20, 725.
75. WARMUZINSKI, K. and TANCZYK, M. Multi-component pressure swing adsorption. *Chemical Engineering & Processing*, **1997**, 36, 89.
76. ZHU, D. Simulation of a PSA process for purifying hydrogen from reforming gas by using an equilibrium model. *Tianranqi Huagong*, **1998**, 23, 36 (in Chinese).
77. PARK, J.H., KIM. J.N., and CHO, S.H. Performance analysis of four bed H_2 PSA process using layered beds. *AIChE Journal*, **2000**, 46, 790.

Chapter 11

Integration of H$_2$/Syngas Production Technologies with Future Energy Systems

WEI WEI, PARAG KULKARNI, AND KE LIU

Energy & Propulsion Technologies, GE Global Research Center

11.1 OVERVIEW OF FUTURE ENERGY SYSTEMS AND CHALLENGES

Meeting the growing energy demand while keeping the environment clean is one of the most challenging issues facing the world in the 21st century. According to International Energy Reports,[1] currently around 86% of the world's energy demand is satisfied with fossil fuels such as natural gas, coal, and petroleum, as shown in Figure 11.1. Furthermore, these fossil fuels are expected to remain the main source of energy for at least the next few decades. The major technologies of producing electricity from fossil fuels include coal power plants using pulverized coal (PC) boiler with steam turbine, Natural Gas Combined Cycle (NGCC) plants using natural gas with both gas and steam turbines, Integrated Gasification Combined Cycle (IGCC) plants with both gas and steam turbine, internal combustion engines (ICEs) for transportation and distributed power generation, and finally, the future hybrid power plants with turbines and fuel cells. Figure 11.2 summarizes the range of fossil fuel-based power generation technologies.[2]

Figure 11.3 shows the electric power generation by fuel type in the U.S. Notice that ~50% of electricity is produced from coal in the U.S. As natural gas prices increase, more power generation will rely on coal. To deal with the environmental issues of today's coal power plants, more IGCC plants with or without CO$_2$ capture could be built in the U.S. in the foreseeable future.

Hydrogen and Syngas Production and Purification Technologies, Edited by Ke Liu, Chunshan Song and Velu Subramani
Copyright © 2010 American Institute of Chemical Engineers

451

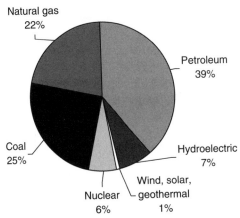

Figure 11.1. Worldwide energy sources distribution. See color insert.

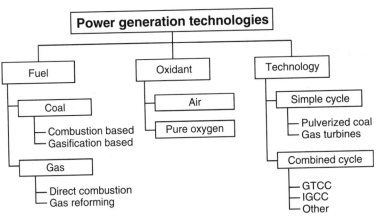

Figure 11.2. Technology options for fossil fuel-based power generation. GTCC, gas turbine combined cycle.

The main technical challenges in utilizing fossil fuels include achieving high efficiency and effective management and control of the pollutants such as CO, NO_x, and SO_x. In addition, the global warming caused by the greenhouse gases (GHGs) such as CO_2 and CH_4 emitted from burning fossil fuels will become a more and more challenging issue in the near future. Several European countries have already started a carbon-tax program, and this trend could gradually spill over to the U.S. and the rest of the world, calling for CO_2-free energy generation in the future.

In the last 150 years, the concentration of the GHGs has increased by ~25%, largely due to human activities and industrialization of the world, as shown in Figure

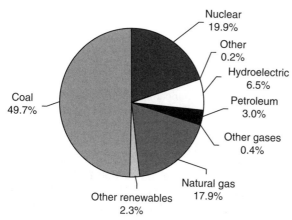

Figure 11.3. U.S. electric power generation by fuel type (2004); http://tonto.eia.doe.gov/ FTPROOT/electricity/034804.pdf.html. See color insert.

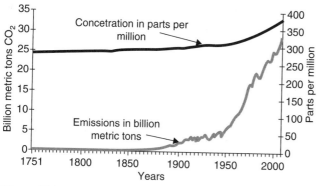

Figure 11.4. Trends in atmospheric concentrations and man-made emissions of CO_2. *Source:* Oak Ridge National Laboratory, Carbon Dioxide Information Analysis Center. http://www.eia.doe.gov/ bookshelf/brochures/greenhouse/Chapter1.htm.

11.4.[3] A majority of these GHG emissions are in the form of CO_2. For example, in the U.S., CO_2 generated by fossil fuel combustion accounts for ~82% of the GHGs. Around 21% of the CO_2 emitted comes from the combustion of natural gas.

The future fossil fuel-based power plants would most likely use high efficiency and low emission technologies such as NGCC power plants, IGCC power plants, and, in long term, various types of hybrid fuel cell power plants. The fuel processing technologies discussed in earlier chapters of this book can be integrated with the future power plants to meet the future demands of efficiency and pollution control. These syngas production and purification technologies are also critical to other fossil fuel energy conversion systems such as coal to liquid (CTL), gas to liquid, and biomass to liquid.

11.2 APPLICATION OF REFORMING-BASED SYNGAS TECHNOLOGY

11.2.1 NGCC Plants

Natural gas is an important fuel in the U.S. It accounts for ~18% of all the electricity generated in the U.S. as shown in Figure 11.3 (U.S. DOE, 2004).[4] Increasing the use of natural gas for electricity production can benefit the environment in many ways compared with electricity generated from other fossil fuels. Since natural gas has a lower sulfur and nitrogen content than coal and NGCC also has higher efficiency than the coal-based boiler power plant, power production from natural gas results in fewer SO_x and NO_x emissions per kilowatt-hour of electricity. Additionally, unlike coal-fired power plants, an NGCC system produces no large solid waste streams.[5]

The basic principle of the NGCC is shown in Figure 11.5 and is described in various sources available in the literature. Air is compressed to high pressure in the compressor and is combusted with high-pressure natural gas in the combustor. The high-pressure and -temperature combustion products are expanded through the expander. The expander is coupled with a generator to produce electric power. The combination of the compressor, combustor, and the expander is referred as the gas turbine (GT). The exhaust of the GT is at fairly high temperature, and it is sent to a heat recovery steam generator (HRSG) to produce high-pressure superheated steam. The steam is sent to a network of high-, medium-, and low-pressure steam turbines (LPSTs) to produce additional electric power.

This setup of GT, HRSG, steam turbines, and generators is called a combined cycle (CC). This type of power plant is being installed in increasing numbers around the world where there is an access to substantial quantities of natural gas. This type of power plant produces high power outputs with high efficiencies and low

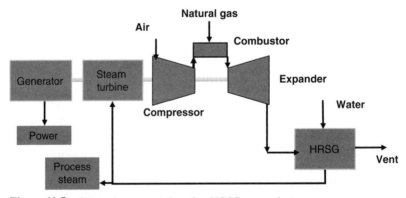

Figure 11.5. Schematic representation of an NGCC power plant.

emissions. NGCC plant efficiencies, depending on the layout and size of the installation, vary from about 40% (now considered poor) to about 60%. For example, GE's H-turbine CC efficiency is ~60%, and F-class CC is in the range of 58.5%, based on lower heating value (LHV), for large new natural gas-fired stations.

11.2.2 Integration of H$_2$/Syngas Production Technologies in NGCC Plants

The hydrogen/syngas production technologies discussed in the earlier chapters of the book such as reforming and water-gas shift (WGS) reactions can be integrated with the advanced NGCC plants to improve the emissions and efficiency. One of the main advantages of such integration is efficient capture of GHGs such as CO_2.

The combustion of natural gas accounts for approximately 21% of the U.S. emission of CO_2. The majority of NGCC plants are not equipped with the CO_2 capture capability. Researchers in both academia and industry recognize the importance of controlling the CO_2 emissions and are working on developing technologies to effectively capture and sequestrate CO_2.

The capture and sequestration of CO_2 reduces the efficiency and increases the capital cost of the NGCC plant significantly.[6] Typically, ~20% of the net power of the NGCC plant is lost in order to separate and sequestrate 90% of CO_2 emissions as shown in Figure 11.6. The addition of CO_2 capture units also results in ~90% increase in the capital cost of the NGCC plant and ~61% increase in the cost of

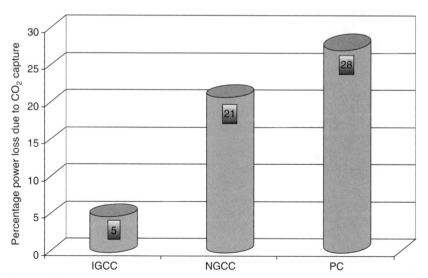

Figure 11.6.   Parasitic power loss for power plants with CO_2 capture (percent of net power plant power) with state-of-the-art scrubbing technologies.

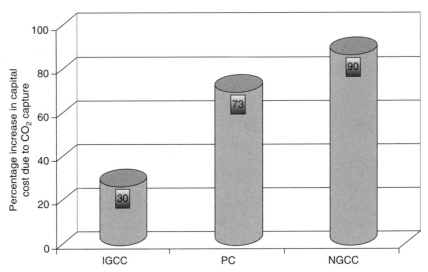

Figure 11.7.   Effect of CO₂ capture on power plant capital cost (percent of increase resulting from CO₂ capture) with state-of-the-art scrubbing technologies.

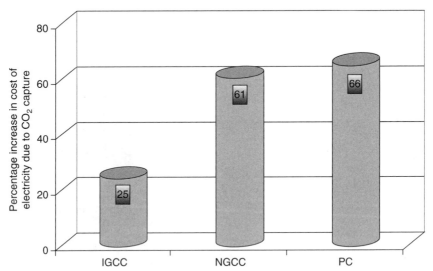

Figure 11.8.   Effect of CO₂ capture on power plant cost of electricity (percent of increase resulting from CO₂ capture) with state-of-the-art scrubbing technologies.

electricity (COE) as shown in Figures 11.7 and 11.8, respectively. It should be noted that these system analysis of the impact of CO₂ sequestration are based on assumptions and designs of specific plant configurations and, therefore, should be used for qualitative comparison only.

There are three main ways to capture CO_2 in the NGCC process: post-combustion CO_2 capture, oxyfuel combustion, and precombustion CO_2 capture.

11.2.2.1 Post-Combustion CO_2 Capture

The principle of post-combustion capture is shown in Figure 11.9. The exhaust gas from the NGCC plant is sent to the CO_2 capture unit, which typically utilizes chemical solvents such as amines or alcohols.[7-13] Chemical absorption of CO_2 using solvent is preferred because of the low concentration of CO_2 (~5%) and low pressure in the flue gas. The solvent loaded with CO_2 is regenerated typically in a reboiler using low-quality steam. The CO_2 released from the solvent can be sent to a sequestration plant.

One advantage of the post-combustion CO_2 capture route is that it does not require any modifications to existing combustion methods, which means current power plants can be retrofitted with this process. In fact, this method has been demonstrated at some small-scale power plants.[14]

The challenges for post-combustion CO_2 capture include degradation of solvents, high capital cost and high energy usage in the solvent regeneration process. It is estimated that in order to capture 90% of CO_2 from the NGCC plant using conventional amine technology, the capital cost of the plant increases by 90% and the COE can increase by ~60%.[6]

Alternative technologies to chemical solvent absorption such as adsorption and membranes have also been explored.[15] Adsorption processes are relatively expensive due to the low CO_2 loading and low lifetime of chemical adsorbents. Cryogenic CO_2 separation is also considered less attractive than absorption because of high-energy requirement. The capital costs savings of a membrane contactor with integrated solvent system are marginal when compared with conventional absorption and desorption units.[16]

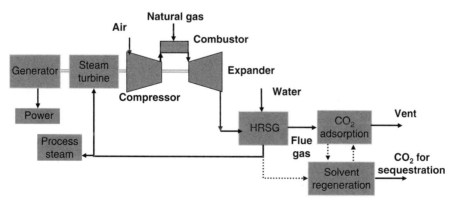

Figure 11.9. Schematic representation of an NGCC power plant with post-combustion CO_2 capture.

11.2.2.2 Oxyfuel Combustion

In oxyfuel combustion, the natural gas is combusted with a concentrated oxygen stream (~95% O_2). The flue gas from the combustion process is not diluted with N_2 but consists of mainly CO_2 and steam. The steam in the flue gas can be condensed, and CO_2 can be sent direcly to the sequestration plant without further separation.

Various technologies have been explored to achieve upfront separation of O_2 from air for oxyfuel combustion. Cycles with a conventional air separation unit (ASU) to produce pure oxygen can be distinguished from novel oxyfuel power cycles such as advanced zero emission power (AZEP) plants that use membrane technologies and chemical looping combustion (CLC) that employs oxygen-carrier particles to enable stoichiometric combustion with oxygen.[16]

11.2.2.2.1 Oxyfuel Cycles with ASU. There are various configurations of oxyfuel cycles with ASU and CO_2/H_2O recycling, including Graz cycle, water cycle, and Matiant (combined) cycle.

The Graz cycle consists of a high-pressure combustor with steam injection and a recuperated GT integrated with a steam cycle.[17] Both CO_2 and steam are recycled to the combustion chamber. It consists of a high-temperature Brayton cycle (CO_2 compressors C_2, C_3, combustion chamber, and high-temperature turbine [HTT]) and a low-temperature Rankin cycle (low-pressure turbine [LPT], condenser, HRSG, and high-pressure turbine [HPT]). The schematic of Graz cycle is shown in Figure 11.10.

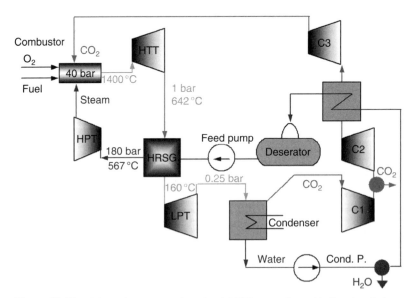

Figure 11.10. Schematic representation of an NGCC power plant with Graz (oxyfuel combustion) cycle.

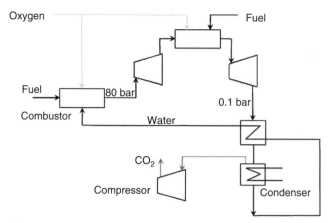

Figure 11.11.   Schematic representation of an NGCC power plant with water (oxyfuel combustion) cycle.

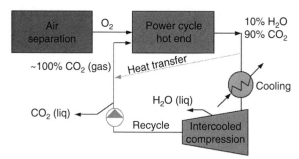

Figure 11.12.   Schematic representation of an NGCC power plant with Matiant (oxyfuel combustion) cycle.

The water cycle is an oxygen-fired cycle with steam recycle, producing a high-pressure, superheated mixture of mainly H_2O and CO_2 in a gas generator. The mixture stream is then expanded in a series of (advanced) turbines. A 20 MWth gas generator fueled with natural gas has been tested and a 500-kWe power plant is being developed by Clean Energy Systems (CES).[16] A schematic of the water cycle is shown in Figure 11.11.[14]

In the Matiant or oxyfuel CC, natural gas is combusted in an O_2/CO_2 atmosphere. The reaction product, consisting principally of CO_2 and H_2O, is expanded in an adapted GT and the heat of the turbine exhaust is used to generate steam, which is expanded in a steam cycle. An example of Matiant cycle is shown in Figure 11.12.[18]

The primary challenge facing the oxyfuel combustion with ASU is the requirement of advanced GTs that use CO_2/H_2O as the working fluid, which has different expansion characteristics than nitrogen used in conventional turbines. Existing GTs cannot directly use CO_2 as working fluid; hence, entirely new GTs (combustor and compressor) need to be developed. Such a development is not likely to occur unless there is a significant market demand for such GTs.

11.2.2.2.2 Advanced Zero Emissions Plant (AZEP).

In the AZEP concept, the conventional combustion chamber in a GT is substituted by a mixed conducting membrane (MCM) reactor that combines oxygen production, fuel combustion, and heat transfer. A sketch of the AZEP concept is shown in Figure 11.13.[19]

Compressed air enters the MCM reactor, where oxygen passes through the membrane and is transported to the combustion chamber. The heat of combustion is transferred to oxygen-depleted air, which is expanded in a conventional turbine. The thermal energies recovered from turbine exhaust and the CO_2/H_2O stream that is generated in the combustion chamber are used to generate high-pressure steam, which is then expanded in a steam turbine. The use of a conventional turbine is an essential advantage of AZEP over oxyfuel concepts using GT with CO_2/H_2O as working fluid. The AZEP concept is highly compatible with current technologies, requiring minor adaptations in GT (the working fluid has lower oxygen content) and HRSG. Obviously, there are still technical challenges in the development of the MCM reactor associated with the membrane developments and other challenges.

A consortium of energy companies is now developing the AZEP concept. The GT selected for test phase is a 64 MWe GTX100, with 53% efficiency in the CC mode. Also, a 400 MWe V94.3A turbine with an efficiency of 57.9% has been studied. Due to the limited temperature in the MCM reactor (~1200 °C), turbine inlet temperature (TIT) is lower than that of the most advanced GTs, which exceed 1400 °C. The efficiency loss can be reduced by installing an afterburner to increase TIT and reducing the CO_2 capture ratio to 85%.[16]

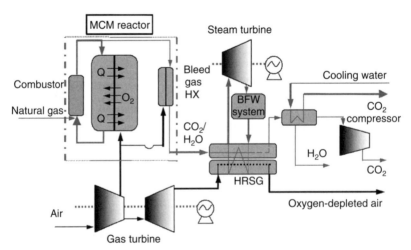

Figure 11.13.   Schematic representation of the AZEP (oxyfuel combustion) concept. HX, heat exchanger; BFW, boiling feed water.

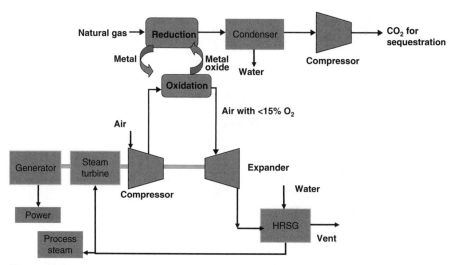

Figure 11.14. Schematic representation of chemical looping combustion (CLC).

11.2.2.2.3 Chemical Looping Combustion (CLC). CLC is an oxyfuel concept where combustion is achieved using a metal oxide as the oxygen carrier without mixing fuel and air. A schematic of CLC concept is shown in Figure 11.14.

In this oxyfuel process, the fuel is combusted using oxygen from an oxygen-carrier material such as a metal oxide (e.g., iron, nickel). The metal oxide undergoes reduction reaction with natural gas by donating the oxygen atom in the metal oxide and producing CO_2 and steam as products. After recovering heat from this product stream to generate power via steam turbines, steam is condensed and CO_2 is sequestered. The reduced oxygen carrier is transferred into a second reactor where it is oxidized by air in an exothermic reaction to produce high-temperature exhaust air suitable for power generation. Both natural gas and syngas can be used as fuel, and the technology can be integrated in various power cycles.[16,20–22]

The CLC process is still in the early stage of research, and there are no satisfactory solutions yet on certain technical issues such as stability of the oxygen-carrier materials under the extreme operating conditions.

11.2.2.3 Precombustion CO_2 Capture

In this route, CO_2 is removed from the fuel before it is combusted in the GT.[16,18] This can be achieved by reforming the fuel to syngas, converting CO to CO_2 through WGS reaction and then removing CO_2 from the syngas at high pressure. The resulting fuel mixture mostly containing H_2 is burned in the GT to produce electricity. Figure 11.15 shows the block diagram for this process.

As shown in Chapters 2 and 3 of this book, reforming of natural gas to syngas can be achieved using partial oxidation (POX)/autothermal reforming (ATR) or by

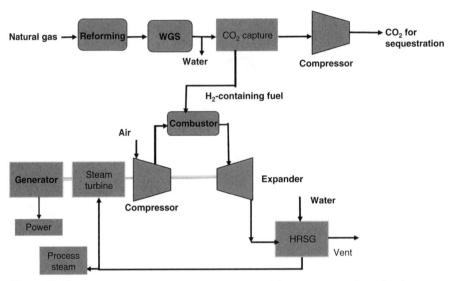

Figure 11.15.   Schematic representation of precombustion CO₂ capture route using reforming.

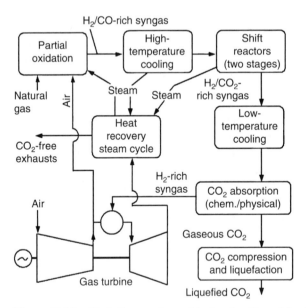

Figure 11.16.   Block diagram of a combined cycle with partial oxidation/ATR.

using steam methane reforming (SMR).[23,24] The block diagram of a precombustion decarbonization concept using POX/ATR is shown in Figure 11.16.[23] Natural gas, steam, and air are mixed and reformed in the POX reactor to produce syngas via the following reactions:

- POX: $CH_4 + \frac{1}{2}O_2 = CO + 2H_2 + 35.67\,kJ/mol$,
- SMR: $CH_4 + H_2O = CO + 3H_2 - 206.158\,kJ/mol$, and
- WGS: $CO + H_2O = CO_2 + H_2 + 41.154\,kJ/mol$.

The POX/oxidation reaction of methane and air/O_2 is exothermic and provides the energy required for the endothermic steam reforming reaction. Conversion is limited by the equilibrium defined by the operating conditions (pressure, temperature) and the residence time. The POX can be carried out catalytically or using a homogeneous burner. Catalytic partial oxygen can be more efficient; however, it is associated with issues such as scale-up, sintering, and uncontrolled reactions. A fuel-rich burner followed by SMR catalyst is a commercially proven ATR route. Alternatively, the reforming can be carried out using steam reforming technology. The block diagram for such integrated reforming combined cycle (IRCC) + SMR plant is shown in Figure 11.17.[24] In this configuration, natural gas and steam are passed through a heat exchanger SMR reactor. The heat required for the endothermic reaction is provided by burning part of the hydrogen-rich syngas and also using the GT exhaust as shown in the figure. The performance comparison indicates that the SMR configuration has lower efficiency and power output than that of the POX case because of the endothermic nature of the steam reforming reaction: $CH_4 + H_2O = CO + 3H_2 - 206.158\,kJ/mol$. Furthermore, the steam reforming needs to be carried

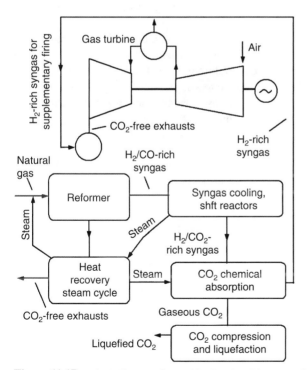

Figure 11.17. Block diagram of a combined cycle with steam reforming.

out at ~900 °C temperature and therefore high-cost alloys are required for the construction of the reformer reactor resulting in increased capital cost for the overall process.

The reforming is usually carried out at high pressure, as reforming is a volume-increase reaction, and it is efficient to compress the reactants compared with the products of the reforming reaction.[25] Separation of CO_2 from syngas by certain methods such as physical adsorption is also favored at high pressure. The hot product gases coming out from the reformer are cooled down and introduced in the shift reactors that convert the CO in the syngas to CO_2 and produce more H_2. The shift reaction is equilibrium limited and is favored at low temperatures. Products of the shift reactors are cooled down to near room temperature for CO_2 separation. The CO_2 separation can be carried out by chemical absorption or physical absorption. The chemical absorption involves a chemical reaction as the name suggests and can be carried out at lower pressures; however, the physical absorption requires higher pressure (>40 bar) for effective CO_2 separation. The high partial pressure of CO_2 after the shift reactor makes physical absorption the most appropriate CO_2 capture technology, because this process is less energy intensive than a chemical absorption process. In physical absorption processes, CO_2 is recovered from the absorbent (e.g., Selexol) by reducing the pressure in flash drums, which is less energy intensive than stripping CO_2 using heat as applied in chemical absorption. The concentrated CO_2 stream can be sent for sequestration. The remaining syngas stream that contains mainly hydrogen can be sent to the power island for the production of electricity.

Several research efforts are ongoing to reduce the cost of precombustion CO_2 capture processes. The overall process efficiency can be increased if the CO_2 capture temperature is increased from room temperature to the GT fuel inlet temperature of ~300 °C. CO_2 separation at higher temperatures is possible using CO_2 adsorbents such as calcium oxides[26] and lithium silicates.[27] Membranes that can capture CO_2 and H_2S from the syngas are under development. The feasibility of the CO_2 sorbents and membranes at large scale is not yet proven; however, if successful, these technologies can lower the cost of CO_2 capture significantly.

A major challenge with precombustion CO_2 capture option is controlling the NO_x emissions from the GT. The fuel sent to GT mainly consists of hydrogen, which may lead to higher NO_x emissions due to higher flame temperatures during hydrogen combustion as shown in Figure 11.18.[28]

For natural gas-based turbines, NO_x is reduced by forcing more air than stoichiometric in the primary combustion zone to achieve lower flame temperatures. This technique of premixing air with natural gas is not suitable for hydrogen GTs due to the much larger flammability limits and the lower ignition temperatures of hydrogen with respect to natural gas.[29] NO_x can be reduced from the exhaust of the GT using techniques such as selective catalytic reduction (SCR). However, these post-combustion treatment options are very costly for hydrogen combustion.[30] For hydrogen GTs, effective NO_x control can be achieved by diluting the combustor with steam or nitrogen. The steam is available from the steam cycle and nitrogen can be available from the ASU if oxygen is used in the ATR/POX reformer reactor.

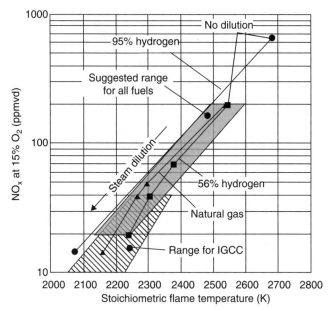

Figure 11.18.   Relation between NOx emission and stoichiometric flame temperature.

A thermodynamic analysis has indicated that the efficiency of CCs is only moderately decreased when diluting the fuel with nitrogen or steam.[31]

Out of the various options available to capture CO_2 from NGCC plants, the precombustion capture route using an autothermal reformer is the most studied and agreed as the feasible route based on today's knowledge. The post-combustion process, which involves processing a large volume of nitrogen-diluted exhaust gas at ambient pressure, is usually more expensive than the precombustion CO_2 capture process. Notice that even the precombustion CO_2 capture process is not economical yet today. Although CO_2 sequestration has been actively debated at numerous energy conferences in recent years, and there is a huge amount of literature in this arena, it seems that few companies are willing to risk the potential environmental liability of CO_2 sequestration. Thus, in addition to the technical challenges related to CO_2 sequestration discussed above, there are still huge economical, environmental, and potential legal challenges related to CO_2 sequestration, which are beyond the scope of this book.

11.3   APPLICATION OF GASIFICATION-BASED SYNGAS TECHNOLOGY

Syngas produced from gasification of coal, biomass, petroleum coke, and other types of feedstock can be used to generate electricity or to produce hydrogen and other liquid fuels or chemicals (ammonia, methanol, dimethyl ether, and diesel fuel) by

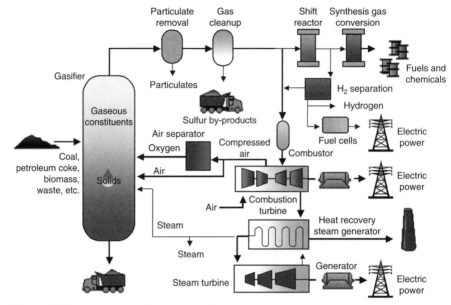

Figure 11.19.   Process flow diagram of gasification-based energy conversion system. See color insert.

synthesis over appropriate catalysts. Figure 11.19 illustrates the process flow diagram of such a gasification-based energy and chemical production system.

According to the 2004 World Gasification Survey sponsored by the U.S. Department of Energy (DOE),[32] "the existing world gasification capacity has grown to 45,001 MWth of syngas output at 117 operating plants with a total of 385 gasifiers." The feedstocks for these operating plants primarily consist of coal (49%) and petroleum residues (37%). Their products include chemicals (37%), Fisher–Tropsch liquid (36%), and electricity (19%). Among the 117 operational gasification plants worldwide, 20 of them are in the U.S., including 14 chemical production plants, 4 power generation plants, and 2 gaseous fuels production plants.

The 2004 World Gasification Survey also shows that additional 38 plants (66 gasifiers) have been announced and are expected to become operational between 2005 and 2010, raising the projected worldwide gasification capacity to 70,000 MWth by 2010.

Figure 11.20 lists the major elements in this syngas conversion system along with corresponding technologies, including feedstock processing, gasification process, syngas cleanup, energy production/conversion, and end-product delivery. Due to the supply and demand gap of NG in most parts of the world, the coal-based IGCC process could play more and more important role in future energy portfolio, especially for countries with abundant coal recourses such as the U.S., China, and India. The CTL process is another important type of syngas-based energy conversion process. With the rising cost of crude oil, CTL process is becoming more and more

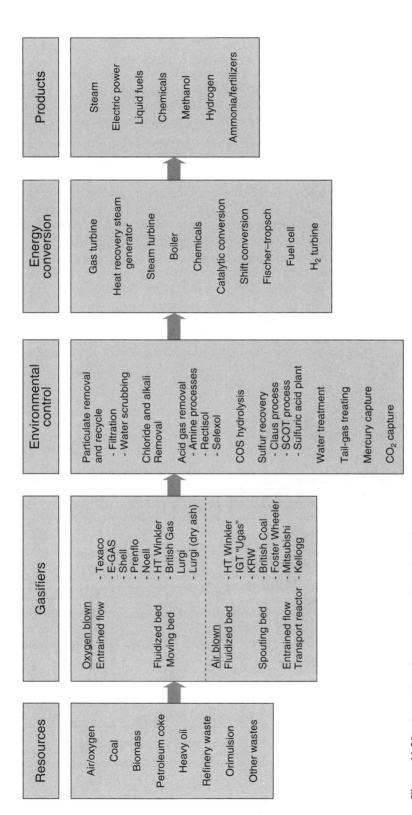

Figure 11.20. Syngas-based energy conversion technologies.

attractive economically where coal, a relatively low-cost feedstock, is converted to high-value chemicals such as methanol, hydrogen, and diesel.

The following sections will briefly describe these syngas conversion systems, focusing on key technologies, system efficiency, and overall economics.

11.3.1 IGCC Plant

Coal, as a low-price and abundant energy source, is an excellent choice for the production of electric power. The coal-fueled IGCC is one of the most promising technologies to convert coal to electricity with not only superior efficiency than PC boiler plants, but also the ability to meet the ever-demanding future environmental regulations including CO_2 emissions.

As shown in Figure 11.21, a typical IGCC plant consists of three major blocks. The first block, coal gasification block, consists of the coal preparation unit, the ASU, a coal gasifier, and the syngas cooling system. The syngas cleanup block includes particulates removal unit, COS hydrolysis unit, acid gas cooling, low-temperature gas cooling (LTGC), acid gas removal (AGR) unit, and sulfur recovery unit (SRU). The final block is the power island that includes GT, HRSG, and steam turbines.

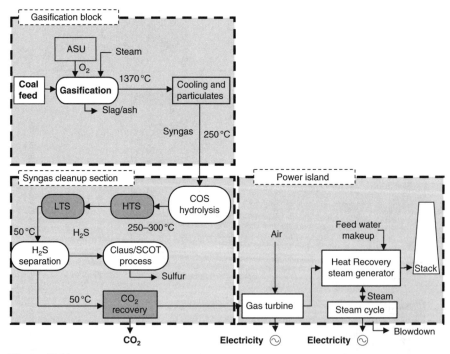

Figure 11.21. Typical IGCC plant process flow diagram (HTS, LTS, and CO_2 recovery blocks are required only for IGCC with CO_2 capture).

11.3.1.1   *Gasification Block*

Coal preparation is the first step in an IGCC process. Depending on the type of gasifier, this process typically includes coal grinding and coal drying (for low-rank coals and/or dry feeding gasifiers) or slurry plant (for slurry feeding gasifier). The coal particle size required is usually between 50 and 150 μm for entrained-flow gasifiers, less than 50 mm for a moving-bed gasifier and less than 6 mm for a fluidized-bed gasifier. There is a trade-off between coal particle sizes and kinetics of coal gasification. From the kinetics point of view, the finer the coal particles, the faster the coal gasification kinetics. However, finer particles require more energy during coal grinding, and it becomes difficult to prepare coal water slurry if coal particles are too small for slurry feeding system.

An ASU is required for the oxygen-blown type of gasifier. Most commonly used air separation technology for IGCC system is the cryogenic process where air is cooled to cryogenic temperature (nearly −185 °C) so that liquid oxygen can be separated from nitrogen via subsequent distillation. Typical oxygen purity for oxygen-blown gasifier is around 95%.

The most critical component in the gasification block is the gasifier for which quite a few designs and technologies were developed worldwide—either air blown or oxygen blown, moving bed, entrained bed, or fluidized bed with slurry or dry feeding. In recent years, most plants were built using the entrained-flow gasifier technology. It partially oxidizes coal particles under high pressure (20~60 bar) and high temperature. Such gasifiers can achieve >90% carbon conversion and produce syngas consisting mainly of CO, H_2, CO_2, and steam.

Table 11.1 lists some of the commercially available gasifier technologies and their typical operation conditions. The gasifier outlet temperature is typically above

Table 11.1.   Gasifier Technology Suppliers and Typical Operation Conditions

	Gasifier Type	Oxidant	Coal Feed Method	Exit Gas Temperature (°C)	Operating Pressure (Bar)	Ash Condition
GE, U.S. (formerly Texaco gasifier)	Entrained flow	O_2	Water slurry	1200~1400	25~60	Slagging
Shell, U.S.	Entrained flow	O_2	Dry	1300~1400	25~28	Slagging
British Gas/ Lurgi	Moving bed	O_2	Dry	420~650	25~28	Dry/slagging
Kellogg–Rust– Westinghouse (KRW)	Fluidized bed	Air	Dry	920~1050	25~28	Dry/ agglomerating

1000 °C; therefore, syngas cooling is required to cool the syngas stream and recover thermal energy before proceeding to downstream processes. Various cooling methods are being adopted for various gasification processes including fire tube, radiant syngas cooling, convective syngas cooling, and quench cooling.

One application of the entrained gasifier is the IGCC plant built in the Tampa Electric Company's Polk Station.[33] It uses an oxygen-blown, slurry-fed GE gasifier, which process about 2100 t of coal per day and operates at 1200~1400 °C and 375 psig. The syngas cooling section consists of a radiant syngas cooler and a convective syngas cooler. The radiant syngas cooler cools the syngas to about 730 °C recovering between 250~300 MMBTU/h heat (12~15%) of the fuel's heating value and generates high-pressure steam at 1650 psig. Convective syngas cooler further cools the syngas down to about 385 °C (recovers 3%–4.5% of the fuel's heating value) and generates additional 1650 psig steam. Tampa Electric Company's IGCC plant uses GE's 7FA GT and a three-pressure HRSG with three-pressure steam turbines. During normal operation on syngas fuel, the GT produces about 192 MW of electricity each, and the steam turbines generate about 120–135 MW of electricity.

11.3.1.2 Syngas Cleanup

The typical steps for syngas gas cleanup include particle removal, COS hydrolysis, acid gas cooling, sulfur removal, and sulfur recovery. CO_2 sequestration can be implemented in an IGCC plant by adding additional units in the syngas cleanup section.

Particulates in syngas, consisting of primarily char and fly ash, can be removed using cyclone filters, ceramic or metal candle filters, or wet scrubbing. Candle filters are typically comprised of an array of candle elements in a pressure vessel. It can be cleaned periodically by back pulsing using syngas. The collected fine ash and char particulates can be recycled back to the gasifier. It is claimed that the plugging of the convective syngas cooler could be one of the major reliability issue of an IGCC plant.

Raw syngas exits the candle filters at around 200~350 °C. A syngas scrubber can be used, in combination with candle filters or stand along, to complete the particulates removal and HCl removal. In Tampa Electric IGCC plant, syngas at 350 °C directly enters a series of three water/gas contact steps to remove particulates and HCl.

If a methyldietheanolamine (MDEA)-based AGR process is used, a COS hydrolysis unit is needed to convert COS to H_2S via the following reaction:

$$COS + H_2O \rightarrow H_2S + CO_2.$$

This step is necessary due to MDEA's inability to absorb COS that was formed in the gasifier. In a conventional hydrolysis process, water-saturated syngas is passed through a fixed-bed catalytic hydrolysis reactor where 85%–95% of the COS is converted to H_2S (COS at 1~10 ppm). Once COS is converted to H_2S, it is readily absorbed by the MDEA process.

Two types of AGR technologies are available: physical absorption and chemical absorption. Selexol, which is a prime example of the physical absorption process developed by Honeywell's UOP LLC utilizes physical solvent (a mixture of dimethyl ethers of polyethylene glycol) to absorb sulfur-containing compounds such as H_2S, COS, and also CO_2. Selexol solvents can be regenerated either thermally, by flashing, or by stripping gas. Chemical solvents, such as MDEA and diethanolamine (DEA), can absorb H_2S at relatively low acid gas partial pressure by forming chemical bonds. Solvent regeneration is typically achieved by reboiling.

Chemical solvent technologies are usually favored at low acid gas partial pressure, while physical solvents are preferred at high acid gas pressure. The MDEA-based AGR generally has a lower capital cost than the Selexol-based AGR process and has been the predominant process for IGCC application up to the late 1990s. However, physical solvent-based AGR, such as Selexol, has emerged as a competitive technology in the recent years. It consumes considerably less amount of energy during solvent regeneration than the MDEA process. Other benefits of the physical solvent-based AGR technologies are primarily attributed to their capability of "deep" sulfur removal and the capability to also remove CO_2. This is particularly important for CTL or coal-to-chemical plants since the downstream catalysis processes usually requires the sulfur level in syngas to be reduced to <1 ppm versus <20 ppm in today's IGCC plant. In addition to sulfur, IGCC plants also need to remove other pollutants in syngas such as Cl, NH_3, Hg, and heavy metals.

Sulfur recovery can be achieved using the industrially proven Claus and SCOT (Shell Claus Off-gas Treating process) processes where the absorbed H_2S is recovered as elemental sulfur via the following reactions:

$$H_2S + O_2 \rightarrow SO_2 + H_2O,$$
$$H_2S + SO_2 \rightarrow S + H_2O.$$

An alternative approach is to combust H_2S to form SO_2 and, after further oxidation into SO_3, produce sulfuric acid as end product.

If CO_2 sequestration is required, an IGCC plant can be easily modified to have such capabilities. Deep decarbonization of syngas requires the addition of a series of WGS reactors and a CO_2 separation unit to the existing IGCC plant as shown in Figure 11.21.

WGS reactors are usually divided into two or more stages, including high-temperature shift (HTS) reactor and low-temperature shift (LTS) reactor. This multi-stage design is necessary due to the strong exothermic nature of the WGS reaction:

$$CO + H_2O \rightarrow CO_2 + H_2 \quad \Delta H = -41.16 \, kJ/mol.$$

Due to adiabatic temperature rise, the reaction quickly reaches chemical equilibrium at the end of the HTS (about 400 °C), and the syngas needs to be cooled to approximately 300 °C before entering LTS to achieve >90% CO conversion. Another method of shifting the WGS reaction equilibrium forward is to inject excess steam into the syngas stream so that an H_2O-to-CO ratio of more than 2.5 is reached prior to WGS reactors. Details of WGS reaction are described in Chapter 6 of this book.

CO_2 can be separated from the shifted syngas stream in a Selexol type of absorber. Multistage flash drums can be used to extract CO_2 from absorbed solvents at various pressure levels typically ranging from atmospheric pressure to 400 psi. The CO_2 product stream is usually compressed to high pressure (110 bar) for geologic storage or enhanced oil recovery.

Note that both physical solvent- and chemical solvent-based desulfurization technologies require operation temperature at or near room temperature (30 °C). The clean syngas is typically reheated before entering GT.

To avoid significant energy loss associated with syngas cooling and reheating, various hot syngas cleanup technologies are currently under development or demonstration phase, including hot candle filters for particulates removal, fluidized beds with solid absorbents for warm gas sulfur removal,[34] and so on.

11.3.1.3   Power Island

As shown in Figure 11.21, the power island of an IGCC plant consisted of GTs that use syngas as fuel, an HRSG, and steam turbines. Additional condensers, feed water heaters, and compressors are also essential parts of the steam power cycles.

The latest GT technology, such as GE's H-class turbine, can achieve ~60% fuel efficiency in CC while maintaining low NO_x emission level. Typical GT's firing temperature is about 1300~1450 °C and can produce exhaust gas at about 560 °C. HRSG utilizes a series of heat exchangers to recover heat from the GT exhaust (560 °C, 1 bar) and raises steam at various pressure levels to generate additional electricity in steam turbines.

Figure 11.22 shows a typical IGCC power island configuration where three-pressure levels of steam are raised corresponding to three levels of steam turbines namely the high-pressure steam turbine (HPST), the intermediate-pressure steam turbine (IPST), and the LPST. In addition to the heat recovery from GT exhaust gas, the high-temperature syngas coolers in the gasification block also raise high-pressure steam to generate electricity in HPST. The typical operation pressure and temperature of these steam turbines are summarized in Table 11.2.

Table 11.2.   Typical Steam Turbine Operation Conditions

	Steam Pressure (psi)	Steam Temperature (°C)
HPST	1400~2600	530~550
IPST	300~600	500~540
LPST	20~80	200~300

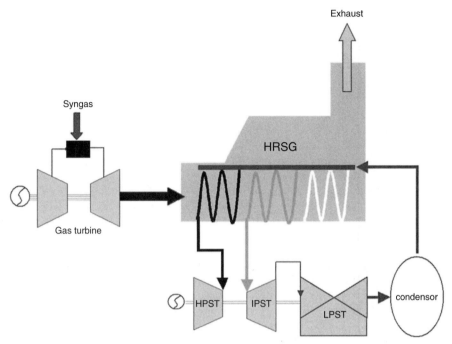

Figure 11.22.   IGCC combined cycle with a three-pressure level HRSG.

11.3.1.4 IGCC System Performance

IGCC plants generally can achieve higher efficiencies than conventional PC power plant. The primary difference lies in the combined use of GT and steam turbine to maximize the amount of energy produced. Current IGCC technology can achieve thermal efficiency of about 38%~43% based on higher heating value of coal.[35–38] Currently, a few commercial-scale IGCC plants are operating worldwide with excellent environmental performances.

Figure 11.23 illustrates the energy conversion and distribution in a typical IGCC process. Assuming 100 MWth of higher heating value of coal fed into an IGCC plant, around 70%~80% of the energy is converted into syngas and another 15% can be recovered in the syngas coolers in the form of high-pressure steam. The GT can generate about 30 MW of electricity while the HRSG/steam turbines can recover another 22 MW of electricity (together achieving about 55% to as much as 60% CC efficiency). After subtracting 9 MW of parasitic power consumptions in the gasification and syngas cleanup blocks (9% of the total energy input), the net energy output of this IGCC plant is about 43 MW. In another word, a typical IGCC plant can achieve an overall efficiency of about 43% on a higher heating value (HHV) base.

Table 11.3. Comparison of Four Different Existing IGCC Plants

Plant Location	Tampa, FL	Wabash River, IN	Buggenum, The Netherlands	Puertollano, Spain
Start-up date	1997	1995	1996	1997
Gasification technology	GE Energy (formerly Texaco)	E-Gas (ConocoPhillips)	Shell	Prenflo
Feeding method	Slurry	Slurry	Dry	Dry
Gasifier type	Entrained flow	Entrained flow	Entrained flow	Entrained flow
Gasifier configuration	Single-stage down flow	Two-stage up flow	Single-stage up flow	Single-stage up flow
Pressure (psi)	300~1000	<500	<600	<600
Hot face wall	Refractory	Refractory	Membrane wall	Membrane wall
Cold gas efficiency (%)	71~76	74~78	80~83	80~83
Carbon conversion (%)	96~98	98	>98	>98
Coal feed rate (TPD)	2200~2400	2500	2000	2600
Gasifier availability (%)	82.0	84.4	81.8	76.0
Plant availability (with fuel backup) (%)	95.0	94.3	96.4	93.0
Net power output (MW)	250	261.6	253	300
Designed net efficiency (%)	41.6 (HHV)	37.8 (HHV)	43 (LHV)	43 (LHV)

TPD, ton per day; MW, mega watt; HHV, higher heating value; LHV, lower heating value.

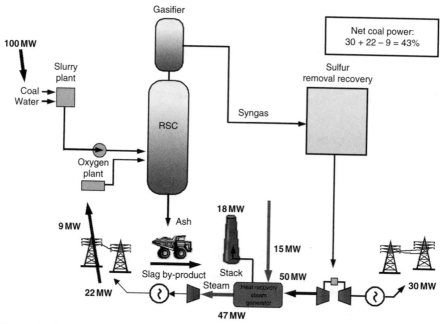

Figure 11.23. Energy conversion and distribution in a typical IGCC process. RSC, radiant syngas cooler. See color insert.

Table 11.3 summarizes some of key operation information and performance data for four existing commercial-scale IGCC plants.[39] It can be seen that, all four types of gasifiers are oxygen blown (95% pure) and can all reach coal feed rate of more than 2000t per day. In all four cases, the reliability and availability of the gasification units limit the overall plant availability. There are many researches focusing on improving refractory life, preventing slag/ash fouling and improving feed injectors, and so on.

Besides its higher system efficiency performance, IGCC plant also exhibits superior environmental performance. Its cleanup process deals with syngas stream at high pressure prior to combustion and, therefore, is far more efficient and cost-effective than most post-combustion cleanup processes, which operate at ambient pressure with large amounts of nitrogen dilution. The pollutant emission of a typical IGCC plant is only 1/10th of a similar size PC power plant. Its sulfur removal level is at 99%; SO_2 emission is around $25\,mg/m^3$ and consumes about 40%~60% less water than do conventional PC power plants.

Another important factor affecting IGCC technology's competitiveness is its economic performances. Over the last two decades, the capital cost of an IGCC plant has reduced dramatically. Figure 11.24 shows a simplified breakdown of the total plant capital cost (TPC) of a typical IGCC plant (excluding piping, electrical and control system, civil construction, engineering fees, project contingency, etc.).

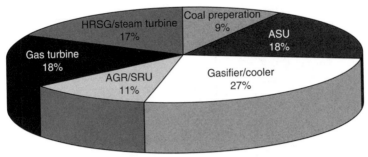

Figure 11.24.   IGCC plant capital cost breakdown. Total TPC = $1438/kWe. See color insert.

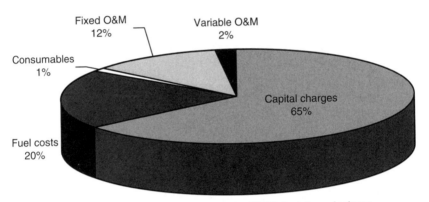

Figure 11.25.   Breakdown of cost of electricity of an IGCC plant. See color insert.

The COE for an IGCC plant mainly includes the levelized capital charges, cost of fuel, consumables, and cost of operating and maintenance (O&M). Typical IGCC COE is around 5–5.5 cents/kWh (2006 U.S. dollars).[40] A breakdown of the IGCC COE is illustrated in Figure 11.25, which shows the capital cost accounts for about 65% of the COE, fuel cost about 20%, and O&M cost about 14%.

With the world's ever-increasing concern for CO_2's potential impact on global warming, IGCC technology has become more and more attractive for its unique advantages in CO_2 capture and sequestration that are not available through conventional combustion technologies. An IGCC plant can be readily designed to capture as much as 90% of the CO_2 for geologic sequestration or enhanced oil recovery. The addition of CO_2 sequestration capability has a significantly less impact on overall system efficiency and capital cost than it would for a PC plant or even an NGCC plant. With the IGCC technology already being increasingly cost competitive, the need for carbon emission control would undoubtedly favor IGCC as the least expensive option.

Table 11.4 illustrates the impact of CO_2 sequestration on IGCC system efficiency and economics.[40] On average, an efficiency loss of about 6%–10% is expected

Table 11.4.   Impact of CO$_2$ Sequestration on the System Efficiency and Capital Cost of an IGCC Plant

Gasification Technology	GE Energy (Texaco) with Quench		E-Gas (ConocoPhillips)		Shell	
CO$_2$ capture	No capture	90% capture	No capture	90% capture	No capture	90% capture
Efficiency (% HHV)	37%	30%	40%	31%	41%	33%
Heat rate (BTU/ kWh)	9300	11,300	8550	11,000	8370	10,350
Capital cost ($/kWe)	$1270	$1620	$1300	$1850	$1470	$2020
COE ($/MWh)	$46	$57	$46	$62	$49	$65

for an IGCC plant with ~90% CO$_2$ sequestration capability, majority of which is due to the extra energy consumption in ASU, CO$_2$ capture, and CO$_2$ compression processes. Similarly, an increase of about 20%~30% in capital cost as well as cost of electricity can be expected with the additional equipment for CO$_2$ capture and compression. On average, it costs about $20~$30 for each ton of CO$_2$ avoided.

11.4 APPLICATION OF H$_2$/SYNGAS GENERATION TECHNOLOGY TO LIQUID FUELS

Hydrogen is not only important for fuel cells but also plays a vital role in today's refining industry for upgrading crude oil to clean fuels. The U.S. consumes 15 million barrels of crude oil daily, of which 9 million barrels are imported (2 million from the Middle East). The main products are gasoline (8.8 million barrels a day), distillates (3.8 million), and petrochemicals (approximately 1 million barrels a day).

Ultimately, human being may shift to non-fossil-fuel-based energy sources such as nuclear, solar, wind, and others; in the near term, however, our reliance on oil imports can be alleviated by modifying the refining process with cheap H$_2$ produced from coal, biomass, waste, or petroleum coke via the well-known gasification and shift processes. These provide a cheaper near-term solution to cut U.S. oil imports while simultaneously reducing GHG emissions.[41]

First, one could increase the hydrogen content of today's fuels. Gasoline and distillates contain a mixture of paraffins ($-CH_2-$)$_n$, naphthenes, and aromatics (CH)$_n$. Paraffins are environmentally superior to aromatics and naphthenes due to its cleaner combustion and lower CO$_2$ emission per BTU (British Thermal Unit) energy output. Today's gasoline contains about 30% aromatics. One can convert aromatics at least partially to paraffins by a hydrogenation reaction and utilize hydrogen from coal, petroleum coke, residual oil, or biomass. Diesel fuel can also be refined via similar hydrogenation process. This is equivalent to the generation of ~0.5 million barrels of additional high-grade liquid fuels using hydrogen.

Second, today's crude oil contains about 30% low boiling fractions (vacuum resid), which in most cases is either used as heavy fuel oil or sent to a coker in the refinery. The coker plant produces, in addition to coke, about 50% low-quality liquid products. There are mature technologies today to hydrocrack these 4.5 million barrels of resid per day, and upgrade them to high-grade liquid products. Again by a simple mass balance, this would be equivalent to about 2.5 million barrels a day of high-grade liquid gasoline and diesel using hydrogen. The amount of hydrogen that can be added in the current U.S. refinery processes is equivalent to 600,000 barrels of oil.[41] The total potential savings in oil imports could be about 5 million barrels a day, of which 2 million are due to lower gasoline consumption, 2 million due to higher yields of gasoline and diesel per barrel crude oil via hydrocracking of heavier feeds, and 1 million barrels due to the use of hydrogen generated from biomass, waste, coal, or other fossil fuels into the gasoline.[41]

Since hydrogen needed for the refineries is usually produced in a central facility, it is easy to achieve near 100% GHG sequestration. Even if only 60% of this potential is realized, it is equal to turning 35% of all cars to hydrogen vehicles. It is feasible to achieve this in 20 years without having to develop H₂ fuel cell cars to reach the same goal.[41] We could look at this method as an improved form of a hydrogen economy. Thus, from this point of view, H₂ economy is not something people can only dream for the future, it is out there today and people are practicing it today.

There is available proven technology for hydrogen production from resid, natural gas, coal, biomass, and waste, as well as for hydrocracking of oil resid. In fact, as shown in Figure 11.26, as natural gas price increases in recent years, it makes more sense to make H₂ from coal and/or resid via gasification followed by WGS reaction.

Coal is one of the most economical and abundant feedstock for the centralized hydrogen production. The environmental concerns associated with coal as an energy

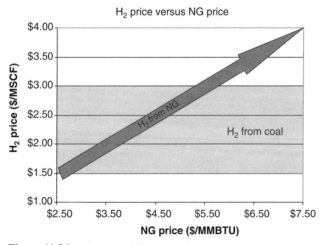

Figure 11.26.   The price of H₂ from coal and natural gas as a function of natural gas prices.[43]

source can be effectively addressed via coal gasification and syngas cleanup processes used in today's IGCC plants. In addition to coal, one can also make H_2 from waste and biomass gasification. The gasification-based hydrogen production technologies, along with the abundant supplies of coal, biomass, and waste in the U.S., will ensure the establishment of hydrogen as one of the primary resources for reducing U.S. oil imports, and CO_2 emissions from transportation fuels.

11.4.1 Coal-to-H₂ Process Description

As one can see from the above session, how to produce H_2 cheaply from different resources is vital to meet the future energy and environmental needs. Coal or biomass gasification could be the cheapest way of making H_2 when natural gas price is high. If one produces H_2 in a centralized plant, it is possible to capture and sequestrate CO_2.

A gasification-based coal-to-hydrogen process, shown in Figure 11.27, largely resembles that of an IGCC plant. They share the same coal gasification and syngas cooling processes.

The primary end product in a coal-to-hydrogen plant, however, is the concentrated hydrogen gas instead of electricity. For this purpose, a series of WGS reactors

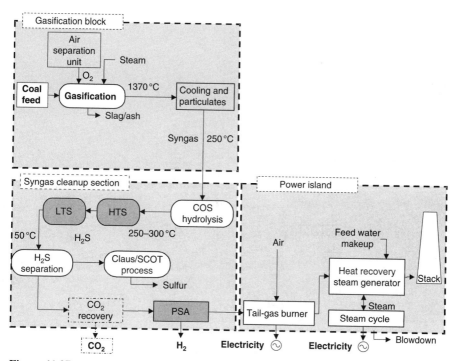

Figure 11.27. A gasification-based coal-to-hydrogen plant.

are required to convert CO into hydrogen. The conversion of CO to H$_2$ via WGS reaction not only maximizes the production of H$_2$ but also creates an excellent opportunity for the isolation and sequestration of CO$_2$. Similar to desulfurization processes, CO$_2$ sequestration can be achieved using either physical- or chemical-based solvents (Selexol or MDEA processes).

In addition to WGS and CO$_2$ recovery units, hydrogen separation and purification units, such as the pressure swing absorber (PSA), are needed to generate high-purity hydrogen product stream. The residue fuel in the PSA-off gas, containing about 10%–15% of the H$_2$ in the original fuel, is sent to a tail-gas boiler to generate additional power through steam turbines.

If both hydrogen and electricity are desired as end products, as in the case of a polygeneration plant, part of the syngas stream can be sent directly to the boiler or even a GT, bypassing the CO$_2$ recovery and PSA units. The split of syngas between H$_2$ production and power generation can be adjusted to accommodate the peak periods for electricity demand by lowering the H$_2$ production, and off-peak periods by sending most of the syngas stream to H$_2$ production. Such flexibility in the plant operations can result in significant economic benefits for a coal gasification-based polygeneration system.

One of the high potential future technologies for the hydrogen production from coal is the integrated WGS–membrane cleanup system, as shown in Figure 11.28.

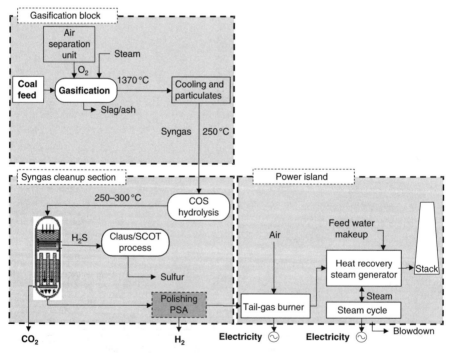

Figure 11.28. Schematic representation of coal-to-hydrogen plant with integrated WGS–membrane cleanup system.

Such integrated system is designed to conduct WGS reaction and the removal of CO_2 or H_2 from the syngas stream simultaneously. In doing so, the equilibrium of the WGS reaction is constantly shifted in favor of the forward reaction, converting more CO into H_2, due to the constant removal of product gas, either CO_2 or H_2, from the syngas stream. As a result, the integrated WGS–membrane system only requires near stoichiometric ratio of H_2O/CO in the syngas feed stream, whereas an H_2O/CO ratio of as high as 2.5 is required for conventional WGS reactor to achieve high (>90%) conversion of CO. Such a significant reduction in the steam requirement represents an overall system efficiency improvement of as much as 5%. On the other hand, such integrated syngas cleanup technology can potentially replace or combine multiple unit operations such as sulfur removal, WGS, and CO_2 separation into one reactor unit, thus reducing the capital cost of an IGCC plant with CO_2 capture by 17%.[40]

11.4.2 Coal-to-Hydrogen System Performance and Economics

Although no coal gasification-based polygeneration plant that coproduces hydrogen and electricity currently exists, the technologies required for such a plant are fairly mature. Many heavy oil-based facilities, which utilize similar technology, have already been in operation. There have been many studies and analysis conducted to evaluate the system performance and economics of such coal-to-hydrogen systems.

A study by Mitretek,[42] sponsored by DOE, compared various configurations of a coal-based hydrogen production plant that uses existing or future technologies for hydrogen separation, with or without CO_2 sequestrations. Some of the major findings of this study are listed in Table 11.5.

It can be seen that even with CO_2 sequestration, the hydrogen produced via coal gasification is reasonably cost competitive with that produced through other means. Future technologies such as hot gas cleanup and advanced membranes have the

Table 11.5. Efficiency and Economic Analysis of Coal-to-Hydrogen Plant

Syngas cleanup technology	Coal to Hydrogen		
	Cold syngas cleanup	Cold syngas cleanup	Hot syngas cleanup
H$_2$/CO$_2$ separation technology	PSA	PSA	Advanced membrane
CO$_2$ sequestration ratio	0	87%	100%
Efficiency (% HHV)	63.90%	59%	75%
Unit capital cost ($×1000/TPD of H$_2$ production)	$1345	$1389	$1066
Cost of hydrogen ($/MMBTU)	$6.83	$8.18	$5.89

TPD, tons per day.

482

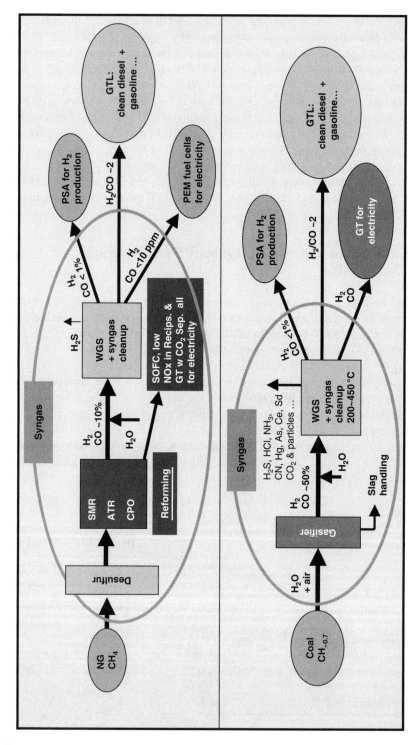

Figure 11.29. Syngas and H₂ production and purification are in the center of future energy conversion processes.

potential to improved the overall system efficiency and therefore significantly reduce the capital cost of such a coal gasification-based hydrogen production plant. Similar studies of polygeneration systems indicated that, in a 50-50 electricity to hydrogen production plant, the overall efficiency can reach about 53.6% and cost of hydrogen can be reduced to as low as $3.98/MMBTU.

11.5 SUMMARY

As you can see from this chapter, the syngas-based energy conversion process will play a more and more important role in the future energy portfolio of the world. Figure 11.29 summarized the gasification-based syngas conversion processes and products, which provide a good summary for this chapter.

As shown in Figure 11.29, syngas and H_2 production and purification technologies are the critical link of future energy conversion processes, such as the H_2 production for future H_2 economy, IGCC with or without CO_2 capture for future coal-based power plants, NGCC with CO_2 capture, natural gas to H_2 and liquids (GTL), and coal to H_2 and liquids (CTL), as well as biomass and waste to H_2 and liquids. All of above-mentioned processes are calling for developments of different H_2 and syngas production and purification technologies discussed in this book. Many R&D activities will continue to receive more and more funding from government and industries in this arena. Although many papers are published in this field, there is no recent book that systematically summarizes the progresses made in this exciting area, and that is the motivation of this book.

REFERENCES

1. Solutions for 21st century. *IEA Technical Report*, p. 2, **2002**. Available at http://www.iea.org/textbase/papers/2002/tsr_layout.pdf.
2. RUBIN, E. and RAO, A. A technical, economic and environmental assessment of amine-based CO_2 capture technology for power plant greenhouse gas control. *Annual Technical Progress Report* (submitted to DOE), October **2002**.
3. http://www.eia.doe.gov/oiaf/1605/ggccebro/chapter1.html.
4. http://tonto.eia.doe.gov/FTPROOT/electricity/034804.pdf.
5. SPATH, P.L. and MANN, M.K. *Life cycle assessment of a natural gas combined-cycle power generation system*. September 2000, NREL/TP-570-27715. http://www.nrel.gov/docs/fy00osti/27715.pdf.
6. KLARA, S.M. Reducing CO_2 emissions from fossil fuel power plants. *EPGA's 3rd Annual Power Generation Conference*. Hershey, PA, October 16–17, **2002**.
7. BOLLAND, O. and UNDRUM, H. A novel methodology for comparing CO_2 capture options for 2 natural gas-fired combined cycle plants. *Advances in Environmental Research*, **2003**, 7, 901.
8. ALLAM, R.J. and SPILSBURY, C.G. A study of extraction of CO from the flue gas of a 500 MW pulverized coal fired boilers. *Proceedings of the First International Conference on Carbon Dioxide Removal*. Amsterdam, **1992**.
9. SUDA, T., FUJII, M., and YOSHIDA, K. Development of flue gas carbon dioxide recovery technology. *Proceedings of the First International Conference on Carbon Dioxide Removal*. Amsterdam, **1992**.
10. ERGA, O., JULIUSSEN, O., and LIDAL, H. Carbon dioxide recovery by means of aqueous amines. *Energy Conversion and Management*, **1995**, 36, 387.
11. CHAKMA, A. Separation of CO_2 and SO_2 from flue gas streams by liquid membranes. *Energy Conversion and Management*, **1995**, 36, 405.
12. MEISEN, A. and SHUAI, X. Research and development issues in CO_2 capture. *Energy Conversion and Management*, **1997**, 38, S37.

13. UNDRUM, H., BOLLAND, O., and AAREBROT, E. Economical assessment of natural gas fired combined cycle power plant with CO capture and sequestration. *Presented at the Fifth International Conference on Greenhouse Gas Control Technologies. Cairns, Australia*, August **2000**.

14. DAVISON, J. CO_2 capture and storage and the IEA capture and storage and the IEA greenhouse as R & D program greenhouse gas R & D program. *Workshop on CO_2 Issues*. Middelfart, Denmark, May 24, **2005**.

15. EIMER, D. Post-combustion CO_2 separation technology summary. In *Carbon Dioxide Capture for Storage in Deep Geological Formations—Results from the CO_2 Capture Project. Capture and Separation of Carbon Dioxide from Combustion Sources* (ed. D.C. Thomas), Vol. 1. Amsterdam: Elsevier, p. 91–97, **2005**.

16. DAMEN, K., van TROOST, M., FAAIJ, A., and TURKENBURG, W. A comparison of electricity and hydrogen production systems with CO_2 capture and storage. Part A: Review and selection of promising conversion and capture technologies. *Progress in Energy and Combustion Science*, **2006**, 32, 215.

17. HEITMEIR, F., SANZ, W., GÖTTLICH, E., and JERICHA, H. The Graz cycle—A zero emission power plant of highest efficiency. *XXXV Kraftwerkstechnisches Kolloquium*. Dresden, Germany, **2003**.

18. BOLLAND, O. Options for oxy-fuels & pre-combustion decarbonisation cycles. *Presentation for Alstom Cross Segment CO_2 Mitigation Group*. October 11, **2002**. Available at http://folk.ntnu.no/obolland/pdf/Alstom_Power_Oct_2002_Bolland.pdf.

19. SUNDKVIST, S.G., KLANG, Å., SJÖDIN, M., WILHELMSEN, K., ÅSEN, K., TINTINELLI, A., MCCAHEY, S., and YE, H. AZEP gas turbine combined cycle power plants: thermal optimisation and LCA analysis. Available at http://uregina.ca/ghgt7/PDF/papers/peer/079.pdf (accessed October 18, 2009).

20. BRANDVOLL, Ø. and BOLLAND, O. Inherent CO_2 capture using chemical looping combustion in a natural gas fired power cycle. *Journal of Engineering for Gas Turbines and Power*, **2002**, 126, 316.

21. YU, J., CORRIPIO, A.B, HARRISON, D.P, and COPELAND, R.J. Analysis of the sorbent energy transfer system (SETS) for power generation and CO_2 capture. *Advances in Environmental Research*, **2003**, 7, 335.

22. BOLLAND, O., KVAMSDAL, H.M, and BODEN, J.C. A comparison of the efficiencies of the oxy-fuel power cycles water cycle, Graz cycle and Matiant cycle. In *Carbon Dioxide Capture for Storage in Deep Geological Formations—Results from the CO_2 Capture Project. Capture and Separation of Carbon Dioxide from Combustion Sources* (ed. D.C. Thomas), Vol. 1. Amsterdam: Elsevier, p. 499–511, **2005**.

23. LOZZA, G. and CHIESA, P. Natural gas decarbonization to reduce CO_2 emission from combined cycles: Part I: Partial oxidation. *Transactions of ASME*, **2002**, 124, 84.

24. LOZZA, G. and CHIESA, P. Natural gas decarbonization to reduce CO_2 emission from combined cycles: Part II: Steam-methane reforming. *Journal of Engineering for Gas Turbines and Power, Transactions of ASME*, **2002**, 124, 89.

25. BORDIGA, G. and CAMPAGNA, M. *Natural gas fired combined cycles with abatement of CO_2 emissions*. Graduation Thesis, Politecnico di Milano, Italy, **1999**.

26. GUPTA, H. and FAN, LS. Carbonation-calcination cycle using high reactivity calcium oxide for carbon dioxide separation from flue gas. *Industrial & Engineering Chemistry Research*, **2002**, 41, 4035–4042.

27. LI, W., GANGWAL, S., GUPTA, R., and TURK, B. Development of fluidizable lithium silicate-based sorbents for high temperature carbon dioxide removal. *Pittsburgh Coal Conference*. Pittsburgh, PA, **2006**.

28. CHIESA, P., LOZZA, G., and MAZZOCCHI, L. Using hydrogen as gas turbine fuel. *Journal of Engineering for Gas Turbines and Power*, **2005**, 127, 73.

29. DRELL, I.L. and BELLES, F.E. Survey of hydrogen combustion properties. *NACA Report 1383, Research Memorandum E57D24*, **1957**.

30. MAJOR, B. and POWERS, B. Cost analysis of NOx control alternatives for stationary gas turbines. *Contract DE-Fc02-97CHIO877*, **1999**.

31. CHIESA, P., LOZZA, G., and MAZZOCCHI, L. Using hydrogen as a gas turbine fuel. *Proceedings of ASME Turbo Expo*. Atlanta, GA, **2003**.

32. JAMES CHILDRESS (GASIFICATION TECHNOLOGIES COUNCIL). Industry in transition—The 2004 World Gasification Survey. *2004 Gasification Technology Conference.* October 3, **2004**.

33. U.S. DEPARTMENT OF ENERGY. Tampa Electric Integrated Gasification Combined-Cycle Demonstration Project: A DOE Assessment. http://www.netl.doe.gov/technologies/coalpower/cctc/cctdp/bibliography/demonstration/pdfs/tampa/TampaPPA8%20Final080904.pdf. August **2004**.

34. GANGWAL, S.K., PORTZER, J.W., TURK, B.S., and GUPTA, R. Advanced sulfur control processing. *Proceedings of the Advanced Coal-Fired Systems Review Meeting* (CD-ROM). Morgantown, WV: U.S. Department of Energy. July 16–19, **1996**.

35. U.S. Department of Energy. Texaco gasifier IGCC base cases. *Process Engineering Division Topical Report PED-IGCC-98-001*, July **1998**.

36. U.S. Department of Energy. Shell gasifier IGCC base cases. *Process Engineering Division Topical Report PED-IGCC-98-002*, July **1998**.

37. U.S. Department of Energy. Destec gasifier IGCC base cases. *Process Engineering Division Topical Report PED-IGCC-98-003*, September **1998**.

38. U.S. Department of Energy. British Gas/Lurgi gasifier IGCC base cases. *Process Engineering Division Topical Report PED-IGCC-98-004*, September **1998**.

39. HOLT, N. Coal-based IGCC plants—Recent operating experience and lessons learned. Presented at the *Gasification Technologies Conference.* Washington, DC, October **2004**.

40. BRACHT, M., ALDERLIESTEN, P.T., KLOSTER, R., PRUSCHEK, R., HAUPT, G., XUE, E., ROSS, J.R.H., KOUKOU, M.K., and PAPAYANNAKOS, N. Water gas shift membrane reactor for CO_2 control in IGCC systems: Techno-economic feasibility study. *Energy Conversion and Management*, 38 (Suppl.), **1997**.

41. SHINNAR, R. The hydrogen economy, fuel cells, and electric cars. *Technology in Society*, **2003**, 25, 455.

42. GRAY, D. and TOMLINSON, G. Hydrogen from coal. *US Department of Energy DE-AM26-99FT40465*, **2001**.   http://www.netl.doe.gov/technologies/hydrogen_clean_fuels/refshelf/pubs/Mitretek%20Report.pdf (accessed November 2001).

43. STURM, K., LIAW, V., and THACKER, S. Beyond the 2002 ChevronTexaco coal IGCC reference plant. *Presented at the Gasification Technologies Conference. San Francisco, CA, October* **2003**. http://www.gasification.org/Docs/Conferences/2003/21STUR.pdf (accessed October 2003).

Chapter 12

Coal and Syngas to Liquids

KE LIU, ZHE CUI, WEI CHEN, AND LINGZHI ZHANG

Energy & Propulsion Technologies, GE Global Research Center

12.1 OVERVIEW AND HISTORY OF COAL TO LIQUIDS (CTL)

The coal and syngas to liquids process is a process whereby coal or other carbonaceous solid feedstocks such as biomass are converted to liquid fuel or chemicals. The production of coal-derived liquids as by-products of coke-making commenced in Germany and the U.K. in the 1840s. Most CTL development derives from the early 1900s, when two distinct approaches were pursued. The earliest process route involved high-temperature and high-pressure dissolution of coal in a solvent to produce high-boiling point liquids. No hydrogen or catalyst was used at this time.

The modern process for converting coal into liquids, indirect liquefaction, was discovered in the 1920s by Franz Fischer and Hans Tropsch. The process was invented in petroleum-poor but coal-rich Germany to produce liquid fuels. Fischer and Tropsch passed synthesis gas—consisting of carbon monoxide and hydrogen—over metallic catalysts and produced pure hydrocarbons. The process was commercialized in the 1930s, initially for the production of chemical feedstocks rather than liquid fuels. The hydrocarbons produced by the Fischer–Tropsch (FT) process proved to be excellent transportation fuels. Most of today's CTL technologies use variants of the FT reactor at the core of their process. In 1935, a commercial-scale direct liquefaction plant was built to process coal and creosote oil to produce a total of 150,000 t/year of gasoline at Billingham in the U.K. Other commercial CTL plants were built in Germany around 1936, and each plant produced about 1500 bbl/day. By the end of World War II, approximately 4,000,000 t/year gasoline was produced in Germany from 9 indirect and 18 direct liquefaction plants.

After World War II, the price of oil fell relative to the price of coal, and the economics of liquefaction became increasingly unattractive. Large oil reserves were being discovered in the Middle East, reducing the perceived need for strategic R&D

Hydrogen and Syngas Production and Purification Technologies, Edited by Ke Liu, Chunshan Song and Velu Subramani

programs and reducing interest in coal liquefaction in all countries, except South Africa. Due to its apartheid policies, South Africa became increasingly isolated politically from the mid-1950s to the mid-1980s. With large coal reserves, South Africa further developed coal liquefaction process because trading oil and oil products was forbidden. Indirect liquefaction process was selected for South African coal liquefaction. The first plant, SASOL 1, was built in the 1950s. In 1980s, two much larger plants were built, using the same basic process chemistry but employing improved catalyst formulations and reactor designs. In the mid-1980s, approximately 10,000,000 t/year of transportation fuels was produced in these plants, which covers 60% of South Africa's requirements. All three plants are currently still in operation. However, interest in the production of transportation fuels from coal remained low in the rest of the world for many years except South Africa, and China in recent years. This is mainly due to the high cost of CTL and the low price of oil in the 1980s and 1990s. However, due to high oil prices in recent years, there has been a renewed interest in CTL. Extensive feasibility studies on CTL plants have been conducted, but not many started construction due to the recent concern on massive CO_2 releases from such plants plus their relatively high cost.

In the 1960s, two direct coal liquefaction processes were under development in the U.S.: the Exxon Donor Solvent (EDS) process and the H-Coal process. The distinguishing feature of the EDS process was a separate solvent hydrogenation step to carefully control the hydrogen donor characteristics of the solvent. The most important feature of the H-Coal process was the emulated bed reactor in the process.

In the 1970s, after the dramatic world oil price fluctuation, considerable amounts of research in the area of CTL production technologies have been prompted. However, none of those processes was commercialized. The sudden collapse of world oil prices in the mid-1980s led to the abandonment of most of the pilot and process development scale CTL facilities built in the U.S.

Except for Germany's experience with technologies that directly transferred coal into fuel oil during World War II, there has been little experience with, or economic analysis of, commercial direct coal liquefaction (direct coal to liquid [DCTL]) until the new DCTL plants currently being built in China.

The U.S. Department of Energy (DOE) had an active coal liquefaction program in the 1990s that was halted in the late 1990s. However, work has been resumed recently on the most critical steps: gasification and catalyst development. In 2003, the DOE announced a 5000 bbl/day demonstration project of advanced FT fuel production. The project was expected to coproduce 35 MW of electricity in Gilberton, PA. The plant incorporates a Shell gasifier and FT liquefaction technology provided by Sasol.

Starting from the late 1990s, China has shown a strong interest in coal liquefaction and has been investing billions of dollars into developing production facilities. Shenhua Group, one of China's largest coal companies, is building a $3.3 billion DCTL project. The process originally was licensed from Hydrocarbon Technologies, Inc. (HTI), and later, Shenhua and China Coal Research Institute (CCRI) spent a lot of its own resources to further improve the processes and direct CTL catalysts. Shenhua has almost completed the construction of the Inner Mongolia coal

Figure 12.1.   Shenhua CTL plant (China). Reprinted from Chen et al.[1] See color insert.

liquefaction project's infrastructure. Operations of its first production line are expected to commence by 2009, and recently, they started the first production line with a capacity of 1 million tons of gasoline and diesel fuel per year. Shenhua is planning a second phase to the project with a total investment of $7.3 billion. It is expected to process 15 million tons of coal to produce 5 million tons of oil products with four more production lines becoming operational in the future. Figure 12.1 shows a picture of the Shenhua CTL projects.[1]

12.2   DIRECT COAL LIQUEFACTION (DCTL)

12.2.1   DCTL Process

The purpose of the DCTL process is to add hydrogen to the organic structure of the coal, breaking it down only as far as it is necessary to produce distillable liquids. Common features are the dissolution of a high proportion of coal in a solvent at elevated temperature and pressure, followed by the hydrocracking of the dissolved coal with H_2 and a catalyst. Direct liquefaction is an efficient route. Liquid yields in excess of 70% by weight of the dry, mineral matter-free coal feed have been demonstrated for some processes in favorable circumstances. Total thermal efficiencies for modern processes are generally in the range of 60%–70% if allowance is made for generating losses and other noncoal energy imports.[2]

The liquid products from direct liquefaction processes do require further upgrading before they can be used directly as transportation fuels. This upgrading utilizes standard petroleum refining industry technologies, allowing the products from a

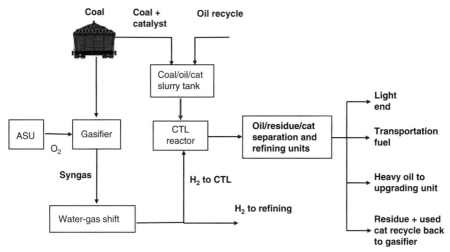

Figure 12.2.   Diagram of the direct coal liquefaction process. ASU, air separation unit.

liquefaction plant to be blended into the feedstock streams of a petroleum refinery.

In a direct coal liquefaction process, coal is reacted with hydrogen produced from coal gasification under high temperature (450 °C) and high pressure (20–30 MPa) conditions to produce crude-like liquids, which can be further hydrocracked to different liquid fuels or chemicals such as high-quality clean gasoline, diesel, jet fuel, and liquefied petroleum gas (LPG). Figure 12.2 shows the simplified process diagram of a direct coal liquefaction process.

The first direct coal liquefaction process was developed and patented by Bergius from Germany in 1913 and, therefore, is often referred to as the Bergius process. The world's first industrial-scale direct coal liquefaction plant was built in Leuna, Germany, in 1927 with an annual fuel production of 10,000 t. By 1939, Germany built 12 direct coal liquefaction plants with a combined annual fuel production of about 4.23 million metric tons, which supplied about 70% of the aviation fuel and 50% of the transportation fuel for German troops during World War II.

A modern direct coal liquefaction plant can produce about 0.5–0.6 t of liquid fuel from every ton of dry coal. In the cases where the hydrogen supply is also produced from coal, the overall system can produce about 1 t of liquid fuel from every 3–4 t of coal. China Shenhua Coal Liquefaction Co. Ltd is currently building a 20,000 bbl/day direct coal liquefaction plant in Inner Mongolia, China. The economic viability of this large-scale direct coal liquefaction remains to be revealed. Economic analysis performed in 1990[3] showed that the process of the two-stage conversion of CTL has a product cost of $38 per barrel. Several areas, such as improved pretreatment and cleaning methods, novel catalysts, and improved hydrocarbon recovery, were identified as ways to lower the cost further.[1]

In a DCTL plant, a coal–oil slurry containing up to about 50% coal is heated to moderately high temperatures in a high-pressured hydrogen atmosphere. A variety of suitable catalysts are used to promote the conversion of CTL products, with most common compounds of iron, nickel, molybdenum, or mixtures thereof. The gasoline-like and diesel-like products are recovered from the partially refined synthetic fuel by distillation. Hydrogen is added to the mixture to increase the hydrogen/carbon ratio and reduce the oxygen, sulfur, and nitrogen in the coal. The addition of hydrogen increases the energy need and thus increases the cost of the process.

DCTL processes aim to add hydrogen to the organic structure of the coal, breaking it down only as far as necessary to produce distillable liquids. Common features are the dissolution of a high proportion of coal in a solvent at elevated temperature and pressure, followed by the hydrocracking of the dissolved coal with H_2 and a catalyst. For example, coal can be treated with hydrogen under pressure (>30 MPa) at 450 °C in the presence of a solvent and an iron oxide catalyst. The liquid products have molecular structures similar to those found in aromatic compounds and need further upgrading to produce specification fuels such as gasoline and fuel oil.

Some processes are designed specifically to coprocess coal with petroleum-derived oils and these may fall into either group. Also, coal liquefaction processes from both groups have been adapted for coprocessing. In the mid- to late 1960s, as interest was growing, all of the available processes were single stage. Most development therefore continued to adopt a single-stage approach. Some developers added a second stage during the 1970s, following the oil crisis, to increase the production of light oils. However, most of these have been superseded and abandoned. Several other less important processes were developed to a modest scale in the U.S. Most of the recent research and development on second-generation liquefaction processes have concentrated on operating under less severe conditions, especially at reduced pressures. This is achieved by enhancing hydrogen transfer rates from the solvent to coal and by recycling liquids to increase conversion. These engineering improvements ultimately translate to lower costs.

12.2.2 The Kohleoel Process

The Kohleoel process with integrated gross oil refining (IGOR+) is a recent development by Ruhrkohle AG and VEBA OEL AG based on the IG (Interessen-Gemeinshaft) process established by a German fuel company I.G. Farbenindustrie in 1927. Initially, a 0.2 t/day continuous unit was evaluated at Bergbau-Forschung (now DMT). In 1981, a 200 t/day industrial pilot plant was built at Bottrop, which remained operational until 1987. During approximately 22,000 h operating lifetime of this plant, it produced an overall of 85,000 t of distillates from 170,000 t of coal. The CCRI signed a 2-year agreement with Ruhrkohle AG in 1997 to conduct a feasibility study for a 5000 t/day demonstration coal liquefaction plant in Yunnan Province.

The Kohleoel process is illustrated in Figure 12.3. Coal and iron oxide catalysts are slurried with process-derived recycle solvents. The specific coal feed rate is in the range of 0.5–0.65 t/m^3/h. H_2 feedstock together with additional H_2 recycled from the process, after mixing with the slurry, is sent to an up-flow tubular reactor. The reactor is typically maintained at 30 MPa and 470 °C. Hot effluents from this reactor

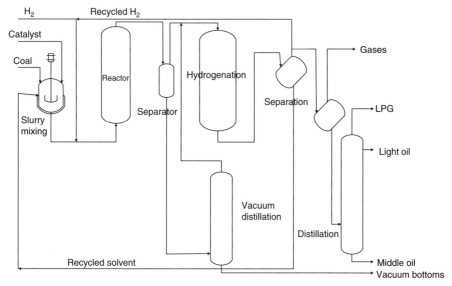

Figure 12.3. Diagram of the Kohleoel process.

are subsequently directed to a high-temperature separator. Vapor products from the separator are introduced to a hydrogenation reactor operating at 30 MPa and 350–420 °C. The liquid portion from the separator passes through a vacuum distillation column to recover distillable liquids as additional feed to the hydrotreating reactor. The rest is collected as vacuum bottom containing pitch, mineral matter, unreacted coal, and catalyst, with possible use as gasifier feedstock for H_2 production. The hydrogenated products are routed to two separators in series, where the products are depressurized and cooled. The liquid from the first separator is recycled as the solvent for slurry mixing. The liquid product from the second separator is sent to an atmospheric distillation column, yielding different oil products: light oil (C_5–200 °C boiling point [BP]) and medium oil (200–325 °C BP). This process can reach a conversion of 90% using bituminous coals, with liquid yields in the range of 50%–60% calculated based on dry ash-free coal. Process yields and quality, when using Prosper, a German bituminous coal, are summarized in Table 12.1.

12.2.3 NEDOL (NEDO Liquefaction) Process

As a response to the first global oil crisis in 1974, Japan launched the Sunshine Project to develop a coal liquefaction technology and alleviate the dependence on imported oil. This project was managed by the New Energy and Industrial Technology Development Organization (NEDO) and received supports from government agencies and many Japanese companies and research institutes for technology advancement. Subbituminous and low-rank bituminous coals were targeted for their process development. By 1983, coal liquefaction had been tested at scales of

Table 12.1. Kohleoel Process and Product Quality with Prosper Coal

Process Yields		Yield
Hydrocarbon gases (C_1–C_4)		19.0
Light oil (C5–200 °C)		25.3
Medium oil (200–325 °C)		32.6
Untreated coal and pitch		22.1
Product Quality	Light Oil	Medium Oil
Hydrogen (%)	13.6	11.9
Nitrogen (ppm)	39.0	174
Oxygen (ppm)	153	84.0
Sulfur (ppm)	12.0	<5
Density (kg/m^3)	772	912

0.1~2.4 t/day. Subsequently, a consortium of 20 companies was founded under the name Nippon Coal Oil Company, Ltd., aiming to design, build, and operate a 250 t/ day pilot plant. Although the project was terminated in 1987 due to budgetary constraints, a 1 t/day process support unit (PSU) was built and put into operation by two joint coal liquefaction companies in Japan in 1988. A pilot plant was designed in 1988 for an operation scale of 150 t/day. Construction started in October 1991 at Kashima and was completed in early 1996. Five operation runs were conducted from March 1997 to September 1998. Three types of coal of different ranks (Tanito Harum and Adaro coals from Indonesia and Keshena coal from Japan) were liquefied without serious problems. This pilot achieved 80 days of continuous coal-charging operation, 58% by weight (dry ash-free coal basis) of oil yield, successful use of slurry at 50% concentration, and 6200 h of cumulative operation. CCRI signed an agreement with NEDO and the Center to carry out a feasibility study for a 5000 t/ day demonstration plant. Several tests have been completed in both bench and process demonstration unit (PDU) scales with Yilan coal. A 62% oil product yield was predicted. NEDO has recently announced coal liquefaction technical support to China to promote coal utilization in Asia and reduce global oil supply tension. Some initial efforts include partnering with Chinese energy companies Datang International Power Generation Co. and Xinwen Mining Group to conduct feasibility tests on the efficiency of their liquefaction processes. NEDO also participates in two liquefaction trials in China, one with Shenhua. Their technology will be licensed for the construction of liquefaction plants in China.

As can be seen in the NEDOL process in Figure 12.4, pulverized coal and 2%–4% by weight iron-based catalysts are first slurried together with recycled solvents. The slurry is mixed with H_2 and preheated before it is delivered to the liquefaction reactor. The tubular up-flow reactor typically operates at temperatures of

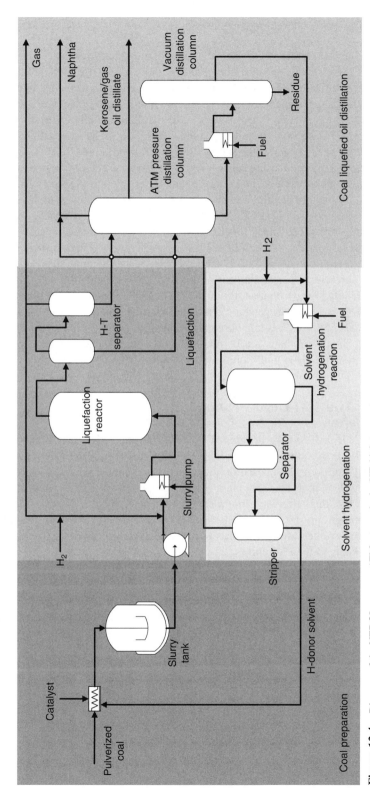

Figure 12.4. Diagram of the NEDOL process. ATM, atmospheric; HT, high temperature.

430–465 °C and pressures of 15–20 MPa. The nominal slurry residence time is 1 h, with actual liquid-phase residence times in the range of 90–150 min. Products are cooled, depressurized, and distilled in an atmospheric column to obtain a light distillate product. The atmospheric column bottoms, after a heating step, are sent to a vacuum distillation column. Most of the middle distillate and all of the heavy distillate are recycled as solvent via hydrogenation. These solvent hydrogenation reactors are down-flow packed catalyst beds operating at 320–400 °C and 10–15 MPa. Residence time is ~1 h. Products are subsequently depressurized at temperature and sent to a flash distillation vessel. The liquid product is recycled to the slurrying step as the solvent. Some light naphtha can also be collected. The vacuum column bottoms, which contain unreacted coal, mineral matter, and catalyst, are discharged and can be used as gasifier feedstock for H_2 production in commercial operation. The maximum solid loading that can be achieved is ~50%. However, in practice, it is understood that a loading of 35% is more typical. Since the pitch discharged with the solids represents a substantial loss of potential product, the process is limited to coal of relatively low ash content. Product yields vary depending on the type of coal, although the primary hydrogenation reaction operating parameters can be adjusted to obtain 50%–55% (dry ash-free basis) distillate product yields. It should be pointed out that the liquid products are of relatively low quality and require substantial upgrading in comparison with liquid fuels produced from other processes.

12.2.4 The HTI-Coal Process

The HTI-Coal process was developed by HTI (originally Hydrocarbon Research Institute [HRI]) from the commercialized H-Oil process used for heavy oil upgrading. A 200 t/day pilot plant based on this process was scaled up and constructed at Catlettsburg, KY, in 1980. This plant remained operational for 3 years, and plant design was modified afterward into a commercial-scale plant at Breckinridge, KY. The Shenhua Group in China collaborated with HTI in 1997, targeting to build a DCTL plant in Inner Mongolia. Feasibility studies were performed for HTI-Coal plants in India and the Philippines.

Figure 12.5 illustrates the HTI-Coal process. Pulverized coal is dissolved in recycled coal-derived heavy process solvent, mixed with hydrogen (~17 MPa) and then delivered to the first stage hydrocracking reactor operated at 435–460 °C. Because of high exothermicity of the hydrocracking reactions, temperature control is crucial for engineering scale-up. A new type of reactor is employed instead of fixed-bed reactors for its advantages in temperature control. This type of reactor offers additional benefits in convenient catalyst replacement during operation to maintain constant performance, which is particularly important with supported catalysts that deactivate relatively rapidly under coal liquefaction conditions. Effluents from this stage are sent to a liquefaction reactor, where smaller molecules derived from coal hydrocracking are hydrogenated in the presence of catalysts. The liquid fuel from this stage requires further upgrading to produce gasoline and jet and diesel fuels that meet or exceed existing fuel specifications.[4] The products from the

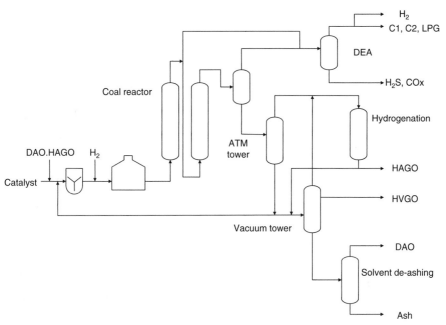

Figure 12.5. Diagram of the H-Coal process. DAO, de-ashed oil; DEA, diethanolamine; HAGO, heavy atmospheric gas oil; HVGO, heavy vacuum gas oil.

liquefaction reactor are injected to a flash separator. Liquids in the overheads are condensed and routed to an atmospheric distillation column, producing naphtha and middle distillate. The flash bottoms are fed to a bank of hydrocyclones. The overheads stream, which contains 1%–2% solids, is recycled to the slurrying stage. The bottom is routed to a vacuum distillation column, where distillable products are collected and solid portion is removed. The overall process yield is dependent on the type of coal. Typical conversion is 95%, with liquid yields up to 50% (dry basis). Table 12.2 summarizes the comparison of different direct coal liquefaction technologies.

12.2.5 Other Single-Stage Processes

The solvent-refined coal (SRC) processes were originally developed to produce cleaner boiler fuels from coal. A 0.5 t/day plant was built in 1965 and scaled up in 1974 into two separate pilot plants located at Wilsonville (SRC-I, 6 t/day) and Fort Lewis, WA (SRC-I, 50 t/day). The Fort Lewis plant was later converted to an SRC-II unit. Due to the more severe conditions required for SRC-II, the capacity was downgraded to about 25 t/day. The objective of the SRC-II process was to produce distillate products. Detailed designs for large-scale plant were subsequently prepared, although these plants were not built. The Wilsonville plant continued to be funded

Table 12.2. Comparisons of Different Direct Coal Liquefaction Technologies

Process	HTI	IGOR	NEDOL
Reactor type	Tubular reactor	Bubble column	Bubble column
Temperature (°C)	435–460	470	430–465
Pressure (MPa)	17	30	15–20
Feed rate (t/m^3/h)	0.24	0.60	0.36
Coal type	Shenhua	Xiangfeng	Shenhua
Conversion rate % (daf coal)	95	90	89.7
Water % (daf coal)	13.8	28.6	7.3
C$_4$+ oil % (daf coal)	67.2	58.6	52.8
Residue % (daf coal)	13.4	11.7	28.1
Hydrogen consumption % (daf coal)	8.7	11.2	6.1

by DOE until 1992 as a pilot-scale test facility for the whole U.S. direct liquefaction development program. The SRC processes have now been abandoned, but elements have been incorporated in more recent U.S. processes.

The development of the Imhausen high-pressure process commenced in 1982, and a 100 kg/day PDU was commissioned in 1984. However, the severe operating conditions (470–505 °C and 60–100 MPa) limit their commercialization potential.

In the late 1970s and early 1980s, Conoco developed a process that used molten zinc chloride to hydrocrack coal directly to give good yields of gasoline in a single step. This process is one of the very few direct liquefaction processes that is not a direct derivative of prewar technology. The process was taken to the 1 t/day pilot plant scale, although this was operated for only a short period with limited success. Major metallurgical difficulties were experienced as a result of the highly corrosive nature of zinc chloride and other chloride salts formed in the system. However, it is believed that if these problems can be resolved, this is one area in which further development might have the potential to make a significant improvement in liquefaction economics. Most two-stage direct liquefaction processes were developed in response to the oil price rises of the early 1970s, often as a development of earlier single-stage processes. Work was carried out in many different countries, but relatively few processes were developed beyond the laboratory scale, and most of these processes were generically very similar.

12.3 INDIRECT COAL TO LIQUID (ICTL)

12.3.1 Introduction

Indirect liquefaction processes were developed in Germany around the same time as direct liquefaction processes. In the early 1920s, Franz Fischer and Hans Tropsch patented a process to produce a mixture of alcohols, aldehydes, fatty acids, and hydrocarbons known as synthol, from a synthesis gas of hydrogen and carbon monoxide. The FT process forms the basis for indirect coal liquefaction. Instead of hydrogenating coal directly, the ICTL process primarily consists of two steps. The first gasification step converts the coal into a "synthesis gas" (or syngas) enriched with hydrogen (H_2) and carbon monoxide (CO). A variety of gasifier designs are in commercial operation worldwide using coal or other dirty, low-value feedstocks. Following gasification, the raw syngas is cooled and cleaned to remove all contaminants. Clean syngas is sent to an appropriate FT catalyst bed to produce predominantly paraffinic liquid hydrocarbons with a wide range of molecular weights.

This method was used to produce motor fuel during World War II. After the war, the FT synthesis technology was used by HRI (predecessor of HTI) to construct a 7000 bbl/day gas-to-liquids (GTL) plant in Brownsville, TX, in 1949. The plant was operated by Cathage Hydrocol from 1950 to 1953 before shutting down due to declining oil prices. The partial oxidation unit, which was used to convert natural gas into synthesis gas to feed the fixed-bed FT reactors at this plant, was the basis for what eventually became the Texaco coal gasification process (currently owned by GE Energy).

In the 1950s, South Africa faced worldwide oil sanctions against its apartheid policy. Lack of petroleum resources in the country greatly motivated indirect coal liquefaction technology development and commercialization efforts to sustain the country's needs for liquid fuels. South Africa Synthetic Oil Limited (SASOL) was founded during this time. Their process is based on Lurgi gasification and FT liquefaction. Substantial improvements in the synthesis step have been achieved over 50 years' continuous development. SASOL is the world's largest indirect coal liquefaction company, with highly profitable business. SASOL's first indirect coal liquefaction plant, SASOL-I, was constructed in 1955. Two more plants (SASOL-II and SASOL-III) were commissioned in 1980 and 1982, after the world oil crisis in the 1970s. Figure 12.6 shows the gasification and syngas cleanup technologies used in an ICTL plant. These technologies are similar to those used in an Integrated Gasification Combined Cycle (IGCC) plant or a coal-to-hydrogen plant, which have been extensively covered in this book.

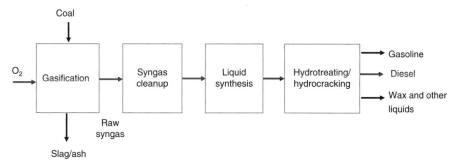

Figure 12.6. Diagram of the indirect coal to liquid (ICTL).

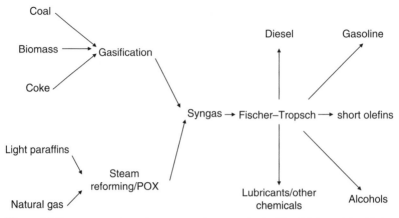

Figure 12.7.   Diagram of the Fischer–Tropsch application in CTL. POX, partial oxidation.

The indirect route yields a large number of by-products, leading to a lower overall thermal efficiency than that of the DCTL. Similar to DCTL, ICTL processing has been restrained commercially by its high capital costs as well as high operational and maintenance costs. Fuels produced from ICTL process include methanol, dimethyl ether, FT diesel- or gasoline-like fuels, and hydrogen (H_2). The availability of CO and H_2 as molecular building blocks at an ICTL facility also provides opportunities for the production of other chemicals. This can be illustrated in Figure 12.7.

12.3.2 FT Synthesis

FT synthesis is one of the most important syngas-based liquid synthesis processes. As presented in Figures 12.6 and 12.7, raw syngas produced from gasification is sent to gas cleanup units to remove contaminants such as H_2S, COS, and HCl, which potentially act as catalyst poisons. Clean syngas reacts over FT catalysts to generate liquid fuels. Those liquid fuels are further upgraded through hydrotreating or hydrocracking reactions. Therefore, the final product fuels from FT synthesis usually exhibit exceptional quality in terms of hydrogen content, product molecule uniformity, freeze point, combustion characteristics, and sulfur contents. These products are of high market values. Hydrocarbons consisting of a chain length of 5–10, after a downstream hydrogenation or isomerization treatment to increase the octane value, are mainly used as gasoline fuel. Shorter products including butane and propane can also be oligomerized into gasoline. However, complex treatments put additional cost and lower profitability. FT diesel featured by high linearity and low aromatic compounds has superior quality compared with diesels derived from other processes. It has very high value of cetane number (as high as 75) and is ultraclean.[5]

Olefins from FT synthesis are important building blocks for petrochemical synthesis to manufacture high-molecular weight chemicals, polymer materials, or alcohols. The linear paraffin products are either recycled to generate more syngas or used as gas fuel, for example, synthetic natural gas (SNG), which can serve as town gas, or be injected directly into a gas turbine to generate electricity.

Wax product is in the heavy end of the FT synthesis product spectrum. Although most wax will be sent to a cracking process to produce shorter-chain products, the FT wax itself has a high quality. It is hard and very clean; more importantly, its fusion temperature is very high (up to 170 °C, compared with less than 100 °C for wax produced from crude oil), and its good feedstocks for lube oil and other high-value products.

12.3.2.1　FT Synthesis Chemistry and Thermodynamics

The main reactions during FT synthesis include paraffin and olefin formation reactions and the water-gas shift (WGS) reaction, as shown correspondingly in Equations 12.1–12.3:

$$(\text{Paraffin formation})(2n+1)H_2 + nCO \rightarrow C_nH_{2n+2} + nH_2O, \quad (12.1)$$

$$(\text{Olefin formation})2nH_2 + nCO \rightarrow C_nH_{2n} + nH_2O, \quad (12.2)$$

$$(\text{WGS})H_2O + CO \leftrightarrow CO_2 + H_2. \quad (12.3)$$

FT synthesis products are primarily a mixture of linear and branched hydrocarbons. Linear paraffins and α-olefins are main products with carbon chain lengths ranging from one carbon atom up to 50. Occurrence of WGS reaction is dependent on operation conditions and catalyst formulations as elaborated in subsequent sections.

Major side reactions that occur during FT synthesis are listed below:

$$(\text{Alcohol formation})2nH_2 + nCO \rightarrow C_nH_{2n+2}O + (n-1)H_2O, \quad (12.4)$$

$$(\text{Catalyst reduction})M_xO_y + yH_2 \rightarrow yH_2O + xM, \quad (12.5)$$

$$(\text{Catalyst oxidation})M_xO_y + yCO \rightarrow yCO_2 + xM, \quad (12.6)$$

$$(\text{Bulk carbide formation})yC + xM \rightarrow M_xC_y, \quad (12.7)$$

$$(\text{Boudouard reaction})2CO \rightarrow CO_2 + C. \quad (12.8)$$

A small amount of alcohols can be produced during FT synthesis. Catalysts may interact with reactants and get reduced or oxidized. Catalysts could be further damaged by carbide formation or coke deposition on the surface from the Boudouard reaction.

According to the above reactions, the ideal H_2/CO ratio in the syngas for the FT process is about 2:1, which is higher than typical H_2/CO ratios produced from coal gasification. Therefore, WGS reactors are used as part of the syngas conditioning unit to adjust the H_2/CO ratio in syngas prior to the FT synthesis process for maximum yields. In the case of fluidized-bed FT reactors, a high H_2/CO ratio also minimizes the formations of waxy hydrocarbons, thus maintaining the fluidization characteristics in the reactor.

An alternative process is to integrate the WGS functionalities into the FT catalyst itself. The WGS reaction can be carried out *in situ* in the FT reactor, with water produced in the chain growth reaction as WGS reactants as shown below:

$$CO + 2H_2 = [-CH_2-] + H_2O, \quad\quad\quad (12.9)$$

Equation 12.3: $CO + H_2O = H_2 + CO_2$.

The overall reaction can be rewritten as:

$$2CO + H_2 = [-CH_2-] + CO_2. \quad\quad\quad (12.10)$$

Therefore, an FT catalyst with WGS function can, in theory, achieve the maximum syngas conversion with an H_2/CO ratio of only 0.5.

Thermodynamic analysis is an important step in FT synthesis process design. The free energy changes for paraffin and olefin formations are largely negative, which suggests favorable reactions. As the hydrocarbon chain length becomes longer, the numbers are more negative, implying the tendency to form heavier hydrocarbons thermodynamically. In contrast, the formation of lower oxygenates are thermodynamically unfavorable. However, longer-chain alcohol formation becomes exothermic.[6]

The main reactions producing paraffins and olefins during FT synthesis are highly exothermic. Although higher temperatures lead to faster reactions and higher conversion rates, from thermodynamic considerations, high-temperature operation should be avoided. Generally, FT reactors are operated at temperatures about 200~400 °C. Additionally, methane is preferentially produced at higher temperatures, which is undesirable for FT liquid fuel synthesis. As can be seen from FT reactions, high pressures favor high conversion rates and formation of long-chained hydrocarbons. However, high pressure raises additional equipment and operational cost. Typical operation pressures are 2~4 MPa.

12.3.2.2 FT Catalysts

The most common FT catalysts are Group VIII metals (Co, Ru, Fe, and Ni).[7] Those metals are dispersed in porous medium such as silica and alumina to achieve high surface area. Nickel catalysts are not widely used because (a) the selectivity shifts rapidly to the methanation reaction when temperature increases and (b) nickel carbonyl is likely to form under high pressure, which causes deactivation. With higher temperature, the tendency of selectivity shifting to methane happens to cobalt and ruthenium too, but with much lower sensitivity. Ruthenium catalysts are very active but expensive. At relatively low pressure (less than 10 MPa), it produces mainly methane. However, at low temperatures and high pressures, the selectivity is toward very heavy products ("polymethylene synthesis"). The molecular weight of the products can reach 1 million at low temperature (100 °C) and high pressure (100 MPa). Due to the price constraints and limited supply of ruthenium, only cobalt and iron catalysts are commercially used.

Cobalt catalysts are reported about three times more active compared with Fe-based catalysts at typical operation conditions of 473K and 2.0 MPa.[8] They demonstrate long lifetime and produce predominantly linear paraffins.[9] This type of catalyst is primarily employed in low-temperature FT (LTFT) processes for the production of middle distillates and high-molecular-weight fuels, achieving high selectivity. At higher temperatures, product quality becomes unacceptable. Another advantage of cobalt catalysts is resistance to the presence of water produced during the reaction. Although cobalt can be deactivated by water, it is usually very stable under low single-pass conversion, which implies low water pressure. However, different from Fe-based catalysts, Co-based catalysts do not promote the WGS reaction, which puts a higher requirement for H_2/CO ratio (about 2.0–2.3[10]) in the feed and consequently increases syngas-conditioning cost. Cobalt catalysts are adopted by the emerging liquid fuels from natural gas process because of favorable syngas compositions generated by this process.

Cobalt catalysts are much more expensive than iron. Therefore, catalyst design is aimed at minimizing the amount used and maximizing the available cobalt surface area for synthesis reactions.[7] There are several directions available in the literature to achieve this.[11] High area and stable supports, such as Al_2O_3, SiO_2, or TiO_2, are commonly used.[12] Other support materials have also been examined, such as carbon nanofiber,[13] ZrO_2,[14] and MgO.[15] Supports can not only influence the metal dispersion but also play an important role in cobalt particle size control, which determines product selectivity as suggested from Borg and colleagues' work.[16] Catalysts can also be promoted with a small amount of noble metal, for example, Pt,[17] Ru,[18] and Re,[12,19] which is claimed to enhance the reduction process and also keep the Co metal surface "clean" during FT. It was found that even 0.3 wt% MnO loading onto Co catalysts can considerably improve the selectivity toward C_{5+} products at 1 bar.[13] Cobalt precursor and pretreatment conditions were investigated for their effects on catalyst reducibility, active site density, and ultimate catalytic performance.[20] In addition to conventional impregnation of cobalt onto porous support, sol-gel method is employed for the preparation of cobalt catalysts for its advantage of creating optimum interaction at the metal/support interface.[19]

Although less active and stable than cobalt catalysts and easily inhibited by the presence of water, the iron catalyst is still a popular option for FTS. Iron catalysts are much cheaper and exhibit strong WGS activity. Therefore, this type of catalyst is favorable for syngas with a relatively low H_2/CO ratio (0.5–1.3) derived from coal or biomass gasification.[8,10] The selectivity to heavy products is not as good as cobalt and ruthenium. However, it is less sensitive to temperature change. Methane selectivity remains low even at temperatures higher than 300 °C. With iron catalysts, two directions have been pursued.[21] One is aimed at the production of light olefins using high-temperature FT (HTFT) process as described in the following section. Fluidized-bed reactors are employed in which no liquid products exist. The other direction is to produce liquid fuels at lower temperatures. Extensive research efforts have been devoted to increase iron-based catalyst active component dispersion and density in order to achieve similar performance range as cobalt catalysts.

It has been shown that surface basicity of iron-based catalysts greatly affects probability of chain growth.[5,7] Alkali metal incorporation promotes the formation of higher carbon number molecules. As described in Dry's review,[7] iron-based catalysts are prepared by precipitation techniques and promoted with Cu, K_2O, and SiO_2 (typical composition: 5 g of K_2O, 5 g of Cu, and 25 g of SiO_2 per 100 g of Fe). This type of catalyst is well suited for heavy liquid fuels produced from a low-temperature operation. For high-temperature application, the iron catalyst is prepared by fusing magnetite together with K_2O and structural promoters such as Al_2O_3 or MgO. For iron catalysts promoted with K and Si together, the active iron oxide and carbide phases can be stabilized during the reaction and sustain the long-term performance.[10] Iglesia et al. synthesized high surface area precursors based on precipitated Fe–Zn oxides and promoted the oxides with Cu, Ru, and K. This catalyst demonstrated high FTS reaction rates and low CH_4 selectivities. When operated at typical conditions used for Co-based catalysts (473K, 2.0 MPa), this catalyst showed similar hydrocarbon synthesis rates per catalyst mass or volume.[8] K, Ru, and Cu promoters were found to increase reduction/carburization rates of Fe–Zn oxide precursors and to enhance active site densities and long-term steady-state performance. The key features of different FT catalysts are shown in Table 12.3. Table 12.4 summarizes the requirements of syngas for FTS catalysts.

Table 12.3. Properties for Major FT Catalysts

Catalyst	Product Chain Length	Olefin Selectivity	Price Ratio	FT Activity	WGS Activity
Fe	Shorter, depends weakly on temperature	Higher, increase with increasing temperature	1	Lowest	High
Co	Longer, depends strongly on temperature	Lower	230	Higher	No
Ru	Depends strongly on pressure	No	31,000	Very high	No

Table 12.4. Requirements of Syngas for FTS Catalysts

Catalyst	Temperature	Sulfur Content	Pressure	H_2/CO	Steam Contents
Co	200–240 °C	<0.1 ppm	2–4 MPa	Around 2, requiring additional WGS	As less as possible
Fe	200–240 °C for low-temperature FT 300–350 °C for high-temperature FT			>0.6, preferably, >1.1	Depends on H_2/CO

12.3.2.3 FT Reaction Mechanism

FTS mechanism study is of considerable academia interest. Although the mechanism investigation started as early as when the FT synthesis reaction was discovered in 1926, controversy still remains as to the mechanism of the chain growth. Initially, Fischer and Tropsch reported the carbide mechanism. Carbides were first formed from the synthesis gas and subsequently hydrogenated to methylene groups. Polymerization of the methylene groups generates hydrocarbon chains that eventually desorb from the surface as saturated and unsaturated hydrocarbons products.[22] The carbide mechanism was further developed by Craxford and Rideal in 1939.[23] During FT reaction, CO is dissociatively adsorbed on the surface in the presence of hydrogen, yielding water or CO_2 together with chemisorbed carbon on catalyst surface. Surface carbon quickly hydrogenated to form CH_2, the growth of which led to a wide range of product distribution. It is suggested that carbon chain polymerization was terminated by hydrogenolysis.

Emmett and coworkers conducted extensive ^{14}C experiments and proposed the widely accepted oxygenate reaction pathway in 1950.[24,25] This mechanism suggested an intermediate species involving adsorbed CO and H_2 with catalyst surface. Growth of the carbon chain was accompanied by condensation and water elimination.

With significant advancement of surface science in recent 40 years, the carbide mechanism regained its popularity. This was supported by experimental observation that essentially no oxygen is present on the catalyst surface during FT reaction. However, carbon species is dominant on the surface. Therefore, the new view is that carbon chain grows from metal carbide species.[26] As reported in Joyner's work,[27] during chain growth, two adsorbed methylenes polymerized to produce an adsorbed ethylidene species. Addition of a third methylene species yields a diadsorbed C_3H_6 entity. Shift of hydrogen may terminate the chain growth and give an olefin. The extent of chain branching is suggested to be determined by the relative stability of secondary and primary carbonium ions.

Davis[26] pointed out in his review on FT synthesis mechanism that the reaction pathway is also dependent on catalyst compositions and operation conditions. Their experimental data on FT synthesis with iron catalysts at low temperatures suggested that instead of carbide mechanism, an oxygenate intermediate, similar as the formate species responsible of the WGS reaction, existed on the surface and initiated carbon chain propagation. The final chain termination step was accompanied by elimination of the oxygen atom.

12.3.2.4 FT Product Selectivity

Concurrent with FTS mechanism studies, product distribution models were developed based on the analysis of product composition. Friedel and Anderson[28,29] in the 1950s published the Anderson–Schulz–Flory (ASF) distribution model to predict the wide range of products yielded from FTS. The equation is shown as follows:

$$W_n = n(1+\alpha)^2 \, \alpha^{-1}.$$

(12.11)

Here, n is the number of carbon atoms in products. W_n represents the weight fraction of products containing n carbon atoms, and α is the chain growth probability. The parameter α is assumed chain-length independent.

The ASF distribution provides a description on the distributions of hydrocarbon in the product mixture. If one plots the logarithm of molar fraction with respect to the carbon number, a straight line will be obtained with α being the slope. A higher α value implies heavier products, while a lower value means lighter compounds. Figure 12.8 shows the product selectivity depending on α. It can be seen that maximum diesel selectivity can be obtained at $\alpha = 0.9$, whereas gasoline selectivity is the highest when α is at 0.76.

Van der Laan and Beenackers[9] summarized the characteristics of the product distribution for FTS:

- The selectivity is the highest for methane and then decreases as the chain length increases, although there is a local maximum at C_3–C_4.

- Most hydrocarbons are linear. Monomethyl-substituted hydrocarbons are present, while dimethyl products are significantly less. None of the branched products contain four carbon atoms on the branch.

- Selectivity of olefins from iron catalyst exceeds 50%, most of which are α-olefins. The selectivity of propene is the highest. For longer carbon chains, the selectivity decreases asymptotically to zero as chain length increases.

- The change in chain growth probability with chain length is observed only for paraffins, but not for olefins and alcohols.

The process conditions as well as the catalyst can influence the selectivity. Table 12.5 shows the effects of main reaction parameters on the product selectivity. It is shown that for the production of heavy fuel, the reaction should occur at lower temperature, higher pressure, and lower H_2/CO ratio. In practice, the reaction rate should also be considered, which probably requires contradictory conditions, for example, higher temperature and higher H_2/CO ratio.

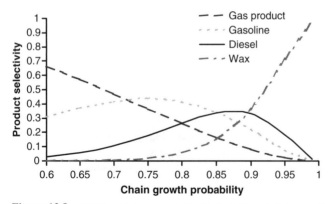

Figure 12.8. Relationship between product selectivity and chain growth probability.

Table 12.5.   Effect of Main FT Reaction Parameters on the Product Selectivity

Parameter	Chain Length	Chain Branching	Olefin Selectivity	Methane Selectivity
Temperature	⇓	⇑	*	⇑
Pressure	⇑	⇓	*	⇓
H_2/CO	⇓	⇑	⇓	⇑

⇑, increase with increasing parameter; ⇓, decrease with increasing parameter; *, complex effect.

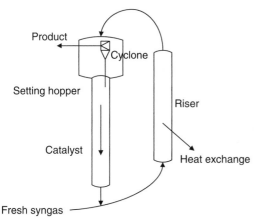

Figure 12.9.   Schematic diagram of a typical CFB reactor.

12.3.2.5 *Typical FT Processes*

As discussed in prior sections, two types of products are produced from FT that are of high market value. At the light end of the product spectrum are the gaseous olefins that can be further converted to various chemicals. At the high end is the high-quality diesel product. Accordingly, there are two types of processes aiming on these two products: HTFT and LTFT.

HTFT typically operates at temperatures higher than 300 °C and produces short α-olefins and gasoline. Cobalt catalysts are very sensitive to temperatures and produce primarily undesired methane under this operation condition. Therefore, for an HTFT process, iron catalysts are more favorable than cobalt catalysts. At high temperatures, iron catalysts preferentially catalyze olefin formation, with a relatively low methane selectivity. The HTFT technology has been applied by Sasol since the 1950s, starting with a circulating fluidized-bed (CFB) reactor in the synthol process, with a capacity of 7500 t/day. Figure 12.9 shows a schematic diagram of a typical CFB reactor. The synthol CFB reactors operate at 2.5 MPa and 340 °C. The syngas is converted in a fast fluidized bed, and then cyclones are used to separate the entrained catalyst particles. These particles are then settled down in a hopper and are recycled into the fluidized-bed reactor by high-velocity fresh syngas. In 1993, a

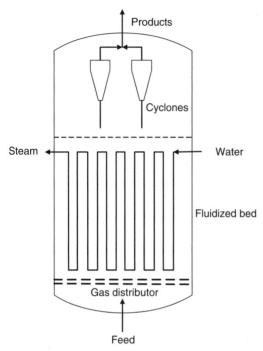

Figure 12.10.   Schematic diagram of a typical SAS reactor.

new fixed fluidized-bed reactor named Sasol Advanced Synthol (SAS) was developed. This reactor was designed to work at 2–4 MPa and 340 °C, with capacities up to 20,000 bbl/day. Figure 12.10 shows a schematic diagram for a typical SAS reactor. The SAS reactor uses a conventional fluidized-bed reactor instead of a circulating bed, with cyclones located on top of the column. All catalysts remain in the reaction region, resulting in a higher catalyst/syngas ratio. In addition, since there is no need to carry catalyst by syngas, a lower velocity and therefore a bigger column diameter can be applied. Typically, SAS reactor has a much larger capacity than a CFB reactor.

LTFT operates at lower temperatures (200–240 °C) and produces mainly heavier liquid fuels such as diesel. Both iron and cobalt catalysts can be used. Two types of reactors are commercially commonplace: fixed-bed reactor and slurry bubble columns (SBCs).

Figure 12.11 shows a typical multitubular fixed-bed reactor. Multiple tubes are packed with large catalyst particles and heat removal is realized by the generation of steam inside the shell. In a fixed-bed reactor, intraparticle diffusion is important because the particle size is relatively large due to pressure drop constraints. The relatively poor heat conductivity and heat transfer to the tube wall give rise to temperature profiles, particularly when it is operated at a high rate once-through conversion, as in classical fixed-bed processes. A better way to achieve higher overall

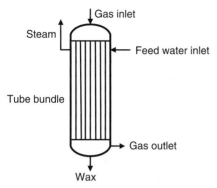

Figure 12.11. Typical multitubular fixed-bed reactor.

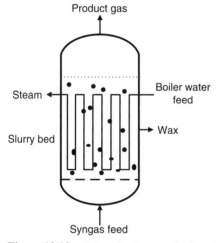

Figure 12.12. Schematic diagram of a slurry bubble column.

conversion is a low per-pass conversion operation with recycle. An improvement of heat transfer can also be obtained by operating in the trickle-flow regime. Typical commercial fixed-bed reactors in operation today are the Sasol Arge tubular fixed-bed reactor (TFBR) (from 1955, 400 bbs/day) and the Shell Middle Distillate Synthesis (SMDS) reactor (from 1993, 3000 bbl/day).

An SBC is a vertical, tubular column in which a three-phase (gas–solid–liquid) mixture is used. The slurry phase consists of FT catalysts and FT wax. The syngas flows though the slurry phase in the form of bubbles, as shown in Figure 12.12. The effective heat and mass transfer, low intraparticle diffusion, low pressure drop, and design simplicity are important advantages of this type of reactor. However, considerable problems arise in separating the liquid-phase synthesis products from the catalyst. With their attractive features, the SBC reactors are receiving extensive investment in both R&D and commercialization. The concept of SBC is not new.

In the 1950s to 1970s, Rheinperussen AG and Koppers GmbH had invested on its development. In the early 1990s, Sasol commercialized its Sasol Slurry Bed Reactor (SSBR) with a capacity of 2500 bbl/day. New research programs have focused on larger-scale SSBR up to 20,000 bbl/day. Other companies are also aggressively investing in this process including Exxon demonstration with 200 bbl/day in 1990 and Conoco demonstration with 400 bbl/day in 2000.

The developments by Sasol have resulted in several changes to the processes. Sasol employs both LTFT and HTFT processes. LTFT is used exclusively at Sasolburg and comprises the older fixed-bed technology as well as the newer-generation slurry-phase FT process. The LTFT process operates at 200–250 °C and 2–3 MPa, and produces paraffins and waxes using an iron-based catalyst. At Secunda, HTFT is the dominant process utilizing the older CFB technology as well as the new-generation SAS technology. A typical Sasol HTFT process is presented in Figure 12.13. The primary reactor is operated at 300–350 °C and 2–3 MPa and packed with iron-based catalysts. Lighter, more olefinic product slate including gasoline, petrochemicals, and oxygenated chemicals can be produced. Product quality can be markedly improved after further upgrading.

Table 12.6 summarizes different FT reactor technologies developed by Sasol and Table 12.7 provides a comparison of the two FT operational modes. In addition to using coal for FT synthesis, Sasol developed a slurry-phase distillate (SSPD) process that comprises of natural gas reforming, slurry-phase FT, and mild hydroprocessing to produce naphtha and diesel from natural gas. The SSPD process uses a specially developed cobalt-based catalyst. Studies carried out by Sasol have shown that Haldor Topsoe autothermal reforming, which is used to reform natural gas with oxygen, is the most appropriate for the FT process. The SSPD process has been demonstrated commercially in a 2500 bbl/day unit at Sasolburg. It is believed that

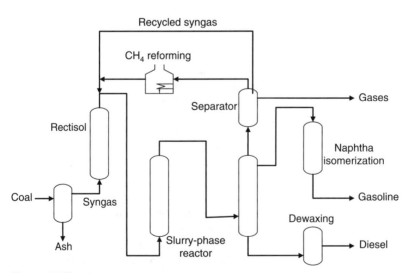

Figure 12.13.   Schematics of the HTFT process.

Table 12.6. Different FT Reactor Technologies Developed by SASOL

Process	Circulating Fluidized Bed (CFB)	Fixed Fluidized Bed	Fixed Bed	Bubble Slurry Column
Typical equipment	Sasol Synthol CFB	Sasol Advanced Synthol (SAS)	Sasol Arge Shell Middle Distillate Synthesis (SMDS)	Sasol Slurry Bed Reactor (SSBR) Exxon mobile demonstration Conoco demonstration
Advantage/disadvantage	Rapid deactivation, high capacity, low selectivity on liquid fuels, solid-gas separation needed		Simple process, high liquid selectivity, heat and mass transfer limitation exists, hot spot exists	High capacity, good mass/heat transfer, simple design, high liquid selectivity, additional liquid-solid separation needed
Operation mode	HTFT	HTFT	LTFT	LTFT
Capacity	7500 bbl/day	20,000 bbl/day	Up to 3000 bbl/day	3000 bbl/day 20,000 bbl/day being developed

Table 12.7.   Key Features of Two FT Operational Modes

Mode	Reaction Conditions	Catalysts	Reactors	Main Products
HTFT	300–350 °C	Fe	Fluidized bed	Short olefins as feedstock for downstream synthesis
LTFT	~200 °C	Fe, Co	Packed bed, bubble slurry column	Liquid fuel, especially high-quality diesel

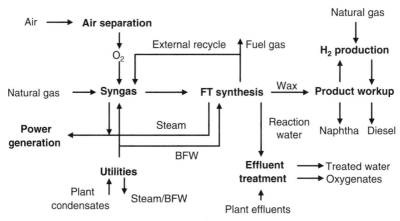

Figure 12.14.   Schematic diagram of the SSPD. BFW, boiler feed water.

this relatively simple three-step process is superior in all aspects and can be built and operated economically. Figure 12.14 shows a schematic for the SSPD process.

12.4   MOBIL METHANOL TO GASOLINE (MTG)

The Mobil MTG process produces gasoline from coal or natural gas in two distinct steps. The process has been taken to a commercial scale in a 12,500 bbl/day plant built in New Zealand to process gas from the Maui field. In the first step, synthesis gas produced by steam reforming of natural gas or by coal gasification is reacted over a copper-based catalyst to produce methanol close to 100% yield. The reaction is carried out at 260–350 °C and 5–6.5 MPa. The second step involves partial dehydration of methanol to dimethyl ether at 300 °C over an activated alumina catalyst, followed by reaction over a fixed-bed zeolite ZSM-5 catalyst. These reactions are strongly exothermic. The feed temperature is 360 °C and product leaves at 415 °C. The reactor pressure is 2.2 MPa. A series of reactions convert methanol and dimethyl ether to olefins and then to saturated hydrocarbons. Approximately 80% of the total products are in the gasoline boiling range. With alkylation of by-products propane and butane, total gasoline yields of 90% at 93.7 research octane number (RON) were achieved at the New Zealand plant. The use of a fluidized-bed reactor offers advantages for temperature control and maintenance of constant catalytic activity over a fixed-bed system. The fluidized-bed reactor operates at an almost isothermal temperature of 410 °C but at a pressure of only 0.3 MPa.

12.5 SMDS

The SMDS process, as shown in Figure 12.15, produces high-quality diesel fuel from natural gas. Although synthesis gas generated from coal gasification would, presumably, be equally suitable, it is a process being considered in many GTL processes in oil production.

During the process, natural gas is first partially oxidized in an oxygen-blown Shell gasifier to produce synthesis gas. This gasification approach is preferred over steam reforming, despite of the considerably higher capital cost and lower thermal efficiency, because it produces a synthesis gas with the ideal CO/H_2 ratio of 1:2. Steam reforming produces excess H_2, which, in a stand-alone operation, can only be used as fuel. The cleaned synthesis gas is then reacted over a proprietary Shell catalyst in a fixed-bed tube bundle reactor that is cooled in boiling water. The product is almost exclusively paraffinic. The catalyst formulation and operating conditions in this step are deliberately controlled to give a much higher-boiling product than usual, since this minimizes the production of hydrocarbon gases.

In the final step, the waxy heavy paraffin is catalytically hydrogenated, isomerized, and hydrocracked in a single trickle-bed reactor over a proprietary catalyst to

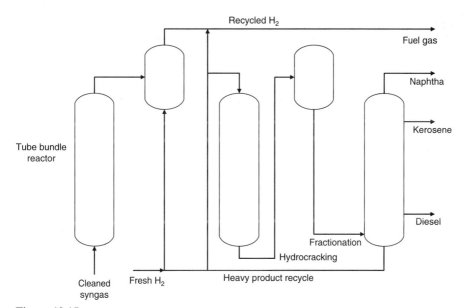

Figure 12.15. Diagram of the Shell Middle Distillate Synthesis process.

Table 12.8.   Key Features of DCTL and ICTL

	DCTL	ICTL (FT)
Feed	H$_2$+ coal/oil slurry	2H$_2$ + CO
T (°C)	440–470	200–300
P (bar)	170	20–40
Liquid fuel quality	Poor (high S and aromatics)	Excellent (low S and aromatics, high paraffin)
Capital expense	Slightly lower than ITCL	~$50K/bbl/day

give products that are mainly middle distillates. The reactor operates at 300–350 °C and 3–5 MPa. A high degree of product recycle is used to minimize the production of light products and to ensure that higher-boiling-point products are largely recycled. By varying the hydrocracking severity and the extent of recycle, the product distribution can be adjusted to give up to 60% diesel, with 25% kerosene and 15% naphtha. Alternatively, up to 50% kerosene can be produced, with 25% each of naphtha and diesel. Table 12.8 summarizes the key features of DCTL and ICTL.

12.6   HYBRID COAL LIQUEFACTION

Hybrid coal liquefaction integrates direct and indirect coal liquefaction into a single plant. This concept takes advantage of the complementary characteristics of the two processes. As mentioned previously, direct coal liquefaction produces high-octane gasoline and low-cetane diesel, while indirect coal liquefaction produces high-cetane diesel and low-octane gasoline. Blending the products in an integrated plant allows the production of premium quality gasoline and diesel with minimal refining.

The concept of a hybrid DCTL/ICTL plant has been discussed for many years. The DOE commissioned MITRE Corporation to study the concept between 1990 and 1991. Initial studies indicated that production costs were slightly lower for a hybrid plant compared with stand-alone direct or indirect plants. No testing has been performed based on this concept to date.

The synergy between the direct and indirect processes, shown in Figure 12.16, improves overall thermal efficiency of an integrated hybrid plant. Higher-quality coal can be fed as feedstock to the direct coal liquefaction reactors. Lower-quality coal can be fed to the gasifier to provide syngas for FT synthesis. The hydrogen-rich FT tail gas can be used to provide hydrogen for product upgrading and for direct coal liquefaction.

Blending the raw distillable products prior to refining takes advantage of their complementary characteristics. High-octane naphtha from direct coal liquefaction is blended with low-octane naphtha from indirect coal liquefaction, and high-cetane diesel from indirect coal liquefaction is blended with low-cetane diesel from direct coal liquefaction. The blended liquids require less refining to meet premium product specifications than if they were refined separately.

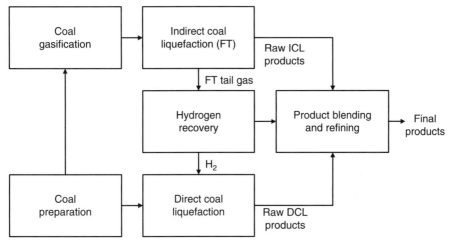

Figure 12.16.   Synergy between direct and indirect process. DCL, direct coal to liquid; ICL, indirect coal to liquid.

12.7   COAL TO METHANOL

12.7.1   Introduction of Methanol Synthesis

The synthesis of methanol from syngas is a well-established technology. Syngas can be produced from oil, natural gas, or coal. Syngas production via different coal gasification technologies is discussed in Chapter 4.

All commercially produced methanol is made by the catalytic conversion of syngas containing carbon monoxide, carbon dioxide, and hydrogen as the main components. The basic reactions involved in methanol synthesis are

$$CO + 2H_2 = CH_3OH \quad \Delta H^{\circ}_{298} = 94.08 \text{ kJ/mol}, \tag{12.12}$$

$$CO_2 + 3H_2 = CH_3OH + H_2O \quad \Delta H^{\circ}_{298} = -52.81 \text{ kJ/mol}, \tag{12.13}$$

$$CO_2 + H_2 = CO + H_2O \quad \Delta H^{\circ}_{298} = 41.27 \text{ kJ/mol}. \tag{12.14}$$

Of all three reactions, only two are stoichiometrically independent. The first reaction produces methanol with low hydrogen consumption but involves significant amounts of heat. The second reaction involves less heat but consumes more hydrogen and produces steam as the by-product. Thermodynamically, low temperature and high pressure favor methanol formation. The cost of methanol production depends on certain variables related to the conversion rate of syngas to crude methanol, and also on the desired grade of the final product. In the conversion of syngas to methanol, only a part of the syngas is transformed to methanol *per pass* through the reactors. The conversion rate is affected by syngas composition, syngas feed pressure, space velocity (SV), and the activity of the catalyst.

12.7.2 Methanol Synthesis Catalysts

In 1923, BASF first found that methanol was the main product of carbon monoxide hydrogenation when ZnO/Cr_2O_3-containing catalysts were used. This catalytic system was used until 1970s and is also known as Zn-based or ZnO-based catalysis. This catalyst is only active at elevated temperatures 350–400 °C. The process is operated under high pressure, about 25–35 MPa. Therefore, it is called high-pressure methanol synthesis. High operating pressure requires complex facilities and results in high power consumption. This process is no longer economical and is being phased out.[30]

In 1963, Imperial Chemical Industries (ICI) PLC announced an innovative process using $Cu/ZnO/Al_2O_3$ catalysts, later called low-pressure synthesis. The major constituents of this catalytic system are Cu (reduced form of CuO) and ZnO on an Al_2O_3 support. The reaction pressure and temperature are 220–270 °C and 5–10 MPa, respectively. The catalysts have been found to be susceptible to sulfur and carbonyl poisoning, sintering, and thermal aging. Nowadays, almost all of commercial methanol syntheses are carried out by low-pressure processes.

Although methanol synthesis from syngas hydrogenation is a proven process for chemical industries, it still poses great challenges. The syngas conversion per pass is severely limited by reaction thermodynamics. To resolve the limitation of thermodynamics on the conversion and methanol yield, one of the most effective ways is timely removal of the reaction heat. Another method is to develop an effective catalyst system that has good performance at low temperatures. Thus far, substantial efforts are still being devoted to the development of novel methanol catalysts.[31,32] While the nature of the active sites and effects of the support and promoter are still under investigation, the reaction mechanism still remains an open question. Only when these issues are truly overcome can the optimum catalyst with perfect activity and selectivity be synthesized.

12.7.3 Methanol Synthesis Reactor Systems

Conventionally, methanol is produced in a two-phase system. The reactants (CO, CO_2, and H_2) and products (CH_3OH and H_2O) form the gas phase, and the catalyst is the solid phase. Two reactor types were most popular: an adiabatic reactor containing a continuous bed of catalyst with quenching by cold gas injections (the ICI system), and a multitubular reactor with an internal heat exchanger (the Lurgi system). Both systems are operated at temperatures from 210 to 280 °C and relatively low pressures of around 5–7 MPa, using $Cu/ZnO/Al_2O_3$ catalysts. Other innovative reactor systems for methanol synthesis currently under development include[30]

- slurry reactor,
- trickle-bed reactor,
- gas–solid–solid trickle-flow reactor (GSSTFR), and
- reactor system with interstage product removal (RSIRP).

Among them, the slurry reactor technology is close to commercial readiness under the DOE Clean Coal Technology Program, which will be discussed below in detail.

The ICI low-pressure methanol (LPM) process accounts for approximately 60% of the world production capacity of methanol. A schematic diagram of the synthesis loop of the process is given in Figure 12.17. Synthesis gas is mixed with recycled unreacted gas, pressurized in a compressor to the operational pressure, then pre-heated in a heat exchanger to reaction temperature, and subsequently fed into the synthesis reactor. The conversion rate of methanol is determined thermodynamically. Depending on the process conditions used, the product gas contains 4–8 vol% methanol. A large part of the condensation heat of methanol is transferred to the reactor feed gas in a feed effluent heat exchanger before the gas is cooled to room temperature in a water cooler. Approximately 95% of the methanol condenses in the water cooler. Separation of the gas phase takes place in a separator. After passing through the recycle compressor, the gas is mixed with fresh synthesis gas and subsequently reintroduced into the reactor. A fixed percentage of the gas leaving the separator is continuously purged. This is to prevent inert gases that do not condense (mainly N_2 and CH_4) from building up in the synthesis loop. The crude methanol is purified by means of distillation. At the top column, gases and impurities with a low boiling point are removed. A second refining column removes the heavier products and water.

The Lurgi LPM process involves the same basic steps as the ICI processes. The two processes differ mainly in their reactor designs and the way in which the produced heat is removed as shown in Figure 12.18. The ICI design consists of a number of adiabatic catalytic beds, and cold gas is used to cool the reactant gases between the beds. The highest temperature is reached in the first catalyst bed. The Lurgi

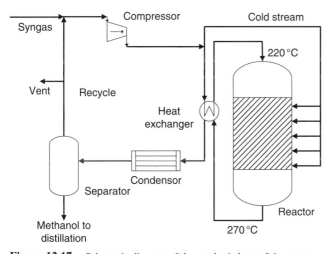

Figure 12.17.   Schematic diagram of the synthesis loop of the process.

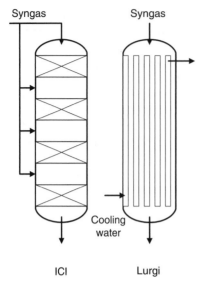

Syngas Syngas

Cooling
water

ICI Lurgi

Figure 12.18. Comparison between the ICI and Lurgi reactor.

reactors have a much flatter temperature profile, as the reactor is almost isothermal owing to its cooling on the shell side by the generation of valuable high-pressure steam. As a result of the flat temperature profile, catalyst deactivation is slowed down. However, in both processes, the catalyst still deactivates, which means that the productivity decreases and the plant capacity will no longer be met. Fortunately, the productivity is a function of the operating pressure, so increasing the pressure can compensate for the loss in catalyst activity. By doing so, the plant can sustain its design value over a longer time (2–3 years). In the present day, large methanol plants exclusively use centrifugal compressors to bring the synthesis gas to the desired operating pressure, which varies from 6 to 9 MPa, depending on the activity of the catalyst, that is, the lower pressure being used with fresh catalyst and the highest pressure at the end of the catalyst lifetime.

12.7.4 Liquid-Phase Methanol (LPMEOH™) Process

The liquid-phase methanol (LPMEOH™) process was developed during the 1980s through a cooperative effort by Air Products and Chemicals, Inc. (AP) and the DOE. This effort resulted in the successful 69-month operation of a commercial-scale 260 t/day demonstration plant at Eastman Chemical Company's coal to chemicals complex in Kingsport, TN. During the 69-month demonstration program since start-up in April 1997, the plant availability averaged 97.5%. The LPMEOH™ demonstration plant produced over 100 million gallons of methanol for Eastman Company's production of its commercial products.

The SBC reactor differentiates the LPMEOH™ process from conventional technology. As mentioned above, conventional methanol reactors use fixed beds of catalyst pellets and operate in the gas phase. The LPMEOH™ process utilizes a bubble column reactor containing a standard methanol synthesis catalyst powder suspended in an inert mineral oil to form a slurry system. The mineral oil acts as a temperature moderator and heat removal medium, transferring the heat of reaction away from the catalyst surface to boiling water in an internal tubular heat exchanger. Since the heat transfer coefficients on both sides of the exchanger are relatively large, the heat exchanger occupies only a small fraction of the cross-sectional area of the reactor. As a result of this capability to remove heat and maintain a constant, highly uniform temperature throughout the entire length of the reactor, the slurry reactor can achieve much higher syngas conversion per pass than its gas-phase counterparts.

Furthermore, because of the LPMEOH™ reactor's unique temperature control capabilities, it can directly process syngas rich in carbon oxides (carbon monoxide and carbon dioxide). Gas-phase methanol technology would require that syngas feedstocks with similar compositions undergo stoichiometry adjustment by the WGS reaction, to increase the hydrogen content, and subsequent CO_2 removal. In a gas-phase reactor, temperature moderation is achieved by recycling large quantities of hydrogen-rich gas, utilizing the higher gas velocities around the catalyst particles and minimizing the conversion per pass. Typically, a gas-phase process is limited to CO concentrations of about 16 vol% in the reactor feed, as a means of constraining the conversion per pass to avoid excess heating. In contrast, for the LPMEOH™ reactor, CO concentrations in excess of 50 vol% have been tested in the laboratory, at the alternative fuels development unit (AFDU) and at Kingsport, without any adverse effect on catalyst activity. As a result, the LPMEOH™ reactor can achieve approximately twice the conversion per pass of the gas-phase process, yielding lower recycle gas compression requirements and capital savings.

A third differentiating feature of the LPMEOH™ process is that a high-quality methanol product is produced directly from syngas rich in carbon oxides. Gas-phase methanol synthesis, which must rely on H_2-rich syngas feedstocks, yields a crude methanol product with 4–20 wt% water. The product from the LPMEOH™ process, using CO-rich syngas, typically contains only 1 wt% water. As a result, raw methanol coproduced in an IGCC facility would be suitable for many applications at substantial savings in purification costs. The steam generated in the LPMEOH™ reactor is suitable for the purification of the methanol product to a higher quality or for use in the IGCC power generation cycle.

Another unique feature of the LPMEOH™ process is the ability to periodically withdraw spent catalyst slurry and add fresh catalyst online. This facilitates uninterrupted operation and also allows perpetuation of high production rate of methanol from the reactor. Furthermore, choice of catalyst replacement rate permits the optimization of methanol production rate versus catalyst replacement cost.

Figure 12.19 shows a simplified process flow diagram of the LPMEOH™ demonstration facility in Kingsport. Three different feed gas streams (hydrogen, carbon monoxide, and the primary syngas feed known as balanced gas) are diverted from

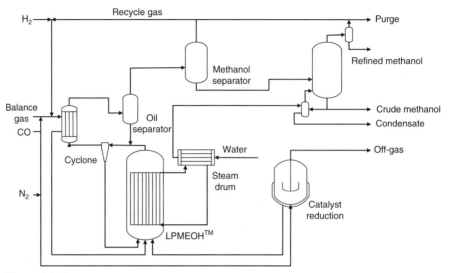

Figure 12.19.   Process diagram of the LPMEOH™ process.

existing operations to the LPMEOH™ demonstration unit, thus providing the range of coal-derived syngas ratio (H_2-to-CO ratio) needed to meet the technical objectives of the demonstration project. Syngas enters at the bottom of the slurry reactor, which contains solid particles of catalyst suspending in liquid mineral oil. The syngas first dissolves through the mineral oil and then is adsorbed on the catalyst surface and reacts to form methanol. The reaction is highly exothermic and the reaction heat is absorbed by the slurry and removed from the reactor by steam coils. The product methanol vapor diffuses from the catalyst surface through the mineral oil and exits the reactor with unreacted syngas. It is then condensed to a liquid and sent to distillation columns for the removal of higher alcohols, water, and other impurities. Most of the unreacted syngas is recycled to the reactor by the syngas recycle compressor.

Because of the above features, the LPMEOH™ process has great potential combined with IGCC power generation. Because heat generated in methanol synthesis can be effectively removed from these reactors, high conversion rate can be realized in a single pass of syngas through the reactor, often making it attractive to processes that limit conversion to what can be realized in a single pass.[33] In such "once-through" processes, the unconverted syngas might be burned in a combined cycle to generate electricity or deliver via pipeline as town gas to distributed users. One typical polygeneration configuration for the combination of methanol manufacture and IGCC is shown in Figure 12.20.

Polygeneration strategies based on use of liquid-phase reactors in once-through configurations can lead to cost savings both by means of scale economies and by exploitation of potential synergisms (i.e., avoiding capital investments associated with syngas recycling to improve conversion). Combined-cycle electricity from

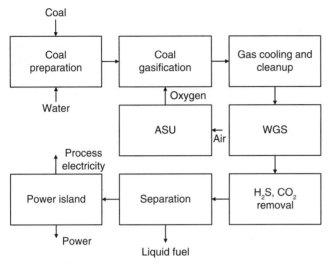

Figure 12.20. Polygeneration configuration for the combination of methanol manufacture and IGCC. ASU, air separation unit.

coal-derived syngas will be less costly when generated in polygeneration configurations than in IGCC-only configurations. As a matter of fact, the electricity from polygeneration can compete today with electricity derived from conventional coal steam-electric plants equipped with end-of-pipe air pollution control equipment. Moreover, the nonintegrated gasifier/combined-cycle designs are less complicated and more reliable than highly integrated IGCC designs that produce electricity only.

12.8 COAL TO DIMETHYL ETHER (DME)

DME (CH_3OCH_3) is a clean fuel that has similar physical properties as LPG. The cetane number of DME is ~55, which is higher than that of diesel. Thus, it can be used as a good substitute of diesel. Currently, DME is made by MeOH dehydration:

$$2CH_3OH = CH_3OCH_3 + H_2O. \qquad (12.15)$$

But DME can also be made, prospectively at lower cost, in a single step by combining the following three reactions in a single reactor:

Equation 12.3: $CO + H_2O = CO_2 + H_2$ (WGS),
Equation 12.12: $CO + 2H_2 = CH_3OH$ (methanol synthesis),
$\qquad 2CH_3OH = CH_3OCH_3 + H_2O$ (methanol dehydration). (12.16)

Haldor Topsoe in Denmark[34] is developing a single-step process for making DME from natural gas. NKK Corporation in Japan[35] and Air Products and Chemicals, Inc. in the U.S.[36] are developing single-step processes for large-scale DME manufacture from coal-derived syngas using slurry-phase reactors.

In China, the Institute of Coal Chemistry (ICC) of the Chinese Academy of Sciences together with the Shanxi New Style Fuel and Stove Company constructed a 500 t/year DME plant in Xi'an based on MeOH dehydration for use as a domestic cooking fuel as an alternative to LPG. Since 1995, the ICC has also been carrying out R&D on one-step DME synthesis based on slurry-phase reactor technology.[37] Also, plants of a total of 5,000,000 t/year DME capability have been proposed.

REFERENCES

1. CHEN, W., QIN, Q., XUE, J., and QIN, Q. White paper: Coal polygeneration for liquid fuels and chemicals. Shanghai, China: Coal Polygeneration Technologies Lab, Global Research, November, **2006**.
2. Department of Trade and Industry UK (DTI). Technological status report: Coal liquefaction. *Technology Status Report 010*, **1999**.
3. HIRSCHON, A.S. and WILSON, R.B. Highly dispersed catalysts for coal liquefaction. *USDOE Phase I Final Report*, **1995**.
4. HTI. *HTI Direct Coal Liquefaction Technology*, http://www.covol.com/data/upfiles/pdfadmin/DCL%206.07.pdf (accessed September 16, 2009).
5. DRY, M.E. Present and future applications of the Fischer-Tropsch process. *Applied Catalysis. A, General*, **2004**, 276, 1.
6. MAITLIS, P.M. Metal catalysed CO hydrogenation: Hetero- or homo-, what is the difference? *Journal of Molecular Catalysis A: Chemical*, **2003**, 204–205, 54.
7. DRY, M.E. The Fischer-Tropsch process: 1950–2000. *Catalysis Today*, **2002**, 71 (3–4), 227.
8. LI, S., KRISHNAMOORTHY, S., LI, A., MEITZNER, G. D., and IGLESIA, E. Promoted iron-based catalysts for the Fischer-Tropsch synthesis: Design, synthesis, site densities, and catalytic properties. *Journal of Catalysis*, **2002**, 206, 202.
9. VAN DER LAAN, G.P. and BEENACKERS, A.A.C.M. Kinetics and selectivity of the Fischer-Tropsch synthesis: A literature review. *Catalysis Reviews—Science and Engineering*, **1999**, 41, 255.
10. DAVIS, B.H. Fischer-Tropsch synthesis: Relationship between iron catalyst composition and process variables. *Catalysis Today*, **2003**, 84, 83.
11. IGLESIA, E. Design, synthesis, and use of cobalt-based Fischer-Tropsch synthesis catalysts. *Applied Catalysis. A, General*, **1997**, 161, 59.
12. BERTOLE, C.J., MIMS, C.A., and KISS, G. Support and rhenium effects on the intrinsic site activity and methane selectivity of cobalt Fischer-Tropsch catalysts. *Journal of Catalysis*, **2004**, 221, 191.
13. BEZEMER, G.L., RADSTAKE, P.B., FALKE, U., OOSTERBEEK, H., KUIPERS, H.P.C.E., VAN DILLEN, A.J., and DE JONG, K.P. Investigation of promoter effects of manganese oxide on carbon nanofiber-supported cobalt catalysts for Fischer-Tropsch synthesis. *Journal of Catalysis*, **2006**, 237, 152.
14. ENACHE, D.I., ROY-AUBERGER, M., and REVEL, R. Differences in the characteristics and catalytic properties of cobalt-based Fischer-Tropsch catalysts supported on zirconia and alumina. *Applied Catalysis. A, General*, **2004**, 268, 51.
15. REUEL, R.C. and BARTHOLOMEW, C.H. Effects of support and dispersion on the CO hydrogenation activity/selectivity properties of cobalt. *Journal of Catalysis*, **1984**, 85, 78.
16. BORG, Ø., DIETZEL, P.D.C., SPJELKAVIK, A.I., TVETEN, E.Z., WALMSLEY, J.C., DIPLAS, S., ERI, S., HOLMEN, A., and RYTTER, E. Fischer-Tropsch synthesis: Cobalt particle size and support effects on intrinsic activity and product distribution. *Journal of Catalysis*, **2008**, 259, 161.
17. CHU, W., CHERNAVSKII, P.A., GENGEMBRE, L., PANKINA, G.A., FONGARLAND, P., and KHODAKOV, A.Y. Cobalt species in promoted cobalt alumina-supported Fischer-Tropsch catalysts. *Journal of Catalysis*, **2007**, 252, 215.
18. HONG, J., CHERNAVSKII, P.A., KHODAKOV, A.Y., and CHU, W. Effect of promotion with ruthenium on the structure and catalytic performance of mesoporous silica (smaller and larger pore) supported cobalt Fischer-Tropsch catalysts. *Catalysis Today*, **2009**, 140, 135.

19. GUCZI, L., STEFLER, G., BORKÓ, L., KOPPÁNY, Z., MIZUKAMI, F., TOBA, M., and NIWA, S. Re-Co bimetallic catalysts prepared by sol/gel technique: Characterization and catalytic properties. *Applied Catalysis. A, General*, **2003**, 246, 79.

20. GIRARDON, J.-S., LERMONTOV, A.S., GENGEMBRE, L., CHERNAVSKII, P.A., GRIBOVAL-CONSTANT, A., and KHODAKOV, A.Y. Effect of cobalt precursor and pretreatment conditions on the structure and catalytic performance of cobalt silica-supported Fischer-Tropsch catalysts. *Journal of Catalysis*, **2005**, 230, 339.

21. SCHULZ, H. Short history and present trends of Fischer-Tropsch synthesis. *Applied Catalysis. A, General*, **1999**, 186, 3.

22. FISCHER, F. and TROPSCH, H. Synthesis of petroleum at atmospheric pressures. *Brennstoff-Chemie*, **1926**, 7, 97.

23. CRAXFORD, S.R. and RIDEAL, E.K. The mechanism of the synthesis of hydrocarbons from water gas. *Journal of Chemical Society*, **1939**, 1604.

24. STORCH, H.H., GOLUMBIC, N., and ANDERSON, R.B. *The Fischer-Tropsch and Related Synthesis*. New York: John Wiley & Sons, **1951**.

25. DAVIS, B.H. Fischer-Tropsch synthesis: Current mechanism and futuristic needs. *Fuel Processing Technology*, **2001**, 71 (1–3), 157.

26. DAVIS, B.H. Fischer-Tropsch synthesis: Reaction mechanisms for iron catalysts. *Catalysis Today*, **2009**, 141 (1–2), 25.

27. JOYNER, R.W. The mechanism of chain growth in the Fischer-Tropsch hydrocarbon synthesis. *Catalysis Letters*, **1988**, 1, 307.

28. FRIEDEL, R.A. and ANDERSON, R.B. Composition of synthetic liquid fuels. I. Product distribution and analysis of C5-C8 paraffin isomers from cobalt catalyst. *Journal of American Chemical Society*, **1950**, 72, 1212.

29. ANDERSON, R.B. *Catalysis-Science and Technology*. New York: John Wiley & Sons, **1956**.

30. LEE, S. *Alternative Fuels*. Washington, DC: Taylor and Francis, **1996**.

31. WU, J., SAITO, M., TAKEUCHI, M., and WATANABE, T. The stability of Cu/ZnO-based catalysts in methanol synthesis from a CO_2-rich feed and from a CO-rich feed. *Applied Catalysis. A, General*, **2001**, 218, 235.

32. LIU, X. Recent advances in catalysts for methanol synthesis via hydrogenation of CO and CO_2. *Industrial & Engineering Chemistry Research*, **2003**, 42, 6518.

33. LARSON, E.D. and REN, T.J. Synthetic fuel production by indirect coal liquefaction. *Energy for Sustainable Development*, **2003**, VII (4), 79.

34. BØGILD-HANSEN, J. Dimethyl ether, transportation fuel and chemical intermediate, possibilities for co-production and multi-process integration. Presented at the *20th World Gas Congress*. Copenhagen, **1997**.

35. ADACHI, Y., KOMOTO, M., WATANABE, I., OHNO, Y., and FUJIMOTO, K. Effective utilization of remote coal through dimethyl ether synthesis. *Fuel*, **2000**, 79, 229.

36. PENG, X.D., TOSELAND, B.A., WANG, A.W., and PARRIS, G.E. Progress in development of PPMDE process: Kinetics and catalysts. Presented at the *1997 Coal Liquefaction and Solid Fuels Contractors' Review Conference*. Pittsburgh, PA, September 3–4, **1997**.

37. NIU, Y. Dimethyl ether (DME)—Clean fuel in the 21st century. Paper presented at the *Workshop on Polygeneration Strategies Based on Oxygen-Blown Gasification—Strategic Energy Thinking for the 10th 5-Year Plan*, convened by the Working Group on Energy Strategies and Technologies, China Council for International Cooperation on Environment and Development. Beijing, May 11–12, **2000**.

Index

ABB Lummus, 242
acid gas removal (AGR), 190, 468
   dietheanolamine (DEA), 471
   methyldietheanolamine (MDEA), 470–471, 480
ADIP, 288–291
adsorption, 418, 420, 423–424
   activated carbon (AC), 262–264
   pressure swing (PSA), 7, 18, 329–330, 357–358, 367, 414–448, 480
      basic concept, 414–415
      for bulk CO_2 removal, 427–429
      for C_1-C_4 hydrocarbon removal, 432–434
      for dilute CH_4 removal, 432
      for dilute CO and N_2 removal, 429–432
      engineering process design, 444–447
      future trends, 435–441
      hybrid adsorbent membrane, 442–444
      improvements, 441–444
      integration, 441–442
      for production of hydrogen only, 418–426
      rapid (RPSA), 426, 436–438
      reliability, 441
      sorption-enhanced reaction process (SERP), 439–441
      structured, 438–439
   selective sulfur (PSU-SARS), 243–244
   zeolite-based, 260–262
Advanced Refining, 236
advanced refining technologies (ART), 239
advanced sulphur removal (ASR), 296
Advantica, 110
AED. *See* atomic emission detector
AFCs. *See* fuel, cells, alkaline

AGR. *See* acid gas removal
Air Liquide S.A., 367
Air Products and Chemicals, Inc., 367, 380–381, 415–417, 422–423, 427, 432, 437, 516, 519
air separation unit (ASU), 458, 468–469
AkzoNobel, 227, 236–238, 241
Alcoa Industrial Chemicals, 253
alumina, 372, 378
Amine Guard FS Process, 281
ammonia synthesis, 14, 128, 156
Anderson–Schultz–Flory (ASF) distribution model, 503–504
Argonne National Laboratory, 137
aromatics, 477
Arrhenius equation, 60
ART. *See* advanced refining technologies
ASF. *See* Anderson–Schultz–Flory distribution model
ash
   content, and fusion temperature of ATF, 184
   -slag chemistry, 166–168
ASR. *See* advanced sulphur removal
ASU. *See* air separation unit
atomic emission detector (AED), 224
ATR. *See* reforming, autothermal
AZEP. *See* power plants, advanced zero emission

BA. *See* basic (BA) model
Ballard, 137
Baltimore Gas Company, 158
BASF Catalysts, LLC, 21, 282, 313, 345
basic (BA) model, 23–24
Battelle Memorial Institute, 36
BDS. *See* desulphurization, bio-

Hydrogen and Syngas Production and Purification Technologies, Edited by Ke Liu, Chunshan Song and Velu Subramani
Copyright © 2010 American Institute of Chemical Engineers